Very Truly Yours,

John J. McLaurin

Sketches in Crude-Oil

SOME ACCIDENTS AND INCIDENTS OF THE PETROLEUM
DEVELOPMENT IN ALL PARTS OF
THE GLOBE

WITH PORTRAITS AND ILLUSTRATIONS

By JOHN J. McLAURIN

*Author of "A Brief History of Petroleum," "The Story of
Johnstown," etc.*

"I will a round, unvarnished tale deliver."—*Shakespeare*

HARRISBURG, PA.
PUBLISHED BY THE AUTHOR
1896

ENGRAVING PRINTING AND BINDING
BY J. HORACE McFARLAND COMPANY
HARRISBURG, PA.

To —
my neighbor and friend for many years,
a man of large heart and earnest purpose,

— HON. CHARLES MILLER —
FRANKLIN, PA.,
whose sterling qualities have achieved the
highest success in life and won the confidence
and esteem of his fellows, This Volume is

— Respectfully Dedicated.

NOTE

Many of the illustrations and portraits in this volume are from photographs by John A. Mather, Titusville, Pa. . . . .

INTRODUCTION

L IFE is too short to compile a book that would cover the subject fully, hence this work is *not* a detailed history of the great petroleum development. Nor is it a mere collection of dry facts and figures, set forth to show that the oil business is a pretty big enterprise. But it *is* a sincere endeavor to print something regarding petroleum, based largely upon personal observation, which may be worth saving from oblivion. The purpose is to give the busy outside world, by anecdote and incident and brief narration, a glimpse of the grandest industry of the ages and of the men chiefly responsible for its origin and growth. Many of the portraits and illustrations, nearly all of them now presented for the first time, will be valuable mementoes of individuals and localities that have passed from mortal sight forever. If the reader shall find that "within is more of relish than of cost" the writer of these "Sketches" will be amply satisfied.

.

PORTRAITS.

CONTENTS.

I.

LOOKING BACKWARD.

PETROLEUM IN ANCIENT TIMES—KNOWN FROM AN EARLY PERIOD IN THE
WORLD'S HISTORY—MENTIONED IN THE SCRIPTURES AND BY PRIMITIVE
WRITERS—SOLOMON SUSTAINED—STUMBLING UPON THE GREASY STAPLE
IN VARIOUS LANDS—INCIDENTS AND ANECDOTES OF DIFFERENT SORTS AND
SIZES—OVER ASIA, AFRICA AND EUROPE.

"Oil out of the flinty rock."—*Deuteronomy xxxii: 13.*
"And the rock poured me out rivers of oil."—*Job xxix: 6.*
"Will the Lord be pleased with * * * ten thousands of rivers of oil?"—*Micah vi: 7.*
"I have myself seen pitch drawn out of the lake and from water in Zacynthus."—*Herodotus.*
"The people of Agrigentum save oil in pits and burn it in lamps."—*Dioscorides, Vol. 1: p. 99.*

ETROLEUM, a name to conjure with and weave romances around, helps out Solomon's oft-misapplied declaration of "Nothing new under the sun." Possibly it filled no place in domestic economy when the race, if the Darwinian theory passes muster, sported as ring-tailed simians, yet the Scriptures and primitive writers mention the article repeatedly. Many intelligent persons, recalling the tallow-dip and lard-oil lamp of their youth, consider the entire petroleum business of very recent date, whereas its history goes back to remotest antiquity. Naturally they are disappointed to find it, in various aspects, "the same thing over again." Men and women in the prime of life have forgotten the flickering pine-knot, the sputtering candle or the smoky sconce hardly long enough to associate rock-oil with "the brave days of old." This idea of newness the host of fresh industries created by oil operations has tended to deepen in the popular mind. Enjoying the brilliant glow of a modern argand-burner, double-wicked, silk-shaded, onyx-mounted and altogether a genuine luxury, it seems hard to realize that the actual basis of this up-to-date elegance has existed from time immemorial. Of derricks, drilling-tools, tank-cars, refineries and pipe-lines our ancestors were blissfully ignorant; but petroleum itself, the foundation of the countless paraphernalia of the oil trade of to-day, flourished "ere Noah's flood had space to dry." Although used to a limited extent in crude form for thousands of years, it was reserved for the present age to introduce the grand illuminant to the world generally. After sixty centuries the game of "hide-and-seek" between Mother Earth and her children has terminated in favor of the latter. They have pierced Nature's internal laboratories, tapping the huge oil-tanks wherein the products of her quiet chemistry had accumulated "in bond," and up came the unctuous fluid in volumes ample to fill all the lamps the universe could manufacture and to grease every wheel on this revolving planet! Edward Bellamy may, perhaps, be imitated pleasantly and profitably in this connection by "Looking Backward."

Precisely how, when, where and by whom petroleum was first discovered and utilized nobody living can, and nobody dead will, tell anxious inquirers. The information has "gone where the woodbine twineth," joining the dodo, the megatherium, the ichthyosaurus and the "lost arts" Wendell Phillips embalmed in fadeless prose. An erratic Joe-Millerite has traced the stuff to the Garden of Eden in a fashion akin to the chopping logic of the Deacon's "Won-

derful One-Horse Shay." Hear him :

"Adam had a fall?"

"Granted."

"He fell very easily?"

"Sure as death and taxes."

"Why did he fall with such neatness and dispatch?"

"May be he took a spring to fall."

"Because everything was greased for the occasion! Unquestionably the only lubricant on this footstool just then was the petroleum brewed in God's own subterranean stills. Therefore, petroleum figured in Eden, which was to be demonstrated. See?"

There is no "irrepressible conflict" between this reasoning, the version of the Pentateuch and the idea of Peck's

THE BAD BOY'S IDEA OF ADAM'S FALL.

Bad Boy that "Adam clumb a appultree to put coal-oil on it to kill the insecks, an' he saw a snaik, an' the oil made the tree slippy, an' he fell bumpety-bump!"

Other wags attribute the longevity of antediluvian veterans to their unstinted use of petroleum for internal and external ailments! Had medical almanacs, patent nostrums and circus-bill testimonials been evolved at that interesting period, the oleum vender would have hit the bull's-eye plump in the center. Guess at the value of recommendations like these, with the latest accompaniment of "before-and-after" pictures in the newspapers :

LAND OF NOD, *April 1, B. C. 5678.*—This is to certify that I keep my strength up to blacksmith pitch by frequent applications of Petroleum Prophylactic and six big drinks of Benzine Bitters daily. Lifting an elephant, with one hand tied behind me, is my favorite trick.

SANDOW TUBAL-CAIN.

MT. ARARAT, *July 4, B. C. 4004.*—Your medicine is out of sight in our family. It relieved papa of an overdose of fire-water, imbibed in honor of his boat distancing Dunraven's barge on this glorious anniversary, and cured Ham of trichina yesterday. Mamma's pug slid off the upper deck into the swim and was fished out in a comatose condition. A solitary whiff of your Pungent Petroleum Pastils revived him instantly, and he was able to howl all night.

SHEM & JAPHETH.

SOMEWHERE IN ASIA, *Dec. 21, B.C. 4019.*—Your incomparable Petroleum Prophylactic, which I first learned about from a college chum, is a daisy-cutter. Thanks to its superlative virtues, I have lived to be a trifle older than the youngest ballet-girl in the "Black Crook." I celebrated my nine-hundred-and-sixty-ninth birthday by walking umsteen miles before luncheon, playing left-tackle with the Y. M. C. A. Football Team in the afternoon and witnessing "Uncle Tom's Cabin"—two Topsys, two Markses, two Evas, two donkeys and four Siberian bloodhounds —in the evening. Next morning's paper flung this ticket to the breeze:

> "For Mayor of Jeroosalum
> We nominate Methoosalum."

By sticking faithfully and fearlessly to your unrivaled elixir I expect to round out my full thousand years and run for a second term. Refer silver sceptics and gold-bug office-seekers to me for particulars as to the proper course of treatment.

GROVER LINGER LONGER METHUSELAH.

PLEASANT VALLEY, *Oct. 30th, B. C. 5555.*—I just want to shout "Eureka," "Excelsior," "Hail Columbia," "E Pluribus Unum," and give three cheers for your Kill-em-off Kerosene! Both my mothers-in-law, who had bossed me seventy decades, tried a can of it on a sick fire this morning. Their funeral is billed for four o'clock p. m. to-morrow. Send me ten gallons more at once.

BRIGHAM YOUNG LAMECH.

ISLES OF GREECE.—I defy the Jersey Lightning to knock me out while your Benzine Bitters are in the ring. "A good thing; push it along."

SULLIVAN AJAX.

Leaving the realm of conjecture, it is quite certain that the "pitch" which coated the ark and the "slime" of the builders of Babel were products of petroleum. Genesis affirms that "the vale of Siddim was full of slime-pits"— language too direct to be dismissed by hinting vaguely at "the mistakes of Moses." Deuteronomy speaks of "oil out of the flinty rock" and Micah puts the pointed query: "Will the Lord be pleased with * * * ten thousands of rivers of oil?" To the three friends who condoled with him in his grievous visitation of boils, the patriarch of Uz asserted: "And the rock poured me out rivers of oil." Whatever his hearers might think of this apparent stretch of fancy, Job's forecast of the oleaginous output was singularly felicitous. Evidently the Old Testament writers, whose wise heads geology had not muddled, knew a good deal about the petroleum situation.

Away back in the fifties, a zealous New-York follower of Voltaire was accustomed to wind up his assaults on inspiration by criticising these oily quotations unmercifully. "Could anything be more absurd," he would ask, "than to talk of 'oil from a flinty rock' and 'rocks pouring forth rivers of oil'? If anything were needed to prove the Bible a fool-book from start to finish, such utterances would settle the matter beyond dispute. Rocks yielding rivers of oil cap the climax of ridiculous nonsense! Next they'll be wanting us to believe that Jonah swallowed the whale! Bah!"

"I'LL BE JIGGERED!"

Months and years passed away swiftly, as they have a habit of doing, and the sturdy agnostic continued arguing pluckily. At length tidings of wells flowing thousands of barrels of oil reached him from William Penn's broad heritage. He came, he saw and, unlike Julius Cæsar, he surrendered unconditionally. Remarking "I'll be jiggered!" the doubter doubted no more. He revised his opinions, humbly accepted the gospel and professed religion, openly and above-board. Hence the petroleum development is entitled to the credit of one notable conversion, at least, and the balance is on the right side of the ledger, assuming that a human soul outweighs the terrestrial globe in the unerring scales of the Infinite.

Whether petroleum, which literally signifies "rock-oil," be of mineral, vegetable or animal origin matters little to the producer or consumer, who views it from a commercial standpoint. In its natural state it is a variable mixture of numerous liquid hydro-carbons, holding in solution paraffine and solid bitumen, or asphaltum. The fountains of Is, on the Euphrates, were familiar to the founders of Babylon, who secured indestructible mortar for the walls of the city by pouring melted asphaltum between the blocks of stone. These famous springs attracted the attention of Alexander, Trajan and Julian. Even now asphaltum procured from them is sold in the adjacent villages. The commodity is skimmed off the saline and sulphurous waters and solidified by evaporation. The ancient Egyptians used another form of the same substance in preparing mummies, probably obtaining their supplies from a spring on the Island of Zante, described by Herodotus. It was flowing in his day, it is flowing to-day, and a citizen of Boston owns the property. Wells drilled near the Suez canal in 1885 found petroleum. So the gay world jogs on. Mummified Pharaohs are burned as fuel to drive locomotives over the Sahara, while the Zantean fount whose oil besmeared "the swathed and bandaged carcasses" is purchased by a Massachusetts bean-eater! Yet victims of "that tired feeling" turn to namby-pamby fiction for real romance!

Asphaltum is found in the Dead Sea, the supposed site of Sodom and Gomorrah, and on the surface of a chain of springs along its banks, far below the level of the ocean. Strabo referred to this remarkable feature two thousand years ago. The destruction of the two ill-fated cities may have been connected with, if not caused by, vast natural stores of this inflammable petroleum. The immense accumulations of hardened rock-oil in the center and on the banks of the sea were oxidized into rosin-like asphalt. Pieces picked up from the waters are frequently carved, in the convents of Jerusalem, into ornaments, which retain an oily flavor. Aristotle, Josephus and Pliny mention similar deposits at Albania, on the shores of the Adriatic. Dioscorides Pedanius, the Greek historian, tells how the citizens of Agrigentum, in Sicily, burned petroleum in rude lamps prior to the birth of Christ. For two centuries it lighted the streets of Genoa and Parma, in northern Italy. Plutarch describes a lake of blazing petroleum near Ecbatana. Persian wells have produced oil liberally for ages, under the name of "naphtha," the descendants of Cyrus, Darius and Xerxes consuming the fluid for its light. The earliest records of China refer to petroleum and small quantities have been found in Thibet. An oil-fountain on one of the Ionian Islands has gushed steadily for over twenty centuries, without once going on a strike or taking a vacation. Austria and France likewise possess oil-springs of considerable importance. Thomas Shirley, in 1667, tested the contents of a shallow pit in Lancashire, England, which burned readily. Rev. John Clayton, who visited it a dozen years later, wrote in 1691 :

"I saw a ditch where the water burned like brandy. Country folk boil eggs and even meat in it."

Near Bitche, a small fort perched on the top of a peak, at the entrance of one of the defiles of Lorraine, opening into the Vosges Mountains—a fort which was of great embarrassment to the Prussians in their last French campaign—and in the valley guarded by this fortress stand the chateau and village of Walsbroun, so named from a strange spring in the forest behind it. In the middle ages this fountain was famous. Inscriptions, ancient coins and the relics of a Roman road attest that it had been celebrated in even earlier times. In the sixteenth century a basin and baths for sick people existed. No record of its abandonment has been preserved. In the last century it was rediscovered by a medical antiquarian, who found the naphtha, or white petroleum, almost exhausted.

Around the volcanic isles of Cape Verde oil floats on the water and to the south of Vesuvius rises through the Mediterranean, exactly as when "the morning stars sang together." Hanover, in Germany, boasts the most north-erly of European "earth-oils." The islands of the Ottoman Archipelago and Syria are richly endowed with the same product. Roumania is literally flowing with petroleum, which oozes from the Carpathians and pollutes the water-springs. Turkish domination has hindered the development of the Roumanian region. Southern Australia is blessed with bituminous shales, resembling those in Scotland, good for sixty gallons of petroleum to the ton. The New-Zealanders obtained a meager supply from the hill-sides, collecting carefully the droppings from the interior rocks, and several test wells have resulted satisfactorily. Verily, "no pent-up Utica confines" petroleum within the narrow compass of a nation or a continent. With John Wesley it may exultingly exclaim: "The world is my parish!"

The Rangoon district of India long yielded four hundred thousand hogs-heads annually, the Hindoos using the oil to heal diseases, to preserve timber and to cremate corpses. Birma has been supplied from this source for an unknown period. The liquid, which is of a greenish-brown color and resem-bles lubricating-oil in density, gathers in pits sunk twenty to ninety feet in beds of sandy clays, overlying slates and sandstones. Clumsy pots or buckets, oper-ated by quaint wind-lasses, hoist the oil slowly to the mouth of the pits, whence it is often carried across the country in leathern bags, borne on men's shoulders, or in earthen jars, packed into carts drawn by oxen. Major Michael Symes, ambassador to the Court of Ava in 1765, published a nar-rative of his sojourn, in which this passage occurs :

OIL-WELLS IN INDIA.

"We rode until two o'clock, at which hour we reached Yaynangheomn, or Petroleum Creek. * * * The smell of the oil is extremely offensive. It was nearly dark when we approached the pits. There seemed to be a great many pits within a small compass. Walking to the nearest, we found the aperture about four feet square and the sides lined, as far as we could see down, with timber. The oil is drawn up in an iron pot, fastened to a rope passed over a wooden cylinder, which revolves on an axis supported by two upright posts. When the pot is filled, two men take hold of the rope by the end and run down a declivity, which is cut in the ground, to a distance equal to the depth of the well. When they reach the end of the track the pot is raised to its proper elevation; the contents, water and oil together, are discharged into a cistern, and the water is afterwards drawn through a hole in the bottom. * * * When a pit yielded as much as came up to the waist of a man, it was deemed tolerably productive; if it reached his neck it was abundant, and that which reached no higher than his knee was accounted indifferent."

Labor-saving machinery has not forged to the front to any great degree in the oil-fields of the East Indies. For the Burmese trade flat-boats ascend the Irrawaddy to Rainanghong, a town inhabited almost exclusively by the potters who make the earthen jars in which the oil is kept for this peculiar traffic. The methods of saving and handling the greasy staple have not changed one iota since John the Baptist wore his suit of camel's-hair and curry-combed the Sadducees in the Judean wilderness. Progress cuts no ice beneath the shadows of the Himalayas, notwithstanding the missionary efforts of Xavier, Judson, Cary, Morrison and Duff.

Petroleum in India occurs in middle or lower tertiary rock. In the Rawalpindi district of the Punjab it is found at sixteen localities. At Gunda a well yielded eleven gallons a day for six months, from a boring eighty feet deep, and one two-hundred feet deep, at Makum, produced a hundred gallons an hour. The coast of Arakan and the adjacent islands have long been famed for mud volcanoes caused by the eruption of hydrocarbon gasses. Forty. thousand gallons a year of petroleum have been exported by the natives from Kyoukpyu. The oil is light and pure. In 1877 European enterprise was attracted to this industry and in 1879 work was undertaken by the Borongo Oil Co. The company started on a large scale and in 1883 had twenty-four wells in operation, ranging from five-hundred to twelve-hundred feet in depth, one yielding for a few weeks one-thousand gallons daily. The total pumped from ten wells during the year was a quarter-million gallons; and in 1884 the company had to suspend payment. Large supplies of high-class petroleum might be obtained from this region, if suitable methods of working were employed.

Japan also takes a position in the oleiferous procession allied to that of the yellow dog under the band-wagon. At the base of Fuji-Yama, a mountain of respectable altitude, the thrifty subjects of the Mikado manage a cluster of oil-pits in the style practiced by their forefathers. The miry holes, the creaking apparatus and the general surroundings are second editions of the Rangoon exhibits. Yum-Yum's countrymen are clever students and they have much to learn concerning petroleum. Twenty-one years ago a Japanese nobleman inspected the Pennsylvania oil-fields, sent thither to report to the government all about the American system of operating the territory. His observations, embodied in an official statement, failed to amend the moss-grown processes of the Fuji-Yamans, who preferred to "fight it out on the old line if it took all summer."

In 1874 S. G. Bayne, now president of the Seaboard Bank of New York and the most successful disposer of oil-well outfits "that e'er the sun shone on," visited these oriental regions. The hard fate of the benighted heathen moved him to briny tears. They had never heard or read of "the annealed steel coupling," "the Palm link," the tubing, casing, engines and boilers the distinguished tourist had planted in every nook and corner of Oildom. With the spirit of a true philanthropist, Bayne determined to "set them on a higher plane." His choicest Hindostanee persiflage was aired in detailing the advantages of the Pennsylvania plan of running the petroleum machine. Tales of fortunes won on Oil Creek and the Allegheny River were garnished with scintillations of Irish wit that ought to have convulsed the listeners. Alas! the supine Asiatics were not built that way and the good seed fell upon barren soil. The story and, despite the finest lacquer and veneer embellishments, the experience were repeated in Japan. What better could be expected of pagans who wore skirts for full-dress, practiced hari-kari and knew not a syllable about Brian Boru? Their conduct was another convincing evidence of "the stern Calvinistic doctrine" of total depravity. The Japs voted to stay in their venerable rut and not monkey with the Yankee buzz-saw. "And the band played on."

Years afterwards two cars of drilling-tools and well-machinery were shipped to Calcutta and a couple of complete rigs to Yeddo— "only this and nothing more." The genial Bayne attempted to square the account by printing his eastern adventures and sending marked copies of translations to the Indo-Japanese press. Doubtless the waste-basket received what the office-cat spared of this un-

S. G. BAYNE.

usual consignment. Mr. Bayne began his prosperous career as an oilman by striking a snug well in 1869, on Pine Creek, near Titusville. His unique advertisements have spread his fame from the Atlantic to the Pacific. Digest these random flecks of heart-foam as samples of originality worthy of John J. Ingalls:

"We never make kite-track records; our speed takes in the full circle."

"The graveyards of the enemy are the monuments of our success."

"We never speak of our goods without glancing at the bust of George Washington which squats on the top of our annealed steel safe; a twenty-five cent plaster cast of George lends an atmosphere of veracity to a trade which in these days it sometimes needs "

"Abdul Aziz, the late Sultan of Morocco, bought a cheap boiler to drill a water well. It bu'st and he is now Abdul Azwas."

"We will never be buried with the ' unknown dead '—we advertise."

"Our patent coupling is the precipitated vapor of fermented progress."

"The intellectual and æsthetic are provided for in consanguinity to their taste."

"Our conversational soloists never descend to orthochromatic photography in their orphean flights; they hug the shore of plain Anglo-Saxon and scoop the doubting Thomas."

"It will never do to shake a man because the lambrequins begin to appear on the bottom of his pants and he wears a 'dickey' with a sinker."

"The Forget-me-nots of to-day are frequently found the Has-beens of to-morrow."

"Our collector bestrides a hearse-horse, and you can bore a well before he reaches your neck of woods."

"Credit is the flower that blooms in life's buttonhole."

"Many a man who now gives dinner-parties in a Queen-Anne front would be nibbling his Frankfurter in a Mary-Ann back had we not given him a helping hand at the right moment."

The classic ground of petroleum is the little peninsula of Okestra, jutting into the Caspian Sea. Extraordinary indications of oil and gas extend over a strip of country twenty-five miles long by a half-mile wide, in porous sandstone. Springs of heavier petroleum flow from hills of volcanic rocks in the vicinity. Open wells, in which the oil settles as it oozes from the rocks, are dug sixteen to twenty feet deep. For countless generations the simple natives dipped up the sticky fluid and carried it great distances on their backs, to burn in its crude state, besides sending a large amount yearly to the Shah's dominions. It is a forbidding spot—rocky, desolate, without a stream or a sign of vegetation. The unfruitful soil is saturated with oil, which exudes from the

2

neighboring hills and sometimes filters into receptacles hewn in the rock at a prehistoric epoch. On gala days it was part of the program to pour the oil into the Caspian and set it ablaze, until the sea and land and sky appeared one

CLASSIC GROUND OF PETROLEUM.

unbroken mass of vivid, lurid, roaring flame. The "pillar of fire" which guided the wandering Israelites by night could scarcely have presented a grander spectacle. The sight might well convey to awe-stricken beholders intensely realistic notions of the place of punishment Col. Ingersoll and Henry Ward Beecher have sought by tongue and pen to abolish. "Old Nick," however, at last advices was still doing a wholesale business at the old stand!

Near Belegan, six miles from the chief village of the Baku District, the grandest of these superb exhibitions was given in 1817. A column of flame, six-hundred yards in diameter, broke out naturally, hurling rocks for days together, until they raised a mound nine-hundred feet high. The roar of steaming brine was terrific. Oil and gas rise wherever a hole is bored. The sides of the mountain are black with dark exudations, while a spring of white oil issues from the foot. A clay pipe or hollow reed, steeped in limewater and set upright in the floor of a dwelling, serves as a sufficient gas-pipe. No wonder such a land as Baku, where in the fissures of the earth and rock the naphtha vapors flicker into flame, where a boiling lake is covered with flame devoid of sensible heat, where after the autumn showers the surrounding country seems wrapped in fire, where the October moon lights up with an azure tint the entire west and Mount Paradise dons a robe of fiery red, where innumerable jets of flame cover the plains on moonless nights, where all the phenomena of distillation and combustion can be studied, should have aroused the religious sentiment of oriental mystics. The adoring Parsee and the cold-blooded chemist might worship side by side. Amidst this devouring element men live and love, are born and die, plant onions and raise sheep, as in more prosaic regions.

At the southern extremity of the peninsula oil and gas shoot upward in a huge pyramid of quenchless light. Here is "the eternal fire of Aaku," burning as when Zoroaster reverently beheld it and flame became the symbol of Deity to the entranced Parsees. Here the poor Gheber gathered the fuel to feed the sacred fire which burns perpetually on his altar. Hither devout pilgrims journeyed even from far-off Cathay, to do homage and bear away a few drops of the precious oil, before the wolf had suckled Romulus or Nebuchadnezzar had been turned out to pasture. At Lourakhanel, not far from Baku, is a temple built by the fire-worshipers. The sea in places has such quantities of gas that it can be lighted and burned on the surface of the water until extinguished by a strong wind. Strange destiny of petroleum, first and last, to be the panderer of idolatry—fire-worship in the olden time, mammon-worship in this era of the "almighty dollar!"

Developments from Baku to the region north of the Black Sea, seven hundred miles westward, have revealed vast deposits of petroleum. Hundreds of wells have been drilled, some flowing one-hundred-thousand barrels a day ! Nobel Brothers' No. 50, which commenced to spout in 1886, kept a stream rising four-hundred feet into the air for seventeen months, yielding three-million barrels. This would fill a ditch five feet wide, six feet deep and a hundred miles long. These monsters eject tons of sand daily, which piles up in high mounds. Stones weighing forty pounds have been thrown out. The common way of obtaining the oil is to raise it by means of long metal cylinders with trap bottoms. Pumps are impossible, on account of the fine sand coming up with the oil. These cylinders will hold from one to four barrels and, on being raised to the surface, are discharged into pipes or ditches. Each trip of the bucket or cylinder takes a minute-and-a-half and the well is worked day and night. The average daily yield of a Russian well is about two-hundred barrels.

Pipe-lines, refineries and railroads have been provided and the three big companies operating the whole field consolidated in 1893. The Rothschilds combined with the Nobels and a prohibitory tariff prevents the importation of foreign oils. Tank-steamers ply the Caspian Sea and the Volga, many of the railways use the crude oil for fuel and the supply is practically unlimited. The petroleum products are carried in these steamers to a point at the mouth of the Volga River called Davit Foot, about four-hundred miles north of Baku and ninety miles from Astrakhan, and transferred into barges. These are towed by small tug-boats to the various distributing points on the Volga, where tanks have been constructed for railway shipments. The chief distributing point upon the Volga is Tsaritzin, but there is also tankage at Saratof, Kazan, and Nijni-Novgorod. From these points it is distributed all over Russia in tank-cars. Some is exported to Germany and to Austria. Russian refined may not be as good an illuminant as the American, but it is made to burn well enough for all purposes and emits no disagreeable odor. After taking from crude thirty per cent. illuminating distillate, about fifteen per cent. is taken from the residuum. It is called "solar oil" and the lubricating-oil distillate is next taken off. From this distillate a very good lubricant is obtained, affected neither by intense heat nor cold. The lubricating oil is made in Baku, but great quantities of the distillate are shipped to England, France, Belgium and Germany and there purified.

Russian competition was for years the chief danger that confronted American producers. Three partial cargoes of petroleum were sent to the United States as an experiment, netting a snug profit. Heaven favors the hustler from Hustlerville, who hoes his own row and doesn't squat on a stump expecting the cow will walk up to be milked, and the Yankee oilmen are not easily downed. They have perfected such improvements in handling, transporting, refining and marketing their product that the major portion of Europe and Asia, outside of the czar's dominions, is their customer. Nailing their colors to the mast and keeping their powder dry, the oil interests of this glorious climate don't propose to quit barking until the last dog is dead !

All these things prove conclusively that petroleum is a veritable antique, always known and prized by millions of people in Asia, Africa and Europe, and not a mushroom upstart. Indeed, its pedigree sizes up to the most exacting Philadelphia requirement. Mineralogists think it was quietly distilling "underneath the ground" when the majestic fiat went forth: "Let there be light !" Happily "age does not wither nor custom stale the infinite variety" of its admirable qualities. Neither is it a hot-house exotic, adapted merely to a single clime or limited to one favored section of any country. It is scattered widely throughout the two hemispheres, its range of usefulness is extending constantly and it is not put up in retail packages, that exhaust speedily. Alike in the tropics and the zones, beneath cloudless Italian skies and the bleak Russian firmament, amid the flowery vales of Cashmere and the snow-crowned heights of the Caucasus, by the banks of the turbid Ganges and the shores of the limpid Danube, this priceless boon has ever contributed to the comfort and convenience of mankind.

A ragged street-Arab, taken to Sunday-school by a kind teacher, heard for the first time the story of Christ's boundless love and sufferings. Big tears coursed down his grimy cheeks, until he could no longer restrain his feelings. Springing upon the seat, the excited urchin threw his tattered cap to the ceiling and screamed "Hurrah for Jesus!" It was an honest, sincere, reverent tribute, which the recording angel must have been delighted to note. In like manner, considering its wondrous past, its glowing present and its prospective future, men, women and children everywhere, while profoundly grateful to the Divine Benefactor for the transcendent gift, may fittingly join in a universal "Hurrah for Petroleum!"

Don't make the mistake that Petroleum,
Like the kodak, the bike, or linoleum,
  Is something decidedly new;
Whereas it was known in the Garden
When Eve, in fig-leaf Dolly Varden,
  Gave Adam an apple to chew.
Nor deem it a human invention,
By reason of newspaper mention
Just lately commanding attention,
  Because it is Nature's own brew.

Repeatedly named in the Bible,
Let none its antiquity libel
  Or seek to explain it away.
It garnished Methuselah's table,
Was used by the builders of Babel
  And pilgrims from distant Cathay;
When Pharaoh and Moses were chummy
It help'd preserve many a mummy,
Still dreadfully life-like and gummy,
  In Egypt's stone tombs from decay!

At Baku Jove's thunderbolts fir'd it,
Devout Zoroaster admir'd it
  As Deity symbol'd in flame;
Parsees from the realms of Darius,
Unweariedly earnest and pious,
  Adoring and worshipping came.
It cur'd Noah's Ham of trichina,
Greas'd babies and pig-tails in China,
Heal'd Arabs from far-off Medina—
  The blind and the halt and the lame!

Herodotus saw it at Zante,
It blazed in the visions of Dante
  And pyres of supine Hindostan;
The tropics and zones have rich fountains,
It bubbles 'mid snow-cover'd mountains
  And flows in the pits of Japan.
Confin'd to no country or nation,
A blessing to God's whole creation
For light, heat and prime lubrication,
  All hail to this grand gift to man!

EXPORTS OF PETROLEUM.

QUANTITY AND VALUE OF PETROLEUM—REFINED REDUCED TO CRUDE EQUIVALENT—EXPORTED FROM THE UNITED STATES FROM SEPTEMBER 1, 1859, TO JUNE 1, 1894, WITH THE AVERAGE YEARLY PRICE OF REFINED AT NEW YORK.

ENDING JUNE 30.	TOTAL GALLONS.	VALUE AT NEW YORK.	REFINED, PER GAL.
1860	1,300	$8,50	. . . .
1861	12,700	5,800	$0 61⅛
1862	400,000	146,000	36⅜
1863	. . . . .	. . . . .	44¾
1864	22,210,369	10,782,689	65
1865	25,496,849	16,563,413	58¾
1866	50,987,341	24,830,887	42⅛
1867	70,255,581	24,407,642	28⅜
1868	79,456,888	21,810,676	29⅛
1869	100,636,684	31,127,433	32¾
1870	113,735,294	32,668,960	26⅜
1871	149,892,691	36,894,810	24¾
1872	145,171,583	34,058,390	23⅝
1873	187,815,187	42,050,756	17⅞
1874	247,806,483	41,245,815	13
1875	221,955,308	30,078,568	13
1876	243,660,152	32,915,786	19⅛
1877	309,198,914	61,789,438	15⅛
1878	338,841,303	46,574,974	10¾
1879	378,310,010	40,305,249	08⅛
1880	423,964,699	36,2.8,625	09
1881	397,660,262	40,315,609	08
1882	559,954,590	51,232,706	07⅜
1883	505,931,622	44,913,079	08
1884	513,660,092	47,103,248	08⅛
1885	574,628,180	50,257,947	08
1886	577,781,752	50,199,844	07⅛
1887	592,803,267	46,824,933	06¾
1888	578,351,638	47,042,409	07⅝
1889	616,195,459	49,913,677	07⅛
1890	664,491,498	51,403,089	07⅜
1891	710,124,077	52,026,734	06⅞
1892	715,471,979	44,805,992	06
1893	804,337,168	42,142,058	05¼
1894	908,281,968	41,499,806	05⅙
Totals	11,829,482,888	$1,224,157,192	

VIEW OF BRIDGEWATER MINE, LUZERNE COUNTY, PA., IN 1863

II.

AMERICA ON DECK.

NUMEROUS INDICATIONS OF OIL ON THIS CONTINENT—LAKE OF ASPHALTUM—
PETROLEUM SPRINGS IN NEW YORK AND PENNSYLVANIA—HOW HISTORY IS
MANUFACTURED—PIONEERS DIPPING AND UTILIZING THE PRECIOUS FLUID—
TOMBSTONE LITERATURE—PATHETIC EPISODE—SINGULAR STRIKE—GEOLOGY
TRIES TO EXPLAIN A KNOTTY POINT.

"Near the Niagara is an oil spring known to the Indians."
—*Joseph de la Roche D'Altion, A. D. 1629.*
"There is a fountain at the head of the Ohio, the water of which is like oil, has a taste of iron
and seems to appease pain."—*Captain de Joncaire, A. D. 1721.*
"It is light bottled-up for tens of thousands of years—light absorbed by plants and vegetables. * * * And now after being buried long ages, that latent light is again brought forth
and liberated and made to work for human purposes."—*Stephenson.*

HE LAND Columbus ran against, by
anticipating Horace Greeley's advice to
"Go West," was not neglected in the
lavish distribution of petroleum. It
abounds in South America, the West
Indies, the United States and Canada.
The most extensive and phenomenal
natural fountain of petroleum ever known
is on the Island of Trinidad. Hot bitumen has filled a basin four miles in cir-
cumference, three-quarters of a mile from
the sea, estimated to contain the equivalent of ten-millions of barrels of crude-
oil. The liquid boils up continually, observing no holidays or Sundays, seething
and foaming at the center of the lake, cooling and thickening as it recedes,
and finally becoming solid asphaltum. The bubbling, hissing, steaming caldron emits a sulphurous odor, perceptible for ten or twelve miles and decidedly suggestive of the orthodox Hades. Humboldt in 1799 reported his
impressions of this spontaneous marvel, in producing which the puny hand of
man had no share. From it is derived the dark, tough, semi-elastic material,
first utilized in Switzerland for this purpose, which paves the streets of scores
of cities. Few stop to reflect, as they glide over the noiseless surface on
whirling bicycles or behind prancing steeds, that the smooth asphaltum
pavements and the clear "water-white" in the piano-lamp have a common

parentage. Yet bloomers and pantaloons, twin creations of the tailor, or diamonds and coal, twin links of carbon, are not related more closely.

The earliest printed reference to petroleum in America is by Joseph de la Roche D'Allion, a Franciscan missionary who crossed the Niagara river from Canada in 1629 and wrote of oil, in what is now New York, known to the Indians and by them given a name signifying "plenty there." Likely this was the petroleum occupying cavities in fossils at Black Rock, below Buffalo, in sufficient abundance to be an object of commerce. Concerning the celebrated oil-spring of the Seneca Indians near Cuba, N. Y., which D'Allion may also have seen, Prof. Benjamin Silliman in 1833 said :

"This is situated in the western part of the county of Alleghany, in the state of New York. This county is the third from Lake Erie on the south line of the state, the counties of Cattaraugus and Chautauqua lying west and forming the southwestern termination of the state of New York. The spring is very near the line which divides Alleghany and Cattaraugus. * * * The country is rather mountainous, but the road running between the ridges is very good and leads through a cultivated region rich in soil and picturesque in scenery. Its geographical formation is the same as that which is known to prevail in the western region; a silicious sandstone with shale, and in some places limestone, is the immediate basis of the country. * * * The oil-spring or fountain rises in the midst of a marshy ground. It is a muddy, dirty pool of about eighteen feet in diameter and is nearly circular in form. There is no outlet above ground, no stream flowing from it, and it is, of course, a stagnant water, with no other circulation than than which springs from the changes in temperature and from the gas and petroleum that are constantly rising through the pool.

"We are told that the odor of petroleum is perceived at a distance in approaching the spring. This may be true in particular states of the wind, but we did not distinguish any peculiar smell until we arrived on the edge of the fountain. Here its peculiar character became very obvious. The water is covered with a thin layer of petroleum or mineral oil, as if coated with dirty molasses, having a yellowish-brown color.

"They collect the petroleum by skimming it like cream from a milk-pan. For this purpose they use a broad, flat board, made thin at one edge like a knife; it is moved flat upon and just under the surface of the water and is soon covered by a coating of petroleum, which is so thick and adhesive that it does not fall off, but is removed by scraping the instrument upon the lip of a cup. It has then a very foul appearance, but it is purified by heating and straining it while hot through flannel. It is used by the people of the vicinity for sprains and rheumatism and for sores on their horses."

The "muddy, dirty pool" was included in an Indian reservation, one mile square, leased in 1860 by Allen, Bradley & Co., who drove a pipe into the bog. At thirty feet oil began to spout to the tune of a-barrel-an-hour, a rhythm not unpleasing to the owners of the venture. The flow continued several weeks and then "stopped short, never to go again." Other wells followed to a greater depth, none of them proving sufficiently large to give the field an orchestra chair in the petroleum arena.

It is told of a jolly Cuban, wearing a skull innocent of garbage as Uncle Ned's, who "had no wool on the top of his head in the place where the wool ought to grow," that he applied oil from the "dirty pool" to an ugly swelling on the apex of his bare cranium. The treatment lasted a month, by which time a crop of brand-new hair had begun to sprout. The welcome growth meant business and eventually thatched the roof of the happy subject with a luxuriant vegetation that would have turned Paderewski, Absalom, or the most ambitious foot-ball kicker green with envy! Tittlebat Titmouse, over whose excruciating experiences with the "Cyanochaitanthropopoion" that dyed his locks a bright emerald readers of "Ten-Thousand a Year" have laughed consumedly, was "not in it" compared with the transformed denizen of the pretty village nestling amid the hills of the Empire State. Those inclined to pronounce this a bald-headed fabrication may see for themselves the precise spot the mud-hole furnishing the oil occupied prior to the advent of the prosaic, unsentimental driving-pipe.

Captain de Joncaire, a French officer in colonial days, who had charge of military operations on the Upper Ohio and its tributaries in 1721, reported "a fountain at the head of a branch of the Ohio, the water of which is like oil." Undoubtedly this was the same "fountain" referred to in the *Massachusetts Magazine* for July, 1791, as follows:

" In the northern part of Pennsylvania is a creek called Oil Creek, which empties into the Allegheny river. It issues from a spring on which floats an oil similar to that called Barbadoes tar, and from which one may gather several gallons a day. The troops sent to guard the western posts halted at this spring, collected some of the oil and bathed their joints with it. This gave them great relief from the rheumatism, with which they were afflicted."

The history of petroleum in America commences with the use the pioneer settlers found the red-men made of it for medicine and for painting their dusky bodies. The settlers adopted its medicinal use and retained for various affluents of the Allegheny the Indian name of Oil Creek. Both natives and whites collected the oil by spreading blankets on the marshy pools along the edges of the bottom-lands at the foot of steep hill-sides or of mountain-walls that hem in the valleys supporting coal-measures above. The remains of ancient pits on Oil Creek—the Oil Creek ordained to become a household word—lined with timbers and provided with notched logs for ladders, show how for generations the aborigines had valued and stored the product. Some of these queer reservoirs, choked with leaves and dirt accumulated during hundreds of years, bore trees two centuries old. Many of them, circular, square, oblong and oval, sunk in the earth fifteen to twenty feet and strongly cribbed, have been excavated. Their number and systematic arrangement attest that petroleum was saved in liberal quantities by a race possessing in some degree the elements of civilization. The oil has preserved the timbers from the ravages of decay, "to point a moral or adorn a tale," and they are as sound to-day as when cut down by hands that crumbled into dust ages ago.

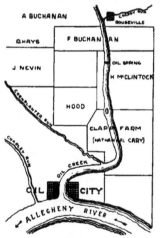

OIL-SPRING ON OIL CREEK.

Scientists worry and perspire over "the mound-builders" and talk glibly of "a superior race anterior to the Indians," while ignoring the relics of a tribe smart enough to construct enduring storehouses for petroleum. People who did such work and filled such receptacles with oil were not slouches who would sell their souls for whiskey and their forest heritage for a string of glass-beads. Did they penetrate the rock for their supply of oil, or skim it drop by drop from the waters of the stream? Who were they, whence came they and whither have they vanished? Surely these are conundrums to tax the ingenuity of imaginative solvers of perplexing riddles. Shall Macaulay's New-Zealand voyager, after viewing the ruins of London and flying across the Atlantic, gaze upon the deserted oil-wells of Venango county a thousand years hence and wonder what strange creatures, in the dim and musty past, could have bored post-holes so deep and so promiscuously? Rip Van Winkle was right in his plaintive wail : " How soon are we forgotten !"

The renowned "spring" which may have supplied these remarkable vats was located in the middle of Oil Creek, on the McClintock farm, three miles above Oil City and a short distance below Rouseville. Oil would escape from the rocks and gravel beneath the creek, appearing like air-bubbles until it reached the surface and spread a thin film reflecting all the colors of the rainbow. From shallow holes, dug and walled sometimes in the bed of the stream, the oil was skimmed and husbanded jealously. The demand was limited and the enterprise to meet it was correspondingly modest. Nathanael Cary, the first

FIRST OIL " SHIPPED " TO PITTSBURG.

tailor in Franklin and owner of the tract adjoining the McClintock, peddled it about the townships early in the century, when the population was sparse and every good housewife laid by a bottle of "Seneca Oil" in case of accident or sickness. Cary would sling two jars or kegs across a faithful horse, belonging to the class of Don Quixote's "Rosinante" and too sedate to scare at anything short of a knickerbockered feminine astride a rubber-tired wheel. Mounting the willing steed, which carried him steadily as "Jess" bore the self-denying physician of "Beside the Bonnie Brier Bush," the tailor-peddler went his rounds at irregular intervals. Occasionally he took a ten-gallon cargo to Pittsburg, riding with it eighty miles on horseback and trading the oil for cloth and groceries. His memory should be cherished as the first "shipper" of petroleum to "the Smoky City," then a mere cluster of log and frame buildings in a patch of cleared ground surrounding Fort Pitt. "Things are different now."

The Augusts, a family living in Cherrytree township and remembered only by a handful of old residents, followed Cary's example. Their stock was procured from springs farther up Oil Creek, especially one near Titusville, which achieved immortality as the real source of the petroleum development that has astounded the civilized world. They sold the oil for "a quarter-dollar a gill" to the inhabitants of neighboring townships. The consumption was extremely moderate, a pint usually sufficing a household for a twelvemonth. Nature's own remedy, it was absolutely pure and unadulterated, a panacea for "the thousand natural shocks that flesh is heir to," and positively refused to mix with water. If milk and water were equally unsocial, would not many a dispenser of the lacteal fluid train with Othello and "find his occupation gone?" Don't "read the answer in the stars ;" let the overworked pumps in thousands of barnyards reply !

No latter-day work on petroleum, no book, pamphlet, sketch or magazine article of any pretensions has failed to reproduce part of a letter purporting to have been sent in 1750 to General Montcalm, the French commander who perished at Quebec nine years later, by the commander of Fort Du Quesne, now Pittsburg. A sherry-cobbler minus the sherry would have been pronounced less insipid than any oil-publication omitting the favorite extract. It has been quoted as throwing light upon the religious character of the Indians and offered as evidence of their affinity with the fire-worshippers of the orient ! Official reports printed and endorsed it, ministers embodied it in missionary sermons and it posed as infallible history. This is the paragraph :

"I would desire to assure you that this is a most delightful land. Some of the most astonishing natural wonders have been discovered by our people. While descending the Allegheny, fifteen leagues below the mouth of the Conewango and three above the Venango, we were invited by the chief of the Senecas to attend a religious ceremony of his tribe. We landed and drew up our canoes on a point where a small stream entered the river. The tribe appeared unusually solemn. We marched up the stream about half-a-league, where the company, a band, it appeared, had arrived some days before us. Gigantic hills begirt us on every side. The scene was really sublime. The great chief then recited the conquests and heroism of their ancestors. The surface of the stream was covered with a thick scum, which upon applying a torch at a given signal, burst into a complete conflagration. At the sight of the flames the Indians gave forth the triumphant shout that made the hills and valleys re-echo again. Here, then, is revived the ancient fire-worship of the East ; here, then, are the Children of the Sun."

The style of this popular composition, in its adaptation to the occasion and circumstances, rivals Chatterton's unsurpassed imitations of the antique. Montcalm was a gallant soldier who lost his life fighting the English under General Wolfe, the hero whose noble eulogy of the poet Gray—" I would rather be the author of the ' Elegy Written in a Country Churchyard ' than the captor of Quebec "—should alone crown him with unfading laurels. The commander of Fort Du Quesne also "lived and moved and had a being." The Allegheny River meanders as of yore, the Conewango empties into it at Warren, the "Venango" is the French Creek which joins the Allegheny at Franklin. The "small stream" up which they marched "about half-a-league" was Oil Creek and the destination was the oil-spring of Joncaire and "Nat" Cary. The "gigantic hills" have not departed, although the "thick scum" is stored in iron tanks. But neither of the French commanders ever wrote or read or heard of the much-quoted correspondence, for the excellent reason that it had not been evolved during their sojourn on this mundane sphere !

Franklin, justly dubbed "the nursery of great men," gave birth to the pretty story. Sixty-six years ago a bright young man was admitted to the bar and opened a law-office in the attractive hamlet at the junction of the Allegheny River and French Creek. He soon ranked high in his profession and in 1839 was appointed judge of a special district-court, created to dispose of accumulated business in Venango, Crawford, Erie and Mercer counties. The same year a talented divinity-student was called to the pastorate of the Presbyterian church in Franklin. The youthful minister and the new judge became warm friends and cultivated their rare literary tastes by writing for the village paper, a six-column weekly. Among others they prepared a series of fictitious articles, based upon the early settlement of Northwestern Pennsylvania, designed to whet the public appetite for historic and legendary lore. In one of these sketches the alleged letter to Montcalm was included. Average readers supposed the minute descriptions and bold narratives were rock-ribbed facts, an opinion the authors did not care to controvert, and at length the "French commander's letter" began to be reprinted as actual, bona-fide, name-blown-in-the-bottle history !

One of the two writers who coined this interesting "fake" was Hon.

James Thompson, the eminent jurist, who learned printing in Butler, practiced law in Venango county, served three terms—the last as speaker—in the Legislature and one in Congress, was district-judge six years and sat on the Supreme

REV. NATHANAEL R. SNOWDEN.

bench fifteen years, five of them as chief-justice of this state. Judge Thompson removed to Erie in 1842 and finally to Philadelphia. He married a daughter of Rev. Nathanael R. Snowden, first pastor of the First Presbyterian church in Harrisburg, in 1794–1803, and afterwards master of a noted academy at Franklin. Mr. Snowden's wife was the daughter of Dr. Gustine, a survivor of the frightful Wyoming massacre. Their son, an eminent Franklin physician of early times, was the father of the late Dr. S. Gustine Snowden and of Major-General George R. Snowden, of Philadelphia, commander of the National-Guard of Pennsylvania. The good minister died in Armstrong county, descending to the grave as a shock of wheat fully ripe for the harvest, leaving to his posterity the precious legacy of a good example. Let his memory and his grave be kept green.

——" What is death
To him who meets it with an upright heart?
A quiet haven, where his shatter'd bark
Harbors secure till the rough storm is past ;
Perhaps a passage overhung with clouds,
But at its entrance, a few leagues beyond,
Opening to kinder skies and milder suns
And seas pacific as the soul that seeks them."

Judge Thompson's literary co-worker was the Rev. Cyrus Dickson, D.D., who resigned his first charge in 1848, settled in the east and gained distinction in the pulpit and as a forcible writer. How thoroughly these kindred spirits, now happily reunited "beyond the smiling and the weeping," must have enjoyed the overwhelming success of their ingenious plot and laughed at the easy credulity which accepted every line of their contributions as gospel truth ! They could not fail to relish the efforts, prompted mainly by their fanciful scene on Oil Creek, to identify as Children of the Sun the savage braves in buckskin and moccasins whose noblest conception of heaven was an eternal surfeit of dog-sausage !

Signs of petroleum in the Keystone state were not confined to Oil Creek. Ten miles westward, in water-wells and in the bed and near the mouth of French Creek, the indications were numerous and unmistakable. The first white man to turn them to account was Marcus Hulings, of Franklin, the original Charon of Venango county. Each summer he would skim a quart or two of "earth-oil" from a tiny pond, formed by damming a bit of the creek, the fluid serving as a liniment and medicine. This was the small beginning of one whose relative and namesake, two generations later, was to rank as a leading oil millionaire. Hulings "ferried" passengers across the unbridged stream in

a bark-canoe and plied a keel-boat to Pittsburg, the round trip frequently requiring four weeks. Passengers were "few and far between," consequently
a book-keeper and a treasurer were not engaged to take care of the receipts.
The proprietor of the canoe-ferry cleared a number of acres, raised corn and
potatoes and lived in a log cabin, not far from the site of the brush-factory,
which stood for fifty years after his death. Probably he was buried in the north-
west corner of the old graveyard, beside his wife and son, of whom two sunken
headstones record :

<table>
<tr><td>In
memory of
Michael Hulings who
departed this life: the 9th
of August, 1797. Aged
27 years, 1 moth &
14 days.</td><td>Massar,
wife of
Marcus Hulings
Died
Feb. 9. 1813.
Aged 67 yrs,
2 ms and 22 ds.</td></tr>
</table>

The once hallowed resting-place of many worthy pioneers sadly needs the
kindly ministrations of some "Old Mortality" to replace broken slabs, restore
illegible inscriptions and brush away the obnoxious weeds. Quaint spelling
and lettering and curious epitaphs are not uncommon. Observe these examples :

| In memory of James and Catherine Ha nne Ho departed this life July 3 1830 JAMES AGED TWO years one months, ten days CETHERIN E AGED two months and 14 days. | JANE consort of DAVID KING who departed this life. April 14 1829 aged 31 years O may I see thy tribes rejoic and aid their triumphs with my voice this all. my Glory Lord to be joined to thy saints and near to. thee | In memory of Samuel Riddle, Esq. Born Aug. 4, 1821 At Scrubgrass. Died May 28, 1853, At Franklin, Venango County, Pa. Here lies an honest lawyer, Honored and respected living, Lamented and mourned dead. |

Trains on the Lake Shore Railroad thunder past the lower end of the quiet
"God's Acre," close to the mounds of the McDowells, the Broadfoots, the Bow-
mans, the Hales and other early settlers, but the peaceful repose of the dead
can be disturbed only by the blast of Gabriel's trumpet on the resurrection
morning.

The venerable William Whitman, familiarly called "Doctor," over whose
grave the snows of twenty-five long years have drifted, often told me how, when
a youngster, he carried water to the masons building Colonel Alexander
McDowell's stone house, on Elk street. He hemmed in a pool on the edge of
French Creek, soaked up the greasy scum with a piece of flannel, wrung out
the cloth and filled several bottles with dark-looking oil. The masons would
swallow doses of it, rub it on their bruised hands and declare it a sovereign
internal and external remedy. In early manhood Mr. Whitman settled in
Canal township, eleven miles northwest of Franklin, cultivated a farm and
reared a large family. It was the dream of his old age to see oil taken from
his own land. In 1866 two wells were drilled on the Williams tract, across the
road from Whitman's, with encouraging prospects. Depressed prices retarded
operations and these wells remained idle. Four years later my uncle, George
Buchanan, and myself drilled on the Whitman farm. The patriarch watched the

progress of the work with feverish interest, spending hours daily about the rig. A string of driving-pipe, up to that time said to be the longest—153 feet—ever needed in an oil-well, had to be forced down. Three feet farther a vein of sparkling water, tinged with sulphur, spouted above the pipe and it has flowed uninterruptedly since. The heavy tools pierced the rock rapidly and the delight of the "Doctor" was unbounded. He felt confident a paying well would result and waited impatiently for the decisive test. A boy longing for

GEORGE BUCHANAN.

Christmas or his first pair of boots could not be more keenly expectant. His fondest wish was not to be gratified. He took sick and died, after a very short illness, in 1870, four days before the well was through the sand and pumping at the rate of fifty barrels per diem !

This singular well merits a brief notice. From the first sand, not a trace of which was met in the two wells on the other side of the road, oil and gas arose through the water so freely that drilling was stopped and tubing inserted. In twenty-four hours the well yielded fifty-eight barrels of the blackest lubricant in America, 28° gravity, the hue of a stack of ebony cats and with plenty of gas to illuminate the neighborhood. Subsiding quickly, the tubing was drawn and the hole drilled in quest of the third sand, the rock which furnished the lighter petroleum on Oil Creek. Eight feet were found seven-hundred feet towards the antipodes, a torpedo was exploded, the tubing was put back and the well produced two barrels a day for a year, divided between the sands about equally, the green and black oils coming out of the pipe side-by-side and positively declining to merge into one. Other wells were drilled years afterwards close by, without finding the jugular. Mr. Whitman sleeps in the Baptist churchyard near Hannaville, the sleep that shall have no awakening until the Judgment Day. Mr. Buchanan, who operated at Rouseville, Scrubgrass, Franklin and Bradford, left the oil-regions six years ago for the Black Hills and he is "adding to his pile" in South Dakota.

Excavating for the Franklin canal in 1832, on the north bank of French Creek, opposite "the infant industry" of Hulings forty years previous, the workmen were annoyed by a persistent seepage of petroleum, execrating it as a nuisance. A well dug on the flats ten years later, for water, encountered such a glut of oil that the disgusted wielder of the spade threw up his job and threw his besmeared clothes into the creek ! When the oil excitement invaded the county-seat the greasy well was drilled to the customary depth and proved hopelessly dry ! At Slippery Rock, in Lawrence county, oil exuded abundantly from the sandy banks and bed of the creek, failing to pan out when wells were put down. A geological expert explains it in this manner :

" 'Surface shows' have been the fascination of many. The places of most copious escape to the surface were regarded as the favored spots where ' the drainage from the coal-measures, in defiance of the laws of gravity and hydro-dynamics, had obligingly deposited itself. Such shows' were always illusory. A great ' surface show ' is a great waste ; where nature plays the spendthrift she retains little treasure in her coffers. The production of petroleum in quantities

Applying this method, the place to find petroleum is where not a symptom of it is visible! An honest Hibernian, asked his opinion of a notorious falsifier, answered that "he must be chock-full ov truth, fur bedad he niver lets any ov it git out!" The above explanation is of this stripe. "Flee to the mountains of Hepsidam" rather than attempt to bore for oil in localities having "shows" of the very thing you are after! These dreadfully deceptive "shows" show that the oil has got out and emptied the "reservoirs in which nature had been hoarding it up!" This is a pretty rough joke on poor deluded nature! How could these "surface shows" have strayed off anyhow, unless connected with reservoirs of genuine petroleum at the outset? The first wells on Oil Creek and at Franklin were drilled beside "surface shows" which revealed the existence of petroleum and supplied Cary, August, McClintock and Hulings with the coveted oil. These wells produced petroleum "in quantities of economical importance," demonstrating that "such shows were *not* always illusory." Is nature buncoing petroleum-seekers by hanging out a Will-o'-the-Wisp signal where there is "little treasure in her coffers?" The failures at Slippery Rock and divers other places resulted from the fact that the seepages had traveled considerable distances to find breaks in the rocks that would permit of the "most copious escape."

Central and South America are fairly stocked with petroleum indications. In the early days of the Panama Railroad and during the construction of the ill-fated canal numerous efforts were made to explore the coal regions of the Atlantic, in proximity to the ports of Colon and Panama. These researches led to the discovery of bituminous shales and lignite near the port of Boca del Toro, on the Caribbean Sea. The map of Colombia shows a great indenture on the Atlantic Coast of the department of Cauca, formed by the Gulfo de Uraba, or Darian del Nord. Into this gulf flow the Atrato, Arboletes, Punta de Piedra and many small streams. Explorations on the Gulf of Uraba and its tributaries disclosed extensive strata of "oil rock" and "oil springs" near the Rio Arboletes. The largest of forty of these springs has a twelve-inch crater, which gushes oil sufficient to fill a six-inch pipe. Near this Brobdignagian spring is a petroleum-pond sixty feet in diameter and from three to ten feet deep. The flow of these oil-springs deserves the attention of geologists and investors. They lie at a distance of one to three miles from the shores of the gulf. The oil is remarkably pure, passing through a bed of coral, which seems to act as a filter and refiner. A proper survey of the oil-region of the Uraba would be interesting from a scientific and an industrial standpoint. The proper development of its possibilities might result in the control of the petroleum market of South America. The climate is too sultry for the display of seal sacques and fur overcoats, a palm-hat constituting the ordinary garb of the average citizen. This providential dispensation eliminates dudes and tailor-made girls, stand-up collars and bifurcated skirts from the domestic economy of the happy Isthmians.

In the canton of Santa Elena, Ecuador, embracing the entire area of country between the hot springs of San Vicente and the Pacific coast, petroleum is found in abundance. It is of a black color, its density varies, it is considered superior to the Pennsylvania product and is entirely free from offensive odor. Little has been done towards working these wells. The people are unacquainted with the proper method of sinking them and no well has exceeded a few feet in depth. Geologists think, when the strata of alumina and rock are pierced, reservoirs will be found in the huge cavities formed by volcanic convulsions of the Andes.

From the Chira to the Fumbes river, a desert waste one-hundred-and-eighty miles in length and fifteen miles in width, lying along the coast between the Pacific ocean and the Andes, the oil-field of Peru is believed to extend. For two centuries oil has been gathered in shallow pits and stored in vats, precisely as in Pennsylvania. The burning sun evaporated the lighter parts, leaving a glutinous substance, which was purified and thickened to the consistency of sealing-wax by boiling. It was shipped to southern ports in boxes and used as glazing for the inside of Aguardiente jars. The Spanish government monopolized the trade until 1830, when M. Lama purchased the land. In 1869 Blanchard C. Dean and Rollin Thorne, Americans, "denounced" the mine, won a lawsuit brought by Lama and drilled four wells two-hundred-and-thirty feet deep, a short distance from the beach. Each well yielded six to ten barrels a day, which deeper drilling in 1871-2 augmented largely. Frederic Prentice, the enterprising Pennsylvania operator, secured an enormous grant in 1870, bored several wells—one a thousand-barreler—erected a refinery, supplied the city of Lima with kerosene and exported considerable quantities to England and Australia. The war with Chili compelled a cessation of operations for some years. Dr. Tweddle, who had established a refinery at Franklin, tried to revive the Peruvian fields in 1887-8. He drilled a number of wells, refined the output, enlisted New York capital and shipped cargoes of the product to San Francisco. The district has been rather quiet of late, but qualified judges have no doubt that "in the sweet-bye-and-bye" the oleaginous goose may hang altitudinum in Peru.

THE BABY HAS GROWN.

The petroleum industry is the one circus bigger inside the canvas than on the posters.

A larger percentage of the oil product of the United States is sent abroad than of any other except cotton, while nearly every home in the land is blessed with petroleum's beneficent light.

> If needs be petroleum may well be defiant;
> The baby has grown to be earth's greatest giant.

Beginning with 1866, the exports of illuminating oils were doubled in 1868, again in 1871, again in 1877 and again in 1891. The average exports per week in 1894 were as much as for the entire year of 1864. The world has reason to be thankful for Pennsylvania petroleum, which has a wider sale than any other American product.

> Breadstuffs and cotton, iron and coal
> All have been distanc'd; oil has the pole.

From 23,000,000 gallons in 1864 the exports of Pennsylvania petroleum have multiplied thirty times. The total exports for the first three months of 1896 foot up 195,637,153 gallons, valued at $12,389,384. For the nine months ending March 31, 1896, they amounted to 650,676,974 gallons, valued at $45,-563,750. For the same period of 1895 the exports were 671,196,133 gallons and their value, $31,554,308. Not less impressive is the marvelous reduction in the price of refined, so that the oil has found a welcome everywhere. Export oil averaged in 1861, 61½ cents per gallon; in 1871, 23⅝ cents per gallon; in 1881, 8 cents per gallon; in 1891, 6⅞ cents per gallon; in 1892, 6 cents per gallon; in 1894, 5 1-6 cents per gallon, or one-twelfth that in 1861. But this decrease, great as it is, does not represent the real reduction in the price of oil, as the cost of the barrel is included in these prices. A gallon of bulk-oil cost in 1861 not less than 58 cents; in 1894, not more than 3½ cents, or hardly one-seventeenth. In January, 1861, the price was 75 cents; in January, 1894, one-twenty-fifth that of thirty-three years before. The consumers have received the benefit of constant improvements and reductions in prices, while twelve-hundred-million dollars have come from abroad to this country for petroleum.

> It takes oil to make a showing
> In the line of rapid growing,
> And make others quit their blowing.

The capital invested in the petroleum-business has increased from one-thousand dollars, raised in 1859 to drill the first well in Pennsylvania, to five-hundred-millions. It is just as easy to say five-hundred-million dollars as five-hundred-million grains of sand, but the possibilities of such a sum of money afford material for endless flights of the imagination. Thirty-thousand miles of pipe-lines handle the output most expeditiously, conveying it to the seaboard at less than teamsters used to receive for hauling it a half-mile. Ten-thousand tank-cars have been engaged in transportation. Seventy-five bulk-steamers and fleets of sailing-vessels carry refined from Philadelphia and New York to the remotest ports in Europe, Africa and Asia. Astral Oil and Standard White have penetrated "wherever a wheel can roll or a camel's foot be planted." In Pennsylvania thirty-five-million barrels have been produced and four-thousand wells drilled in a single year. Add to this the results of operations in Ohio, West Virginia, Indiana, Colorado, Kansas, Wyoming and California, and the whole race must acknowledge that "Petroleum is King."

3

VIEW IN OIL CITY, PA., AFTER THE FLOOD, MARCH 17, 1913.

III.

NEARING THE DAWN.

Salt-Water Helping Solve the Problem—Kier's Important Experiments
—Remarkable Shaft at Tarentum—West Virginia and Ohio to the
Front—The Lantern Fiend—What an Old Map Showed—Kentucky
Plays Trumps—The Father of Flowing Wells—Sundry Experiences
and Observations at Various Points.

"A salt-well dug in 1814, to the depth of four hundred feet, near Marietta, discharged oil peri-
odically at intervals of two to four days."—*Dr. Hildredth, A. D. 1819.*
" Nearly all the Kanawha salt-wells have contained more or less petroleum."—*Dr. J. P. Hale,
A. D. 1825.*
" There are numerous springs of this mineral oil in various regions of the West and South."—
Prof. B. Silliman, A. D. 1833.

AT RUFFNER BROS 1ST WELL

HILE cannel coal in the western end of Pennsylvania and other sections of the country,
bitumen and shales from the Gulf of Mexico to Lake Huron, chapapote or mineral
pitch in Cuba and San Domingo, oozings in
Peru and Ecuador, asphaltum in Canada
and oil-springs in Columbia and a half-
dozen states of the Union from California
to New York denoted the presence of petroleum over the greater part of this hemisphere, wells bored for salt were leading
factors in bringing about its full development. Scores of these wells pumped more
or less oil long before it " entered into the
mind of man" to utilize the unwelcome intruder. Indeed, so often were brine and
petroleum found in the same geological formation that scientists ascribed to
them a kindred origin. The first borings to establish this peculiarity were on
the Kanawha River, in West Virginia, a state destined to play an important part
in oleaginous affairs. Dr. J. P. Hale, a reputable authority, claims oil appeared
in the deep salt-wells of Ruffner Brothers, who began operations in 1806, causing
much annoyance. His account continues:

" Neary all the Kanawha salt-wells have contained more or less petroleum, and some of the
deeper wells a considerable flow. Many persons now think, trusting to their recollections, that
some of the wells afforded as much as twenty-five to fifty barrels per day. This was allowed to
flow over from the top of the salt cisterns to the river, where, from its specific gravity, it spread
over a large surface, and by its beautiful iridescent hues and not very savory odor could be traced
for many miles down the stream. It was from this that the river received the nickname of 'Old
Greasy,' by which it was long known by Kanawha boatmen and others."

25

At the mouth of Hawkinberry Run, three miles north of Fairmount, in Marion county, a well for salt was put down in 1829 to the depth of six hundred feet. "A stinking substance gave great trouble," an owner reported, "forming three or four inches on the salt-water tank, which was four feet wide and sixteen feet long." They discovered the stuff would burn, dipped it off with buckets and consumed it for fuel under the salt-pan. J. J. Burns in 1865 leased the farm, drilled the abandoned well deeper, stuck the tools in the hole and had to quit after penetrating sixty feet of "a fine grit oil rock." Mr. Burns wrote in 1871 regarding operations in Marion:

" The second well put down in this county was about the year 1835, on the West Fork River, just below what is now known as the Gaston mines. The well was sunk by a Mr. Hill, of Armstrong county, Pa., who found salt water of the purest quality and in a great quantity, same as in the first well. He died just after the well was finished, so nothing was done with it. About the time this well was completed one was drilled in the Morgan settlement, just below Rivesville. Salt water was ound with great quantities of gas. Twenty-five years since the farmers on Little Bingamon Creek formed a company and drilled a well—I think to a depth of eight-hundred feet—in which they claimed to have found oil in paying quantities. You can go to it to-day and get oil out of it. The president told me he saw oil spout out of the tubing forty or fifty feet, just as they started the pump to test it. The company got to quarreling among themselves, some of the stockholders died and part of the stock got into the hands of minor heirs, so nothing was done with the enterprise."

Similar results attended other salt-wells in West Virginia. The first oil-speculators were Bosworth, Wells & Co., of Marietta, Ohio, who as early as 1843 bought shipments of two to five barrels of crude from Virginians who secured it on the Hughes River, a tributary of the Little Kanawha. This was sold for medical purposes in Pittsburg, Baltimore, New York and Philadelphia. Two of the Pittsburg firms handling it were Schoomaker & Co. and B. A. Fahnestock.

Notable instances of this kind occurred on the Allegheny River, opposite Tarentum, twenty miles above Pittsburg, as early as 1809. Wells sunk for brine to supply the salt-works were troubled with what the owners called "odd, mysterious grease." Samuel M. Kier, a Pittsburg druggist, whose father worked some of these wells, conceived the idea of saving the "grease," which for years had run waste, and in 1846 he bottled it as a medicine. He knew it had commercial and medicinal value and spared no exertions to introduce it widely. He believed implicitly in the greenish fluid taken from his salt wells, at first as a healing agent and farther on as an illuminant. A bottle of the oil, corked and labeled by Kier's own hands, lies on my desk at this moment, in a wrapper dingy with age and redolent of crude. A four-page circular inside recites the good qualities of the specific in gorgeous language P. T. Barnum himself would not have scorned to father. For example:

" Kier's Petroleum, or Rock Oil, Celebrated for its Wonderful Curative Powers. A Natural Remedy ! Procured from a Well in Allegany Co., Pa., Four Hundred Feet below the Earth's Surface. Put up and Sold by Samuel M. Kier, 363 Liberty Street, Pittsburgh, Pa.

> The healthful balm, from Nature's secret spring,
> The bloom of health and life to man will bring;
> As from her depths this magic liquid flows,
> To calm our sufferings and assuage our woes.

" The Petroleum has been fully tested ! It was placed before the public as A REMEDY OF WONDERFUL EFFICACY. Every one not acquainted with its virtues doubted its healing qualities. The cry of humbug was raised against it. It had some friends—those who were cured through its wonderful agency. Those spoke in its favor. The lame through its instrumentality were made to walk—the blind to see. Those who had suffered for years under the torturing pains of RHEUMATISM, GOUT AND NEURALGIA were restored to health and usefulness. Several who were blind were made to see. If you still have doubts, go and ask those who have been cured ! * * * We have the witnesses, crowds of them, who will testify in terms stronger than we can write them to the efficacy of this remedy ; cases abandoned by physicians of unquestionable celebrity have been made to exclaim, " This Is the Most Wonderful Remedy Yet Discovered !" * * * Its transcendent power to heal must and will become known and appreciated. * * * The Petroleum is a Natural Remedy ; it is put up as it flows from the bosom of the earth, without anything being added to or taken from it. It gets its ingredients from the beds of substances which it passes over in its secret channel. They are blended together in such a way as to defy all human competition. * * * Petroleum will continue to be used and applied as a Remedy as long as man continues to be afflicted with disease. Its discovery is a new era in medicine."

A host of certificates of astonishing cases of curable and incurable ailments, from blindness to colds, followed this preliminary announcement. The "remedy" was trundled about by agents in vehicles elaborately gilt and painted with representations of the Good Samaritan ministering to a wounded Hebrew writhing in agony under a palm-tree. Two barrels of oil a day were sold at fifty cents a half-pint. The expense of bottling and peddling it consumed the bulk of the profits. Kier experimented with it for light, about 1848, burning it at his wells and racking his fertile brain for some means to get rid of the offensive smoke and odor. To be entirely successful the oil must have some other than this crude form. The tireless experimenter went to Philadelphia to consult a chemist, who advised distillation, without a hint as to the necessary apparatus. Fitting a kettle with a cover and a worm, the first outcome of the embryo refiner's one-barrel still was a dark substance little superior to the crude. Learning to manage the fires so as not to send the oil over too rapidly, by twice distilling he produced an article the color of cider, which had a horrible smell, as he knew nothing of the treatment with acids.

Slight changes in the camphene lamp enabled him to burn the distillate without smoke. Improvements in the lamp, especially the addition of the " Virna burner," and in the quality of the fluid brought the " carbon oil," as it was usually termed, to a goodly measure of perfection. One lot of oil used in these experiments was a purchase of six barrels in March, 1853, from Charles Lockart, now an officer of the Standard Oil Company in Pittsburg. It came from the Huff well, across the river from Tarentum. " Carbon oil " sold readily for a dollar-fifty per gallon and provided a market for all the petroleum the salt-wells in the vicinity could produce. The day was dawning and the great light of the nineteenth century had been foreshadowed in the broad commonwealth that was to send it forth on its shining mission.

SAMUEL M. KIER.

Samuel M. Kier slumbers in Allegheny cemetery, resting in peace " after life's fitful fever." He was the first to appreciate the value of petroleum and to purify it by ordinary refining. His product was in brisk demand for illuminating purposes. He invented a lamp with a four-pronged burner, arranged to admit air and give a steady light. If he failed to reap the highest advantage from his researches, to patent his process and to sink wells for petroleum alone, he paved the way for others, enlarged the field of the product's usefulness and by his labors suggested its extensive development. Has not he earned a monument more enduring than brass or marble?

These operations at Tarentum and Pittsburg led to an extraordinary attempt to fathom the petroleum basin by *digging* to the oil-bearing rock! Through Kier's experiments the crude had become worth from fifty cents to one dollar a gallon. Among the owners of Tarentum's salt-wells was Thomas Donnelly, who sold his well on the Humes farm to Peterson & Irwin. The senior partner, ex-Mayor Louis Peterson, of Allegheny, lived until recently to recount his inter. esting experiences with the coming light. He thought the Donnelly well, which produced salt-water only, if enlarged and pumped vigorously, would produce oil. Humes received twenty-thousand dollars for his farm. The hole was reamed out and yielded five barrels of petroleum a day. This was in 1856. A specimen sent to Baltimore was used successfully in oiling wool at the carding-mills and the total production was shipped to that city for eight years. Eastern capitalists bought the farm and well in 1864, organized "The Tarentum Salt and Oil Company" and determined to dig a shaft down to the source of supply! The wells were four-hundred and five-hundred feet deep. The officers of the company argued that it was feasible to reach that far into the bowels of the earth with pick and shovel and discover a monstrous cave of brine and oil! They picked a spot twenty rods from the Donnelly well, sent to England for skilled miners and started a shaft about eight feet square. Over two years were employed and forty thousand dollars spent in sinking this shaft. Heavy timbers walled the upper portion, the hard rock below needing none. The water was pumped through iron pipes, nine men formed each shift and the work progressed merrily to the depth of four-hundred feet. Then the salt-water in the Donnelly well was affected by the fresh-water in the shaft, losing half its strength whenever the latter was let stand a few hours, showing their intimate connection by veins or crevices. Mr. Peterson said of it :

" The digging of the shaft was finally abandoned in the darkest period of the war, from the necessities of the time. A New Yorker named Ferris, and Wm. McKeown, of Pittsburg, bought the property, shaft and all. The daring piece of engineering was neglected and finally commenced to fill up with cinders and dirt, until at last it was level again with the surface of the ground. You may walk over it to-day and I could point it out to you if I was up there. Dig it out and you will find those iron pipes and timbers still there, just as they were originally put in."

Dyed-in-the-wool Tarentumites insist that natural gas caused the suspension of work, flowing into the shaft at such a gait that the miners refused to risk the chances of a speedy trip to Kingdom Come by suffocation or the ignition of the subtile vapor. This was the case with two shafts at Tidioute and Petroleum Centre, neither of them nearly the depth of "the daring piece of engineering" which "set the pace" for enterprises of this novel brand. The New York Enterprise and Mining Company projected the former, intending to sink a shaft eight feet by twelve to the third sand and tunnel the rock for petroleum by wholesale. The shaft reached oil-producing sand at one-hundred-and-sixty feet. The miners worked in squads, eight-hour turns. Holes had been drilled into the rock at various angles and a lot of conglomerate brought to the surface. Once a short delay occurred in changing squads, during which the air-pump, employed to exhaust the gases from the pit and supply pure ozone from above, was let stand idle. Mr. Hart was seated on a timber across the shaft when the men were ready to go down. As was the custom, a man dropped a taper into the opening to test the air. Natural gas had filled the shaft and it ignited from the burning torch, causing a terrific explosion. The workmen were thrown in all directions and lay stunned and burned. When they regained conciousness Hart was nowhere to be seen and flames rose from the mouth of the pit to the tree-tops. Hart's body was eventually recovered from the bottom of the shaft, horribly mangled and charred. Work was abandoned and the hole was partly filled up and covered, none caring to pry farther into the petroleum secrets of nature. Were meddlers who seek to poke their noses into the secrets of other people dealt with thus summarily, what a thinning out of the population there would be!

Peterson & Irwin's treatment of the Donnelly well brings out clearly that the sole object was to procure oil. This is important, in view of the claim that the first well drilled exclusively for petroleum was put down in 1859. Practically the two Pittsburgers anticipated this by three years, a circumstance to remember when considering the varied events which led up to the petroleum development.

On an old map of the United States, printed in England in 1787, the word "petroleum" is marked twice, indicating that the "surface shows" of oil had attracted the notice of the earliest explorers of Southern Ohio and Northwestern Pennsylvania a century ago. In one instance it is placed at the mouth of the stream since famed the world over as Oil Creek, where Oil City is situated; in the other on a stream represented as emptying into the Ohio river, close to the site of what is now the village of Macksburg. When that section of Ohio was first settled, various symptoms of greasiness were detected, thin films of oil floating on the waters of Duck Creek and its tributaries, globules rising in different springs and seepings occurring frequently in the same manner as in Pennsylvania and West Virginia. Thirty miles north of Marietta, on Duck Creek, a salt well sunk by Mr. McKee, in 1814, to the depth of four-hundred-and-seventy-five feet, discharged "periodically, at intervals of from two to four days and from three to six hours' duration, thirty to sixty gallons of petroleum at each inception." Eighteen years afterwards the discharges were less frequent and the yield of oil diminished to one barrel a week, finally ceasing altogether. Once thirty or forty barrels stored in a cistern took fire from the gas

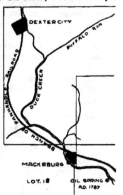

AN OLD OIL-SPRING IN OHIO.

at the well having been ignited by a workman carrying a light. The burning oil ran into the creek, blazed to the tops of the trees and exhibited for hours to the amazed settlers the novelty of a rivulet on fire. Ten miles above McConnellsville, on the Muskingum River, results almost identical attended the boring of salt-wells in 1819. Dr. S. P. Hildreth, of Marietta, in an account of the region, written that year, says of the borings for salt water:

"They have sunk two wells more than four hundred feet; one of them affords a strong and pure water, but not in great quantity; the other discharges such vast quantities of petroleum, or as it is vulgarly called "Seneca Oil," and besides is subject to such tremendous explosions of gas * * * that they make little or no salt. Nevertheless, the petroleum affords considerable profit and is beginning to be in demand for workshops and manufactories. It affords a clear, brisk light, when burned in this way, and will be a valuable article for lighting the street-lamps in the future cities of Ohio."

The last sentence has the force of a prophecy. Writing about the year

1832 the same observant author directs attention to another notable feature:

"Since the first settlement of the regions west of the Appalachian range the hunters and pioneers have been acquainted with this oil. Rising in a hidden and mysterious manner from the bowels of the earth, it soon arrested their attention and acquired great value in the eyes of the simple sons of the forest. * * * From its success in rheumatism, burns, coughs, sprains etc., it was justly entitled to its celebrity. * * * It is also well adapted to *prevent friction in machinery*, for, being free of gluten, so common to animal and vegetable oils, it preserves the parts to which it is applied for a long time in free motion; where a heavy vertical shaft runs in a socket it is preferable to all or any other articles. This oil rises in greater or less abundance in most of the salt-wells and, collecting where it rises, is removed from time to time with a ladle."

Is it not strange that, with the sources of supply thus pointed out in different counties and states and the useful applications of petroleum fairly understood, its real value should have remained unappreciated and unrecognized for more than thirty years and be at last determined through experiments upon the distillation of bituminous shales and coals? Wells sunk hundreds of feet for salt water produced oil in abundance, yet it occurred to no one that, if bored expressly for petroleum, it could be found in paying quantity! Hamilton McClintock, owner of the "oil-spring" famed in history and romance, when somebody ventured to suggest that he should *dig* into the rock a short distance, instead of skimming the petroleum with a flannel-cloth, retorted hotly: "I'm no blanked fool to dig a hole for the oil to get away through the bottom!"

If West Virginia, Pennsylvania and Ohio played trumps in the exciting game of Brine vs. Oil, Kentucky held the bowers. The home of James Harrod and Daniel Boone, Henry Clay and George D. Prentice was noted for other things besides backwoods fighters, statesmanship, sparkling journalism, thoroughbred horses, superb women and moonshine whiskey. Off in the southeast corner of Wayne county, near the northeast corner of a six-thousand-acre tract

KENTUCKY'S FIRST OIL-WELL.

of wild land, David Beatty bored a well for salt about the year 1818. The land extended four miles westward from the Big South Fork of the Cumberland River, its eastern boundary, and three miles down the Fork from Tennessee, its southern line. The well was located on a strip of flat ground between the stream and a rocky bluff, streaked with veins of coal and limestone. Five yards from the water a hole nine feet square was dug ten feet to the rock and timbered. The well, barely three inches in diameter, was punched one-hundred-and-seventy feet by manual labor, steam-engines not having penetrated the trackless forests of Wayne at that period. To the intense disgust of the workmen a black, sticky, viscid liquid persisted in coming up with the salt-water and a new location was chosen two miles farther down the creek. Extra care not to drill too deep averted an influx of the disagreeable fluid which spoiled the first venture. Salt-works were established and flourished for years, a simon-pure oasis in the interminable wilderness.

Beatty was elected to Congress, serving his constituents faithfully and illustrating the Mulberry-Sellers policy of "the old flag and an appropriation." He secured a liberal grant for a road to his property on the South Fork and constructed a passable thoroughfare. Traces of deep cuttings, log culverts and blasted rocks, still discernible amid the underbrush that well-nigh hides them from view, are convincing evidences of the magnitude and difficulty of the task. "The rocky road to Dublin" was a mere bagatelle in comparison with this long-deserted pathway. "Jordan is a hard road to travel," says an old song, and the sentiment would fit equally well in this case. At one rugged point holes were cut in a rock as steep as the roof of a house, to afford footing for the mules engaged in drawing salt from the works ! Considering the roughness of the country, the height of the hills, the depth of the chasms and the scanty facilities available, Beatty's road was quite as remarkable a feat as Bonaparte's passage across the Alps or Ben Butler's "Dutch-Gap Canal." Its spirited projector lived and died at Monticello, the county-seat, where his descendants resided until recently.

The abandoned well did not propose to be snuffed out unceremoniously or to enact the role of "Leah the Forsaken." In its bright lexicon the word fail was not to be inserted merely because it was too fresh to participate in the salt-trade. Far from retiring permanently, it spouted petroleum at a Nancy-Hanks quickstep, filling the pit, running into the Fork and covering mile after mile of the water with a top-dressing of oil. Somehow the floating mass caught fire and mammoth pyrotechnics ensued. The stream blazed and boiled and sizzled from the well to the Cumberland River, thirty-five miles northward, calcining rocks and licking up babbling brooks on its fiery march ! Trees on its banks burned and blistered and charred to their deepest roots. Iron-pans at the salt-wells got red-hot, shriveled, warped, twisted and joined the junk-pile ! Was not that a sweet revenge for plucky No. 1, the well its owner "had no use for" and devoutly wished at the bottom of the sea ?

The Chicago fire "couldn't hold a candle" to this rural conflagration, which originated the expressive phrase of "hell with the lid off," applied sixty years afterwards by James Parton to the flaming furnaces at Pittsburg. Unluckily, the region was populated so sparsely that few spectators had front seats at "the greatest show on earth." The deluge of oil ceased eventually, the fire following suit. Anon the salt industry began to languish and the works were dismantled. No more the forest road echoed the sharp crack of the teamster's whip or heard his lusty oaths. The district along the South Fork was left as silent as "the harp that once through Tara's halls the soul of music shed," ready to be labeled "Ichabod," and tradition alone preserved the name and record of the "Beatty Well," THE FIRST REAL OIL-SPOUTER IN AMERICA !

Accompanied by Dr. W. G. Hunter, and a native as guide, it was my good fortune to visit this memorable locality in 1877. The start was from Burksville, Cumberland county, the doctor's home and my headquarters for a twelvemonth. At Albany, Clinton county, sure-footed mules, the only animals that could be ridden safely through the rough country, took the place of our horses. Soon the last signs of civilization disappeared and we plunged into the thick woods, a crooked, tortuous trail pointing the way. Hills, rocks, ravines, fallen trees and mountain streams by turns impeded our progress, as we rode in Indian file for thirty miles. Birds twittered and snakes hissed at the invasion of their solitudes. Several times the path touched the line of Beatty's forgotten road and once a ruined cabin, with three grave-like mounds in a corner of the small clearing, met our gaze. The guide explained how, twenty years before, the poor family tenanting the wretched hovel had been poisoned by eating some kind of berries, the parents and their only child dying alone and unattended.   No human eye beheld their struggles, no soft hand cooled the fevered brows of the sufferers whose lives went out in that desolate waste.

Provisions in our saddle-bags, a clear brook and evergreen boughs supplied us with food, drink and an open-air bed.   Next morning we traversed a broad plateau, ending abruptly at the top of a precipitous bluff a hundred feet high. Beneath us lay a stretch of bottom-land, with the Big South Fork on its east side and the Cumberland Mountains rearing their bold crests five miles away.   In the center of a patch of cleared ground stood a shanty, built of poles and roofed with split slabs of oak.   From an open space in one end smoke escaped freely, showing that the place was inhabited.   Tethering the mules and throwing the saddles upon the grass, we crawled down a slope formed by the collapse of a portion of the bluff.   A shot from my revolver—everybody carried a pistol— shattered the atmosphere and brought the inmates to the side of the dwelling. The father, mother, a child in arms and two boys entering their teens watched our approach.   As we drew nigh they scampered into the shanty and took refuge under a queer structure of rails, straw and blankets that did duty as a bed for the household !   A blanket hung over the space cut for a door.   Drawing this aside, the frightened family could be seen crouching on the bare soil, for the abode had neither door, window, floor nor chinking between the logs.   It was quite unfit to shelter a decent porker.   Not a chair, table, stove, looking-glass, bureau or any of the articles of furniture deemed necessary for modern comfort was in sight !   A bench hewn out of timber with an axe, two metal pots, some tin-dishes and knives and forks composed the domestic outfit !   Yet it was "home" to the squalid beings huddled in the dark, damp, musty angle farthest from the intruders who had dropped in upon them so unexpectedly.

Calling them to come out and speak with us a moment, the woman appeared, bearing the inevitable baby.   She was truly a revelation, with unkempt brindle hair and sallow skin to match.   Her raiment consisted of a single jean garment, dirty and tattered beyond description, too narrow to encircle her waist and too short to reach within a dozen inches of her naked feet.   Compared with the flimsy toilet of "a living picture," this costume was simplicity itself.   The poor creature smoked a cob-pipe viciously.   A request to see her husband evoked the command : "Old man, I reckon you best git out hyer !"   The "old man" heeded this summons and emerged from his hiding-place, trembling vio- lently.   His attire was in harmony with his wife's, threadbare jean-pants and shirt comprising it.   Head and feet were bare.   His trembling ceased the instant he saw our guide, whom he knew and greeted cordially.   Introductions followed and we asked if he could show us the way to the Beatty Well.   He answered in perfect English, with the grace of a Chesterfield : "It will be the greatest pleas- ure I have known for many a day to go with you to the exact spot and give you all the information in my power !"

A brisk walk brought us to the well.   Dirt and leaves had filled the pit nearly level, forming a depression which one might pass without special notice. Scraping away the rubbish, blackened fragments of the timbered walls ap- peared.   But not a drop of oil had issued from the veteran well for scores of years.   One man alone survived of those who had gazed upon the flow of pe- troleum previous to the fire which checked the greasian tide forever.   He lived ten miles northwest and his short story was learned on the return trip by another route.   The scattered rustics were accustomed to go to the well once or twice a year and dip enough oil to medicate and lubricate whoever or whatever needed it. The fluid was dark and heavy and for years rose to within a few feet of the surface. At length the well clogged up and was almost obliterated. The dim eyes of the aged narrator sparkled as he recalled the big blaze, concluding with the emphatic words: "It jes' looked ez if the devil had hitched up the hull bottomless pit fur a torchlight percession!"

AT THE BEATTY WELL IN 1877.

Except the squatter on the tract of land, which Dr. Hunter and myself had secured the winter of our visit, the nearest settler lived five miles distant! The Cincinnati Southern Railroad, now the Queen & Crescent route, had not crossed the Kentucky River and the country was practically inaccessible. Men and women grew up without ever hearing of a church, a school, a book, a newspaper, a preacher, a doctor, a wheeled vehicle or a lucifer-match! The heathen of Bariaboola-Gha were as well informed concerning God and a future state. They herded in miserable cabins, lived on "corn-dodgers and sow-belly," drank home-made whiskey and never wandered ten miles from their own fireside. Of the great outside world, of moral obligations, of religious conviction and of current events they were profoundly ignorant. Think of people fifty, sixty, seventy years old, born and reared in the United States, who never saw a loaf of wheat-bread, a wagon, a cart or a baby-carriage, to say nothing of a plum-pudding, railway-coach, a trolley-car or a tandem-bicycle! It seems incredible, in this advanced age and bang-up nation, that such conditions should be possible, yet they existed in Southeastern Kentucky. And the American eagle flaps his wings, while Americans boast of their culture and send barrels of cold cash to buy flannel-shirts for perspiring Hottentots and goody-goody tracts for jolly cannibals!

Small need of barbed-wire fences to shut out the cattle and chickens of neighbors five miles apart! Their children did not quarrel and sulk and yell "You can't play in our yard!" Our host, who took us over the property and told us all he knew about it, had not seen a strange face for twenty-nine weary months! Then the neighbor five miles off had come in the vain search of a cruse of oil from the old well to rub on an afflicted hog! Three years had rolled by since his last expedition to the cross-roads, fourteen miles away, to trade "coon skins" for jeans and groceries. Could isolation be more complete? Was Alexander Selkirk less blessed with companionship on his secluded island? Had Coleridge's Ancient Mariner, "on a wide, wide sea," greater cause for an attack of the blues?

The steel-track and the iron-horse are prime civilizers and eighteen years have wrought a wondrous change in the section bordering upon the Cumberland

Mountains. The schoolmaster has come in with the railroad and improvement is the prevailing order. Farmers have turned their forests into cultivated fields and bought the latest implements. Their boys read the papers, yearn for the city, smoke cigarettes, dabble in politics and dream of unbounded wealth. The girls, no longer content with homespun frocks and sunbonnets, dress in silk and velvet, wear stylish hats, devour French novels, sport high-heeled shoes and balloon-sleeves, play Beethoven and Chopin, waltz divinely and are altogether lovable!

An apparition muttering "I am thy father's ghost" would not have surprised us so much as the politeness of our half-clad, barefooted, bareheaded pilot to the neglected well. His manners and his language were faultless. Not a coarse word or grammatical error marred his fluent speech. At noon he invited us to share his humble dinner, apologizing with royal dignity for the poverty of his surroundings. "Gentlemen," he said, "I regret that parched corn and fat bacon are all I can offer, but I beg you to honor me with your presence at my table!" Remembering the cabin and its presiding divinity, we felt obliged to decline and requested him to lunch with us. It was a positive pleasure to see with what relish he ate the baked chicken, biscuit and good things Mrs. Hunter had packed in our saddle-bags. After the meal we prepared to depart. The end of a Louisville paper under the flap of my saddle attracted the old man's attention.

"Is that a newspaper?" he inquired.

"Yes, do you want it?"

"Oh, thank you a thousand times! It is fifteen years since I have seen a paper and this will be such a treat!"

He seized the sheet eagerly, dropped upon the grass and glanced over the printed page. In an instant he jumped to his feet and tears coursed down his wrinkled cheeks.

"I did not mean to be rude," he said earnestly, "but you cannot imagine how my feelings mastered me, after so many years of separation from the world, at sight of a paper from the city of my birth!"

The next moment the good-byes were uttered and we had left the hermit of South Fork, to meet no more this side of eternity. He stood peering after us until the woods shut us from his wistful gaze. Six years later death, the grim detective no vigilance can elude, claimed the guardian of the Beatty Well. His family removed to parts unknown. He rests in an unmarked grave, beneath a spreading oak, near the murmuring stream. The lonely exile has reached home at last!

Who on earth was this educated, courteous, gentlemanly personage, and how did he drift into such a place? This perplexing problem beat the fifteen-puzzle, "Pigs in Clover," or the confusing dogma of Freewill and Predestination. Our guide enlightened us. The old man was reared in Louisville, graduated from college and entered an office to study law. In a bar-room row one night a young man, with whom he had some trouble, was stabbed fatally. Fearing he would be accused of the deed, the student fled to the woods. For years he shunned mankind, subsisting on game and fruit and sleeping in a cave. Every rustling leaf or snapping twig terrified him with the idea that officers were at his heels. Ultimately he gained courage and sought the acquaintance of the few settlers in his vicinity. Striving to forget the past, he cohabited with the woman he called his wife, erected a shanty and brought up three children. Fire destroyed his hut and its contents, leaving him destitute, and he located where we met him. The fear of arrest could not be shaken off and he supposed we had come to take him a prisoner, after twenty-five years of hiding, for a crime of which he was innocent. This explained his retreat under the bed and violent trembling. He carried his secret in his own bosom until 1873, when he was believed to be dying and disclosed it to a friend, our guide, with a sealed letter giving his true name. He recovered, the letter was handed back unopened and the fugitive's identity was never revealed. What an existence for a man of refinement and collegiate training! What volumes of unwritten, unsuspected tragedies environ us, could we but pierce the outward mask and read the tablets of the heart!

Eight or ten years ago J. O. Marshall, a Pennsylvania oil-operator, cleaned out the Beatty Well and drilled another a half-mile north. Neither yielded any oil, although the second was put down nine-hundred feet. Mr. Marshall leased a great deal of land in Wayne and adjacent counties, expecting to operate extensively, but he died without seeing his purposes accomplished. He was a genial, enterprising, whole-souled fellow, whose faith in Kentucky as an oil-field never faltered.

DR. W. G. HUNTER.

Dr. Hunter, my esteemed associate on many a delightful trip, was practising at Newcastle, Pa., when the civil war broke out. He sold his drug-store, offered his services to the Government and was placed in charge of a medical department, where he made a first-class record. He amputated the leg of General James A. Beaver, subsequently Governor of Pennsylvania. At the close of the war he settled in Cumberland county, married a prominent young lady, built up an immense practice and acquired a competence. He served with signal ability and credit in the Legislature and Congress, elected time and again in a district overwhelmingly against his party. He is chairman of the Republican State Committee, member of Congress, ought to be the successor of Senator Blackburn and ranks with the leading men of Kentucky in character and influence.

Such is a brief sketch of the father of American flowing-wells and some of the individuals at different times connected with it more or less directly.

> Let future generations tell
> The story of the Beatty Well,
> The father of oil-spouters!
> In spite of quips and jibes and sneers
> Of arrant cranks and doubters,
> Whose forte is flinging wretched jeers,
> It richly merits hearty cheers
> From true petroleum-shouters.

NOTABLE WELLS ON OIL CREEK, CURRIE IN IRELAND.

IV.

A TALE OF TWO STATES.

Interesting Petroleum Developments in Kentucky and Tennessee—
The Famous American Well—A Boston Company Takes Hold—Provi-
dential Escape—Regular Mountain Vendetta—A Sunday Lynching
Party—Peculiar Phases of Piety—An Old Woman's Welcome—Warm
Reception—Stories of Rustic Simplicity.

"Coming events cast their shadows before."—*Thomas Campbell.*
"In Cumberland county, Kentucky, a run of pure oil was struck."—*Niles' Register, A. D. 1829.*
"Indications of oil are plentiful in the neighborhood of Chattanooga, Tennessee."—*Robert B. Roosevelt, A. D., 1863.*
"Ever since the first settlement of the country oil has been gathered and used for medical pur-
poses."—*Cattelsburg, Ky., Letter, A. D. 1884.*

NTERESTING and unexpected results from borings for salt-water
in Kentucky were not exhausted
by the initial experiment on South
Fork. Special peculiarities invest
that venture with a romantic halo
essentially its own, but " there · re
others." Wayne county was not
to monopolize the petroleum feature of salt-wells by a large majority. "Westward the star of empire takes its way" affirmed Bishop Berkeley two - hundred years
ago, with the instinct of a born
prophet, and it was so with the
petroleum star of Kentucky, however it might be with brilliant Henri Watter-
son's "star-eyed goddess of Reform."

The storm-center next shifted to Cumberland county, the second west of
Wayne, Clinton separating them. Hardy breadwinners, braving the hardships
and privations of pioneer life in the backwoods, early in this century settled
much of the country along the Cumberland River. Upon one section of irregular shape, its southern end bordered by Tennessee, the state of Davy Crockett
and Andrew Jackson, the name of the winding river intersecting it was appropriately bestowed. A central location, between the west bank of the Cumberland and the foot of a lordly hill, was selected for the county-seat and christ-
ened Burksville, in honor of a respected citizen who owned the site of the embryo
hamlet. From a cross-roads tavern and blacksmith-shop the place expanded
gradually into an inviting village of one-thousand population. It has fine stores,

37

good churches and schools, a brick court-house, and for years it boasted the only college in Kentucky for the education of girls.

Burksville pursued "the even tenor of its way" slowly and surely. Forty miles from a railroad or a telegraph-wire, its principal outlet is the river during the season of navigation. The Cumberland retains the fashion of rising sixty to eighty feet above its summer level when the winter rains set in and dwindling to a mere brooklet in the dry, hot months. Old-timers speak of "the flood of 1826" as the greatest in the history of the community. The rampant waters overflowed fields and streets, invaded the ground-floors of houses and did a lot of unpleasant things, the memory of which tradition has kept green. In January of 1877 the moist experience was repeated almost to high-water mark. Saw-logs floated into kitchens and parlors and improvised skiffs navigated back-yards and gardens. Seldom has the town cut a wide swath in the metropolitan press, because it avoided gross scandals and attended strictly to home-affairs. The chief dissipation is a trip by boat to Nashville or Point-Burnside, or a drive overland to Glasgow, the terminus of a branch of the Louisville & Nashville Railroad.

The first great event to stir the hearts of the good people of Cumberland county occurred in 1829. A half-mile from the mouth of Rennix Creek, a minor stream that empties into the Cumberland two miles north of the county town, a well was sunk one-hundred-and-eighty feet for salt-water. Niles' *Register*, published the same year, told the tale tersely:

"Some months since, in the act of boring for salt-water on the land of Mr. Lemuel Stockton, situated in the county of Cumberland, Kentucky, a run of pure oil was struck, from which it is almost incredible what quantities of the substance issued. The discharges were by floods, at intervals of from two to five minutes, at each flow vomiting forth many barrels of pure oil. I witnessed myself, on a shaft that stood upright by the aperture in the rock from which it issued, marks of oil 25 or 30 feet perpendicularly above the rock. These floods continued for three or four weeks, when they subsided to a constant stream, affording many thousand gallons per day. This well is between a quarter and a half-mile from the bank of the Cumberland River, on a small rill (creek) down which it runs to the Cumberland. It was traced as far down the Cumberland as Gallatin, in Sumner county, Tennessee, nearly a hundred miles. For many miles it covered the whole surface of the river and its marks are now found on the rocks on each bank.

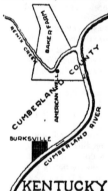

FAMOUS "AMERICAN WELL."

"About two miles below the point on which it touched the river, it was set on fire by a boy, and the effect was grand beyond description. An old gentleman who witnessed it says he has seen several cities on fire, but that he never beheld anything like the flames which rose from the bosom of the Cumberland to touch the very clouds."

This was the beginning of what was afterwards known from the equator to the poles as the "American Well." The flow of oil spoiled the well for salt and the owners quitted it in disgust, sinking another with better success in an adjacent field. For years it remained forsaken, an object of more or less curiosity to travelers who passed close by on their way to or from Burksville. It was very near the edge of the creek, on flat ground most of which has been washed away. Neighboring farmers dipped oil occasionally for medicine, for axle-grease and—"tell it not in Gath, publish it not in Ascalon"—to kill vermin on swine!

Job Moses, a resident of Buffalo, N. Y., visited the locality about the year 1848. He had read of the oil-springs in New York, Pennsylvania, West Virginia and Ohio and he decided that the hole on Rennix Creek ought to be a prize-package. His moderate offer for the well was accepted by the Bakers, into whose hands the Stockton tract had come. He drilled the well to four-hundred feet and erected a pumping-rig. The five or six barrels a day of greenish-amber fluid, 42° gravity, he put up in half-pint bottles, labeled "American Rock Oil" and sold at fifty cents, commending it as a specific for numberless complaints. He reaped a harvest for several years, until trade languished and the well was abandoned.

With the proceeds of his enterprise Moses bought a large block of land at Limestone, N. Y., adjoining the northern boundary of McKean county, Pa., and built a mansion big enough for a castle. He farmed extensively, raised herds of cattle, employed legions of laborers and dispensed a bountiful hospitality. In 1862-3 he drilled three wells near his dwelling, finding a trifling amount of gas and oil. Had he drilled deeper he would inevitably have opened up the phenomenal Bradford field a dozen years in advance of its actual development. Wells twelve-hundred to three-thousand feet deep had not been dreamt of in petroleum philosophy at that date, else Job Moses might have diverted the whole current of oil operations northward and postponed indefinitely the advent of the Clarion and Butler districts! Boring a four-inch hole a few hundred feet farther would have done it!

On what small causes great effects sometimes depend! Believing a snake-story induced our first parents to sample "the fruit of that forbidden tree whose mortal taste brought sin into our world and all our woe." Ambition to be a boss precipitated Lucifer "from the battlements of heaven to the nethermost abyss." A dream released Joseph from prison to be "ruler over Egypt." The smiles of a wanton plunged Greece into war and wiped Troy from the face of the earth. A prod on the heel slew Achilles, a nail—driven by a woman at that—finished Sisera and a pebble ended Goliath. The cackling of a goose saved Rome from the barbarous hordes of Brennus. A cobweb across the mouth of the cave secreting him preserved Mahomet from his pursuers and gave Arabia and Turkey a new religion. The scorching of a cake in a goatherd's hut aroused King Alfred and restored the Saxon monarchy in England. The movements of a spider inspired Robert Bruce to renewed exertions and secured the independence of Scotland. An infected rag in a bundle of Asiatic goods scourged Europe with the plague. The fall of an apple from a tree resulted in Sir Isaac Newton's sublime theory of gravitation. The vibrations of a tea-kettle lid suggested to the Marquis of Worcester the first conception of the steam-engine. A woman's chance remark led Eli Whitney to invent the cotton-gin. The twitching of a frog's muscles revealed galvanism. A diamond necklace hastened the French Revolution and consigned Marie Antoinette to the guillotine. Hacking a cherry-tree with a hatchet earned George Washington greater glory than the victory of Monmouth or the overthrow of Cornwallis. A headache helped cost Napoleon the battle of Waterloo and change the destiny of twenty kingdoms. An affront to an ambassador drove Germany to arms, exiled Louis Napoleon and made France a republic. Mrs. O'Leary's kicking cow laid Chicago in ashes and burst up no end of insurance companies. An alliterative phrase defeated James G. Blaine for President of the United States. An epigram, a couplet or a line has been known to confer immortality. A new bonnet has disrupted a sewing-society, split a congregation and put devout members on the toboggan in their hurry to backslide. An onion-breath has severed doting lovers, cheated parsons of their wedding-fees and played hob with Cupid's calculations. Statistics

4

fail to disclose the awful havoc wrought in millions of homes by such observations, on the part of thoughtless young husbands, as "this isn't the way mother baked," or "mother's coffee didn't taste like this!"

Moses lived to produce oil from his farms and to witness, five miles south of Limestone, the grandest petroleum development of any age or nation. He was built on the broad-gauge plan, physically and mentally, and "the light went out" peacefully at last. The Kentucky well was never revived. The rig decayed and disappeared, a timber or two lingering until carried off by the flood in 1877.

In the autumn of 1876 Frederic Prentice, a leading operator, engaged me to go to Kentucky to lease and purchase lands for oil purposes. Shortly before Christmas he wished me to meet him in New York and go from there to Boston, to give information to parties he expected to associate with him in his Kentucky projects. Together we journeyed to the city of culture and baked-beans and met the gentlemen in the office of the Union Pacific Railroad Company. The gathering was quite notable. Besides Mr. Prentice, who had long been prominent in petroleum affairs, Stephen Weld, Oliver Ames, Sen., Oliver Ames,

GOVERNOR AMES.

Jun., Frederick Ames, F. Gordon Dexter and one or two others were present. Mr. Weld was the richest citizen of New England, his estate at his death inventorying twenty-two millions. The elder Oliver Ames, head of the giant shovel manufacturing firm of Oliver Ames & Sons, was a brother of Oakes Ames, the creator of the Pacific Railroads, whom the Credit-Mobilier engulfed in its ruthless destruction of statesmen and politicians. His nephew and namesake was a son of Oakes Ames and Governor of Massachusetts in 1887-8-9. He began his career in the shovel-works, learning the trade as an employé, and at thirty-five had amassed a fortune of ten millions. He occupied the finest house in Boston, entertained lavishly, spent immense sums for

paintings and bric-a-brac and died in October of 1895. Frederick Ames, son of the senior Oliver, has inherited his father's executive talent and he maintains the family's reputation for sagacity and the acquisition of wealth. F. Gordon Dexter is a multi-millionaire, a power in the railroad world and a resident of Beacon street, the swell avenue of the Hub.

Such were the men who heard the reports concerning Kentucky. They did not squirm and hesitate and wonder where they were at. Thirty-five minutes after entering the room the "Boston Oil Company" was organized, the capital was paid in, officers were elected, a lawyer had started to get the charter and authority was given me to draw at sight for whatever cash was needed up to one-hundred-thousand dollars! This record-breaking achievement was about as expeditious as the Chicago grocer, who closed his store one forenoon and pasted on the door a placard inscribed in bold characters : "At my wife's funeral—back in twenty minutes!"

Oliver Ames, the future governor, invited the party to lunch at the Parker House, Boston's noted hostelry. An hour sped quickly. My return trip had been arranged by way of Buffalo and the Lake-Shore Road to Franklin. The time to start arrived, the sleigh to take me to the depot was at the door, the good-byes were said, the driver tucked in the robes and grasped the lines. At that instant Oliver Ames, Sen., called: "Please come into the hotel one moment; I want to jot down something you told us about the American Well." The other gentlemen looked on, the explanation was penciled rapidly, my seat in the sleigh was resumed and Mr. Dexter jokingly said to the Jehu: "You'll have to hustle, or your fare will miss his train!"

Through the narrow, twisted, crowded streets the horses trotted briskly. Rushing into the station, the train was pulling out and the ticket-examiner was shutting the iron-gates. He refused to let me attempt to catch the rear car and my disappointment was extreme. A train for New York and Pittsburg left in fifteen minutes. It bore me, an unwilling passenger, safely and satisfactorily to the "Smoky City." There the news reached me of the frightful railway disaster at Ashtabula, in which P. P. Bliss and fourscore fellow-mortals, filled with fond anticipations of New-Year reunions, perished in the icy waters ninety feet beneath the treacherous bridge that dropped them into the yawning chasm! The doomed train was the same that would have borne me to Ashtabula and—to death, had not Mr. Ames detained me to make the entry in his memorandum-book! Call it Providence, Luck, Chance, what you will, an incident of this stamp is apt to beget "a heap of tall thinking."

Returning to Burksville in January, the work of leasing went ahead merrily. The lands around the American Well were taken at one-eighth royalty. Forty rods northeast of the American, in a small ravine, a well was drilled eight-hundred feet. At two-hundred feet some gas and oil appeared, but the well proved a failure. While it was under way the gas in a deserted salt-well twenty rods northwest of the American burst forth violently, sending frozen earth, water and pieces of rock high into the air. The derrick at the Boston Well, rising to the height of seventy-two feet, was a perennial delight to the natives. Youths, boys and old men ascended the ladder to the topmost round to enjoy the beautiful view. Pretty girls longed to try the experiment and it was whispered that six of them, one night when only the man in the moon was peeping, performed the perilous feat. Certain it is that a winsome teacher at the college, who climbed the celestial stair years ago, succeeded in the effort and wrecked her dress on the way back to solid ground. A dining-room girl at Petrolia, in 1873, stood on top of a derrick, to win a pair of shoes banteringly offered by a jovial oilman to the first fair maiden entitled to the prize. Lovely woman and Banquo's ghost will not "down!"

GIRL CLIMBING A DERRICK.

Three miles northeast of the American Well, at the mouth of Crocus Creek, C. H. English drilled eight shallow wells in 1865. They were bunched closely and one flowed nine-hundred barrels a day. Transportation was lacking, the product could not be marketed and the promising field was deserted. Twelve years later the Boston Oil Company drilled in the midst of English's cluster, to discover the quality of the strata, and could not exhaust the surface water by the most incessant pumping. The company also drilled on the Gilreath farm, across the Cumberland from Burksville, where Captain Phelps found heavy oil in paying quantity back in the sixties. The well produced nicely and would have paid handsomely had a railroad or a pipe-line been within reach. A well two miles west of the American, drilled in 1891, had plenty of sand and showed for a fifty-barreler.

Six miles south-west of Burksville, at Cloyd's Landing, J. W. Sherman, of Oil-Creek celebrity, drilled a well in 1865 which spouted a thousand barrels of 40° gravity oil in twenty-four hours. He loaded a barge with oil in bulk, intending to ship it to Nashville. The ill-fated craft struck a rock in the river and the oil floated off on its own hook. Sherman threw up the sponge and returned to Pennsylvania. Three others on the Cloyd tract started finely, but the wonderful excitement at Pithole was breaking out and operations elsewhere received a cold chill. Dr. Hunter purchased the Cloyd farm and leased it in 1877 to Peter Christie, of Petrolia, who did not operate on any of the lands he secured in Kentucky and Tennessee. Micawber-like, Cumberland county is " waiting for something to turn up " in the shape of facilities for handling oil. When these are assured the music of the walking-beam will tickle the ears of expectant believers in Kentucky as the coming oil-field.

Wayne and Cumberland had been heard from and Clinton county was the third to have its inning. On the west bank of Otter Creek, a sparkling tributary of Beaver Creek, a well bored for salt fifty or more years ago yielded considerable oil. Instead of giving up the job, the owners pumped the water and oil into a tank, over the side of which the lighter fluid was permitted to empty at its leisure. The salt-works came to a full stop eventually and the well relapsed into "innocuous desuetude." L. D. Carter, of Aurora, Ill., sojourning temporarily in Clinton for his health, saw the old well in 1864. He dipped a jugful of oil, took it to Aurora, tested it on the Chicago, Burlington & Quincy Railroad, found it a good lubricant and concluded to give the well a square trial. The railroad company agreed to buy the oil at a fair price. Carter pumped six or eight barrels a day, hauled it in wagons over the hills to the Cumberland River and saved money. He granted Mr. Prentice an option on the property in 1877. The day the option expired J. O. Marshall bought the well, farm and ten-thousand acres of leases conditionally, for a Butler operator who "didn't have the price," and the deal fell through.

The well stood idle until 1892, when J. Hovey, an ex-broker from New York and relative of a late Governor of Indiana, drilled a short distance down the creek. The result was a strike which produced twenty-four hundred barrels of dark, heavy, lubricating oil in fifty days. It was shut down for want of tankage and means to transport the product to market. The Carter again yielded nicely, as did three more wells in this neighborhood. Last year the Standard Oil Company was given a refusal of the Hovey and surrounding interests, in order to test the territory fully and lay a pipe-line to Glasgow or Louisville, should the production warrant the expenditure. Wells have been sunk east of the Carter nearly to Monticello, eighteen miles off, finding gas and indications of oil. Every true Clintonite is positive an ocean of petroleum underlies his particular neck of woods, impatient to be relieved and burden landholders and operators alike with excessive wealth!

A hard-headed youth, out walking with his best girl in the dog-days, told her a fairy story of the dire effects of ice-cream upon the feminine constitution.

"I knew a girl," he declared, "who ate six plates of the dreadful stuff and died next day!" The shrewd damsel exclaimed rapturously: "Oh, wouldn't it be sweet to die that way? Let us begin on six plates now!" And wouldn't it be nice to be loaded with riches, not gained by freezing out some other fellow, by looting a bank, by wedding an unloved bride, by grinding the poor, by manipulating stocks, by cornering grain or by practices that make the angels weep, but by bringing oil honestly from the bowels of the earth?

About the year 1839 a salt-well in Lincoln county, eight miles from the pretty town of Stanford, struck a vein of oil unexpectedly. The inflammable liquid gushed out with great force, took fire and burned furiously for weeks. The owner was a grim joker in his way and he aptly remarked, upon viewing the conflagration: "I reckon I've got a little hell of my own!" Four more wells were drilled farther up the stream, two getting a show of oil. One was plugged and the other, put down by the late Marcus Hulings, the wealthy Pennsylvania operator, proved dry. Surface indications in many quarters gave rise to the belief that oil would be found over a wide area, and in 1861 a well was bored at Glasgow, Barren county, one-hundred-and-ten miles below Louisville. It was a success and a hundred have followed since, most of which are producing moderately. Col. J. C. Adams, formerly of Tidioute, Pa., was the principal operator for twenty years. A suburban town, happily termed Oil City, is "flourishing like a green bay horse." The oil, dark and ill-flavored, smelling worse than "the thousand odors of Cologne," is refined at Glasgow and Louisville. It can be deodorized and converted into respectable kerosene. Sixteen miles south of Glasgow, on Green River, four shallow wells were bored thirty years ago, one flowing at the rate of six-hundred barrels, so that Barren county is by no means barren of interest to the oil fraternity.

At Bowling Green a well was sunk two-hundred feet, a few gallons of green-oil bowling to the surface. Torpedoing was unknown, or the fate of many Kentucky wells might have been reversed. John Jackson, of Mercer, Pa., in 1866 drilled a well in Edmonson county, twenty-five miles north-west of Glasgow. The tools dropped through a crevice of the Mammoth Cave, but neither eyeless fish nor slippery petroleum repaid the outlay of muscle and greenbacks. As if to add insult to injury, the well hatched a mammoth cave that buried the tools eight-hundred feet out of sight!

Loyal to his early training and hungry for appetizing slapjacks, Jackson once imported a sack of the flour from Louisville and asked the obliging landlady of his boarding-house to have buckwheat-cakes for breakfast. He was on hand in the morning, ready to do justice to the savory dish. The "cakes" were brought in smoking hot, baked into biscuit, heavy as lead and irredeemably unpalatable! The sack of flour went to fatten the denizens of a neighbor's pig-pen. Jackson was a pioneer in the Bradford region, head of the firm of Jackson & Walker, clever and generous. The grass and the flowers have grown on his grave for ten years, "the insatiate archer" striking him down in the prime of vigorous manhood.

Sandy Valley, in the north-eastern section of the state, contributed its quota to the stock of Kentucky petroleum. From the first settlement of Boyd, Greenup, Carter, Johnson and Lawrence counties oil had been gathered for medical purposes by skimming it from the streams. About 1855 Cummings & Dixon collected a half-dozen barrels from Paint Creek and treated it at their coal-oil refinery in Cincinnati, with results similar to those attained by Kier in Pittsburg. They continued to collect oil from Paint Creek and Oil-Spring Fork until the war, at times saving a hundred barrels a month. In 1861 they drilled a well three-hundred feet on Mud Lick, a branch of Paint Creek, penetrating shale and sandstone and getting light shows of oil and gas. Surface-oil was found on the Big Sandy River, from its source to its mouth, and in considerable quantities on Paint, Blaine, Abbott, Middle, John's and Wolf Creeks. Large springs on Oil-Spring Fork, a feeder of Paint Creek, yielded a barrel a day. At the mouth of the Fork, in 1860, Lyon & Co. drilled a well two-hundred feet,

NORTH-EASTERN KENTUCKY.

tapping three veins of heavy oil and retiring from the scene when "the late unpleasantness" began to shake up the country. The same year a well was sunk one-hundred-and-seventy feet, on the headwaters of Licking River, near the Great Burning Spring. Gas and oil burst out for days, but the low price of crude and the impending conflict prevented further work. What an innumerable array of nice calculations this cruel war nipped in the bud !

J. Hinkley bored two-hundred feet in 1860, on Paint Creek, eight miles above Paintville, meeting a six-inch crevice of heavy-oil, for which there was no demand, and the capacity of the well was not tested. Salt-borers on a multitude of streams had much difficulty, fifty or sixty years ago, in getting rid of oil that persisted in coming to the surface. These old wells have been filled with dirt, although in some the oil works to the top and can be seen during the dry seasons. The Paint-Creek region had a severe attack of oil-fever in 1864-5. Hundreds of wells were drilled, boats were crowded, the hotels were thronged and the one subject of conversation was "oil—oil—oil !" Various causes, especially the extraordinary developments in Pennsylvania, compelled the plucky operators to abandon the district, notwithstanding encouraging symptoms of an important field. Indeed, so common was it to find petroleum in ten or fifteen counties of Kentucky that land-owners ran a serious risk in selling their farms before boring them full of holes, lest they should unawares part with prospective oil-territory at corn-fodder prices !

Tennessee did not draw a blank in the awards of petroleum indications. Along Spring Creek many wells, located in 1864-5 because of "surface-shows," responded nobly, at a depth comparatively shallow, to the magic touch of the drill. The product was lighter in color and gravity than the Kentucky brand. Twelve miles above Nashville, on the Cumberland River, wells have been pumped at a profit. Around Gallatin, Sumner county, decisive tests demonstrated the presence of petroleum in liberal measure. On Obey Creek, Fentress county, sufficient drilling has been done to justify the expectation of a rich district. Near Chattanooga, on the southern border of the state, oil seepages are "too numerous to mention." The Lacy Well, eighteen miles south of the Beatty, drilled in 1893, is good for thirty barrels every day in the week. The oil is of superior quality, but the cost of marketing it is too great. A dozen wells are going down in Fentress, Overton, Scott and Putnam. Some fine day the tidal wave of development will sweep over the Cumberland-River region, with improved appliances and complete equipment, and give the country a rattling "show for its white alley!" Surely all these spouting-wells, oil-springs and greasy oozings mean something. To quote a practical oilman, who knows both states from a to z : "Twenty counties in Kentucky and Tennessee are sweating petroleum!"

Picking up a million acres of supposed oil-lands in the Blue-Grass and Volunteer States had its serio-comic features. The ignorant squatters in remote latitudes were suspicious of strangers, imagining them to be revenue-officers on the trail of "moonshiners," as makers of untaxed whisky were generally called. More than one northern oilman narrowly escaped premature death on this conjecture. J. A. Satterfield, the successful Butler operator, went to Kentucky in the winter of 1877 to superintend the leasing of territory for his firm, between which and the Prentice combination a lively scramble had been inaugurated. Somebody thought he must be a Government agent and passed the word to the lawless mountaineers. The second night of his stay a shower of bullets riddled the window, two lodging in the bed in which Satterfield lay asleep! Daylight saw him galloping to the railroad at a pace eclipsing Sheridan's ride to Winchester, eager to "get back to God's country." "Once was enough for him" to figure as the target of shooters who seldom failed to score "a hit, a palpable hit." Alas! the grim archer didn't miss him in 1894.

Arriving late one Saturday at Mt. Vernon, the county seat of Rockcastle, the colored waiter on Sunday morning inquired: "Hes yo done gone an' seen em?" Asking what he meant, he informed me that three men were dangling from a tree in the court-house yard, lynched by an infuriated mob during the night on suspicion of horse-stealing, "the unpardonable sin" in Kentucky. A party of citizens had started for the cabin of a notorious outlaw, observed skulking homeward under cover of darkness, intending to string him up. The desperado was alert. He fired one shot,

THREE DANGLING FROM A TREE.

which killed a man and stampeded the assailants. They returned to the village, broke into the jail, dragged out three cowering wretches and hanged them in short metre! The bodies swung in the air all day, a significant warning to whoever might think of "walkin' off with a hoss critter."

On that trip to Rockcastle county the train stopped at a wayside station bearing the pretentious epithet of Chicago. A tall, gaunt, unshaven, uncombed man, with gnarled hands that appealed perpetually for soap and water, high cheekbones, imperfect teeth and homespun clothes of the toughest description, stood on the platform in a pool of tobacco-juice. A rustic behind me stuck his head through the car-window and addressed the hard-looking citizen as "Jedge." Honors are easy in Kentucky, where "colonels," "majors" and "judges" are "thick as leaves in Vallambrosa," but the title in this instance seemed too absurd to pass unheeded. When the train started, in reply to my question whether the man on the platform was a real judge, his friendly acquaintance took the pains to say : "Wal, I can't swar es he's zackly, but las' year he wuz jedge ov a chicken-fight down ter Si Mason's an' we calls 'im jedge ever sence !"

Kentucky vendettas have often figured in thrilling narratives Business took me to the upper end of Laurel county one week. Litigants, witnesses and hangers-on crowded the village, for a suit of unusual interest was pending before the " 'squar." The principals were farmers from the hilly region, whose fathers and grandfathers had been at loggerheads and transmitted the quarrel to their posterity. Blood had been shed and hatred reigned supreme. The important case was about to begin. Two shots rang out so closely together as to be almost simultaneous, followed by a regular fusilade. Everybody ran into the street, where four men lay dead, a fifth was gasping his last breath and two others had ugly wounds. The tragedy was soon explained. The two parties to the suit had met on their way to the justice's house. Both were armed, both drew pistols and both dropped in their tracks, one a corpse

A MOUNTAIN VENDETTA.

and the second ready for the coroner in a few moments. Relatives and adherents continued the dreadful work and five lives paid the penalty of ungovernable passion. The dead were wrapped in horse-blankets and carted home. The case was not called. It had been settled "out of court."

The spectators of this dreadful scene manifested no uncommon concern. "It's what might be expected," echoed the local oracle ; "when them mountain fellers gets whiskey inside them they don't care fur nuthin' !" Within an hour of the shooting a young man stopped me on the street-corner, where stood a wagon containing two bodies. "Kunnel," he went on to say, "I've h'ard es yo's th' man es got our farm fur oil. Dad an' Cousin Bill's 'n that ar wagon, an' I want yo ter giv' me a job haulin' wood agin yo starts work up our way." He mounted the vehicle and drove off with his ghastly freight without a quiver of emotion.

At Crab Orchard, one beautiful Sunday, the clerk chatted with me on the hotel porch. A stalwart individual approached and my companion ejaculated : "Thar's a bigger man 'n Gen'ral Grant !"

"A BIGGER MAN 'N GEN'RAL GRANT."

Next instant Col. Kennedy was added to my list of Kentucky acquaintances. He was very affable, wished oil-operations in the neighborhood success and, with characteristic Southern hospitality, invited me to visit him. After he left us the clerk, in answer to my desire to learn the basis of Kennedy's greatness, naively said : "Why, he's killed eight men !"

Politics and religion were staple wares, the susceptible negroes inclining strongly to the latter. Their spasms of piety were extremely inconvenient at times. News of a "bush meetin' " would be circulated and swarms of darkeys would flock to the appointed place, taking provisions for a protracted siege. No matter if it were the middle of harvest and rain threatening, they dropped everything and went to the meeting. "Doant 'magine dis niggah's gwine ter lose his 'mo'tal soul fer no load uv cow-feed" was the conclusive rejoinder of a colored hand to his employer, who besought him to stay and finish the haying.

Rev. George O. Barnes, the gifted evangelist, who resigned a five-thousand-dollar Presbyterian pastorate in Chicago to assist Moody, was reared in Kentucky and lived near Stanford. He would traverse the country to hold revivals, staying three to six weeks in a place. His personal magnetism, rare eloquence, apostolic zeal, fine education, intense fervor and catholic spirit made him a wonderful power. Converts he numbered by thousands. He preferred Calvary to Sinai, the gentle pleadings of infinite mercy to the harsh threats of endless torment. His daughter Marie, with the voice of a Nilsson and the face of a Madonna, accompanied her father in his wanderings, singing gospel-hymns in a manner that distanced Sankey and Philip Phillips. Her rendering of " Too Late," "Almost-Persuaded," and "Only a Step to Jesus," electrified and thrilled the auditors as no stage-song could have done. Raymon Moore's hackneyed verses had not been written, yet the boys called Miss Barnes "Sweet Marie" and thronged to the penitent bench. The evangelist and his daughter tried to convert New York, but the Tammany stronghold refused to budge an inch. They invaded England and enrolled hosts of recruits for Zion. The Prince of Wales is said to have attended one of their meetings in the suburbs of London. Mr. Barnes finally proposed to cure diseases by "anointing with oil and laying on of hands." His pink cottage became a refuge for cranks and cripples and patients, until a mortgage on the premises was foreclosed and the queer aggregation scattered to the winds.

Albany, the county-seat of Clinton, experienced a Barnes revival of the tip-top order. Business with Major Brentz, the company's attorney, landed me in the cosy town on a bright March forenoon. Not a person was visible. Stores were shut and corner-loungers absent. What could have happened? Halting my team in front of the hotel, nobody appeared. Ringing the quaint, old-fashioned bell attached to a post near the pump, a lame, bent colored man shuffled out of the barn.

"Pow'ful glad ter see yer, Massa," he mumbled slowly," a'l put up de hosses."

"Where is the landlord ?"

"Done gone ter meetin'."

"Will dinner soon be ready ?"

"Soon's de folkses gits back frum meetin'."

" All right, take good care of the horses and I'll go over to the court-house."

"No good gwine dar, dey's at the meetin'."

It was true. Mr. Barnes was holding three services a day and the village emptied itself to get within sound of his voice. For five weeks this kept up. Lawyers quit their desks, merchants locked their stores, women deserted their houses and young and old thought only of the meetings. Hardly a sinner was left to work upon, even the village-editor and the disciples of Blackstone joining the hallelujah band !

An African congregation at Stanford had a preacher black as the ace of spades and wholly illiterate, whom many whites liked to hear. " Brudders an sisters, niggahs and white folks," he closed an exhortation by saying, " dar's no use 'temptin' ter sneak outen de wah 'tween the good Lawd an' de bad debbil, 'cos dar's on'y two armies in dis worl' an' bofe am a-fitin' eberlastingly ! So 'list en de army ob light, ef yer want ter gib ole Satan er black eye an' not roast fureber an' eber in de burnin' lake whar watah-millions on ice am nebber se'ved for dinnah ! " Could the most astute theological hair-splitter have presented the issue more concisely and forcibly to the hearers of the sable Demosthenes?

The first and only circus that exhibited at Burksville produced an immense sensation. It was " Bartholomew's Equescurriculum," with gymnastics and ring exercises to round out the bill. Barns, shops and trees for miles bore gorgeous posters. Nast's cartoons, which the most ignorant voters could understand, did more to overthrow Boss Tweed than the masterly editorials of the New-York *Times*. The flaming pictures aroused the Cumberlanders, hundreds of whom could not read, to the highest pitch of expectation. Monday was the day set for the show. On Saturday evening country-patrons began to camp in the woods outside the village. A couple from Overton county, Tennessee, and their four children rode twenty-eight miles on two mules, bringing food for three days and lodging under the trees ! A Burksville character of the stripe Miss Ophelia styled "shiftless" sold his cooking-stove for four dollars to get funds to attend ! "Alf," the ebony-hued choreman at Alexander College, who built my fires and blacked my shoes, was worked up to fever-heat. "Befo' de Lawd," he sobbed, "dis chile's er gone coon, 'less yer len' er helpin' han' ! Mah wife's axed her mudder an' sister ter th' ci'cus an' dar's no munny ter take 'em an' mah sister !" Giving him the currency for admission dried the mourner's tears and "pushed them clouds away."

At noon on Sunday the circus arrived by boat from Nashville. Service was

AN AFRICAN TALE OF WOE.

in progress in one church, when an unearthly sound startled the worshippers. The wail of a lost soul could not be more alarming. Simon Legree, scared out of his boots by the mocking shriek of the wind blowing through the bottle-neck Cassy fixed in the garret knot-hole, had numerous imitators. Again and again the ozone was rent and cracked and shivered. The congregation broke for the door, the minister jerking out a sawed-off benediction and retreating with the rest. A half-mile down the river a boat was rounding the bend. A steam-calliope, distracting, discordant and unlovely, belched forth a torrent of paralyzing notes. The whole population was on the bank by the time the boat stopped. The crowd watched the landing of the animals and belongings of the circus with unflinching eagerness. Few of the surging mass had seen a theatre, a circus, or a show of any sort except the Sunday-school Christmas performance. They were bound to take in every detail and that Sunday was badly splintered in the peaceful, orderly settlement.

With the earliest streak of dawn the excitement was renewed. Groups of adults and children, of all ages and sizes and complexions, were on hand to see the tents put up. By eleven o'clock the town was packed. A merry-go-round, the first Burksville ever saw, raked in a bushel of nickels. The college domestics skipped, leaving the breakfast dishes on the table and the dinner to shift for itself. A party of friends went with me to enjoy the fun. Beside a gap in the fence, to let wagons into the field, sat "Alf," the image of despair. Four weeping females—his wife, sister, mother-in-law and sister-in-law—crouched at his feet. As our party drew near he beckoned to us and unfolded his tale of woe. "Dem fool-wimmin," he exclaimed bitterly, "hes done spended de free dollars yer guv me on de flyin'-hosses! Dey woodn't stay off nohow an' now dey caint see de ci'cus! Oh, Lawd! Oh, Lawd!" The purchase of tickets poured oil on the troubled waters. The Niobes wiped their eyes on their jean-aprons and "Richard was himself again." How the antics of the clowns and the tricks of the ponies pleased the motley assemblage! Buck Fanshaw's funeral did not arouse half the enthusiasm in Virginia City the first circus did in Burksville.

It was necessary for me to visit Williamsburg, the county-seat of Whitley, to record a stack of leases. Somerset was then the nearest railway-point and the trip of fifty miles on horseback required a guide. The arrival of a Northerner raised a regular commotion in the well-nigh inaccessible settlement of four-hundred population. The landlord of the public-house slaughtered his fattest chickens and set up a bed in the front parlor to be sure of my comfort. The jailer's fair daughter, who was to be wedded that evening, kindly sent me an invitation to attend the nuptials. By nine o'clock at night nearly every business-man and official in the place had called to bid me welcome. Before noon next day seventeen farmers, whose lands had been leased, rode into town to greet me and learn when drilling would likely begin. Each insisted upon my staying with him a week, "or es much longer es yo kin," and fourteen of them brought gallon-jugs

A WELCOME IN JUGS.

of apple-jack, their own straight goods, for my acceptance! Such a reception a king might envy, because it was entirely unselfish, hearty and spontaneous. Williamsburg has got out of swaddling-clothes, the railway putting it in touch with the balance of creation.

Thirteen miles of land, in an unbroken line, on a meandering stream, had been tied-up, with the exception of a single farm. The owner was obdurate and refused to lease on any terms. Often lands not regarded favorably as oil territory were taken to secure the right-of-way for pipe-lines, as the leases conveyed this privilege. Driving past the stubborn farmer's homestead one afternoon, he was chopping wood in the yard and strode to the gate to talk. His bright-eyed daughter of four summers endeavored to clamber into the buggy. Handing the cute fairy in coarse jeans a new silver-dollar, fresh from the Philadelphia mint, the father caught sight of the shining coin.

"Hev yo mo' ov 'em 'ar dollars about yo ?" he asked.

"Plenty more."

"Make out leases fur my three farms an' me an' the old woman 'll sign 'em ! I want three ov 'em kines, for they be th' slickest Demmycratic money my eyes hes sot onto sence I fit with John Morgan !"

The documents were filled up, signed, sealed and delivered in fifteen minutes. The chain of leased lands along Fanny Creek was intact, with the "missing link" missing at last.

The simplicity of these dwellers in the wilderness was equaled only by their apathy to the world beyond and around them. Parents loved their children and husbands loved their wives in a quiet, unobtrusive fashion. "She wuz a hard-workin' woman," moaned a middle-aged widower in Fentress county, telling me of his deceased spouse, "an' she allers wore a frock five year, an' she hed 'leven chil'ren, an' she died right in corn-shuckin'!" He was not stony-hearted, but twenty-five years of married companionship meant to him just so many days' work, so many cheap frocks, child-bearing, corn-cake and bacon always ready on time. Among these people woman was a drudge, who knew nothing of the higher relations of life. Children were huddled into the hills to track game, to follow the plough or to drop corn over many a weary acre. Reading and writing were unknown accomplishments. Jackson, "the great tradition of the uninformed American mind," and Lincoln, whose name the tumult of a mighty struggle had rendered familiar, were the only Presidents they had ever heard of. "Where ignorance is bliss 'tis folly to be wise" may be a sound poetical sentiment, but it was decidedly overdone in South-eastern Kentucky and North-eastern Tennessee so recently as the year of the Philadelphia Centennial.

Opposite the Hovey and Carter wells in Clinton county lives a portly farmer who "is a good man and weighs two-hundred-and-fifty pounds." He is known far and wide as "Uncle John" and his wife, a pleasant-faced little matron, is affectionately called "Aunt Rachel." A log-church a mile from "Uncle John's" is situated on a pretty hill. There the young folks are married, the children are baptized and the dead are buried. The "June meetin'," when services are held for a week, is the grand incident of the year to the people for a score of miles. In December of 1893 Dr. Phillips, of Monticello, drove me to the wells. We stopped at "Uncle John's." As we neared the house a dog barked and the hospitable farmer came out to meet us. Behind him walked a man who greeted the Doctor cordially. He glanced at me, recognition was mutual and we clasped hands warmly. He was Alfred Murray, formerly connected with the Pennsylvania Consolidated Land and Petroleum Company in Butler and at Bradford. Fourteen years had glided away since we met and there were many questions to ask and answer. He had been in the neighborhood a twelvemonth, keeping tab on oil movements and indications, hoping, longing and praying for the speedy advent of the petroleum millenium. We pumped the Hovey Well one hour, rambled over the hills and talked until midnight about persons and things in Pennsylvania. Meeting in so dreary a place, under such circumstances, was as thorough a surprise as Stanley's discovery of Livingstone in Darkest Africa. During our conversation regarding the roughest portion of the county, bleak, sterile and altogether repellant, selected by a hermit as his lonely retreat, my friend remarked : "I have heard that the poor devil was troubled with remorse and, as a sort of penance, vowed to live as near Sheol as possible until he died!"

The stage that bore me from Monticello to Point Burnside on my home-ward journey stopped half-way to take up a countryman and an aged woman. Room was found inside for the latter, a stout, motherly old creature, into whose beaming face it did jaded mortals good to look. She said "howdy" to the three passengers, a local trader, a farmer's young wife and myself, sat down solidly and fixed her gaze upon me intently. It was evident the dear soul was fairly bursting with impatience to find out about the stranger. Not a word was spoken until she could restrain her inquisitive impulse no longer.

"Yo don't liv' eroun' these air parts?" she interrogated.

"No, madam, my home is in Pennsylvania."

"Land sakes! Be yo one ov 'em air ile-fellers?"

"Yes."

"Wal, I be orful glad ter see yo!" and she stretched out her hand and shook mine vigorously. "Hope yo're right peart, but yo' be a long way from home! Did yo see 'em wells over thar by Aunt Rachel's?"

"Oh, yes, I saw the wells and stayed at Aunt Rachel's all night."

"I ain't seed Aunt Rachel for nigh a year an' a half. My old man hed roomatiz and we couldn't get ter meetin' this summer. He sez thar's ile onto our farm. I be seventy-four an' him on the ruf be my son'n-law. Yo see he married, Jess did, my darter Sally an' tha moved ter a place tha call Kansas. Tha's bin thar seventeen year an' hes six chil'ren. Jess he cum back las' week ter see his fokeses an' he be takin' me ter Kansas ter see Sally an' the babies. I never seed 'em things Jess calls cyars, an' he sez tha ain't drord by no hoss nuther! I wuz bo'n eight mile down hyar an' never wuz from home more'n eighteen mile, when we goes ter June meetin'. But I be ter Monticeller six times."

Truly this was a natural specimen, bubbling over with kindness, unspoiled by fashion and envy and frivolity and superficial pretense. Here was the counterpart of Cowper's humble hero-

"I BE ORFUL GLAD TER SEE YO!"

ine, who "knew, and knew no more, her Bible true." The wheezy stage was brighter for her presence. She told of her family, her cows, her pigs, her spin-ning and her neighbors. She lived four miles from the Cumberland River, yet never went to see a steamboat! When we alighted at the Burnside station and the train dashed up she looked sorely perplexed. "Jess" helped her up the steps and the "cyars" started. The whistle screeched, daylight vanished and the train had entered the tunnel below the depot. A fearful scream pierced the ears of the passengers. The good woman seventy-four years old, who "never seed 'em things" before was terribly frightened. We tried to reassure her, but she begged to be let off. How "Jess" managed to get her to Kansas safely may be imagined. But what a story she would have to tell about the "cyars" and "Sally an' the babies" when she returned to her quiet home after such a trip! Bless her old heart!

Although the broad hills and sweeping streams which grouped many sweet panoramas might be dull and meaningless to the average Kentuckian of former days, through some brains glowing visions flitted. Two miles south of Columbia, Adair county, on the road to Burksville, a heap of stones and pieces of rotting timber may still be seen. Fifty-five years ago the man who owned the farm constructed a huge wheel, loaded with rocks of different weights on its strong arms. Neighbors jeered and ridiculed, just as scoffers laughed at Noah's ark and thought it wouldn't be much of a shower anyway. The hour to start the wheel arrived and its builder stood by. A rock on an arm of the structure slipped off and struck him a fatal blow, felling him lifeless to the earth! He was a victim of the craze to solve the problem of Perpetual Motion. Who can tell what dreams and plans and fancies and struggles beset this obscure genius, cut off at the moment he anticipated a triumph? The wheel was permitted to crumble and decay, no human hand touching it more. The heap of stones is a pathetic memento of a sad tragedy. Not far from the spot Mark

"EF YO KNOW'D COUSIN JIM."

Twain was born and John Fitch whittled out the rough model of the first steamboat.

Riding in Scott county, Tennessee, at full gallop on a rainy afternoon, a cadaverous man emerged from a miserable hut and hailed me. The dialogue was not prolonged unduly.

"Gen'ral," he queried, "air yo th' oilman frum Pennsylvany?"

"Yes, what can I do for you?"

"I jes' wanted ter ax ef yo know'd my cousin Jim!"

"Who is your cousin Jim?"

"Law, Jim Sickles!" I tho't ez how evr'ybody know'd Jim! He went up No'th arter th' wah an' ain't cum back yit. Ef yo see 'im tell 'im yo seed me!"

A promise to look out for "Jim" satisfied the verdant backwoodsman, who probably had never been ten miles from his shanty and deemed "up No'th" a place about the size of a Tennessee hunting-ground!

Fair women, pure Bourbon and men extra plucky,
No wonder blue-grass folks esteem themselves lucky—
But wait till the oil boom gets down to Kentucky!

Let Fortune assume forms and fancies Protean,
No matter for that, there will rise a loud pæan
So long as oil gladdens the proud Tennesseean!

FIGURES THAT COUNT.

PRODUCTION AND PRICES OF CRUDE PETROLEUM IN PENNSYLVANIA FROM SEPTEMBER 1, 1859, TO DECEMBER 31, 1895.

YEARS.	BARRELS PRODUCED.	LOWEST MONTHLY AVERAGE PRICE.	HIGHEST MONTHLY AVERAGE PRICE.	YEARLY AVERAGE PRICE.	WELLS COMPLETED.
1859	1,873	$20 00	$20 00	$20 00	
1860	547,439	2 75	19 25	9 60	
1861	2,119,045	10	1 00	49	
1862	3,153,183	10	2 25	1 05	
1863	2,667,543	2 25	3 37½	3 15	
1864	2,215,150	4 00	12 12½	9 87½	
1865	2,560,200	4 62½	8 25	6 59	
1866	3,385,105	2 12½	4 50	3 74	
1867	3,458,113	1 75	3 55	41	
1868	3,540,670	1 95	5 12½	3 62½	
1869	4,186,475	4 95	6 95	5 63¾	
1870	5,308,046	3 15	4 52½	3 89	
1871	5,278,072	3 82½	4 62½	4 34	
1872	6,505,774	3 15	4 92½	3 64	1,183
1873	9,849,508	1 00	2 60	1 83	1,263
1874	11,102,114	55	1 90	1 17	1,317
1875	8,948,749	1 03	1 75	1 35	2,398
1876	9,142,940	1 80	3 81	2 56¼	2,920
1877	13,230,330	1 80	3 53¼	2 42	3,939
1878	15,272,491	82½	1 65¼	1 19	3,064
1879	19,835,903	67⅛	1 18⅛	85⅞	3,048
1880	26,027,631	80	1 10¼	94½	4,217
1881	27,376,509	81¼	95½	85¼	3,880
1882	30,053,500	54½	1 27⅝	78⅝	3,304
1883	23,128,389	92½	1 16⅞	1 06⅝	2,847
1884	23,772,209	63¼	1 11¾	83¼	2,265
1885	20,776,041	70⅞	1 05½	88½	2,761
1886	25,798,000	62⅛	88¼	71¼	3,478
1887	21,478,883	59¼	80	66⅝	1,660
1888	16,488,668	76	93¾	87	1,515
1889	21,487,435	83⅛	1 08⅛	94	5,434
1890	30,065,867	67⅞	1 05	86½	6,435
1891	35,742,152	59	77¼	66¼	3,390
1892	33,332,306	52	64⅛	55½	1,954
1893	31,362,890	53½	78⅛	64	1,980
1894	27,597,614	80	91⅜	84	3,756
1895	....	98⅝	1 79⅝	1 35¼	...

EARLY OPERATORS ON OIL CREEK.

V

A HOLE IN THE GROUND.

"I have tapped the mine."—*Colonel E. L. Drake.*
"Come quick, there's oceans of oil."—"*Uncle Billy*" *Smith.*
"Petroleum has come to be king."—*Professor W. D. Gunning.*
"It is our mission to illuminate all creation."—*New York Ledger.*
"Behold, how great a matter a little fire kindleth."—*St. James III: 5.*

ATURE certainly spared no effort to bring petroleum into general notice ages before James Young manufactured paraffine-oil in Scotland or Samuel M. Kier fired-up his miniature refinery at Pittsburg. North and south, east and west the presence of the greasy staple was manifested positively and extensively. The hump of a dromedary, the kick of a mule or the ruby blossom on a toper's nose could not be more apparent. It bubbled in fountains, floated on rivulets, escaped from crevices, collected in pools, blazed on the plains, gurgled down the mountains, clogged the ozone with vapor, smelled and sputtered, trickled and seeped for thousands of years in vain attempts to divert attention towards the *source* of this prodigal display. Mankind accepted it as a liniment and lubricant, gulped it down, rubbed it in, smeared it on and never thought of seeking whence it came or how much of it might be procured. Even after salt-wells had produced the stuff none stopped to reflect that the golden grease must be imprisoned far beneath the earth's surface, only awaiting release to bless the dullards callous to the strongest hints respecting its headquarters. The dunce who heard Sydney Smith's side-splitting story and sat as solemn as the sphinx, because he couldn't see any point until the next day and then got it heels over head, was less obtuse. Puck was right in his little pleasantry : "What fools these mortals be!"

Dr. Abraham Gesner obtained oil from coal in 1846 and in 1854 patented an illuminator styled "Kerosene," which the North American Kerosene Gaslight Company of New York manufactured at its works on Long Island. The excel-

WILLIAM A SMITH

5   55

lence of the new light—the smoke and odor were eliminated gradually—caused a brisk demand that froze the marrow of the animal-oil industry. Capitalists invested largely in Virginia, Kentucky and Missouri coal-lands, saving the expense of transporting the "raw material" by erecting oil-works at the mines. Exactly in the ratio that mining coal was cheaper than catching whales mineral-oil had the advantage in competing for a market. Realizing this, men owning fish-oil works preserved them from extinction by manufacturing the mineral-product Young and Gesner had introduced. Thus Samuel Downer's half-million-dollar works near Boston and colossal plant at Portland were utilized. Downer had expanded ideas and remarked with characteristic emphasis, in reply to a friend who criticised him for the risk he ran in putting up an enormous refinery at Corry, as the oil-production might exhaust: "The Almighty never does a picayune business !" Fifty or sixty of these works were turning out oil from bituminous shales in 1859, when the influx of petroleum compelled their conversion into refineries to avert overwhelming loss. Maine had one, Massachusetts five, New York five, Pennsylvania eight, Ohio twenty-five, Kentucky six, Virginia eight, Missouri one and one was starting in McKean county, near Kinzua village. The Carbon Oil Company, 184 Water street, New York City, was the chief dealer in the illuminant. The entire petroleum traffic in 1858 was barely eleven-hundred barrels, most of it obtained from Tarentum. A shipment of twelve barrels to New York in November, 1857, may be considered the beginning of the history of petroleum as an illuminator. How the baby has grown !

The price of "kerosene" or "carbon oil," always high, advanced to two dollars a gallon ! Nowadays people grudge ten cents a gallon for oil vastly clearer, purer, better and safer ! One good result of the high prices was an exhaustive scrutiny by the foremost scientific authorities into all the varieties cf coal and bitumen, out of which comparisons with petroleum developed incidentally. Belief in its identity with coal-oil prompted the investigations which finally determined the economic value of petroleum. Professor B. Silliman, Jun., Professor of Chemistry in Yale College, in the spring of 1855 concluded a thorough analysis of petroleum from a "spring" on Oil Creek, nearly two miles south of Titusville, where traces of pits cribbed with rough timber still remained and the sticky fluid had been skimmed for two generations. In the course of his report Professor Silliman observed :

" It is understood and represented that this product exists in great abundance on the property; that it can be gathered wherever a well is sunk, over a great number of acres, and that it is unfailing in its yield from year to year. The question naturally arises, Of what value is it in the arts and for what uses can it be employed ? * * * The Crude Oil was tried as a means of illumination. For this purpose a weighed quantity was decomposed by passing it through a wrought-iron retort filled with carbon and ignited to redness. It produced nearly pure carburetted hydrogen gas, the most highly illuminating of all carbon gases. In fact, the oil may be regarded as chemically identical with illuminating gas in a liquid form. It burned with an intense flame. * * * The light from the rectified Naphtha is pure and white, without odor, and the rate of consumption less than half that of Camphene or Rosin-Oil. * * * Compared with Gas, the Rock-Oil gave more light than any burner, except the costly Argand, consuming two feet of gas per hour. These photometric experiments have given the Oil a much higher value as an illuminator than I had dared to hope. * * * As this oil does not gum or become acid or rancid by exposure, it possesses in that, as well as in its wonderful resistance to extreme cold, important qualities for a lubricator. * * * It is worthy of note that my experiments prove that nearly the *whole* of the raw product may be manufactured without waste, solely by one of the most simple of all chemical processes."

Notwithstanding these researches, which he spent five months in prosecuting, the idea of artesian boring for petroleum—naturally suggested by the oil in the salines of the Muskingum, Kanawha, Cumberland and Allegheny—never occurred to the learned Professor of Chemistry in Yale! If he had been the Yale football, with Hickok swatting it five-hundred pounds to the square inch, the idea might have been pummeled into the man of crucibles and pigments! Once more was nature frustrated in the endeavor to "bring out" a favorite child. The faithful dog that attempted to drag a fat man by the seat of his pants to the rescue of a drowning master, or Diogenes in his protracted quest for an honest Athenian, had an easier task. The "spring" which furnished the material for Silliman's experiments was on the Willard farm, part of the lands of Brewer, Watson & Co.—Ebenezer Brewer and James Rynd, Pittsburg, Jonathan Watson, Rexford Pierce and Elijah Newberry, Titusville—extensive lumbermen on Oil Creek. They ran a sawmill on an island near the east bank of the creek, at a bend in the stream, a few rods south of the boundary-line between Venango and Crawford counties. Close to the mill was the rusty-looking "spring" from which the oil to burn in rude lamps, smoky and chimneyless, and to lubricate the circular saw was derived. The following document explains the first action regarding the care and development of the "spring :"

"Agreed this fourth day of July, A. D. 1853, with J. D. Angier, of Cherrytree Township, in the County of Venango, Pa., that he shall repair up and keep in order the old oil-spring on land in said Cherrytree township, or dig and make new springs, and the expenses to be deducted out of the proceeds of the oil and the balance, if any, to be equally divided, the one-half to J. D. Angier and the other half to Brewer, Watson & Co., for the full term of five years from this date, if profitable."

All parties signed this agreement, pursuant to which Angier, for many years a resident of Titusville, dug trenches centering in a basin from which a pump connected with the sawmill raised the water into shallow troughs that sloped to the ground. Small skimmers, nicely adjusted to skim the oil, collected three or four gallons a day, but the experiment did not pay and it was dropped. In the summer of 1854 Dr. F. B. Brewer, son of the senior member of the firm owning the mill and "spring," visited relatives at Hanover, New Hampshire, carrying with him a bottle of the oil as a gift to Professor Crosby, of Dartmouth College. Shortly after George H. Bissell, a graduate of the college, practicing law in New York with Jonathan G. Eveleth, while on a visit to Hanover called to see Professor Crosby, who showed him the bottle of petroleum. Crosby's son induced Bissell to pay the expenses of a trip to inspect the "spring" and to agree, in case of a satisfactory report, to organize a company with a capital of a quarter-million dollars to purchase lands and erect such machinery as might be required to collect all the oil in the vicinity.

Complications and misunderstandings retarded matters. Everything was adjusted at last. Brewer, Watson & Co. conveyed in fee simple to George H. Bissell and Jonathan G. Eveleth one-hundred-and-five acres of land in Cherry-tree township, embracing the island at the junction of Pine Creek and Oil Creek, on which the mill of the firm and the Angier ditches were situated. The deed was formally executed on January first, 1855. Eveleth and Bissell gave their own notes for the purchase-money—five-thousand dollars—less five-hundred dollars paid in cash. The consideration mentioned in the deed was twenty-five-thousand dollars, five times the actual sum, in order not to appear such a small fraction of the total capital—two-hundred-and-fifty-thousand dollars—as to injure the sale of stock. On December thirtieth, 1854, articles of incorporation of The Pennsylvania Rock Oil Company were filed in New York and Albany. The stock did not sell, owing to the prostration of the money-market and the fact that the company had been organized in New York, by the laws of which state each shareholder in a joint-stock company was liable for its debts to the amount of the par value of the stock he held.   New Haven parties agreed to subscribe for large blocks of stock if the company were reorganized under the laws of Connecticut.   A new company was formed with a nominal capital of three-hundred-thousand dollars, to take the name and property of the one to be dissolved and levy an assessment to develop the island "by trenching" on a wholesale plan.

Eveleth & Bissell retained a controlling interest and Ashael Pierpont, James M. Townsend and William A. Ives were three of the New Haven stockholders. Bissell visited Titusville to complete the transfer.   On January sixteenth he and his partner had given a deed, which was not recorded, to the trustees of the original company.   At Titusville he learned that lands of corporations organized outside of Pennsylvania would be forfeited to the state.   The new company was notified of this law and to avoid trouble, on September twentieth, 1855, Eveleth & Bissell executed a deed to Pierpont and Ives, who gave a bond for the value of the property and leased it for ninety-nine years to a company formed two days before under certain articles of association.   It really seemed that something definite would be done.   The first oil company in the history of nations had been organized.   Pierpont, an eminent mechanic, was sent to examine the "spring," with a view to improve Angier's machinery.   Silliman's reports had a stimulating effect and the Professor was president of the company. But the monkey-and-parrot time was renewed.   Dissensions broke out, Angier was fired and the enterprise looked to be "as dead as Julius Cæsar," ready to bury "a hundred fathoms deep."

One scorching day in the summer of 1856 Mr. Bissell, standing beneath the awning of a Broadway drug-store for a moment's shade, noticed a bottle of Kier's Petroleum and a queer show-bill, or label, in the window.   It struck him as rather odd that a four-hundred-dollar bill—such it appeared—should be displayed

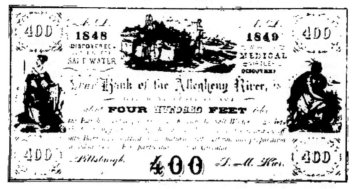

FAC-SIMILE OF LABEL ON KIER'S PETROLEUM.

in that manner.   A second glance proved that it was an advertisement of a substance that concerned him deeply.   He stepped inside and requested permission to scan the label.   The druggist told him to "take it along."   For an instant he gazed at the derricks and the figures—four-hundred feet!   A thought flashed upon him—bore artesian wells for oil!   Artesian wells!   Artesian wells!   rang in his ears like the Trinity chimes down the street, the bells of London telling "Dick" Whittington to return or the pibroch of the Highlanders at Lucknow The idea that meant so much was born at last. Patient nature must have felt in the mood to turn somersaults, blow a tin-horn and dance the fandango. It was a simple thought—merely to bore a hole in the rock—with no frills and furbelows and fustian, but pregnant with astounding consequences. It has added untold millions to the wealth of the country and conferred incalculable benefits upon humanity. To-day refined petroleum lights more dwellings in America, Europe, Asia, Africa and Australia than all other agencies combined.

To put the idea to the test was the next wrinkle. Mr. Eveleth agreed with Bissell's theory. Their first impulse was to bore a well themselves. Reflection cooled their ardor, as this course would involve the loss of their practice for an uncertainty. Mr. Havens, a Wall-street broker, whom they consulted, offered them five-hundred dollars for a lease from the Pennsylvania Rock-Oil Company. A contract with Havens, by the terms of which he was to pay "twelve cents a gallon for all oil raised for fifteen years," financial reverses prevented his carrying out. The idea of artesian boring was too fascinating to lie dormant. Mr. Townsend, president of the company, Silliman having resigned, employed Edwin L. Drake, to whom in the darker days of its existence he had sold two-hundred-dollars' worth of his own stock, to visit the property and report his impressions. Mrs. Brewer and Mrs. Rynd had not joined in the power-of-attorney by which the agent conveyed the Brewer-Watson lands to the company, hence they would be entitled to dower in case the husbands died. Drake was instructed to return by way of Pittsburg and procure their signatures. Illness had forced him to quit work—he was conductor on the New York & New Haven Railroad—for some months and the opportunity for change of air and scene was embraced gladly. Shrewd, far-seeing Townsend, who still lives in New Haven and has been credited with "the discovery" of petroleum, addressed legal documents and letters to "Colonel" Drake, no doubt supposing this would enhance the importance of his representative in the eyes of the Oil-Creek backwoodsmen. The military title stuck to the diffident civilian whose name is interwoven with the great events of the nineteenth century.

Stopping on his way from New Haven to view the salt-wells at Syracuse, about the middle of December, 1857, Colonel Drake was trundled into Titusville—named from Jonathan Titus—on the mail-wagon from Erie. The villagers received him cordially. He lodged at the American Hotel, the home-like inn "Billy" Robinson, the first boniface, and Major Mills, king of landlords, rendered famous by their bountiful hospitality. The old caravansary was torn down in 1880 to furnish a site for the Oil Exchange. Drake stayed a few days to transact legal business, to examine the lands and the indications of oil and to become familiar with the general details. Proceeding to Pittsburg, he visited

JONATHAN TITUS.

the salt-wells at Tarentum, the picture of which on Kier's label suggested boring for oil, and hastened back to Connecticut to conclude a scheme of operating the property. On December thirtieth the three New Haven directors executed a lease to Edwin E. Bowditch and Edwin L. Drake, who were to pay the Pennsylvania Rock-Oil Company "five-and-a-half cents a gallon for the oil raised for fifteen years." Eight days later, at the annual meeting of the directors, the lease was ratified, George H. Bissell and Jonathan Watson, representing two-thirds of the stock, protesting. Thereupon the consideration was placed at "one-eighth of all oil, salt or paint produced." The lease was sent to Franklin and recorded in Deed Book P, page 357. A supplemental lease, extending the time to forty-five years on the conditions of the grant to Havens, was recorded, and on March twenty-third, 1858, the Seneca Oil Company was organized, with Colonel Drake as president and owner of one-fortieth of the "stock." No stock was issued, for the company was in reality a partnership working under the laws governing joint-stock associations.

Provided with a fund of one-thousand dollars as a starter, Drake was engaged at one-thousand dollars a year to begin operations. Early in May, 1858, he and his family arrived in Titusville and were quartered at the American Hotel, which

THE FIRST DRAKE WELL, ITS DRILLERS AND ITS COMPLETE RIG.

boarded the Colonel, Mrs. Drake, two children and a horse for six-dollars-and-a-half per week ! Money was scarce, provisions were cheap and the quiet village put on no extravagant airs. Not a pick or shovel was to be had in any store short of Meadville, whither Drake was obliged to send for these useful tools ! Behold, then, "the man who was to revolutionize the light of the world," his mind full of a grand purpose and his pockets full of cash, snugly ensconced in the comfortable hostelry. Surely the curtain would soon rise and the drama of "A Petroleum Hunt" proceed without further vexatious delays.

Drake's first step was to repair and start up Angier's system of trenches, troughs and skimmers. By the end of June he had dug a shallow well on the island and was saving ten gallons of oil a day. He found it difficult to get a practical "borer" to sink an artesian well. In August he shipped two barrels of oil to New Haven and bargained for a steam-engine to furnish power for drilling. The engine was not furnished as agreed, the "borer" Dr. Brewer hired at Pittsburg had another contract and operations were suspended for the winter. In February, 1859, Drake went to Tarentum and engaged a driller to come in March. The driller failed to materialize and Drake drove to Tarentum in a sleigh to lasso another. F. N. Humes, who was cleaning out salt-wells for Peterson, informed him that the tools were made by William A. Smith, whom he might be able to secure for the job. Smith accepted the offer to manufacture tools and bore the well. Kim Hibbard, favorably known in Franklin, was dispatched with his team, when the tools were completed, for Smith, his two sons and the outfit. On May twentieth the men and tools were at the spot selected for the hole. A "pump-house" had been framed and a derrick built. A room for "boarding the hands" almost joined the rig and the sawmill. The accompanying illustration shows the well as it was at first, with the original derrick enclosed to the top, the "grasshopper walking-beam," the "boarding-house" and part of the mill-shed. "Uncle Billy" Smith is seated on a wheelbarrow in the foreground. His sons, James and William, are standing on either side of the "pump-house" entrance. Back of James his two young sisters are sitting on a board. Elbridge Lock stands to the right of the Smiths. "Uncle Billy's" brother is leaning on a plank at the corner of the derrick and his wife may be discerned in the doorway of the "boarding-house." This interesting and historic picture has never been printed until now. The one with which the second rig, with Peter Wilson, a Titusville druggist, facing Drake. In like manner, the portrait of Colonel Drake in this volume is from the first photograph for which he ever sat. The well and the portrait are the work of John A. Mather, the veteran artist and Drake's bosom-friend, who ought to receive a pension and no end of gratitude for preserving "counterfeit presentments" of a host of petroleum-scenes and personages that have passed from mortal sight.

Delays and tribulations had not retreated from the field. In artesian boring it is necessary to drill in rock. Mrs. Glasse's old-time cook-book gained celebrity by starting a recipe for rabbit-pie: "First catch your hare." The principle applies to artesian drilling: "First catch your rock." The ordinary rule was to dig a pit or well-hole to the rock and crib it with timber. The Smiths dug a few feet, but the hole filled with water and caved-in persistently. It was a fight-to-a-finish between three men and what Stow of Girard—he was Barnum's hot-stuff advance agent—wittily termed "the cussedness of inanimate things." The latter won and a council of war was summoned, at which Drake recommended driving an iron-tube through the clay and quicksand to the rock. This was effectual. Colonel Drake should have patented the process, which was his exclusive device and decidedly valuable. The pipe was driven thirty-six feet to hard-pan and the drill started on August fourteenth. The workmen averaged three feet a day, resting at night and on Sundays. Indications of oil were met as the tools pierced the rock. Everybody figured that the well would be down to the Tarentum level in time to celebrate Christmas. The company, tired of repeated postponements, did not deluge Drake with money. Losing speculations and sickness had drained his own meagre savings. R. D. Fletcher, the well-known Titusville merchant, and Peter Wilson endorsed his paper for six-hundred dollars to tide over a crisis. The tools pursued the downward road with the eagerness of a sinner headed for perdition, while expectation stood on tiptoe to watch the progress of events.

On Saturday afternoon, August twenty-eighth, 1859, the well had reached the depth of sixty-nine feet, in a coarse sand. Smith and his sons concluded to "lay off" until Monday morning. As they were about to quit the drill dropped six inches into a crevice such as was common in salt-wells. Nothing was thought of this circumstance, the tools were drawn out and all hands adjourned to Titusville. Mr. Smith went to the well on Sunday afternoon to see if it had moved away or been purloined during the night. Peering into the hole he saw fluid within eight or ten feet. A piece of tin-spouting was lying outside. He plugged one end of the spout, let it down by a string and pulled it up. Muddy water? No! It was filled with PETROLEUM!

That was the proudest hour in "Uncle Billy" Smith's forty-seven years' pilgrimage. Not daring to leave the spot, he ran the spout again and again, each time bringing it to the surface full of oil. A straggler out for a stroll approached, heard the story, sniffed the oil and bore the tidings to the village. Darkness was setting in, but the Smith boys sprinted to the scene. When Colonel Drake came down, bright and early next morning, they and their father were guarding three barrels of the precious liquid. The pumping apparatus was adjusted and by noon the well commenced producing at the rate of twenty barrels a day! The problem of the ages was solved, the agony ended and petroleum fairly launched on its astonishing career.

The news flew like a Dakota cyclone. Villagers and country-folk flocked to the wonderful well. Smith wrote to Peterson, his former employer: "Come quick, there's oceans of oil!" Jonathan Watson jumped on a horse and galloped down the creek to lease the McClintock farm, where Nathanael Cary dipped oil and a timbered crib had been constructed. Henry Potter, still a citizen of Titusville, tied up the lands for miles along the stream, hoping to interest New York capital. William Barnsdall secured the farm north of the Willard. George H. Bissell, who had arranged to be posted by telegraph, bought all the Pennsylvania Rock-Oil stock he could find and in four days was at the well. He leased farm after farm on Oil Creek and the Allegheny river, regardless of surface indications or the admonitions of meddling wiseacres.

The rush for property resembled the wild scramble of the children when the Pied Piper of Hamelin blew his fatal reed. Titusville was in a whirlpool of excitement. Buildings arose as if by magic, the hamlet became a borough and the borough a city of fifteen-thousand inhabitants. Maxwell Titus sold lots at two-hundred dollars, people acquired homes that doubled in value and speculation held undisputed sway. Jonathan Titus, from whom it was named, lived to witness the farm he cleared transformed into "The Queen City," noted for its tasteful residences, excellent schools, manufactories, refineries and active population. One of his neighbors in the bush was Samuel Kerr, whose son Michael went to Congress and served as Speaker of the House. Many enterprising men settled in Titusville for the sake of their families. They paved the streets, planted shade-trees, fostered local industries, promoted culture and believed in public improvements. When Christine Nillson enraptured sixteen-hundred well-dressed, appreciative listeners in the Parshall Opera-House, the peerless songstress could not refrain from saying that she never saw an audience so keen to note the finer points of her performance and so discriminating in its applause.

" Praise from Sir Hubert is praise indeed " and the compliment of the Swedish Nightingale compressed a whole encyclopedia into a sentence. Titusville has had its ups and downs, but there is no more desirable abiding place in the Keystone State.

Matches are supposed to be made in Heaven and the inspiration that led to the choice of such a site for the future city must have been derived from the same source. Healthfulness and beauty of location attest the wisdom of the selection. Folks don't have to climb precipitous hills or risk life and limb

MAIN STREET, TITUSVILLE, IN 1861.

crossing railway-tracks whenever they wish to exercise their fast nags. Driving is a favorite pastime in fine weather, the leading thoroughfares often reminding strangers of Central Park on a coaching-day. Main, Walnut and Perry streets are lined with trees and residences worthy of Philadelphia or Baltimore. Comfortable homes are the crowning glory of a community and in this respect Titusville does not require to take a back-seat. Near the lower end of Main street is Ex-Mayor Caldwell's elegant mansion, built by Jonathan Watson in the days of his prosperity. Farther up are John Fertig's, the late Marcus Brownson's, Mrs. David Emery's and Mrs. A. N. Perrin's. Franklin S. Tarbell, a former resident of Rouseville, occupies an attractive house. Joseph Seep, who has not changed an iota since the halcyon period of Parker and Foxburg, shows his faith in the town by building a home that would adorn Cleveland's aristocratic Euclid Avenue. The host is the cordial Seep of yore, quick to make a point and not a bit backward in helping a friend. David McKelvy, whom everybody knew in the lower oil-fields, remodeled the Chase homestead, a symphony in red brick. Close by is W. T. Scheide's natty dwelling, finished in a style befitting the ex-superintendent of the National-Transit Pipe-Lines. Byron D. Benson—he died in 1889—nine times elected president of the Tidewater Pipe-Line Company, lived on the corner of Oak and Perry streets. Opposite is John L. McKinney's luxurious residence, a credit to the liberal owner and the city. J. C. McKinney's is "one of the finest." James Parshall, W. B. Sterrett, O. D. Harrington,

J. P. Thomas, W. W. Thompson, Charles Archbold and hundreds more erected dwellings that belong to the palatial tribe. Dr. Roberts—he's in the cemetery —had a spacious place on Washington street, with the costliest stable in seventeen counties. E. O. Emerson's house and grounds are the admiration of visitors. The grand fountain, velvet lawns, smooth walks, tropical plants, profusion of flowers, mammoth conservatory and Marechal-Niel rose-bushes bewilder the novice whose knowledge of floral affairs stops at button-hole bouquets. George K. Anderson—dead, too—constructed this delightful retreat. Col. J. J. Carter, whose record as a military officer, merchant, railroad-president

DANIEL CADY.

and oil-operator will stand inspection, has an ideal home, purchased from John D. Archbold and refitted throughout. It was built and furnished extravagantly by Daniel Cady, once a leading spirit in the business and social life of Titusville. He was a man of imposing presence and indomitable pluck, the confidant of Jay Gould and "Jim" Fisk, dashing, speculative and popular. For years whatever he touched seemed to turn into gold and he computed his dollars by hundreds of thousands. Days of adversity overtook him, the splendid home was sacrificed and he died poor. To men of the stamp of Watson, Anderson, Abbott, Emery, Fertig and Cady Titusville owes its real start in the direction of greatness. Much of the froth and fume of former days is missing, but the baser elements have been eliminated, trade is on a solid basis and important manufactures have been established. There are big refineries, Holly water-works, a race-track, ball-grounds, top-notch hotels, inviting newspapers, inviting churches and a lovely cemetery in which to plant good citizens when they pass in their checks. Pilgrims who expect to find Titusville dead or dying will be as badly fooled as the lover whose girl eloped with the other fellow.

Unluckily for himself, Colonel Drake took a narrow view of affairs. Complacently assuming that he had "tapped the mine"—to quote his own phrase— and that paying territory would not be found outside the company's lease, he pumped the well serenely, told funny stories and secured not one foot of ground ! Had he possessed a particle of the prophetic instinct, had he grasped the magnitude of the issues at stake, had he appreciated the importance of petroleum as a commercial product, had he been able to "see an inch beyond his nose," he would have gone forth that August morning and become "Master of the Oil Country !" "The world was all before him where to choose," he was literally "monarch of all he surveyed," but he didn't move a peg ! Money was not needed, the promise of one-eighth or one-quarter royalty satisfying the easygoing farmers, consequently he might have gathered in any quantity of land. Friends urged him to "get into the game ;" he rejected their counsel and never realized his mistake until other wells sent prices skyward and it was everlastingly too late for his short pole to knock the persimmons. Yet this is the man whom numerous writers have proclaimed "the discoverer of petroleum !" Times without number it has been said and written and printed that he was "the first man to advise boring for oil," that "his was the first mind to conceive the idea of penetrating the rock in search of a larger deposit of oil than was dreamed of by any one," that "he alone unlocked one of nature's vast storehouses" and "had visions of a revolution in light and lubrication." Considering what Kier, Peterson, Bissell and Watson had done years before Drake ever saw—perhaps ever heard of—a drop of petroleum, the absurdity of these claims is "so plain that he who runs may read." Couple with this his incredible failure to secure lands after the well was drilled — wholly inexcusable if he supposed oil-operations would ever be important—and the man who thinks Colonel Drake was "the first man with a clear conception of the future of petroleum" could swallow the fish that swallowed Jonah!

Above all else history should be truthful and "hew to the line, let chips fall where they may." Mindful that "the agent is but the instrument of the principal," why should Colonel Drake wear the laurels in this instance? Paid a salary to carry out Bissell's plan of boring an artesian well, he spent sixteen months getting the hole down seventy feet. For a man who "had visions" and "a clear conception" his movements were inexplicably slow. He encountered obstacles, but salt-wells had been drilled hundreds of feet without either a steam-engine or professional "borer." The credit of suggesting the driving-pipe to overcome the quicksand is justly his due. Quite as justly the credit of suggesting the boring of the well belongs to George H. Bissell. The company hired Drake, Drake hired Smith, Smith did the work. Back of the man who possessed the skill to fashion the tools and sink the hole, back of the man who acted for the company and disbursed its money, back of the company itself is the originator of the idea these were the means employed to put into effect. Was George Stephenson, or the foreman of the shop where the "Rocket" was built, the inventor of the locomotive? Was Columbus, or the man whose name it bears, the discoverer of America? In a conversation on the subject Mr. Bissell remarked : "Let Colonel Drake enjoy the pleasure of giving the well his name ; history will set us all right." So it will and this is a step in that direction. If the long-talked-of monument to commemorate the advent of the petroleum-era ever be erected, it should bear in boldest capitals the names of Samuel M. Kier and George H. Bissell.

Edwin L. Drake, who is linked inseparably with the first oil-well in Pennsylvania, was born on March eleventh, 1819, at Greenville, Greene county, New York. His father, a farmer, moved to Vermont in 1825. At eighteen Edwin left home to begin the struggle with the world. He was night clerk of a boat running between Buffalo and Detroit, worked one year on a farm in the Wolverine state, clerked two years in a Michigan hotel, returned east and clerked in a dry-goods store at New Haven, clerked and married in New York, removed to Massachusetts, was express-agent on the Boston & Albany railroad and resigned in 1849 to become conductor on the New York & New Haven. His younger brother died in the west and his wife at New Haven, in 1854, leaving one child. While boarding at a hotel in New Haven he met James M. Townsend, who persuaded him to draw his savings of two-hundred dollars from the bank and buy stock of the Pennsylvania Rock-Oil Company, his first connection with the business that was to make him famous. Early in 1857 he married Miss Laura Dow, sickness in the summer compelled him to cease punching tickets and his memorable visit to Titusville followed in December. In 1860 he was elected justice-of-the-peace, an office worth twenty-five-hundred dollars that year, because of the enormous number of property-transfers to prepare and acknowledge. Buying oil on commission for Shefflin Brothers, New York, swelled his income to five-thousand dollars for a year or two. He also bought twenty-five acres of land from Jonathan Watson, east of Martin street and through the center of which Drake street now runs, for two-thousand dollars. Unable to meet the mortgage given for part of the payment, he sold the block in 1863 to Dr. A. D. Atkinson for twelve-thousand dollars. Forty times this sum would not have bought it in 1867! With the profits of this transaction and his savings for five years, in all about sixteen-thousand dollars, in the summer of 1863 Colonel Drake left the oil-regions forever.

Entering into partnership with a Wall-street broker, he wrecked his small fortune speculating in oil-stocks, his health broke down and he removed to Vermont. Physicians ordered him to the seaside as the only remedy for his disease, neuralgic affection of the spine, which threatened paralysis of the limbs and caused intense suffering. Near Long Branch, in a cottage offered by a friend, Mr. and Mrs. Drake drank the bitter cup to the dregs. Their funds were exhausted, the patient needed constant attention and helpless children cried for bread. The devoted wife and mother attempted to earn a pittance with her needle, but could not keep the wolf of hunger from the door. Medicine for the sick man was out of the question. All this time men in the region the Drake well had opened to the world were piling up millions of dollars! One day in 1869, with eighty cents to pay his fare, Colonel Drake struggled into New York to seek a place for his twelve-year-old boy. The errand was fruitless. The distressed father was walking painfully on the street to the railway-station, to board the train for home, when he met "Zeb" Martin of Titusville, afterwards proprietor of the Hotel Brunswick. Mr. Martin noted his forlorn condition, inquired as to his circumstances, learned the sad story of actual privation, procured dinner, gave the poor fellow twenty dollars and cheered him with the assurance that he would raise a fund for his relief. The promise was redeemed.

At a meeting in Titusville the case was stated and forty-two hundred dollars were subscribed. The money was forwarded to Mrs. Drake, who husbanded it carefully. The terrible recital aroused such a feeling that the Legislature, in 1873, granted Colonel Drake an annuity of fifteen-hundred dollars during his life and his heroic wife's. California had set a good example by giving Colonel Sutter, the discoverer of gold in the mill-race, thirty-five-hundred dollars a year. The late Thaddeus Stevens, "the Great Commoner," hearing that Drake was actually in want, prepared a bill, found among his papers after his death, intending to present it before Congress for an appropriation of two-hundred-and-fifty-thousand dollars for Colonel Drake. In 1870 the family removed to Bethlehem, Pennsylvania. Years of suffering, borne with sublime resignation, closed on the evening of November ninth, 1881, with the release of Edwin L. Drake from this vale of tears. A faithful wife and four children survived the petroleum-pioneer. They lived at Bethlehem until the spring of 1895 and then moved to New England. Colonel Drake was a man of pronounced individuality, affable, genial and kindly. He had few superiors as a story-teller, neither caroused nor swore, and was of unblemished character. He wore a full beard, dressed well, liked a good horse, looked every man straight in the face and his dark eyes sparkled when he talked. Gladly he laid down the heavy burden of a checkered life, with its afflictions and vicissitudes, for the peaceful rest of a humble grave.

> " Nothing in his life
> Became him like the leaving it ; he died
> As one who had been studied in his death,
> To throw away the dearest thing he owed,
> As 'twere a careless trifle."

George H. Bissell, honorably identified with the petroleum-development from its inception, was a New Hampshire boy. Thrown upon his own resources at twelve, by the death of his father, he gained education and fortune unaided. At school and college he supported himself by teaching and writing for magazines. Graduating from Dartmouth College in 1845, he was professor of Greek and Latin in Norwich University a short time, went to Washington and Cuba, did editorial work for the New Orleans *Delta* and was chosen superintendent of the public schools. Impaired health forced him to return north in 1853, when his connection with petroleum began. From 1859 to 1863 he resided at Franklin, Venango county, to be near his oil-interests. He operated largely on Oil Creek, on the Allegheny river and at Franklin, where he erected a barrel-factory. He removed to New York in 1863, established the Bissell Bank at Petroleum Centre in 1866, developed oil-lands in Peru and was prominent in financial circles. His wife died in 1867 and long since he followed her to the tomb. Mr. Bissell was a brilliant, scholarly man, positive in his convictions and sure to make his influence felt in any community. His son and daughter reside in New York.

William A. Smith, born in Butler county in 1812, at the age of twelve was apprenticed at Freeport to learn blacksmithing. In 1827 he went to Pittsburg and in 1842 opened a blacksmith-shop at Salina, below Tarentum. Samuel M. Kier employed him to drill salt-wells and manufacture drilling-tools. After finishing the Drake well, he drilled in various sections of the oil-regions, retiring to his farm in Butler a few years prior to his death, on October twenty-third, 1890. "Uncle Billy," as the boys affectionately called him, was no small factor in giving to mankind the illuminator that enlightens every quarter of the globe. The farm he owned in 1859 and on which he died proved to be good oil-territory.

Dr. Francis B. Brewer was born in New Hampshire, studied medicine in Philadelphia and practiced in Vermont. His father in 1840 purchased several thousand acres of land on Oil Creek for lumbering, and the firm of Brewer, Watson & Co. was promptly organized. Oil from the "spring" on the island at the mouth of Pine Creek was sent to the young physician in 1848 and used in his practice. He visited the locality in 1850 and was admitted to the firm. Upon the completion of the Drake well he devoted his time to the extensive oil-operations of the partnership for four years. In 1864 Brewer, Watson & Co. sold the bulk of their oil-territory and the doctor, who had settled at Westfield, Chautauqua county, N. Y., instituted the First National Bank, of which he was chosen president. A man of solid worth and solid wealth, he has served as a Member of Assembly and is deservedly respected for his integrity and benevolence.

Jonathan Watson, whose connection with petroleum goes back to the beginning of developments, arrived at Titusville in 1845 to manage the lumbering and mercantile business of his firm. The hamlet contained ten families and three stores. Deer and wild-turkeys abounded in the woods. John Robinson was postmaster and Rev. George O. Hampson the only minister. Mr. Watson's views of petroleum were of the broadest and his transactions the boldest. He hastened to secure lands when oil appeared in the Drake well. At eight o'clock on that historic Monday morning he stood at Hamilton McClintock's door, resolved to buy or lease his three-hundred-acre farm. A lease was taken and others along the stream followed during the day. Brewer, Watson & Co. operated on a wholesale scale until 1864, after which Watson continued alone. Riches poured upon him. He erected the finest residence in Titusville, lavished money on the grounds and stocked a fifty-thousand dollar conservatory with choicest plants and flowers. A million dollars in gold he is credited with "putting by for a rainy day." He went miles ahead, bought huge blocks of land and drilled scores of test-wells. In this way he barely missed opening the Bradford field and the Bullion district years before these productive sections were brought into line. His well on the Dalzell farm, Petroleum Centre, in 1869, renewed interest in that quarter long after it was supposed to be sucked dry. An Oil City clairvoyant indicated the spot to sink the hole, promising a three-hundred-barrel strike. Crude was six dollars a barrel and Watson readily proffered the woman the first day's production for her services. A check for two-thousand dollars was her reward, as the well yielded three-hundred-and-thirty-three barrels the first twenty-four hours. Mrs. Watson was an ardent medium and her husband humored her by consulting the "spirits" occasionally. She became a lecturer and removed to California long since. The tide of Watson's prosperity ebbed. Bad investments and dry-holes ate into his splendid fortune. The gold reserve was drawn upon and spent. The beautiful home went to satisfy creditors. In old age the brave, hardy, indefatigable oil-pioneer, who had led the way for others to acquire wealth, was stripped of his possessions. Hope and courage remained. He operated at Warren and revived some of the old wells around the Drake, which afforded him subsistence. Advanced years and anxiety enfeebled the stalwart fame. His steps faltered, and in 1893 protracted sickness closed the busy, eventful life of the man who, more than any other, fostered and developed the petroleum-industry.

> " What tragic tears bedew the eye,
> What deaths we suffer ere we die !"

The Drake well declined almost imperceptibly, yielding twelve barrels a day by the close of the year. It stood idle on Sundays and for a week in December. Smith had a light near a tank of oil, the gas from which caught fire and burned the entire rig, This was the first "oil-fire" in Pennsylvania, but it was destined to have many successors. Possibly it brought back vividly to Colonel Drake the remembrance of his childish dream, in which he and his brother had set a heap of stubble ablaze and could not extinguish the flames. His mother interpreted it : "My son, you have set the world on fire."

The total output of the well in 1859 was under eighteen-hundred barrels. One-third of the oil was sold at sixty-five cents a gallon for shipment to Pittsburg. George M. Mowbray, the accomplished chemist, who came to Titusville in 1860 and played a prominent part in early refining, disposed of a thousand barrels in New York. The well produced moderately for two or three years from the first sand, until shut down by low prices, which made it ruinous to pay the royalty of twelve-and-a-half cents a gallon. A compromise was effected in 1860, by which the Seneca Oil Company retained a part of the land as fee and surrendered the lease to the Pennsylvania Rock-Oil Company. Mr. Bissell purchased the stock of the other shareholders in the latter company for fifty-thousand dollars. He drilled ten wells, six of which for months yielded eighty barrels a day, on the tract known thenceforth as the Bissell farm, selling it eventually to the Original Petroleum Company. The Drake was deepened to five-hundred feet and two others, drilled beneath the roof of the sawmill in 1862, were pumped by water.

The Drake machinery was stolen or scattered piecemeal. In 1876 J. J. Ashbaugh, of St. Petersburg, and Thomas O'Donnell, of Foxburg, conveyed the neglected derrick and engine-house to the Centennial at Philadelphia, believing crowds would wish to look at the mementoes. The exhibition was a fizzle and the lumber was carted off as rubbish. Lewis F. Emery, Jun., saved the drilling-tools and he has them in his private museum at Bradford. They are pigmies compared with the giants of to-day. A man could walk away with them as readily as Samson skipped with the gates of Gaza. Sandow and Cyril Cyr done up in a single package couldn't do that with a modern set. The late David Emery, a man of heart and brain, contemplated reviving the old well—the land had come into his possession—and bottling the oil in tiny vials, the proceeds to be applied to a Drake monument. He put up a temporary rig and pumped a half-barrel a week. Death interrupted his generous purpose. Except that the trees and the saw-mill have disappeared, the neighborhood of the Drake well is substantially the same as in the days when lumbering was at its height and the two-hundred honest denizens of Titusville slept without locking their doors.

LOCATION AND SURROUNDINGS OF THE DRAKE WELL IN 1895.

There is nothing to suggest to strangers or travelers that the spot deserves to be remembered. How transitory is human achievement!

William Barnsdall, Boone Meade and Henry R. Rouse started the second well in the vicinity, on the James Parker farm, formerly the Kerr tract and now the home of Ex-Mayor J. H. Caldwell. The location was north and within a stone's throw of the Drake. In November, at the depth of eighty feet, the well was pumped three days, yielding only five barrels of oil. The outlook had an indigo tinge and operations ceased for a week or two. Resuming work in December, at one-hundred-and-sixty feet indications were satisfactory. Tubing was put in on February nineteenth, 1860, and the well responded at the rate of fifty barrels a day! In the language of a Hoosier dialect poet: "Things wuz gettin' inter-restin'!" William H. Abbott, a gentleman of wealth, reached Titusville on February ninth and bought an interest in the Parker tract the same month. David Crossley's well, a short distance south of the Drake and the third finished on Oil Creek, began pumping sixty barrels a day on March fourth. Local dealers, overwhelmed with an "embarrassment of riches," could not handle such a glut of oil. Schefflin Brothers arranged to market it in New York. Fifty-six-thousand gallons from the Barnsdall well were sold for seventeen-thousand dollars by June first, 1860. J. D. Angier contracted to "stamp down a hole" for Brewer, Watson & Co., in a pit fourteen feet deep, dug and cribbed to garner oil dipped from the "spring" on the Hamilton-McClintock farm. Piercing the rock by "hand-power" was a tedious process. December of 1860 dawned without a symptom of greasiness in the well, from which wondrous results were anticipated on account of the "spring." One day's hand-pumping produced twelve barrels of oil and so much water that an engine was required to pump steadily. By January twentieth, 1861, the engine was puffing and the well producing moderately, the influx of water diminishing the yield of oil. These four, with two getting under way on the Buchanan farm, north of the McClintock, and one on the J. W. McClintock tract, the site of Petroleum Centre, summed up all the wells actually begun on Oil Creek in 1859.

Three of the four were "kicked down" by the aid of spring-poles, as were hundreds later in shallow territory. This method afforded a mode of development to men of limited means, with heavy muscles and light purses, although totally inadequate for deep drilling. An elastic pole of ash or hickory, twelve to twenty feet long, was fastened at one end to work over a fulcrum. To the other end stirrups were attached, or a tilting platform was secured by which two or three men produced a jerking motion that drew down the pole, its elasticity pulling it back with sufficient force, when the men slackened their hold, to raise

" KICKING DOWN " A WELL.

the tools a few inches. The principle resembled that of the treadle-board of a sewing-machine, operating which moves the needle up and down. The tools were swung in the driving-pipe or the "conductor"—a wooden tube eight or ten inches square, placed endwise in a hole dug to the rock—and fixed by a rope to the spring-pole two or three feet from the workmen. The strokes were rapid and a sand-pump—a spout three inches in diameter, with a hinged bottom opening inward and a valve working on a sliding-rod, somewhat in the manner of a syringe—removed the borings mainly by sucking them into the spout as it was drawn out quickly. Horse-power, in its general features precisely the kind still used with threshing-machines, was the next step forward. Steam-engines, employed for drilling at Tidioute in September of 1860, reduced labor and expedited work. The first pole-derricks, twenty-five to thirty-five feet high, have been superseded by structures that tower seventy-two to ninety feet.

Drilling-tools, the chief novelty of which are the "jars"—a pair of sliding bars moving within each other—have increased from two-hundred pounds to three-thousand in weight. George Smith, at Rouseville, forged the first steel-lined jars in 1866, for H. Leo Nelson, but the steel could not be welded firmly. Nelson also adopted the "Pleasantville Rig" on the Meade lease, Rouseville, in 1866, discarding the "Grasshopper." In the former the walking-beam is fastened in the centre to the "samson-post," with one end attached to the rods in the well and the other to the band-wheel crank, exactly as in side-wheel steamboats. George Koch, of East Sandy, Pa., patented numerous improvements on pumping-rigs, drilling-tools and gas-rigs, for which he asked no remuneration. Primitive wells had a bore of three or four inches, half the present size. To exclude surface-water a "seed-bag"—a leather bag the diameter of the hole —was tied tightly to the tubing, filled with flax-seed and let down to the proper depth. The top was left open and in a few hours the flax swelled so that the space between the tubing and the walls of the well was impervious to water. Drilling "wet holes" was slow and uncertain, as the tools were apt to break and the chances of a paying well could not be decided until the pump exhausted the water. It is surprising that over five-thousand wells were sunk with the rude appliances in vogue up to 1868, when "casing"—a larger pipe inserted usually to the top of the first sand—was introduced. This was the greatest improvement ever devised in oil-developments and drilling has reached such perfection that holes can be put down five-thousand feet safely and expeditiously. Devices multiplied as experience was gained.

The tools that drilled the Barnsdall, Crossley and Watson wells were the handiwork of Jonathan Lock, a Titusville blacksmith. Mr. Lock attained his eighty-third year, died at Bradford in March of 1895 and was buried at Titusville, the city in which he passed much of his active life. He was a worthy type of the intelligent, industrious American mechanics, a class of men to whom civilization is indebted for unnumbered comforts and conveniences. John Bryan, who built the first steam-engine in Warren county, started the first foundry and machine-shop in Oildom and organized the firm of Bryan, Dillingham & Co., began the manufacture of drilling-tools in Titusville in 1860.

JONATHAN LOCK.

Of the partners in the second well William Barnsdall survives. He has lived in Titusville sixty-three years, served as mayor and operated extensively. His son Theodore, who pumped wells on the Parker and Weed farms, adjoining the Barnsdall homestead, is among the largest and wealthiest producers. Crossley's sons rebuilt the rig at their father's well in 1873, drilled the hole deeper and obtained considerable oil. Other wells around the Drake were treated similarly, paying a fair profit. In 1875 this spasmodic revival of the earliest territory died out. Machinery was removed and the derricks rotted. Jonathan Watson, in 1889, drilled shallow wells, cleaned out several of the old ones and awakened brief interest in the cradle of developments. Gas burning and wells pumping, thirty years after the first strike, seemed indeed strange. Not a trace of these repeated operations remains. The Parker and neighboring farms north-west and north of Titusville proved disappointing, owing to the absence of the third sand,

6

which a hole drilled two-thousand feet by Jonathan Watson failed to reveal. The Parker-Farm Petroleum Company of Philadelphia bought the land in 1863 and in 1870 twelve wells were producing moderately. West and south-west the Octave Oil Company has operated profitably for twenty years and Church Run has produced generously. Probably two-hundred wells were sunk above Titusville, at Hydetown, Clappville, Tryonville, Centerville, Riceville, Lincolnville and to Oil-Creek Lake, in vain attempts to discover an extension of juicy territory.

Ex-Mayor William Barnsdall is the oldest living pioneer of Titusville. Not only has he seen the town grow from a few houses to its present proportions, but he is one of its most esteemed citizens. Born at Biggleswade, Bedfordshire, England, on February sixth, 1810, he lived there until 1831, when he came to America. In 1832 he arrived at what is known as the English Settlement, seven miles north of Titusville. The Barnsdalls founded the settlement, Joseph, a brother of William, clearing a farm in the wilderness that then covered the country. Remaining in the settlement a year, in 1833 William Barnsdall came to the hamlet of Titusville, where he has ever since resided. He established a small shop to manufacture boots and shoes, continuing at the business until the discovery of oil in 1859. Immediately after the completion of the Drake strike he began drilling the second well on Oil Creek. Before this well produced oil, in February of 1860, he sold a part interest to William H. Abbott for ten-thousand dollars. He associated himself with Abbott and James Parker and, early in 1860, commenced the first oil-refinery on Oil Creek. It was sold to Jonathan Watson for twenty-five-thousand dollars. From those early days to the present Mr. Barnsdall has been identified with the production of petroleum. Although thus engaged, he found time to identify himself with many other affairs. Always benevolent, he has made a record that will live long after he has passed from earth. In 1878 the people elected him mayor by a very large majority over two opposing candidates. In 1880 he was elected treasurer of the city. He served both positions in a manner that won for him the gratitude of his fellow-citizens. During his administration as mayor he effected many reforms that accrued to the benefit of the taxpayers. At the ripe age of eighty-six years, respected as few men are in any community and enjoying an unusual measure of mental and physical strength, he calmly awaits "the inevitable hour."

Born in England in 1818, David Crossley ran away from home and came to America as a stowaway in 1828. He found relatives at Paterson, N. J., and lived with them until about 1835, when he bound himself out to learn blacksmithing. On March seventeenth, 1839, he married Jane Alston and in the winter of 1841-2 walked from New York to Titusville, walking back in the spring. The following autumn he brought his family to Titusville. For a few years he tried farming, but gave it up and went back to his trade until 1859, when he formed a partnership with William Barnsdall, William H. Abbott and P. T. Witherop, under the firm-name of Crossley, Witherop & Co., and began drilling the third well put down on Oil Creek. The well was completed on March tenth, 1860, having been drilled one-hundred-and-forty feet with a spring-pole. It produced at the rate of seventy-five barrels per day for a short time. The next autumn the property was abandoned on account of decline in production. In 1865 Crossley bought out his partners and drilled the well to a depth of five-hundred-and-fifty feet, but again abandoned it because of water. In 1872 he and his son drilled other wells upon the same property and in a short time had so reduced the water that the investment became a paying one. In 1873 he and

William Barnsdall and others drilled the first producing well in the Bradford oil-field. His health failed in 1875 and he died on October eleventh, 1880, respected by all for his manliness and integrity.

Hon. David Emery, the last owner of the Drake well, was for many years a successful oil-operator. At Pioneer he drilled a number of prime wells, following the course of developments along Oil Creek. He organized the Octave Oil Company and was its chief officer. Removing to Titusville, he erected a fine residence and took a prominent part in public affairs. His purse was ever open to forward a good cause. Had the Republican party, of which he was an active member, been properly alive to the interests of the Commonwealth, he would have been Auditor-General of Pennsylvania. In all the relations and duties of life David Emery was a model citizen. Called hence in the vigor of stalwart manhood, multitudes of attached friends cherish his memory as that "of one who loved his fellow-men."

DAVID EMERY.

Thus dawned the petroleum-day that could not be hidden under myriads of bushels. The report of the Drake well traveled "from Greenland's icy mountains" to "India's coral strands," causing unlimited guessing as to the possible outcome. Crude petroleum was useful for various things, but a farmer who visited the newest wonder hit a fresh lead. Begging a jug of the oil, he paralyzed Colonel Drake by observing as he strode off: "This'll be durned good tew spread onto buckwheat-cakes!"

Bishop Simpson once delivered his lecture on "American Progress," in which he did not mention petroleum, before an immense Washington audience. President Lincoln heard it and said, as he and the eloquent speaker came out of the hall : "Bishop, you didn't 'strike ile'!"

When the Barnsdall well, on the Parker farm, produced hardly any oil from the first sand, the coming Mayor of Titusville quietly clinched the argument in favor of drilling it deeper by remarking : "It's a long way from the bottom of that hole to China and I'm bound to bore for tea-leaves if we don't get the grease sooner!"

"De Lawd thinks heaps ob Pennsylvany," said a colored exhorter in Pittsburg, "fur jes' ez whales iz gettin' sca'ce he pints outen de way fur Kunnel Drake ter 'scoveh petroleum!" A solemn preacher in Crawford county held a different opinion. One day he tramped into Titusville to relieve his burdened mind. He cornered Drake on the street and warned him to quit taking oil from the ground. "Do you know," he hissed, "that you're interfering with the Almighty Creator of the universe? God put that oil in the bowels of the earth to burn the world at the last day and you, poor worm of the dust, are trying to thwart His plans!" No wonder the loud check in the Colonel's barred pantaloons wilted at this unexpected outburst, which Drake often recounted with extreme gusto.

The night "Uncle Billy" Smith's lantern ignited the tanks at the Drake well the blaze and smoke of the first oil-fire in Pennsylvania ascended high. A loud-mouthed professor of religion, whose piety was of the brand that needed close watching in a horse-trade, saw the sight and scampered to the hills shout-ing: "It's the day of judgment!" How he proposed to dodge the reckoning, had his surmise been correct, the terrified victim could not explain when his fright subsided and friends rallied him on the scare.

The Drake well blazed the path in the wilderness that set petroleum on its triumphant march. This nation, already the most enlightened, was to be the most enlightening under the sun. An Atlantic of oil lay beneath its feet. America, its young, plump sister, could laugh at lean Europe. War raged and the old world sought to drain the republic of its gold. The United States ex-ported mineral-fat and kept the yellow dross at home. Petroleum was crowned king, dethroning cotton and yielding a revenue, within four years of Drake's modest strike, exceeding that from coal and iron combined! Talk of Califor-nia's gold-fever, Colorado's silver-furore and Barney Barnato's Caffir-mania.

American petroleum is a leading article of commerce, requiring hundreds of vessels to transport it to distant lands. Its refined product is known all over the civilized world. It has found its way to every part of Europe and the remotest portions of Asia. It shines on the western prairie, burns in the homes of New England and illumines miles of princely warehouses in the great cities of America. Everywhere is it to be met with, in the Levant and the Orient, in the hovel of the Russian peasant and the harem of the Turkish pasha. It is the one article imported from the United States and sold in the bazaars of Bagdad, the "City of the Thousand-and-One-Nights." It lights the dwellings, the tem-ples and the mosques amid the ruins of Babylon and Nineveh. It is the light of Abraham's birthplace and of the hoary city of Damascus. It burns in the Grotto of the Nativity at Bethlehem, in the Church of the Holy Sepulchre at Jerusalem, on the Acropolis of Athens and the plains of Troy, in cottage and palace along the banks of the Bosphorus, the Euphrates, the Tigris and the Golden Horn. It has penetrated China and Japan, invaded the fastnesses of Tartary, reached the wilds of Australia and shed its radiance over African wastes. Pennsylvania petroleum is the true cosmopolite, omnipresent and omnipotent in fulfilling its mission of illuminating the universe! A product of nature that is such a controlling influence in the affairs of men may well chal-lenge attention to its origin, its history and its economic uses.

All this from a hole seventy feet in the ground!

A grape-seed is a small affair,
Yet, swallow'd when you sup,
In your appendix it may stick
Till doctors carve you up.

A coral-insect is not large,
Still it can build a reef
On which the biggest ship that floats
May quickly come to grief.

A hint, a word, a look, a breath
May bear envenom'd stings,
From all of which the moral learn:
Despise not little things!

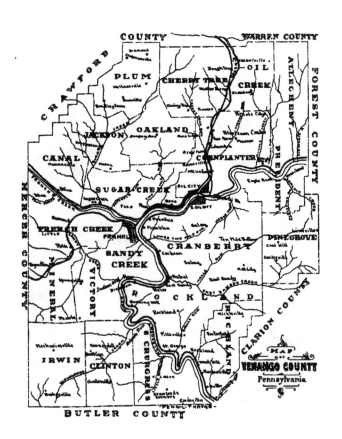

MAJ. W. T. BAUM.

JACOB SHEASLEY.

HENRY F. JAMES.

JAMES EVANS.

W. R. CRAWFORD.

COL. JAMES P. HOOVER.

PENN-L-THRONE

DANIEL GRIMM.

VI.

THE WORLD'S LUBRICANT.

A GLANCE AT A PRETTY SETTLEMENT—EVANS AND HIS WONDERFUL WELL—HEAVY
OIL AT FRANKLIN TO GREASE ALL THE WHEELS IN CREATION—ORIGIN OF A POP-
ULAR PHRASE—OPERATIONS ON FRENCH CREEK—EXCITEMENT AT FEVER HEAT—
GALENA AND SIGNAL OIL-WORKS—RISE AND PROGRESS OF A GREAT INDUSTRY—
CRUMBS SWEPT UP FOR GENERAL CONSUMPTION.

"A cargo of petroleum may cross the ocean in a vessel propelled by steam it has generated, acting
 upon an engine it lubricates and directed by an engineer who may grease his hair, limber his
 joints and freshen his liver with the same article."—*Petrolia, A. D. 1870.*
"Franklin wells produce the finest lubricant on earth."—*Brief History of Petroleum, A. D. 1885.*
"Friction, not motion, is the great destroyer of machinery."—*Engineering Journal.*

BIG ROCK BELOW FRANKLIN.

CHEAP and abundant light the island-well on Oil Creek assured the nations sitting in darkness. If there are "tongues in trees" and "sermons in stones" the trickling stream of greenish liquid murmured: "Bring on your lamps—we can fill them!" The *second* oil-well in Pennsylvania, eighteen miles from Col. Drake's, changed the strain to: "Bring on your wheels—we can grease them!" America was to be the world's illuminator and lubricator—not merely to dispel gloom and chase hobgoblins, but to increase the power of machinery by decreasing the impediments to easy motion. Friction has cost enough for extra wear and stoppages and breakages "to buy every darkey forty acres and a mule." The first coal-oil for sale in this country was manufactured at Waltham, Mass., in 1852, by Luther Atwood, who called it "Coup Oil," from the recent *coup* of Louis Napoleon. Although highly esteemed as a lubricator, its offensive odor and poor quality would render it unmerchantable to-day. Samuel Downer's hydro-carbon oils in 1856 were marked improvements, yet they would cut a sorry figure beside the unrivaled lubricant produced from the wells at Franklin,

77

the county-seat of Venango. It is a coincidence that the petroleum era should have introduced light and lubrication almost simultaneously, one on Oil Creek, the other on French Creek, and both in a region comparatively isolated. "Misfortunes never come singly," said the astounded father of twins, in a paroxysm of bewilderment; but happily blessings often come treading closely on each other's heels.

Pleasantly situated on French Creek and the Allegheny River, Franklin is an interesting town, with a history dating from the middle of the eighteenth century. John Frazer, a gunsmith, occupied a hut and traded with the Indians in 1747. Four forts, one French, one British and two American, were erected in 1754, 1760, 1787 and 1796. Captain Joncaire commanded the French forces. George Washington, a British lieutenant, with no premonition of fathering a great country, visited the spot in 1753. The north-west was a wilderness and Pittsburg had not been laid out. Franklin was surveyed in 1795, created a borough in 1829 and a city in 1869, deriving its chief importance from petroleum. Lofty hills and winding streams are conspicuous. Spring-water is abundant, the air is invigorating and healthfulness is proverbial. James Johnston, a negro-farmer of Frenchcreek township, stuck it out for one-hundred-and-nine summers, lamenting that death got around six months too soon for him to attend the Philadelphia Centennial. Angus McKenzie, of Sugarcreek, whose strong-box served as a bank in early days, reached one-hundred-and-eight. Mrs. McDowell, a pioneer, was bright and nimble three years beyond the century-mark. Galbraith McMullen, of Waterloo, touched par. John Morrison, the first court-crier, rounded out ninety-eight. A successor, Robert Lytle, was summoned at eighty-seven, his widow living to celebrate her ninety-fourth birthday.

J. B NICKLIN.

David Smith succumbed at ninety-nine and William Raymond at ninety-three. Mr. Raymond was straight as an arrow, walked smartly and in youth was the close friend of John J. Pearson, who began to practice law at Franklin and was President Judge of Dauphin county thirty-three years. J. B. Nicklin, fifty years a respected citizen, died in 1890 at eighty-nine. To the end he retained his mental and physical strength, kept the accounts of the Baptist church, was at his desk regularly and could hit the bullseye with the crack shots of the military company. William Hilands, county-surveyor, was a familiar figure on the streets at eighty-seven. Rev. Dr. Crane, who went to his reward a few weeks since, preached, lectured, visited the sick and continued to do good at eighty-six. Ready for the call to "come up higher," not willing to be idle and rust out, of him it could truly be said that "his last days were his best days." At eighty-five "Uncle Billy" Grove, of Canal, would hunt deer in Forest county and walk farther and faster than any man in the township. The people who have rubbed fourscore would fill a ten-acre patch. Of course, some get sick and die young, or the doctors would starve, heaven would be short of youthful tenants and the theories of Malthus might have to be tried on.

Franklin boasts the finest stone side-walks in the State. There are imposing churches, shady parks, broad streets, cosy homes, spacious stores, first-class

schools, fine hotels and inviting drives. For years the Baptist quartette has not
been surpassed in New York or Philadelphia. The opera-house is a gem. Three
railroads—a fourth is coming that will lop off sixty-five miles between New York
and Chicago—and electric street-cars supply rapid transit. Five substantial
banks, a half-dozen millionaires, two-dozen hundred-thousand-dollar-citizens
and multitudes of well-to-do property-holders give the place financial backbone.
Manufactures flourish, wages are liberal and many workmen own their snug
houses. Probably no town in the United States, of seven-thousand population,
has greater wealth, better society and a kindlier feeling clear through the community.

On the south bank of French Creek, at Twelfth and Otter streets, James
Evans, blacksmith, had lived twenty years. A baby when his parents settled
farther up in 1802, he removed to Franklin in 1839. His house stood near the
"spring" from which Hulings and Whitman wrung out the viscid scum. In dry
weather the well he dug seventeen feet for water smelled and tasted of petroleum. Tidings of Drake's success set the blacksmith thinking. Drake had
bored into the well close to the "spring" and found oil. Why not try the experiment at Franklin? Evans was not flush of cash, but the hardware-dealer
trusted him for the iron and he hammered out rough drilling-tools. He and
his son Henry rigged a spring-pole and bounced the drill in the water-well.
At seventy-two feet a crevice was encountered. The tools dropped, breaking
off a fragment of iron, which obstinately refused to be fished out. Pumping by
hand would determine whether a prize or a blank was to be drawn in the greasian lottery. Two men plied the pump vigorously. A stream of dark-green fluid
gushed forth at the rate of twenty-five barrels a day. It was heavy oil, about
thirty degrees gravity, free from grit and smooth as silk. The greatest lubricant
on earth had been unearthed!

Picture the pandemonium that followed. Franklin had no such convulsion
since the William B. Duncan, the first steamboat, landed one Sunday evening
in January, 1828. The villagers speeded to the well as though all the imps of
sheol were in pursuit. November court adjourned in half the number of seconds
Sut Lovingood's nest of hornets broke up the African camp-meeting. Judge
John S. McCalmont, whose able opinions the Supreme Court liked to adopt,
decided there was ample cause for action. A doctor rushed to the scene hatless,
coatless and shoeless. Women deserted their households without fixing their
back-hair or getting inside their dress-parade toggery. Babies cried, children
screamed, dogs barked, bells rang and two horses ran away. At prayer-meeting
a ruling elder, whom the events of the day had wrought to fever-heat, raised a
hilarious snicker by imploring God to "send a shower of blessings—yea, Lord,
twenty-five barrels of blessings!" Altogether it was a red-letter forenoon, for
twenty-five barrels a day of thirty-dollar oil none felt inclined to sneeze at.

That night a limb of the law, "dressed in his best suit of clothes," called at
the Evans domicile. Miss Anna, one of the fair daughters of the house, greeted
him at the door and said jokingly: "Dad's struck ile!" The expression caught
the town, making a bigger hit than the well itself. It spread far and wide, was
printed everywhere and enshrined permanently in the petroleum-vernacular.
The young lady married Miles Smith, the eminent furniture-dealer, still trading
on Thirteenth street. In 1875 Mr. Smith revisited his native England, after many
years' absence. Meeting a party of gentlemen at a friend's house, the conversation turned upon Pennsylvania. "May I awsk, Mr. Smith," a Londoner inquired, "if you hever 'eard in your 'ome about 'dad's stwuck ile'? I wead it

in the papahs, doncherknow, but I fawncied it nevah weally appeahed." Mr. Smith *had* "'eard" it and the delight of the company, when he recited the circumstances and told of marrying the girl, may be conceived. The phrase is billed for immortality.

Sufficient oil to pay for an engine was soon pumped. Steam-power increased the yield to seventy barrels! Franklin became the Mecca of speculators, traders, dealers and monied men. Frederic Prentice, a leader in aggressive enterprises, offered forty-thousand dollars for the well and lot. Evans rejected the bid and kept the well, which declined to ten or twelve barrels within six months. The price of oil shrank like a flannel-shirt, but the lucky disciple of Vulcan realized a nice competence. He enjoyed his good fortune some years before journeying to "that bourne from which no traveler e'er returns." Mrs. Evans long survived, dying at eighty-six. The son removed to Kansas, three daughters died and one resides at Franklin. The old well experienced its complement of fluctuations. Mosely & Co., of Philadelphia, leased it. It stood idle, the engine was taken away, the rig tumbled and the hole filled up partially with dirt and wreckage. Prices spurted and the well was hitched to a pumping-rig operating others around it. Captain S. A. Hull ran a group of the wells on the flats and a dozen three miles down the Allegheny. He was a man of generous impulses, finely educated and exceedingly companionable. His death, in 1893, resulted in dismantling most of these wells, hardly a vestige remaining to tell that the Evans and its neighbors ever existed.

> "Remorseless Time!
> Fierce spirit of the glass and scythe—what power
> Can stay him in his silent course, or melt
> His iron heart with pity?"

James Evans was not "left blooming alone" in the search for oily worlds to conquer. Companies were organized while he was yanking the tools in the well that "set 'em all crazy." The first of these—The Franklin Oil and Mining

COL. JAMES BLEAKLEY.

Company—started work on October fifth, twenty rods below Evans, finding oil at two-hundred-and-forty feet on January twelfth, 1860. The well pumped about one-half as much as the Evans for several months, but did not die of old age. The forty-two shares of stock advanced tenfold in one week, selling at a thousand dollars each. Three or four wells were put down, the company dissolving and members operating on their own hook. It was strongly officered, with Arnold Plumer as president; J. P. Hoover, vice president; Aaron W. Raymond, secretary; James Bleakley, Robert Lamberton, R. A. Brashear, J. L. Hanna and Thomas Hoge, executive committee. Mr. Plumer was a dominant factor in Democratic politics, largely instrumental in the nomination of James Buchanan for President, twice a member of Congress, twice State Treasurer, Canal Commissioner and founder of the First National Bank. At his death, in 1869, he devised his family the largest estate in Venango county. Judge Lamberton opened the first bank in the oil-regions, owned hundreds of houses and in 1885 bequeathed each of his eight children a handsome fortune. Colonel Bleakley rose by his own exertions, keen foresight and skillful management. He invested in productive realty, drilled scores of wells around Franklin, built iron-tanks and brick-blocks, established a bank, held thousands of acres of lands and in 1884 left a very large inheritance to his sons and daughters. Mr. Raymond developed the Raymilton district—it was named from him—in which hundreds of fair wells have rewarded Franklin operators, and at eighty-nine was exceedingly quick in his movements. Mr. Brashear, a civil engineer and exemplary citizen, has been in the grave twenty years. Mr. Hanna operated heavily in oil, acquired numerous farms and erected the biggest block—it contained the first opera-house—in the city. He is handling real-estate, but his former partner, John Duffield, slumbers in the cemetery. Mr. Hoge, an influential politician, elected to the Legislature two terms and Mayor one term, has also joined the silent majority.

In February, 1860, Caldwell & Co., a block south-east of Evans, finished a paying well at two-hundred feet. The Farmers' and Mechanics' Company, Levi Dodd, president, drilled a medium producer at the foot of High street, on the bank of the creek. Mr. Dodd was an old settler, originator of the first Sabbath-school in Franklin and a ruling-elder for over fifty years. Numerous companies and individuals pushed work in the spring. Holes were sunk in front yards, gardens and water-wells. Derricks dotted the landscape thickly. Franklin was the objective point of immense crowds of people. The earliest wells were shallow, seldom exceeding two-hundred feet. The Mammoth, near a huge walnut-tree back of the Evans lot, began flowing on May fifteenth to the tune of a hundred barrels. This was the first "spouter" in the district and it quadrupled the excitement. Four-hundred barrels of oil were shipped to Pittsburg, by the steamboat Venango, on April twenty-seventh. Twenty-two wells were drilling and twenty producing on July first. Farms for miles up French Creek had been bought at high prices and the noise of the drill burdened the summer air. Sugar Creek, emptying into French Creek three miles west of Franklin, shared in the activity. Then prices "came down like a thousand of brick." Pumping was expensive, lands were scarce and dear, hauling the oil to a railroad cost half its value and hosts of small wells were abandoned. On November first, within the borough limits, fifteen were yielding one-hundred and-forty barrels. Curtz & Strain had bored five-hundred feet in October, the deepest well in the neighborhood, without finding additional oil-bearing rock. The Presidential election foreboded trouble, threats of war clouded the sky and the year closed gloomily.

The advantages of Franklin heavy-oil as a lubricant were quickly recognized. It possessed a "body" that artificial oils could not rival. In the crude state it withstood a cold-test twenty degrees below zero. Here is where it "had the bulge" on alleged lubricants which solidify into a sort of liver with every twitch of frost. The producing-area of heavy-oil is restricted to a limited section, where the first sand is thirty to sixty feet thick and the lower sands were entirely omitted in the original distribution of strata. For years operators hugged the banks of the streams and the low grounds, keeping off the hills more willingly than General Coxey kept off the Washington grass. The famous "Point Hill," across French Creek from the Evans well, went begging for a purchaser. At its southern base Mason & Lane, Cook & Co., Welsby & Smith, Shuster, Andrews, Green and others had profitable wells, but nobody dreamed of boring through the steep "Point" for oil. J. Lowry Dewoody offered the lordly hill, with its forty acres of dense evergreen-brush, to Charles Miller for fifteen-hundred dollars. He wanted the money to drill on the flats and the hill was an elephant on his hands.

During the Columbian Exposition an aged man alighted from a western train at the union-depot at Chicago. His rifle and his buckskin-suit indicated the Kit-Carson brand of hunter. He gazed about him in amazement and a crowd assembled. "Wal," ejaculated the white-haired Nimrod, "this be Chicago, eh? Sixty years ago I killed lots ov game right whar we stan' an' old man Kinzey fell all over hisse'f to trade me a hunnerd acre ov land fur a pair ov cowhide boots! I might hev took him up, but, consarn it, I didn't hev the boots!"

Something of this kind would apply to Mr. Miller and the Dewoody proposition. He had embarked in the business that was to bring him wealth and honor, but just at that time " didn't hev" the fifteen-hundred to spare from his working-capital for the fun of owning a hill presumed to be worthless except for scenery. Colonel Bleakley and Dr. A. G. Egbert bought it later at a low figure. Operations scaled the slopes and hills and the first well on the "Point" was of the kind to whet the appetite for more. Bleakley & Egbert pocketed a keg of cold-cash from their wells and the royalty paid by lessees. Daniel Grimm's production put him in the van of Franklin oilmen. He came to the town in 1861, had a dry-goods store in partnership with the late William A. Horton and in 1869 drilled his first well. W. J. Mattern and Edward Rial & Son had a rich slice. The foundation of a dozen fortunes was laid on the "Point," which yields a few barrels daily, although only a shadow of its former self. From the western end of the hill thousands of tons of a peculiar shale have been manufactured into paving-brick, the hardest and toughest in America. A million dollars would not pay for the oil taken from the hill that found no takers at fifteen-hundred!

Dewoody, over whose grave the storms of a dozen winters have blown, was a singular character. He cared not a continental for style and was independent in speech and behavior. Bagging a term in the Legislature as a Democratic-Greenbacker, his rugged honesty was proof against the allurements of the lobbyists, jobbers and heelers who disgrace common decency. His most remarkable act was a violent assault on the Tramp-Bill, a measure cruel as the laws of Draco, which Rhoads of Carlisle contrived to pass. He paced the central aisle, spoke in the loudest key and gesticulated fiercely. Tossing his long auburn hair like a lion's mane, he wound up his torrent of denunciation with terrible emphasis: "If Jesus Christ were on earth this monstrous bill would jerk him as a vagrant and dump him into the lock-up!"

Gradually developments crept north and east. The Galloway—its Dolly Varden well was a daisy—Lamberton and McCalmont farms were riddled with holes that repaid the outlay lavishly. Henry F. James drilled scores of paying wells on these tracts. In his youth he circled the globe on whaling voyages and learned coopering. Spending a few months at Pithole in 1865, he returned to Venango county in 1871, superintended the Franklin Pipe-Line five years and operated judiciously. He was active in agriculture and served three terms in the Legislature with distinguished fidelity. He defeated measures inimical to the oil-industry and promoted the passage of the Marshall Bill, by which pipe-lines were permitted to buy, sell or consolidate. This sensible law relieves pipe-lines in the older districts, where the production is very light, from the necessity of maintaining separate equipments at a loss or ruining hundreds of well-owners by tearing up the pipes for junk and depriving operators of transportation. The late Casper Frank, William Painter—he was killed at his wells —Dr. Fee, the Harpers, E. D. Yates and others extended the field into Sugarcreek township. Elliott, Nesbett & Bell's first well on the Snyder farm, starting at thirty barrels and settling down to regular work at fifteen, elongated the Galloway pool and brought adjoining lands into play. Kunkel & Newhouse, Stock & Co., Mitchell & Parker, Crawford & Dickey, Dr. Galbraith and M. O'Connor kept many sets of tools from rusting. The extension to the Carter and frontierfarms developed oil of lighter gravity, but a prime lubricator. Mrs. Harold, a Chicago lady, dreamed a certain plot, which she beheld distinctly, would yield heavy-oil in abundance. She visited Franklin, traversed the district a mile in advance of developed territory, saw the land of her dream, bargained for it, drilled wells and obtained "lashin's of oil!" Still there are bipeds in bifurcated garments who declare woman's "sphere" is the kitchen, with dish-washing, sock-darning and meal-getting as her highest "rights!"

Jacob Sheasley, who came from Dauphin county in 1860 and branched into oil in 1864, is the largest operator in the bailiwick. He drilled at Pithole, Parker, Bradford, on all sides of Franklin and put down a hundred wells the last two years. He enlarged the boundaries of the lubricating section by leasing lands previously condemned and sinking test-wells in 1893-4, with gratifying results. Rarely missing his guess on territory, he has been almost invariably fortunate. His son, George R., has operated in Venango and Butler counties and owns a bunch of desirable wells on Bully Hill, with his brother Charles as partner. The father and two sons are "three of a kind" hard to beat.

A mile north of Franklin, in February of 1870, the Surprise well on Patchel Run, a streamlet bearing the name of the earliest hat-maker, surprised everybody by its output. It foamed and gassed and frothed excessively, filling the pipe with oil and water. Throngs tramped the turnpike over the toilsome hill to look at the boiling, fuming tank into which the well belched its contents. "Good for four-hundred barrels" was the verdict. A party of us hurried from Oil Creek to judge for ourselves. Although the estimate was six times too great, a lease of adjacent lands would not be bad to take. Rev. Mr. Johns, retired pastor of the Presbyterian church at Spartansburg, Crawford county, had charge of the property. My acquaintance with Mr. Johns devolved upon me the duty of negotiating for the tract. He received me graciously and would be pleased to lease twenty acres for one-half the oil and one-thousand dollars an acre bonus! Br'er John's exalted notions soared far too high to be entertained seriously. The Surprise fizzled down to four or five barrels in a week and the good minister— for twenty years he has been enjoying his treasure in heaven—never fingered a penny from his land save the royalty of two or three small wells.

Major W. T. Baum has operated in the heavy-oil field thirty-two years, beginning in 1864. He passed through the Pithole excitement and drilled largely at Foster, Pleasantville, Scrubgrass, Bullion, Gas City, Clarion, Butler and Tarkiln. His faith in Scrubgrass territory has been recompensed richly. In 1894 he sank a well on the west bank of the Allegheny, opposite Kennerdell Station, in hope of a ten-barrel strike. It pumped one-hundred-and-fifty barrels a day for months and it is doing fifty barrels to-day, with three more of similar caliber to keep it company! The Major's persevering enterprise deserves the reward Dame Fortune is bestowing. He owns the wells and lands on Patchel Run, which yield a pleasant revenue. Colonel J. H. Cain, Colonel L. H. Fassett and J. W. Grant, all successful operators, have their wells in the vicinity. Modern devices connect wells far apart, by coupling them with rods two to ten feet

above ground, so that a single engine can pump thirty or forty in shallow territory. The downward stroke of one helps the upward stroke of the other, each
pair nearly balancing. This enables the owners of small wells to pump them at
the least expense. Heavy-oil has sold for years at three-sixty to four dollars a
barrel, consequently a quarter-barrel apiece from forty wells, handled by one
man and engine, would exceed the income from a quarter-million dollars salted
down in government bonds. It is worth traveling a long distance to stand on
the hill and watch the pumping of Baum's, Grimm's, Cain's, Grant's, Sheasley's
and James's wells, some of them a mile from the
power that sets the strings of connecting rods in
motion.

COL. JOHN H. CAIN.

On Two-Mile Run, up the Allegheny two
miles, W. S. McMullan drilled several wells in
1871-2. The product was the blackest of black
oils, indicating a deposit separated from the main
reservoir of the lubricating region. Subsequent
operations demonstrated that a dry streak inter-
vened. Captain L. L. Ray put down fair wells
near the river in 1894. Mr. McMullan resided
at Rouseville and had valuable interests on Oil
Creek. He served a term in the State Senate,
reflecting honor upon himself and his constituents. A man of integrity and capacity, he
could be trusted implicitly. Fifteen years ago he
removed to Missouri to engage in lumbering.

Senator McMullan, Captain William Hasson, member of Assembly, and Judge
Trunkey, who presided over the court and later graced the Supreme Bench,
were three Venango-county men in public life whom railroad-passes never
swerved from the path of duty. They refused all such favors and paid their way
like gentlemen. If lawgivers and judges of their noble impress were the rule
rather than the exception — "a consummation devoutly to be wished" —
grasping corporations would not own legislatures and courts and "drive a
coach and four" through any enactment with impunity.

George P. Smith's tract of land between Franklin and Two-Mile Run netted
him a competence in oil and then sold for one-hundred-thousand dollars. Mr.
Smith dispenses liberally to charitable objects, assists his friends and uses his
wealth properly. He owns his money, instead of letting it own him. He has
traveled much, observed closely and profited by what he has seen and read. He
is verging on fourscore, his home is in Philadelphia and "the world will be the
better for his having lived in it."

The production of heavy-oil in 1875 aggregated one-hundred-and-thirty-
thousand barrels. In 1877 it dropped to eighty-eight-thousand barrels and in
1878 to seventy thousand. Thirteen-hundred wells produced sixty-thousand
barrels in 1883. Taft & Payn's pipe-line was laid in 1870 from the Egbert and
Dewoody tracts to the river, extended to Galloway in 1872 and combined with
the Franklin line in 1878. The Producers' Pipe-Line Company began to transport oil in 1883. J. A. Harris, who died in 1894, had the first refinery in the oil-
regions in 1860. His plant was extremely primitive. Colonel J. P. Hoover
built the first refinery of note, which burned in the autumn of 1861. Sims &
Whitney had one in 1861 and the Norfolk Oil-Works were established the same
year, below the Allegheny bridge. Samuel Spencer, of Scranton, expended

thirty-thousand dollars on the Keystone Oil-Works, near the cemetery, in 1864. Nine refineries, most of them running the lighter oils, were operated in 1864-5, after which the business collapsed for years. Dr. Tweddle, a Pittsburg refiner who had suffered by fire, organized a company in 1872 to start the Eclipse Works. At different periods many of the local operators have been interested in refining, now the leading Franklin industry.

For some time heavy-oil was used principally in its natural state. At length improvements of great value were devised, out of which have grown the oil-works devoted solely to the manufacture of lubricants. Among these the most important and successful was that adopted in 1869 by Charles Miller, of Franklin, protected by letters-patent of the United States and since by patents covering his complete method. Besides improvements in the method of manufacturing, he discovered the value of Galena, a lead oxide, as an ingredient in lubricating oils and patented the process. The Great Northern Oil-Company had built a refinery in 1865 on the north bank of French Creek, below the Evans Well, and leased it in 1868 to Colonel Street. In May of 1869 Mr. Miller and John Coon purchased the Point Lookout Oil-Works, as the refinery was called, Street retiring. The total tankage was one-thousand barrels and the daily manufacturing capacity scarcely one-hundred. The new firm, of which R. L. Cochran became a member in July, pushed the business with characteristic energy, doubling the plant and extending the trade in all directions. Mr. Cochran withdrew in January of 1870, R. H. Austin buying his interest. The following August fire destroyed the works, entailing severe loss. A calamity that would have disheartened most men seemed only to imbue the partners with fresh vigor. Colonel Henry B. Plumer, a wealthy citizen of Franklin, entered the firm and the Dale light-oil refinery, a half-mile up the creek, was bought and remodeled throughout. Reorganized on a solid basis as the "Galena Oil-Works," a name destined to gain world-wide reputation, within one month from the fire the new establishment, its buildings and entire equipment changed and adapted to the treatment of heavy-oil, was running to its full capacity night and day! Such enterprise and pluck augured happily for the future and they have been rewarded abundantly.

Orders poured in more rapidly than ever. The local demand spread to the adjoining districts. Customers once secured were sure to stay. In addition to the excellence of the product, there was a vim about the business and its management that inspired confidence and won patronage. Messrs. Coon, Austin and Plumer disposed of their interest, at a handsome figure, to the Standard Oil Company in 1878. The Galena Oil-Works, Limited, was chartered and continued the business, with Mr. Miller as president. Increasing demands necessitated frequent enlargements of the works, which now occupy five acres of ground. Every appliance that ingenuity and experience can suggest has been provided, securing uniform grades of oil with unfailing precision.

The machinery and appurtenances are the best money and skill can supply. The same sterling traits that distinguished the smaller firm have all along marked the progress of the newer and larger enterprise. The standard of its products is always strictly first-class, hence patrons are never disappointed in the quality of any of the celebrated Galena brands of "Engine," "Coach," "Car," "Machinery," or "Lubricating" oils. Steadfast adherence to this cardinal principle has borne its legitimate fruit. Railway oils are manufactured exclusively. The daily capacity is three-thousand barrels. "Galena Oils" are used on *seventy-five per cent.* of the railway mileage of America! This includes three distinct lines from Boston and New York to the Pacific coast, the Northern Pacific to Puget Sound, the Vanderbilt system, leading roads in Mexico and dozens of others. Such patronage has never before been gained by one establishment and it is the result of positive merit. The Franklin district furnishes more and better lubricating oil than all the rest of the continent and the Galena treatment brings it to the highest measure of perfection. Reflect for a moment upon the enormous expansion of the Galena Works and see what earnest, faithful, intelligent effort and straightforward dealing may accomplish.

The first three railroads that tried the "Galena Oils" in 1869 have used none other since. Could stronger proof of their excellence be desired? It was a pleasing novelty for railway managers to find a lubricant that would neither freeze in winter nor dissipate in summer and they made haste to profit by the experience. The severest tests served but to place it far beyond all competition. At twenty degrees below zero it would not congeal, while the fiercest heat of the tropical sun affected it hardly a particle. As the natural consequence it speedily superseded all others on the principal railroads of the country. The axles of the magnificent Pullman and Wagner coaches on the leading lines have their friction reduced to the minimum by "Galena Oil." It adds immeasurably to the smoothness and speed of railway travel between the Atlantic and the Pacific, from Maine to the Isthmus, from British Columbia to Florida. Passengers detained by a "hot box" and annoyed by the fumes of rancid grease frying in the trucks beneath their feet may be certain that the offending railways *do not* use "Galena Oil." The "Galena" is not constructed on that plan, but stands alone and unapproachable as the finest lubricator of the nineteenth century.

This is a record-breaking age. The world's record for fast time on a railroad was again captured from the English on September eleventh, 1895. The New-York-Central train, which left New York that morning, accomplished the trip to Buffalo at the greatest speed for a continuous journey of any train over any railroad in the world. The distance—four-hundred-and-thirty-six miles—was covered in four-hundred-and-seven minutes, a rate of sixty-four-and-one-third miles per hour. Until that feat the English record of sixty-three-and-one-fifth miles an hour for five-hundred miles was the fastest. In other words, the American train of four heavy cars, hauled to Albany by engine No. 999, the famous World's Fair locomotive, smashed the English record more than a mile an hour, in the teeth of a stiff head-wind. Father Time, who has insisted for many years that travelers spend at least twenty-four hours on the journey between Chicago and New York, received a fatal shock on October twenty-fourth, 1895. Two men who left Chicago at three-thirty in the morning visited five theatres in New York that night! A special New-York-Central train, with Vice-President Webb and a small party of Lake-Shore officials, ran the nine-hundred-and-eighty miles in seventeen-and-three-quarter hours, averaging sixty-five miles an hour to Buffalo, beating all previous long-distance runs. For the first time copies of Chicago newspapers, brought by gentlemen on the train, were seen in New York on the day of their publication. Every axle, every journal, every box, every wheel of both these trains, from the front of the locomotive to the rear of the hind-coach, was lubricated with "Galena Oil."

The works are situated in the very heart of the heavy-oil district. Two railroads, with a third in prospect, and a paved street front the spacious premises. The main building is of brick, covering about an acre and devoted chiefly to the handling of oil for manufacture or in course of preparation, the repairing and painting of barrels and the accommodation of the engines and machinery.

To the rear stands a substantial brick-structure, containing the steam-boilers, the electric-light outfit and the huge agitators in which the oil is treated. Big pumps next force the fluid into large vessels, where it is submitted to a variety of special processes, which finally leave it ready for the consumer. A dozen iron-tanks, each holding many thousand barrels, receive and store crude to supply the works for months. As this is piped directly from the wells the largest orders are filled with the utmost dispatch. Nothing is lacking that can ensure superiority. The highest wages are paid and every employee is an American citizen or proposes to become one. The men are regarded as rational, responsible beings, with souls to save and bodies to nourish, and treated in accordance with the Golden Rule. They are well-fed, well-housed, prosperous and contented. A strike, or a demand for higher wages or shorter hours, is unknown in the history of this model institution. Is it surprising that each year adds to its vast trade and wonderful popularity? The "Galena Oil-Works, Limited," of Franklin, Venango county, Pennsylvania, must be ranked among the most noteworthy representative industries of Uncle Sam's splendid domain.

Have you a somewhat cranky wife,
  Whose temper's apt to broil?
To ease the matrimonial strife
Just lubricate when trouble's rife—
  Pour on Galena Oil!

Has life some rusty hinge or joint
  That vexes like a boil,
And always sure to disappoint?
The hindrance to success anoint
  As with Galena Oil!

Does business seem to jar and creak,
  Despite long years of toil,
Till wasted strength has left you weak?
Reduce the friction, so to speak—
  Apply Galena Oil!

Are your affairs all run aground,
  The cause of sad turmoil?
To see again "the wheels go 'wound,"
Smooth the rough spots wherever found—
  Soak in Galena Oil!

The Signal Oil-Works, Limited, manufacture Sibley's Perfection Valve-Oil for locomotive-cylinders and Perfection Signal Oil. More than twenty-five years ago Joseph C. Sibley commenced experimenting with petroleum-oils for use in steam-cylinders under high pressure. He found that where the boiler-pressure was not in excess of sixty pounds the proper lubrication of a steam-cylinder with petroleum was a matter of little or no difficulty. With increase in pressure came increase in temperature. As a result the oil vaporized and passed through the exhaust. The destruction of steam-chests and cylinders through fatty acids incident to tallow, or tallow and lard-oils, cost millions of dollars annually; but it was held as a cardinal point in mechanical engineering that these were the only proper steam-lubricants. Mr. Sibley carried on his experiments for years. He conversed with leading superintendents-of-machinery in the United States and with leading chemists. Almost invariably he was laughed at when asserting his determination to produce a product of petroleum, free from fatty acids, capable of better lubrication even than the tallow then in use. Many of his friends in the oil-business, who thought they understood the nature of petroleum, expressed the deepest sympathy for Mr. Sibley's hallucination. Amid partial successes, interspersed with many failures, he continued the experiments. So incredulous were chemists and superintendents-of-machinery, so fearful of disasters to their machinery through the use of such a compound, that he had in many instances to guarantee to assume any damages which might occur to a locomotive through its use. He rode thousands of miles upon locomotives, watching the use of the oil, daily doubling the distance made by engineers. Success at last crowned his efforts and the Perfection Valve-Oil has been for nearly twenty years the standard lubricant of valves and cylinders. To-day there is scarcely a locomotive in the United States that does not use some preparation of petroleum and the steam-chests and cylinders of *more than three-fourths* of the locomotives in the United States are lubricated with Perfection Valve-Oil.

The results have been astounding. Destruction of steam-joints by fatty acids from valve-lubricants is now an unknown thing. Not only this, but as a lubricant the Perfection Valve-Oil has proved itself so much superior that, where valve-seats required facing on an average once in sixty days, they do not now require facing on an average once in two years. The steam-pressure carried upon the boilers at that time rarely exceeded one-hundred-and-twenty pounds. With the increase of pressure and the corresponding increase of temperature it was found next to impossible to properly lubricate the valves and cylinders to prevent cutting. The superintendent-of-machinery of a leading American railway sent for Mr. Sibley at one time, told him that he proposed to build passenger-locomotives carrying one-hundred-and-eighty pounds pressure and asked if he would undertake to lubricate the valves and cylinders under that pressure. The reply was: "Go ahead. We will guarantee perfect lubrication to a pressure very much higher than that." And to-day the majority of the higher type of passenger-locomotives carry one-hundred-and-eighty pounds pressure regularly.

When it was clearly demonstrated that the Perfection Valve-Oil was a success, oil-men who had pronounced it impossible and had been backed in their opinion by noted chemists commenced to make oils similar to it in appearance. While many of them may have much confidence in their own product, the highest testimonial ever paid to Perfection Valve-Oil is that no competitor claims he has its superior. Some urge their product with the assurance that it is the equal of Perfection Valve-Oil, thus unconsciously paying the highest tribute possible to the latter.

The works also make Perfection Signal-Oil for use in railway-lamps and lanterns. Since 1869 this oil has been before the public. It is in daily use in more than three-fourths of the railway-lanterns of the United States and it is the proud boast of Mr. Sibley that, during that time, there has never occurred an accident which has cost either a human limb or life or the destruction of one penny's worth of property, through the failure of this oil to perform its work perfectly. Making but the two products, Valve and Signal-Oils, catering to no other than railroad-trade, studying carefully the demands of the service, keeping in touch with the latest developments of locomotive-engineering and thoroughly acquainted with the properties of all petroleum in Pennsylvania, the company may well believe that, granted the possession of equal natural abilities with competitors, under the circumstances it is entitled to lead all others in the production of these two grades of oils for railroad use.

Hon. Charles Miller, president of the Galena Oil-Works, and Hon. Joseph C. Sibley, president of the Signal Oil-Works, are brothers-in-law and proprietors of the great stock-farms of Miller & Sibley. Mr. Miller is of Huguenot ancestry, born in Alsace, France, in 1843. The family came to this country in 1854, settling on a farm near Boston, Erie county, New York. At thirteen Charles clerked one year in the village-store for thirty-five dollars and board. He clerked in Buffalo at seventeen for one-hundred-and-seventy-five dollars, without board. In 1861 he enlisted in the New-York National Guard. In 1863 he was mustered into the United States service and married at Springville, N. Y., to Miss Ann Adelaide Sibley, eldest child of Dr. Joseph C. Sibley. In 1864 he commenced business for himself, in the store in which he had first clerked, with his own savings of two-hundred dollars and a loan of two-thousand from Dr. Sibley. In 1866 he sold the store and removed to Franklin. Forming a partnership with John

CHARLES MILLER.   JOSEPH C. SIBLEY.

Coon of Buffalo, the firm carried on a large dry-goods house until 1869, when a patent for lubricating oil and a refinery were purchased and the store was closed out at heavy loss. The refinery burned down the next year, new partners were taken in and in 1878 the business was organized in its present form as "The Galena Oil-Works, Limited." The entire management was given Mr. Miller, who had built up an immense trade and retained his interest in the works. He deals directly with consumers. Since 1870 his business-trips have averaged five days a week and fifty-thousand miles a year of travel. No man has a wider acquaintance and more personal friends among railroad officials. His journeys cover the United States and Mexico. Wherever he may be, in New Orleans or San Francisco, on the train or in the hotel, conferring with a Vanderbilt or the humblest manager of an obscure road, receiving huge orders or aiding a deserving cause, he is always the same genial, magnetic, generous exemplar of practical belief in "the fatherhood of God and the brotherhood of man."

Major Miller is one whom money does not spoil. He is the master, not the servant, of his wealth. He uses it to extend business, to foster enterprise, to further philanthropy, to alleviate distress and to promote the comfort and happiness of all about him. His benefactions keep pace with his increasing prosperity. He is ever foremost in good deeds. He gives thousands of dollars yearly to worthy objects, to the needy, to churches, to schools, to missions and to advance the general welfare. In 1889 he established a free night-school for his employés and the youth of Franklin, furnishing spacious rooms with desks and apparatus and engaging four capable teachers. This school has trained hundreds of young men for positions as accountants, book-keepers, stenographers and clerks. The First Baptist church, which he assisted in organizing, is the object of his special regard. He bore a large share of the cost of the brick-edifice, the lecture-room and the parsonage. He and Mr. Sibley have donated the massive pipe-organ, maintained the superb choir, paid a good part of the pastor's salary, erected a branch-church and supported the only services in the Third Ward. For twenty-five years Mr. Miller has been superintendent of the Sabbath-school, which has grown to a membership of six-hundred. His Bible-class of three-hundred men is equalled in the state only by John Wanamaker's, in Philadelphia, and James McCormick's, in Harrisburg. The instruction is scriptural, pointed and business-like, with no taint of bigotry or sectarianism. No matter how far away Saturday may find him, the faithful teacher never misses the class that is "the apple of his eye," if it be possible to reach home. Often he has hired an engine to bring him through on Saturday night, in order to meet the adult pupils of all denominations who flock to hear his words of wisdom and encouragement. Alike in conversation, teaching and public-speaking he possesses the faculty of interesting his listeners and imparting something new. He has raised the fallen, picked poor fellows out of the gutter, rescued the perishing and set many wanderers in the straight path. Not a few souls, "plucked as brands from the burning," owe their salvation to the kindly sympathy and assistance of this earnest layman. Eternity alone will reveal the incalculable benefit of his night-school, his Bible-class, his church-work, his acts of charity, his personal appeals to the erring and his unselfish life to the community and the world.

Twice Mr. Miller served as mayor of Franklin. Repeatedly has he declined nominations to high offices, private affairs demanding his time and attention. He is president or director of a score of commercial and industrial companies, with factories, mines and works in eight states. He has been president time after time of the Northwestern Association of Pennsylvania of the Grand Army of the Republic, Ordnance-Officer and Assistant Adjutant-General of the Second Brigade of Pennsylvania and Commander of Mays Post. He is a leading spirit in local enterprises. He enjoys his beautiful home and the society of his wife and children and friends. He prizes good horses, smokes good cigars and tells good stories. In him the wage-earner and the breadwinner have a steadfast helper, willing to lighten their burden and to better their condition. In short, Charles Miller is a typical American, plucky, progressive, energetic and invincible, with a heart to feel for suffering humanity, genius to plan and talent to execute the noblest designs.

Hon. Joseph Crocker Sibley, eldest son of Dr. Joseph Crocker Sibley, was born at Friendship, N. Y., in 1850. His father's death obliging him to give up a college-course for which he had prepared, in 1866 he came to Franklin to clerk in Miller & Coon's dry-goods store. From that time his business interests and Mr. Miller's were closely allied. In 1870 he married Miss Metta E. Babcock, daughter of Simon M. Babcock, of Friendship. He was agent of the Galena Oil-Works at Chicago for two years, losing his effects and nearly losing his life in the terrible fire that devastated that city. His business-success may be said to date from 1873, when he returned to Franklin. After many experiments he produced a signal-oil superior in light, safety and cold-test to any in use. The Signal Oil-Works were established, with Mr. Sibley as president and the proprietors of the Galena Oil-Works, whose plant manufactured the new product, as partners. Next he compounded a valve-oil for locomotives, free from the bad qualities of animal-oils, which is now used on three-fourths of the railway mileage of the United States.

Every newspaper-reader in the land has heard of the remarkable Congressional fight of 1892 in the Erie–Crawford district. Both counties were overwhelmingly Republican. People learned with surprise that Hon. Joseph C. Sibley, a resident of another district, had accepted the invitation of a host of good citizens, by whom he was selected as the only man who could lead them to victory over the ring, to try conclusions with the nominee of the ruling party, who had stacks of money, the entire machine, extensive social connections, religious associations—he was a preacher—and a regular majority of five-thousand to bank upon. Some wiseacres shook their heads gravely and predicted disaster. Such persons understood neither the resistless force of quickened public sentiment nor the sterling qualities of the candidate from Venango county. Democrats, Populists and Prohibitionists endorsed Sibley. He conducted a campaign worthy of Henry Clay. Multitudes crowded to hear and see a man candid enough to deliver his honest opinions with the boldness of "Old Hickory." The masses knew of Mr. Sibley's courage, sagacity and success in business, but they were unprepared to find so sturdy a defender of their rights. His manly independence, ringing denunciations of wrong, grand simplicity and incisive logic aroused unbounded enthusiasm. The tide in favor of the fearless advocate of fair-play for the lowliest creature no earthly power could stem. His opponent was buried out of sight and Sibley was elected by a sweeping majority.

Mr. Sibley's course in Congress amply met the expectations of his most ardent supporters. The prestige of his great victory, added to his personal magnetism and rare geniality, at the very outset gave him a measure of influence few members ever attain. During the extra-session he expressed his views with characteristic vigor. A natural leader, close student and keen observer, he did not wait for somebody to give him the cue before putting his ideas on record. In the silver-discussion he bore a prominent part, opposing resolutely the repeal of the Sherman act. His wonderful speech "set the ball rolling" for those who declined to follow the administration program. The House was electrified by Sibley's effort. Throughout his speech of three hours he was honored with the largest Congressional audience of the decade. Aisles, halls, galleries and corridors were densely packed. Senators came from the other end of the Capitol to listen to the brave Pennsylvanian who dared plead for the white metal. For many years Mr. Sibley has been a close student of political and social economics and he so grouped his facts as to command the undivided attention and the highest respect of those who honestly differed from him in his conclusions. Satire, pathos, bright wit and pungent repartee awoke in his hearers the strongest emotions, entrancing the bimetalists and giving their enemies a cold chill, as the stream of eloquence flowed from lips "untrained to flatter, to dissemble or to play the hypocrite." Thenceforth the position of the representative of the Twenty-sixth district was assured, despite the assaults of hireling journals and discomfited worshippers of the golden calf.

He took advanced ground on the Chinese question, delivering a speech replete with patriotism and common-sense. An American by birth, habit and education, he prefers his own country to any other under the blue vault of heaven. The American workman he would protect from pauper immigration and refuse to put on the European or Asiatic level. He stands up for American skill, American ingenuity, American labor and American wages. Tariff for revenue he approves of, not a tariff to diminish revenue or to enrich one class at the expense of all. The tiller of the soil, the mechanic, the coal-miner, the coke-burner and the day-laborer have found him an outspoken champion of their cause. Small wonder is it that good men and women of all creeds and parties have abiding faith in Joseph C. Sibley and would fain bestow on him the highest office in the nation's gift.

Human nature is a queer medley and sometimes manifests streaks of envy and meanness in queer ways. Mr. Sibley's motives have been impugned, his efforts belittled, his methods assailed and his neckties criticised by men who could not understand his lofty character and purposes. The generous ex-Congressman must plead guilty to the charge of wearing clothes that fit him, of smoking decent cigars, of driving fine horses and of living comfortably. Of course it would be cheaper to buy hand-me-down misfits, to indulge in loud-smelling tobies, to walk or ride muleback, to curry his own horses and let his wife do the washing instead of hiring competent helpers. But he goes right ahead increasing his business, improving his farms, developing American trotters and furnishing work at the highest wages to willing hands in his factories, at his oil-wells, on his lands, in his barns and his hospitable home. He dispenses large sums in charity. His benevolence and enterprise reach far beyond Pennsylvania. He does not hoard up money to loan it at exorbitant rates. As a matter of fact, from the hundreds of men he has helped pecuniarily he never accepted one penny of interest. He has been mayor of Franklin, president of the Pennsylvania State-Dairymen's Association, director of the American Jersey-Cattle Club and member of the State Board of Agriculture. He is a brilliant talker, a profound thinker, a capital story-teller and a loyal friend. "May he live long and prosper!"

Miller & Sibley's Prospect-Hill Stock-Farm is one of the largest, best equipped and most favorably known in the world. Different farms comprising the establishment include a thousand acres of land adjacent to Franklin and a farm, with stabling for two-hundred horses and the finest kite-track in the United States, at Meadville. On one of these farms is the first silo built west of the Allegheny mountains. Trotting stock, Jersey cattle, Shetland ponies and Angora goats of the highest grades are bred. For Michael Angelo, when a calf six weeks old, twelve-thousand-five-hundred dollars in cash were paid A. B. Darling, proprietor of the Fifth Avenue Hotel, New York City. Animals of the best strain were purchased, regardless of cost. In 1886 Mr. Sibley bought from Senator Leland Stanford, of California, for ten-thousand dollars, the four-year-old trotting-stallion St. Bel. Seventy-five thousand were offered for him a few weeks before the famous sire of numerous prize-winners died. Cows that have broken all records for milk and butter and horses that have won the biggest purses on the leading race-tracks of the country are the results of the liberal policy pursued at Prospect-Hill. Charles Marvin, the prince of horsemen, superintends the trotting department and E. H. Sibley is manager of all the Miller & Sibley interests. Hundreds of the choicest animals are raised every year. Prospect-Hill Farm is one of the sights of Franklin and the enterprise represents an investment not far short of one-million dollars. Wouldn't men like Charles Miller and Joseph C. Sibley sweep away the cobwebs, give business an impetus and infuse new life and new ideas into any community?

Franklin had tallied one for heavy oil, but its resources were not exhausted. On October seventeenth, 1859, Colonel James P. Hoover, C. M. Hoover and Vance Stewart began to drill on the Robert-Brandon—now the Hoover—farm of three-hundred acres, in Sandycreek township, on the west bank of the Allegheny river, three miles south of Franklin. They found oil on December twenty-first, the well yielding one-hundred barrels a day! This pretty Christmas gift was another surprise. Owing to its distance from "springs" and the two wells—Drake and Evans—already producing, the stay-in-the-rut element felt confident that the Hoover Well would not "amount to a hill of beans." It was "piling Ossa on Pelion" for the well to produce, from the *second sand*, oil with properties adapted to illumination and lubrication. The Drake was for light, the Evans for grease and the Hoover combined the two in part. Where and when was this variegated dissimilarity to cease? Perhaps its latest phase is to come shortly. Henry F. James is beginning a well south-west of town, on the N. B. Myers tract, between a sweet and a sour spring. Savans, scientists, beer-drinkers, tee-totalers and oil-operators are on the ragged edge of suspense, some hoping, some fearing, some praying that James may tap a perennial fount of creamy 'alf-and-'alf.

Once at a drilling-well on the "Point" the tools dropped suddenly. The driller relieved the tension on his rope and let the tools down slowly. They descended six or eight feet! The bare thought of a crevice of such dimensions paralyzed the knight of the temper-screw, all the more that the hole was not to the first sand. What a lake of oil must underlie that derrick! He drew up the tools. They were dripping amber fluid, which had a flavor quite unlike petroleum. Did his nose deceive him? It was the aroma of beer! A lick of the stuff confirmed the nasal diagnosis—it had the taste of beer! The alarm was sounded and the sand-pump run down. It came up brimming over with beer! Ten times the trip was repeated with the same result. Think of an ocean of the delicious, foamy, appetizing German beverage! Word was sent to the owners of the well, who ordered the tubing to be put in. They tried to figure how many breweries the production of their well would retire. Pumping was about to begin, in presence of a party of impatient, thirsty spectators, when an excited Teuton, blowing and puffing, was seen approaching at a breakneck pace. Evidently he had something on his mind. "Gott in Himmel!" he shrieked, "you vas proke mit Grossman's vault!" The mystery was quickly explained. Philip Grossman, the brewer, had cut a tunnel a hundred feet into the hill-side to store his liquid-stock in a cool place. The well chanced to be squarely over this tunnel, the roof of which the tools pierced and stove in the head of a tun of beer! Workmen who came for a load were astonished to discover one end of a string of tubing dangling in the tun. It dawned upon them that the drillers three-hundred feet above must have imagined they struck a crevice and a messenger speeded to the well. The saddened crowd slinked off, muttering words that would not look nice in print. The tubing was withdrawn, the hogshead was shoved aside, the tools were again swung and two weeks later the well was pumping thirty barrels a day of unmistakable heavy-oil.

The Hoover strike fed the flame the Evans Well had kindled. Lands in the neighborhood were in demand on any terms the owners might impose. From Franklin to the new well, on both sides of the Allegheny, was the favorite choice, on a theory that a pool connected the deposits. Leases were snapped up at one-half royalty and a cash-bonus. Additional wells on the Hoover rivaled No. 1, which produced gamely for four years. The tools were stuck in cleaning it out and a new well beside it started at sixty barrels. The "Big-Emma Vein" was really an artery to which for years "whoa, Emma!" did not apply. Bissell & Co. and the Cameron Petroleum Company secured control of the property, on which fifteen wells were producing two-hundred barrels ten years from the advent of the Hoover & Vance. Harry Smith, a city-father, is operating on the tract and drilling paying wells at reasonable intervals. Colonel James P. Hoover died on February fourth, 1871, aged sixty-nine. Born in Centre county, he settled in the southern part of Clarion, was appointed by Governor Porter in 1839 Prothonotary of Venango county and removed to Franklin. The people elected him to the same office for three years and State Senator in 1844. The Canal-Commissioners in 1851 appointed him collector of the tolls at Hollidaysburg, Blair county, for five years. He filled these positions efficiently, strict adherence to principle and a high sense of duty marking his whole career. The esteem and confidence he enjoyed all through his useful life were attested by universal

B. E. SWAN.

regret at his death and the largest funeral ever witnessed in Franklin. His estimable widow survived Colonel Hoover twenty years, dying at the residence of her son-in-law, Arnold Plumer, in Minnesota. Their son, C. M. Hoover, ex-sheriff of the county, is now interested in the street railway. Vance Stewart, who owned the farm near the lower river-bridge, removed to Greenville and preceded his wife and several children, one of them Rev. Orlando V. Stewart, to the tomb. Another son, James Stewart, was a prominent member of the Erie bar.

The opening months of 1860 were decidedly lively on the Cochran Farm, in Cranberry township, opposite the Hoover. The first well, the Keystone, on the flats above where the station now stands, was a second-sander of the hundred-barrel class. The first oil sold for fourteen dollars a barrel, at which rate land-owners and operators were not in danger of bankruptcy or the poor-house. Fourteen-hundred dollars a day from a three-inch hole would have seemed too preposterous for Munchausen before the Pennsylvania oil-regions demonstrated that "truth is stranger than fiction." The Monitor, Raymond, Williams, McCutcheon and other wells kept the production at a satisfactory figure. Dale & Morrow, Horton & Son, Hoover & Co., George R. Hobby, Cornelius Fulkerson and George S. McCartney were early operators. B. E. Swan located on the farm in May of 1865 and drilled numerous fair wells. He has operated there for thirty-one years, sticking to the second-sand territory with a tenacity equal to the "perseverance of the saints." When thousands of producers, imitating the dog that let go the bone to grasp the shadow in the water, quit their enduring small wells to take their chance of larger ones in costlier fields, he did not lose his head and add another to the financial wrecks that strewed the greasian shore. Appreciating his moral stamina, his steadfastness and ability, Mr. Swan's friends insist that he shall serve the public in some important office. Walter Pennell—his father made the first car-wheels—and W. P. Smith drilled several snug wells on the uplands, Sweet & Shaffer following with six or eight. Eighteen wells are producing on the tract, which contains one-hundred-and forty acres and has had only two dry-holes in its thirty-six years of active developments.

Alexander Cochran, for forty years owner of the well-known farm bearing his name, is one of the oldest citizens of Franklin. Winning his way in the world by sheer force of character, scrupulous integrity and a fixed determination to succeed, he is in the highest and best sense a self-made man. Working hard in boyhood to secure an education, he taught school, clerked in general stores, studied law and was twice elected Prothonotary without asking one voter for his support. In these days of button-holing, log-rolling, wire-pulling, buying and soliciting votes this is a record to recall with pride. Marrying Miss Mary Bole—her father removed from Lewistown to Franklin seventy-five years ago—he built the home at "Cochran Spring" that is one of the land-marks of the town and established a large dry-goods store. As his means permitted he bought city-lots, put up dwelling-houses and about 1852 paid sixteen-hundred dollars for the farm in Cranberry township for which in 1863, after it had yielded a fortune, he refused seven-hundred-thousand! The farm was in two blocks. A neighbor expostulated with him for buying the second piece, saying it was "foolish to waste money that way." In 1861, when the same neighbor wished to mortgage his land for a loan, he naively remarked: "Well, Aleck, I guess I was the fool, not you, in 1852." A man of broad views, Mr. Cochran freely grants to others the liberality of thought he claims for himself. A hater of cant and sham and hollow pretense, he believes less in musty creeds than kindly deeds, more in giving loaves than tracts to the hungry and

ALEXANDER COCHRAN.

takes no stock in religion that thinks only of dodging punishment in the next world and fails to help humanity in this. In the dark days of low-priced oil and depressed trade, he would accept neither interest from his debtors nor royalty from the operators who had little wells on his farm. He never hounded the sheriff on a hapless borrower, foreclosed a mortgage to grab a coveted property or seized the chattels of a struggling victim to satisfy a shirt-tail note. There is no shred of the Pecksniff, the Shylock or the Uriah-Heep in his anatomy. At fourscore he is hale and hearty, rides on horseback, cultivates his garden, attends to business, likes a good play and keeps up with the literature of the day. The productive oil-farm is now owned by his daughters, Mrs. J. J. McLaurin, of Harrisburg, and Mrs. George R. Sheasley, of Franklin. The proudest eulogy he could desire is Alexander Cochran's just desert: "The Poor Man's Friend."

Down to Sandy Creek many wells were drilled from 1860 to 1865, producing fairly at an average depth of four-hundred-and-fifty to five-hundred feet.

These operations included the Miller, Smith and Pope farms, on the west side of the river, and the Rice, Nicklin, Martin and Harmon, on the east side, all second-sand territory. North of the Cochran and the Hoover work was pushed actively. George H. Bissell and Vance Stewart bored twelve or fifteen medium wells on the Stewart farm of two-hundred acres, which the Cameron Petroleum Company purchased in 1865 and Joseph Dale operated for some years. It lies below the lower bridge, opposite the Bleakley tract, from which a light production is still derived. Above the Stewart are the Fuller and the Chambers farms, the latter extending to the Allegheny-Valley depot. Scores of eager operators thronged the streets of Franklin and drilled along the Allegheny. Joseph Powley and Charles Cowgill entered the lists in the Cranberry district. Henry M. Wilson and George Piagett veered into the township and sank a bevy of dry-holes to vary the monotony. That was a horse on Wilson, but he got ahead of the game by a deal that won him the nicest territory on Horse Creek. Stirling Bonsall and Colonel Lewis—they're dead now—were in the thickest of the fray, with Captain Goddard, Philip Montgomery, Boyd, Roberts, Foster, Brown, Murphy and many more whom old-timers remember pleasantly. Thomas King, whole-souled, genial "Tom"—no squarer man e'er owned a well or handled oil-certificates—and Captain Griffith were "a good pair to draw to." King has "crossed over," as have most of the kindred spirits that dispelled the gloom in the sixties.

Colonel W. T. Pelton, nephew of Samuel J. Tilden, participated in the scenes of that exciting period. He lived at Franklin and drilled wells on French Creek. He was a royal entertainer, shrewd in business, finely educated and polished in manner and address. He and his wife—a lovely and accomplished woman—were fond of society and gained hosts of friends. They boarded at the United-States Hotel, where Mrs. Pelton died suddenly. This affliction led Colonel Pelton to sell his oil-properties and abandon the oil-regions. Returning to New York, when next he came into view as the active agent of his uncle in the secret negotiations that grew out of the election of 1876, it was with a national fame. His death in 1880 closed a busy, promising career.

In the spring of 1864 a young man, black-haired, dark-eyed, an Apollo in form and strikingly handsome, arrived at Franklin and engaged rooms at Mrs. Webber's, on Buffalo street. The stranger had money, wore good clothes and presented a letter of introduction to Joseph H. Simonds, dealer in real-estate, oil-wells and leases. He looked around a few days and concluded to invest in sixty acres of the Fuller farm, Cranberry township, fronting on the Allegheny river. The block was sliced off the north end of the farm, a short distance below the upper bridge and the Valley station. Mr. Simonds consented to be a partner in the transaction. The transfer was effected, the deed recorded and a well started. It was situated on the hill, had twenty feet of second-sand and pumped twenty barrels a day. The owner drilled two others on the bluff, the three yielding twenty barrels for months. The ranks of the oil-producers had received an addition in the person of—John Wilkes Booth.

The firm prospered, each of the members speculating and trading individually. M. J. Colman, a capital fellow, was interested with one or both in various deals. Men generally liked Booth and women admired him immensely. His lustrous orbs, "twin-windows of the soul," could look so sad and pensive as to awaken the tenderest pity, or fascinate like "the glittering eye" of the Ancient Mariner or the gaze of the basilisk. "Trilby" had not come to light, or he might have enacted the hypnotic role of Svengali. His moods were variable and uncertain. At times he seemed morose and petulant, tired of everybody and "unsocial as a clam." Again he would court society, attend parties, dance, recite and be "the life of the company." He belonged to a select circle that exchanged visits with a coterie of young folks in Oil City. A Confederate sympathizer and enemy of the government, his closest intimates were staunch Republicans and loyal citizens. William J. Wallis, the veteran actor who died in December of 1895, in a Philadelphia theater slapped him on the mouth for calling President Lincoln a foul name. Booth's acting, while inferior to his brother Edwin's, evinced much dramatic power. He controlled his voice admirably, his movements were graceful and he spoke distinctly, as Franklinites whom he sometimes favored with a reading can testify.

J. WILKES BOOTH.

One morning in April, 1865, he left Franklin, telling Mr. Simonds he was going east a few days. He carried a satchel, which indicated that he did not expect his stay to be prolonged. His wardrobe, books and papers remained in his room. Nothing was heard of him until the crime of the century stilled all hearts and the wires flashed the horrible news of Abraham Lincoln's assassination. The excitement in Franklin, the murderer's latest home, was intense. Crowds gathered to learn the dread particulars and discuss Booth's conduct and utterances. Not a word or act previous to his departure pointed to deliberate preparation for the frightful deed that plunged the nation in grief. That he contemplated it before leaving Franklin the weight of evidence tended to disprove. He made no attempt to sell any of his property, to convert his lands and wells into cash, to settle his partnership accounts or to pack his effects. He had money in the bank, wells bringing a good income and important business pending. All these things went to show that, if not a sudden impulse, the killing of Lincoln was prompted by some occurrence in Washington that fired the passionate nature John Wilkes Booth inherited from his father. The world is familiar with the closing chapters of the dark tragedy—the assassin's flight, the pursuit into Virginia, the burning barn, Sergeant Corbett's fatal bullet, the pathetic death-scene on the Garrett porch and the last message, just as the dawn was breaking on the glassy eyes that opened feebly for a moment: "Tell my mother I died for my country. I did what I thought was best."

The wells and the land on the river were held by Booth's heirs until 1869, when the tract changed hands. The farm is producing no oil and the Simonds-Booth wells have disappeared. Had he not intended to return to Franklin, Booth would certainly have disposed of these interests and given the proceeds to his mother. "Joe" Simonds removed to Bradford to keep books for Whitney & Wheeler, bankers and oil-operators, and died there years ago. He was an expert accountant, quick, accurate and neat in his work and most fastidious in his attire. A blot on his paper, a figure not exactly formed, a line one hair-breadth crooked, a spot on his linen or a speck of dust on his coat was simply intolerable. He was correct in language and deportment and honorable in his dealings. Colman continued his oil-operations and, in company with W. R. Crawford, a real-estate agency until the eighties. He married Miss Ella

Hull, the finest vocalist Franklin ever boasted, daughter of Captain S. A. Hull, and removed to Boston. For years paralytic trouble has confined him to his house. He is "one of nature's noblemen."

Over the hills to the interior of the townships developments spread. Bredinsburg, Milton and Tarkiln loomed up in Cranberry, where Taylor & Torrey, S. P. McCalmont, Jacob Sheasley, B. W. Bredin and E. W. Echols have sugar-plums. In Sandycreek, between Franklin and Foster, Angell & Prentice brought Bully Hill and Mount Hope to the front. The biggest well in the package was a two-hundred barreler on Mount Hope, which created a mount of hopes that were not fully realized. George V. Forman counted out one-hundred-and-fifty-thousand dollars for the Mount Hope corner. The territory lasted well and averaged fairly. Bully Hill merited its somewhat slangy title. Dr. C. D. Galbraith, George R. Sheasley and Mattern & Son are among

ANGELL & PRENTICE'S WELLS BELOW FRANKLIN IN 1873.

its present operators. Angell and Prentice parted company, each to engage in opening up the Butler region. Prentice, Crawford, Barbour & Co. did not let the grass grow under their feet. They "knew a good thing at sight" and pumped tens-of-thousands of barrels of oil from the country south of Franklin. The firm was notable in the seventies. Considerable drilling was done at Polk, where the state is providing a half-million-dollar Home for Feeble-Minded Children, and in the latitude of Utica, with about enough oil to be an aggravation. The Shippen wells, a mile north of the county poor-house, have produced for thirty years. West of them, on the Russell farm, the Twin wells, joined as tightly as the derricks could be placed, pumped for years. This was the verge of productive territory, test wells on the lands of William Sanders, William Bean, A. Reynolds, John McKenzie, Alexander Frazier and W. Booth, clear to Cooperstown, finding a trifle of sand and scarcely a vestige of oil. The Raymonds, S. Ramage, John J. Doyle and Daniel Grimm had a very tidy offshoot at Raymilton. On this wise lubricating and second-sand oils were revealed for the benefit of mankind generally. The fly in the ointment was the clerical crank who wrote to President Lincoln to demand that the producing of heavy-oil be stopped peremptorily, as it had been stored in the ground to grease the axletree of the earth in its diurnal revolution! This communication reminded Lincoln of a "little story," which he fired at the fellow with such effect that the candidate for a strait-jacket was perpetually squelched.

Hon. William Reid Crawford, a member of the firm of Prentice, Crawford, Barbour & Co., lives in Franklin. His parents were early settlers in north-western Pennsylvania. Alexander Grant, his maternal grandfather, built the first stone-house in Lancaster county, removed to Butler county and located finally in Armstrong county, where he died sixty-five years ago. In 1854 William R. and four of his brothers went to California and spent some time mining gold. Upon his return he settled on a farm in Scrubgrass township, Venango county, of which section the Crawfords had been prominent citizens from the beginning of its history. Removing to Franklin in 1865, Mr. Crawford engaged actively in the production of petroleum, operating extensively in various portions of the oil-regions for twenty years. He acquired a high reputation for enterprise and integrity, was twice a city-councillor, served three terms as mayor, was long president of the school-board, was elected sheriff in 1887 and State-Senator in 1890. Untiring fidelity to the interests of the people and uncompromising hostility to whatever he believed detrimental to the general welfare distinguished his public career. Genial and kindly to all, the friend of humanity and benefactor of the poor, no man stands better in popular estimation or is more deserving of confidence and respect. His friends could not be crowded into the Coliseum without bulging out the walls.

Captain John K. Barbour, a man of imposing presence and admirable qualities, removed to Philadelphia after the dissolution of the firm. The Standard Oil-Company gave him charge of the right-of-way department of its pipe-line service and he returned to Franklin. Two years ago, during a business visit to Ohio, he died unexpectedly, to the deep regret of the entire community. S. A. Wheeler operated largely in the Bradford field and organized the Tuna-Valley Bank of Whitney & Wheeler. For a dozen years he has resided at Toledo, his early home. Like Captain Barbour, "Fred," as he was commonly called, had an exhaustless mine of bright stories and a liberal share of the elements of popularity. One afternoon in 1875, three days before the fire that wiped out the town, a party of us chanced to meet at St. Joe, Butler county, then the centre of oil-developments. An itinerating artist had his car moored opposite the drugstore. Somebody proposed to have a group-picture. The motion carried unanimously and a toss-up decided that L. H. Smith was to foot the bill. The photographer brought out his camera, positions were taken on the store-platform and the pictures were mailed an hour ahead of the blaze that destroyed most of the buildings and compelled the artist to hustle off his car on the double-quick. Samuel R. Reed, at the extreme right, operated in the Clarion field. He had a hardware-store in company with the late Dr. Durrant and his home is in Franklin. James Orr, between whom and Reed a telegraph-pole is seen, was connected with the Central Hotel at Petrolia and later was a broker in the Producers' Exchange at Bradford. On the step is Thomas McLaughlin, now oil-buyer at Lima, once captain of a talented base-ball club at Oil City and an active oil-broker. Back of him is "Fred" Wheeler, with Captain Barbour on his right and L. H. Smith sitting comfortably in front. Mr. Smith figured largely at Pithole, operated satisfactorily around Petrolia and removed years ago to New York. Cast in a giant mould, he weighs three-hundred pounds and does credit to the illustrious legions of Smiths. He is a millionaire and has an office over the Seaboard Bank, at the lower end of Broadway. Joseph Seep, the king-bee of good fellows, sits beside Smith. Pratt S. Crosby, formerly a jolly broker at Parker and Oil City, stands behind Seep. Next him is "Tom" King, who has "gone to the land of the leal," J. J. McLaurin ending the line. James Amm, who went from an Oil-City clerksnip to coin a fortune at Bradford—a street

bears his name—sits on the platform. Every man, woman, child and baby in Oil-City knew and liked "Jamie" Amm, who is enjoying his wealth in Buffalo. Two of the eleven in the group have "passed beyond the last scene" and the other nine are scattered widely. Frederic Prentice, one of the pluckiest operators in petroleum-annals, was the

GROUP AT ST. JOE, BUTLER COUNTY, IN 1875.

first white child born on the site of Toledo, when Indians were the neighbors of the pioneers of Northern Ohio. His father left a fine estate, which the son increased greatly by extensive lumbering, in which he employed three-thousand men. Losses in the panic of 1857 retired him from the business. He retrieved his fortune and paid his creditors their claims in full, with ten per cent. interest, an act indicative of his sterling character. Reading in a newspaper about the Drake well, he decided to see for himself whether the story was fast colors. Journeying to Venango county by way of Pittsburg, he met and engaged William Reed to accompany him. Reed had worked at the Tarentum salt-wells and knew a thing or two about artesian-boring. The two arrived at Franklin on the afternoon of the day Evans's well turned the settlement topsy-turvy. Next morning Prentice offered Evans forty-thousand dollars for a controlling interest in the well, one-fourth down and the balance in thirty, sixty and ninety days. Evans declining to sell, the Toledo visitor bought from Martin & Epley an acre of ground on the north bank of French Creek, at the base of the hill, and contracted with Reed to "kick down" a well, the third in the district. Prentice and Reed tramped over the country for days, locating oil-deposits by means of the witch-hazel, which the Tarentumite handled skillfully. This was a forked stick, which it was claimed turned in the hands of the holder at spots where oil existed. Various causes delayed the completion of the well, which at last proved disappointingly small. Meanwhile Mr. Prentice leased the Neeley farm, two miles up the Allegheny, in Cranberry township, and bored several paying wells. A railroad station on the tract is named after him and R. G. Lamberton has converted the property into a first-class stock-farm. Favorable reports from Little Kanawha River took him to West Virginia, where he leased and purchased immense blocks of land. Among them was the Oil-Springs tract, on the Hughes River, from which oil had been skimmed for generations. Two of his wells on the Kanawha yielded six-hundred barrels a day, which had to be stored in ponds or lakes for want of tankage. Confederate raiders burned the wells, oil and machinery and drove off the workmen, putting an extinguisher on operations until the Grant-Lee episode beneath the apple-tree at Appomattox.

Assuming that the general direction of profitable developments would be north-east and south-west, Mr. Prentice surveyed a line from Venango county through West Virginia, Kentucky and Tennessee. This idea, really the foundation of "the belt theory," he spent thousands of dollars to establish. Personal investigation and careful surveys confirmed his opinion, which was based upon observations in the Pennsylvania fields. The line run thirty years ago touched numerous "springs" and "surface shows" and recent tests prove its remarkable accuracy. On this theory he drilled at Mount Hope and Foster, opening a section that has produced several-million barrels of oil. C. D. Angell applied the principle in Clarion and Butler counties, mapping out the probable course of the "belt" and leasing much prolific territory. His success led others to adopt the same plan, developing a number of pools in four states, although nature's lines are seldom straight and the oil-bearing strata are deposited in curves and beds at irregular intervals.

In company with W. W. Clark of New York, to whom he had traded a portion of his West-Virginia lands, Mr. Prentice secured a quarter-interest in the Tarr farm, on Oil Creek, shortly before the sinking of the Phillips well, and began shipping oil to New York. They paid three dollars apiece for barrels, four dollars a barrel for hauling to the railroad and enormous freights to the east. The price dropping below the cost of freights and barrels, the firm dug acres of pits to put tanks under ground, covering them with planks and earth to prevent evaporation. Traces of these storage-vats remain on the east bank of Oil Creek. Crude fell to twenty-five cents a barrel at the wells and the outlook was discouraging. Clark & Prentice stopped drilling and turned their attention to finding a market. They constructed neat wooden packages that would hold two cans of refined-oil, two oil-lamps and a dozen chimneys and sent one to each United-States Consul in Europe. Orders soon rushed in from foreign countries, especially Germany, France and England, stimulating the erection of refineries and creating a large export trade. Clark & Summer, who also owned an interest in

FREDERIC PRENTICE.

the Tarr farm, built the Standard Refinery at Pittsburg and agreed to take from Clark & Prentice one-hundred-thousand barrels of crude at a dollar a barrel, to be delivered as required during the year. Before the delivery of the first twenty-five-thousand barrels the price climbed to one-fifty and to six dollars before the completion of the contract, which was carried out to the letter. The advance continued to fourteen dollars a barrel, lasting only one day at this figure. These were vivifying days in oleaginous circles, never to be repeated while Chronos wields his trusty blade.

When crude reached two dollars Mr. Prentice bought the Washington-McClintock farm, on which Petroleum Centre was afterwards located, for three-

hundred-thousand dollars. Five New-Yorkers, one of them the president of the
Shoe and Leather Bank and another the proprietor of the Brevoort House, advanced fifty-thousand dollars for the first payment. Within sixty days Prentice
sold three-quarters of his interest for nine-hundred-thousand dollars and organized the Central Petroleum Oil-Company, with a capital of five-millions!
Wishing to repay the New York loan, the Brevoort landlord desired him to retain his share of the money and invest it as he pleased. For his ten-thousand
dollars mine host received eighty-thousand in six months, a return that leaves
government-bond syndicates and Cripple-Creek speculations out in the latitude
of Nansen's north-pole. The company netted fifty-thousand dollars a month in
dividends for years and lessees cleared three or four millions from their operations on the farm. Greenbacks circulated like waste-paper, Jules Verne's fancies were surpassed constantly by actual occurrences and everybody had money
to burn.

Prentice and his associates purchased many tracts along Oil Creek, including
the lands where Oil City stands and the Blood farm of five-hundred acres. In
the Butler district he drilled hundreds of wells and built the Relief Pipe-Line.
Organizing The Producers' Consolidated Land and Petroleum Company, with
a capital of two-and-a-half millions, he managed it efficiently and had a prominent part in the Bradford development. Boston capitalists paid in twelve-hundred-thousand dollars, Prentice keeping a share in his oil-properties representing thirteen-hundred-thousand more. The company is now controlled by the
Standard, with L. B. Lockard as superintendent. Its indefatigable founder also
organized the Boston Oil Company to operate in Kentucky and Tennessee, put
down oil-wells in Peru and gas-wells in West Virginia, produced and piped
thousands of barrels of crude daily and was a vital force in petroleum-affairs
for eighteen years. The confidence and esteem of his compatriots were attested
by his unanimous election to the presidency of the Oilmen's League, a secret-
society formed to resist the proposed encroachments of the South Improvement
Company. The League accomplished its mission and then quietly melted out
of existence.

Since 1877 Mr. Prentice has devoted his attention chiefly to lumbering in
West Virginia and to his brown-stone quarries at Ashland, Wisconsin. The
death of his son, Frederick A., by accidental shooting, was a sad bereavement
to the aged father. His suits to get possession of the site of Duluth, the city of
Proctor Knott's impassioned eulogy, included in a huge grant of land deeded to
him by the Indians, were scarcely less famous than Mrs. Gaines's protracted litigation to recover a slice of New Orleans. The claim involved the title to property valued at twelve-millions of dollars. From his Ashland quarries the owner
took out a monolith, designed for the Columbian Exposition in 1893, forty yards
long and ten feet square at the base. Beside this monster stone Cleopatra's
Needle, disintegrating in Central Park, Pompey's Pillar and the biggest blocks
in the pyramids are Tom-Thumb pigmies. At seventy-four Mr. Prentice, foremost in energy and enterprise, retains much of his youthful vigor. Earnest and
sincere, a master of business, his word as good as gold, Frederic Prentice holds
an honored place in the ranks of representative oil-producers, "nobles of
nature's own creating."

A native of Chautauqua county, N, Y., where he was born in 1826, Cyrus
D. Angell received a liberal education, served as School-Commissioner and
engaged in mercantile pursuits at Forestville. Forced through treachery and
the monetary stringency of the times to compromise with his creditors, he recov-

ered his financial standing and paid every cent of his indebtedness, principal and interest. In 1867 he came to the oil regions with a loan of one-thousand dollars and purchased an interest in property at Petroleum Centre that paid handsomely. Prior to this, in connection with Buffalo capitalists, he had bought

Belle Island, in the Allegheny River at Scrubgrass, upon which soon after his arrival he drilled three wells that averaged one-hundred barrels each for two years, netting the owners over two-hundred-thousand dollars. Operations below Franklin, in company with Frederic Prentice, also proved highly profitable. His observations of the course of developments along Oil Creek and the Allegheny led Mr. Angell to the conclusion that petroleum would be found in "belts" or regular lines. He adopted the theory that two "belts" existed, one running from Petroleum Centre to Scrubgrass and the other from St. Petersburg through Butler county. Satisfied of the correctness of this view, he leased or purchased all the lands within the probable boundaries of the "belt"

CYRUS D. ANGELL.

from Foster to Belle Island, a distance of six miles. The result justified his expectations, ninety per cent. of the wells yielding abundantly. With "the belt theory," which he followed up with equal success farther south, Mr. Angell's name is linked indissolubly. His researches enriched him and were of vast benefit to the producers generally. He did much to extend the Butler region, drilling far ahead of tested territory. The town of Angelica owed its creation to his fortunate operations in the neighborhood, conducted on a comprehensive scale. Reverses could not crush his manly spirit. He did a large real-estate business at Bradford for some years, opening an office at Pittsburg when the Washington field began to loom up. Failing health compelling him to seek relief in foreign travel, last year he went to Mexico and Europe. He is still abroad. Mr. Angell is endowed with boundless energy, fine intellectual powers and rare social acquirements. During his career in Oildom he was an excellent sample of the courageous, unconquerable men who have made petroleum the commercial wonder of the world.

An old couple in Cranberry township, who eked out a scanty living on a rocky farm near the river, sold their land for sixty-thousand dollars at the highest pitch of the oil-excitement around Foster. This was more money than the pair had ever before seen, much less expected to handle and own. It was paid in bank-notes at noon and the log-house was to be vacated next day. Towards evening the poor old woman burst into tears and insisted that her husband should give back the money to the man that "wanted to rob them of their home." She was inconsolable, declaring they would be "turned out to starve, without a roof to cover them." The idea that sixty-thousand dollars would buy an ideal home brought no comfort to the simple-minded creature, whose hopes and ambitions were confined to the lowly abode that had sheltered her for a half-century. A promise to settle near her brother in Ohio reconciled her somewhat, but it almost broke her faithful heart to leave a spot endeared by many tender associations. John Howard Payne, himself a homeless wanderer,

8

whose song has been sung in every tongue and echoed in every soul, was right in asserting :

"Be it ever so humble, there's no place like home."

The refusal of his wife to sign the deed conveying the property enabled a wealthy Franklinite to gather a heap of money. The tract was rough and unproductive and the owner proposed to accept for it the small sum offered by a neighboring farmer, who wanted more pasture for his cattle. For the first time in her life the wife declined to sign a paper at her husband's request, saying she had a notion the farm would be valuable some day. The purchaser refused to take it subject to a dower and the land lay idle. At length oil-developments indicated that the "belt" ran through the farm. Scores of wells yielded freely, netting the land-owner a fortune and convincing him that womanly intuition is a sure winner.

A citizen of Franklin, noted for his conscientiousness and liberality, was interested in a test-well at the beginning of the Scrubgrass development. He vowed to set aside one-fourth of his portion of the output of the well "for the Lord," as he expressed it. To the delight of the owners, who thought the venture hazardous, the well showed for a hundred barrels when the tubing was put in. On his way back from the scene the Franklin gentleman did a little figuring, which proved that the Lord's percentage of the oil might foot up fifty dollars a day. This was a good deal of money for religious purposes. The maker of the vow reflected that the Lord could get along without so much cash and he decided to clip the one-fourth down to one-tenth, arguing that the latter was the scripture limit. Talking it over with his wife, she advised him to stick to his original determination and not trifle with the Lord. The husband took his own way, as husbands are prone to do, and revisited the well next day. Something had gone wrong with the working-valve, the tubing had to be drawn out and the

well never pumped a barrel of oil! The chagrined operator concluded, as he charged two thousand dollars to his profit-and-loss account, that it was not the Lord who came out at the small end of the horn in the transaction.

Rev. Clarence A. Adams, the eloquent ex-pastor of the First Baptist Church at Franklin, is the lucky owner of a patch of

REV. C. A. ADAMS, D.D.

REV. EZRA F. CRANE, D.D.

paying territory at Raymilton. Recently he finished a well which pumped considerable salt-water with the oil. Contrary to Cavendish and the ordinary custom, another operator drilled very close to the boundary of the Adams lease and torpedoed the well heavily. Instead of sucking the oil from the preacher's nice pumper, the new well took away most of the salt-water and doubled the production of petroleum! Commonly it would seem rather mean to rob a Baptist minister of water, but in this case Dr. Adams is perfectly resigned to the loss of aqueous fluid and gain of dollar-fifty crude. A profound student of Shakespeare, Browning and the Bible, a brilliant lecturer and master of pulpit-oratory, may he also stand on a lofty rung of the greasian ladder and attain the goodly age of Franklin's "grand old man," Rev. Dr. Crane. This "father in Israel," whose death in February of 1896 the whole community mourned, left a record of devoted service as a physician and clergyman for over sixty years that has seldom been equaled. He healed the sick, smoothed the pillow of the dying, relieved the distressed, reclaimed the erring, comforted the bereaved, turned the faces of the straying Zionward and found the passage to the tomb "a gentle wafting to immortal life."

> "Though old, he still retained
> His manly sense and energy of mind,
> Virtuous and wise he was, but not severe,
> For he remembered that he once was young;
> His kindly presence checked no decent joy,
> Him e'en the dissolute admired. Can he be dead
> Whose spiritual influence is upon his kind?"

Miss Lizzie Raymond, daughter of the pioneer who founded Raymilton and erected the first grist-mill at Utica, has long taught the infant-class of the Presbyterian Sunday-school at Franklin. Once the lesson was about the wise and the foolish virgins, the good teacher explaining the subject in a style adapted to the juvenile mind. A cute little tot, impressed by the sad plight of the virgins who had no oil in their lamps, innocently inquired: "Miss 'Aymond, tan't oo tell 'em dirls to tum to our house an' my papa 'll div' 'em oil f'um his wells?" Heaven bless the children that come as sunbeams to lighten our pathway, to teach us lessons of unselfishness and prevent the rough world from turning our hearts as hard as the mill-stone.

THOMAS M'DONOUGH.

Another youngster prayed every morning and night that a well her father was drilling on Bully Hill would be a good one. It was a hopeless failure, finished the day before Christmas. The result disturbed the child exceedingly. That night, as the loving mother was preparing her for bed, the little girl observed: "I dess it's no use prayin' 'till after Kismas, 'cos God's so busy helpin' Santa Claus He hasn't time for nobody else!"

The late Thomas McDonough, a loyal-hearted son of the Emerald Isle, was also an operator in the lubricating region. He had an abundance of rollicking wit, "the pupil of the soul's clear eye," and an unfailing supply of droll stories. Desiring to lease a farm in Sandy-Creek township, supposed to be squarely "on the belt," he started at daybreak to interview the owner, feeling sure his mission would succeed. An unexpected sight presented itself through the open door, as the visitor stepped upon the porch of the dwelling. The farmer's wife was setting the table for breakfast and Frederic Prentice was folding a paper carefully. McDonough realized in a twinkling that Prentice had secured the lease and his trip was fruitless. "I am looking for John Smith" he stammered, as the farmer invited him to enter, and beat a hasty retreat. For years his friends rallied the Colonel on his search and would ask with becoming solemnity whether he had discovered John Smith. The last time we met in Philadelphia this incident was revived and the query repeated jocularly. The jovial McDonough died in 1894.

It is safe to assume that he will easily find numerous John Smiths in the land of perpetual reunion.   One day he told a story in an office on Thirteenth street, Franklin, which tickled the hearers immensely.   A full-fledged African, who had been sweeping the back-room, broke into a tumultuous laugh.   At that moment a small boy was riding a donkey directly in front of the premises. The jackass heard the peculiar laugh and elevated his capacious ears more fully to take in the complete volume of sound.   He must have thought the mel-ody familiar and believed he had stumbled upon a relative.   Despite the frantic exertions of the boy, the donkey rushed towards the building whence the bois-terous guffaw proceeded, shoved his head inside the door and launched a ter-rific bray.   The bystanders were convulsed at this evidence of mistaken identity, which the jolly story-teller frequently rehearsed for the delectation of his hosts of friends.

Looking over the Milton diggings one July day, Col. McDonough met an amateur-operator who was superintending the removal of a wooden-tank from a position beside his first and only well.   A discussion started regarding the combustibility of the thick sediment collected on the bottom of the tank.   The amateur maintained the stuff would not burn and McDonough laughingly re-plied, "Well, just try it and see!"   The fellow lighted a match and applied it to the viscid mass before McDonough could interfere, saying with a grin that he proposed to wait patiently for the result.   He didn't have to wait "until Orcus would freeze over and the boys play shinny on the ice."   In the ninetieth frac-tion of a second the deposit blazed with intense enthusiasm, quickly enveloping the well-rig and the surroundings in flames.   Clouds of smoke filled the air, suggesting fancies of Pittsburg or Sheol.   Charred fragments of the derrick, engine-house and tank, with an acre of blackened territory over which the burn-ing sediment had spread, demonstrated that the amateur's idea had been de-cidedly at fault.   The experiment convinced him as searchingly as a Roentgen ray that McDonough had the right side of the argument.   "If the 'b. s.' had been as green as the blamed fool, it wouldn't have burned," was the Colonel's appropriate comment.

> In your wide peregrinations from the poles to the equator,
> Should you hear some ignoramus—let out of his incubator—
> Say the heavy oil of Franklin is not earth's best lubricator,
> Do as did renown'd Tom Corwin, the great Buckeye legislator,
> When a jabberwock in Congress sought to brand him as a traitor:
> Just "deny the allegation and defy the allegator!"

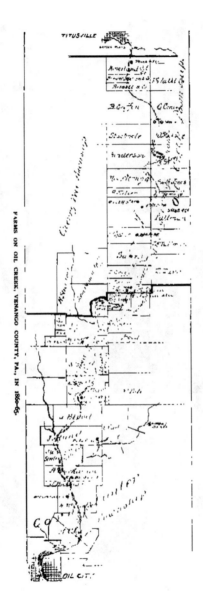

FARMS ON OIL CREEK, VENANGO COUNTY, PA., IN 1860-65.

ORIGINAL C. W. KENNEY'S ALLEMAGOOZELUM-CITY WELL No. 2.

VII.

THE VALLEY OF PETROLEUM.

Wonderful Scenes on Oil Creek—Mud and Grease Galore—Rise and Fall of Phenomenal Towns—Shaffer, Pioneer and Petroleum Centre—Fortune's Queer Vagaries—Wells Flowing Thousands of Barrels—Sherman, Delamater and "Coal-Oil Johnnie"—From Penury to Wealth and Back—Truthful Recitals That Discount Fairy-Tales.

"Some ships come into port that are not steered."—*Seneca.*
"God has placed in his great bank—mother earth—untold wealth and many a poor man's check has been honored here for large amounts of oil."—*T. S. Scoville, A. D. 1861.*
"Ain't that well spittin' oil?"—*Small Boy, A. D. 1863.*
"Wonderful, most wonderful, marvelous, most marvelous, are the stories told of the oil region. It is another California."—*John W. Forney, A. D. 1863.*
"Coal-oil Johnnie is my name."—*Popular Song.*
"Derricks peered up behind the houses of Oil City, like dismounted steeples, and oil was pumping in the back-yards."—*London Post, A. D. 1865.*

ORTY-THREE farms of manifold shapes and sizes lay along the stream from the Drake well to the mouth of Oil Creek, sixteen miles southward. For sixty years the occupants of these tracts had forced a bare subsistence from the reluctant soil. "Content to live, to propagate and die," their requirements and their resources were alike scanty. They knew nothing of the artificial necessities and extravagances of fashionable life. To most of them the great, busy, plodding world was a sealed book, which they had neither the means nor the inclination to unclasp. The world reciprocated by wagging in its customary groove, blissfully unconscious of the scattered settlers on the banks of the Allegheny's tributary. A trip on a raft to Pittsburg, with the privilege of walking back, was the limit of their journeyings from the hills and rocks of Venango. Hunting, fishing and hauling saw-logs in winter aided in replenishing the domestic larder. None imagined the unproductive valley would become the cradle of an industry before which cotton and coal and iron must "hide their diminished heads." No prophet had proclaimed that lands on Oil Creek would sell for more than corner-lots in London or New York. Who could have conceived that these bold cliffs and patches of clearing would enlist ambitious mortals from every quarter of the globe in a mad race to secure a foothold on the coveted acres? What seventh son of a seventh son could foresee that a thousand dollars spent on the Willard farm would yield innumerable millions? Who could predict that a tiny stream of greenish fluid, pumped from a hole on an island too insignificant to have a name, would swell into the vast ocean of petroleum that is the miracle of the nineteenth century? Fortune has played many pranks, but the queerest of them all were the vagaries incidental to the petroleum-development on Oil Creek.

The Bissell, Griffin, Conley, two Stackpole, Pott, Shreve, two Fleming, Henderson and Jones farms, comprising the four miles between the Drake well and the Miller tract, were not especially prolific. Traces of a hundred oil-pits, in some of which oak-trees had grown to enormous size, are visible on the Bissell plot of eighty acres. A large dam, used for pond-freshets, was located on Oliver Stackpole's farm. Two refineries of small capacity were built on the Stackpole and Fletcher lands, where eighteen or twenty wells produced moderately. The owner of a flowing well on the lower Fleming farm, imitating the man who killed the goose that la:d the golden eggs, sought to increase its output by putting the tubing and seed-bagging farther down. The well resented the interference, refusing to yield another drop and pointing the obvious moral: "Let well enough alone!" The Miller farm of four-hundred acres, on both sides of the creek, was purchased in 1863 from Robert Miller by the Indian Rock-Oil Company of New York. Now a railroad station and formerly the principal shipping-point for oil, refineries were started, wells were drilled and the stirring town of Meredith blossomed for a little space. The Lincoln well turned out sixty barrels a day, the Boston fifty, the Bobtail forty, the Hemlock thirty and others from ten to twenty-five, at an average depth of six-hundred feet. The Barnsdall Oil-Company operated on the Miller and the Shreve farms, drilling extensively on Hemlock Run, and George Bartlett ran the Sunshine Oil-Works. The village, the refineries and the derricks have disappeared as completely as Herculaneum or Sir John Franklin.

George Shaffer owned fifty acres below the Miller farm, divided by Oil Creek into two blocks, one in Cherrytree township and the other in Allegheny. Twenty-four wells, eight of them failures, were put down on the flats and the abrupt hill bordering the eastern shore of the stream. Samuel Downer's Rangoon and three of Watson & Brewer's were the largest, ranking in the fifty-barrel list. In July of 1864 the Oil-Creek Railroad was finished to Shaffer farm, which immediately became a station of great importance. From one house and barn the place expanded in sixty days to a town of three-thousand population. And such a town! Sixteen-hundred teams, mainly employed to draw oil from the wells down the creek, supported the stables, boarding-houses and hotels that sprang up in a night. Every second door opened into a bar-room. The buildings were "balloon frames," constructed entirely of boards, erected in a few hours and liable to collapse on the slightest pretext. Houses of cards would be about as comfortable and substantial. Outdo Hezekiah, by rolling back time's dial thirty-one years, and in fancy join the crowd headed for Shaffer six months after the advent of the railway.

Start from Corry, "the city of stumps," with the Downer refinery and a jumble of houses thrown around the fields. Here the Atlantic & Great-Western, the Philadelphia & Erie and the Oil-Creek Railroads meet. The station will not shelter one-half the motley assemblage bound for Oildom. "Mother Cary is plucking her geese" and snow-flakes are dropping thickly. Speculators from the eastern cities, westerners in quest of "a good thing," men going to work at the wells, capitalists and farmers, adventurers and drummers clamor for tickets. It is the reverse of "an Adamless Eden," for only three women are to be seen. At last the train backs to the rickety depot and a wild struggle commences. Scrambling for the elevated cars in New York or Chicago is a feeble movement compared with this frantic onslaught. Courtesy and chivalry are forgotten in the rush. Men swarm upon the steps, clog the platforms, pack the baggage-car, thrust the women aside, stick to the cowcatcher and clamber on the roofs of the coaches. Over the roughest track on earth, which winds and twists and skirts the creek most of the way, the train rattles and jolts and pitches. The conductor's job is no sinecure, as he squeezes through the dense mass that leaves him without sufficient elbow-room to "punch in the presence of the passenjare." Derricks—tall, gaunt skeletons, pickets of the advancing army—keep solemn watch here and there, the number increasing as Titusville comes in sight.

A hundred people get off and two-hundred manage somehow to get on. Past the Drake well, past a forest of derricks, past steep cliffs and tortuous ravines the engineer speeds the train. Did you ever think what a weight of responsibility rests upon the brave fellow in the locomotive-cab, whose clear eye looks straight along the track and whose steady hand grasps the throttle? Should he relax his vigilance or lose his nerve one moment, scores of lives might be the fearful penalty. A short stop at Miller Farm, a whiff of refinery-smells and in five minutes Shaffer is reached. The board-station is on the right hand, landings on the left form a semi-circle hundreds of feet in length, freight-cars jam the double track and warehouses dot the bank. The flat—about thirty rods wide—contains the mushroom town, bristling with the undiluted essence of petroleum-activity. Three-hundred teamsters are unloading barrels of oil from wagons dragged by patient, abused horses and mules through miles of greasy, clayey mud. Everything reeks with oil. It pervades the air, saturates clothes and conversation, floats on the muddy scum and fills lungs and nostrils with its peculiar odor. One cannot step a yard without sinking knee-deep in deceptive mire that performs the office of a boot-jack if given "a ghost of a show." Christian's Slough of Despond wasn't a circumstance to this adhesive paste, which engulfs unwary travels to their trouser-pockets and begets a dreadful craving for roads not

> " Wholly unclassable,
> Almost impassable,
> Scarcely jackassable."

The trip of thirty-five miles has shaken breakfast clear down to the pilgrims' boots. Out of the cars the hungry passengers tumble as frantically as they had clambered in and break for the hotels and restaurants. A dollar pays for a dinner more nearly first-class in price than in quality. The narrow hall leading to the dining-room is crammed with men—Person's Hotel fed four-hundred a day—waiting their turn for vacant chairs at the tables. Bolting the meal hurriedly, the next inquiry is how to get down the creek. There are no coupés, no prancing steeds, no stages, no carriages for hire. The hoarse voice of a hackman would be sweeter music than Beethoven's "Moonlight Sonata" or Mendelssohn's "Wedding March." Horseback-riding is impracticable and

walking seems the only alternative. To wade and flounder twelve miles—Oil
City is that far off—is the dreary prospect that freezes the blood. Hark! In
strident tones a fierce-looking fellow is shouting: "Packet-boat for Oil City!
This way for the packet-boat! Packet-boat! Packet-boat!" Visions of a
pleasant jaunt in a snug cabin lure you to the landing. The "packet-boat"
proves to be an oily scow, without sail, engine, awning or chair, which horses
have drawn up the stream from Oil City. It will float back at the rate of three
miles an hour and the fare is three-fifty! The name and picture of "Pomeroy's
Express," the best of these nondescript Oil-Creek vessels, will bring a smile
and warm the cockles of many an old-timer's heart!

"POMEROY'S EXPRESS" BETWEEN SHAFFER AND OIL CITY.

Perhaps you decide to stay all night at Shaffer and start on foot early in
the morning. A chair in a room thick with tobacco-smoke, or a quilt in a
corner of the bar, is the best you can expect. By rare luck you may happen
to pre-empt a half-interest in a small bed, tucked with two or three more in a
closet-like apartment. Your room-mates talk of "flowing wells—five-hundred-
thousand dollars—third sand—big strike—rich in a week—thousand-dollars a
day," until you fall asleep to dream of wells spouting seas of mud and hapless
wights wading in greenbacks to their waists. Awaking cold and unrefreshed,
your brain fuddled and your thoughts confused, you gulp a breakfast of "ham
'n eggs 'n fried potatoes 'n coffee" and prepare to strike out boldly. Encased
in rubber-boots that reach above the thighs, you choose one of the two paths—
each worse than the other—pray for sustaining grace and begin the toilsome
journey. Having seen the tips of the elephant's ears, you mean to see the end
of his tail and be able to estimate the bulk of the animal. Night is closing in
as you round up at your destination, exhausted and mud-coated to the chin.
But you have traversed a region that has no duplicate "in heaven above, or
in the earth beneath, or in the waters under the earth," and feel recompensed a

thousand-fold for the fatigue and exposure. Were your years to exceed Thomas Parr's and Methuselah's combined, you will never again behold such a scene as the Oil-Creek valley presented in the days of "the middle passage' between Shaffer and Oil City. Rake it over with a fine-comb, turn on the X-rays, dig and scrape and root and to-day you couldn't find a particle of Shaffer as big as a toothpick ! When the railroad was extended the buildings were torn down and carted to the next station.

Widow Sanney's hundred-acre farm, south of the Shaffer, had three refin eries and a score of unremunerative wells. David Gregg's two-hundred acres on the west side of Oil Creek, followed suit with forty non-paying wells, three that yielded oil and the Victoria and Continental refineries. The McCoy well, the first put down below the Drake, at two-hundred feet averaged fifteen barrels a day from March until July, 1860. Fire burning the rig, the well was drilled to five-hundred feet and proved dry. R. P. Beatty sold his two-hundred acres on Oil Creek and Hemlock Run to the Clinton Oil-Company of New York, a bunch of medium wells repaying the investment. James Farrell, a teamster, for two-hundred dollars purchased a thirty-acre bit of rough land south of Beatty, on the east side of Oil Creek and Bull Run, the extreme south-west corner of Allegheny—now Oil Creek—township. In the spring of 1860 Orange Noble leased sixteen acres for six hundred dollars and one-quarter royalty. Jerking a "spring-pole" five months sank a hole one-hundred-and-thirty feet, without a symptom of greasiness, and the well was neglected nearly three years. The "third sand" having been found on the creek, the holders of the Farrell lease decided to drill the old hole deeper. George B. Delamater and L. L. Lamb were associated with Noble in the venture. They contracted with Samuel S. Fertig, of Titusville, whose energy and reliability had gained the good-will of operators, to drill about five-hundred feet. Fertig went to work in April of 1863, using a ten-horse boiler and engine and agreeing to take one-sixteenth of the working-interest as part payment. He had lots of the push that long since placed him in the van as a successful producer, enjoying a well-earned competence. Early in May, at four-hundred-and-fifty feet, a "crevice" of unusual size was encountered. Fearing to lose his tools, the contractor shut down for consultation with the well-owners. Noble was at Pittsburg on a hunt for tubing, which he ordered from Philadelphia. The well stood idle two weeks, waiting for the tubing, surface-water vainly trying to fill the hole.

On the afternoon of May twenty-seventh, 1863, everything was ready. "Start her slowly," Noble shouted from the derrick to Fertig, who stood beside the engine and turned on the steam. The rods moved up and down with steady stroke, bringing a stream of fresh water, which it was hoped a day's pumping might exhaust. Then it

SAMUEL S. FERTIG.

would be known whether two of the owners—Noble and Delamater—had acted wisely on May fifteenth in rejecting one-hundred-thousand dollars for one-half of the well. Noble went to an eating-house near by for a lunch. He was munching a sandwich when a boy at the door bawled : "Golly ! Ain't that well spittin' oil?" Turning around, he saw a column of oil and water rising a hundred feet, enveloping the trees and the derrick in dense spray! The gas roared, the ground fairly shook and the workmen hastened to extinguish the fire beneath the boiler. The "Noble well," destined to be the most profitable ever known, had begun its dazzling career at the dizzy figure of three-thousand barrels a day!

Crude was four dollars a barrel, rose to six, to ten, to thirteen! Compute the receipts from the Noble well at these quotations—twelve-thousand, eighteen-thousand, thirty-thousand, thirty-nine-thousand dollars a day! Sinbad's fabled Valley of Diamonds was a ten-cent side-show in comparison with the actual realities of the valley of Oil Creek.

Soon the foaming volume filled the hollow close to the well and ran into the creek. What was to be done? In the forcible jargon of a driller: "The divil wuz to pay an' no pitch hot!" For two-hundred dollars three men crawled through the blinding shower and contrived to attach a stop-cock device to the pipe. By sunset a seven-hundred-barrel tank was overflowing. Boatmen down the creek, notified to come at once for all they wanted at two dollars a barrel, by midnight took the oil directly from the well. Next morning the stream was turned into a three-thousand-barrel tank, filling it in twenty-one hours! Sixty-two-thousand barrels were shipped and fifteen-thousand tanked, exclusive of leakage and waste, in thirty days. Week after week the flow continued, declining to six-hundred barrels a day in eighteen months. The superintendent of the Noble & Delamater Oil Company—organized in 1864 with a million capital—in February of 1865 recommended pulling out the tubing and cleaning the well. Learning of this intention, Noble and Delamater unloaded their stock at or above par. The tubing was drawn, the well pumped fifteen barrels in two days, came to a full stop and was abandoned as a dry-hole!

The production of this marvelous gusher—over seven-hundred-thousand barrels—netted upwards of four-million dollars! One-fourth of this lordly sum went to the children of James Farrell—he did not live to see his land developed—James, John, Nelson and their sister, now Mrs. William B. Sterrett, of Titusville. Noble and Delamater owned one-half the working-interest, less the sixteenth assigned to S. S. Fertig, who bought another sixteenth from John Farrell while drilling the well and sold both to William H. Abbott for twenty-seven-thousand dollars. Ten persons—L. L. Lamb, Solomon and W. H. Noble, Rev. L. Reed, James and L. H. Hall, Charles and Thomas Delamater, G. T. Churchhill and Rollin Thompson—held almost one-quarter. Even this fractional claim gave each a splendid income. The total outlay for the lease and well—not quite four-thousand dollars—was repaid one-thousand times in twenty months! Is it surprising that men plunged into speculations which completely eclipsed the South-Sea Bubble and Law's Mississippi-Scheme? Is it any wonder that multitudes were eager to stake their last dollar, their health, their lives, their very souls on the chance of such winnings?

Thirteen wells were drilled on the Farrell strip. The Craft had yielded a hundred-thousand barrels and was doing two-hundred a day when the seed-bag burst, flooding the well with water and driving the oil away. The Mulligan and the Commercial did their share towards making the territory the finest property in Oildom, with third sand on the flats and in the ravine of Bull Run forty feet thick. Not a fragment of tanks or derricks is left to indicate that twenty fortunes were acquired on the desolate spot, once the scene of tremendous activity, more coveted than Naboth's vineyard or Jason's Golden Fleece. On the Caldwell farm of two-hundred acres, south of the Farrell, twenty-five or thirty wells yielded largely. The Caldwell, finished in March of 1863, at the north-west corner of the tract, flowed twelve-hundred barrels a day for six weeks. Evidently deriving its supply from the same pool, the Noble well cut this down to four-hundred barrels. A demand for one-fourth the output of the Noble, enforced by a threat to pull the tubing and destroy the two, was settled by paying one-hundred-and-forty-five-thousand dollars for the Caldwell well and an acre of ground. "Growing smaller by degrees and beautifully less," within a month of the transfer the Caldwell quit forever, drained as dry as the bones in Ezekiel's vision!

Hon. Orange Noble, the son of a New-York farmer, dealt in sheep and cattle, married in 1841 and in 1852 removed to Randolph, Crawford county, Pa. He farmed, manufactured "shooks" and in 1855 opened a store at Townville in partnership with George B. Delamater. The partners and L. L. Lamb inspected the Drake well in October of 1859, secured leases on the Stackpole and Jones farms and drilled two dry-holes. Other wells on different farms in 1860-1 resulted similarly, but the Noble compensated richly for these failures. The firm wound up the establishment at Townville in 1863, squared petroleum-accounts, and in 1864 Mr. Noble located at Erie. There he organized banks, erected massive blocks, served as mayor three terms, built the first grain elevator and contributed greatly to the prosperity of the city. Blessed with ample wealth—the Noble well paid him eight-hundred-thousand dollars—a vigorous constitution and the regard of his fellows, he has lived to a ripe age to enjoy the fruits of his patient industry and remarkable success.

Hon. George B. Delamater, whose parents settled in Crawford county in 1822, studied law and was admitted to the Meadville bar in 1847. He published a newspaper at Youngsville, Warren county, two years and in 1852 started in business at Townville. Clients were not plentiful in the quiet village, where a lawsuit was a luxury, and the young attorney found boring juries much less remunerative than he afterwards found boring oil-wells. Returning to Meadville in 1864, with seven-hundred-thousand dollars and some real-estate at his command, he built the magnificent Delamater Block, opened a bank, promoted many important enterprises and engaged actively in politics. Selected to oppose George K. Anderson—he, too, had a bar'l—for the State Senate in 1869, Delamater carried off the prize. It was a case of Greek meeting Greek. Money flowed like water, Anderson spending thirty-thousand dollars and his opponent twenty-eight-thousand on the primaries alone! This was the beginning of the depletion of the Delamater fortune and the political demoralization that scandalized Crawford county for years. Mr. Delamater

GEORGE W. DELAMATER.

served one term, declined to run again and Anderson succeeded him. His son, George W., a young lawyer of ability and superior address, entered the lists and was elected Mayor of Meadville and State Senator. He married an accomplished lady, occupied a brick-mansion, operated at Petrolia, practiced law and assisted in running the bank. Samuel B. Dick headed a faction that opposed the Delamaters bitterly. Nominated for Governor of Pennsylvania in 1890, George W. Delamater was defeated by Robert E. Pattison. He conducted an aggressive campaign, visiting every section of the state and winning friends by his frank courtesy and manly bearing. Ruined by politics, unable longer to stand the drain that had been sapping its resources, the Delamater Bank suspended two weeks after the gubernatorial election. The brick-block, the homes of the parents and the sons, the assets of the concern—mere drops in the bucket—met a trifling percentage of the liabilities. Property was sacrificed, suits were entered and dismissed, savings of depositors were swept away and the failure entailed a host of serious losses. The senior Delamater went to Ohio to start life anew at seventy-one, George W. journeyed to Seattle and quickly gathered a law-practice. That he will regain wealth and honor, pay off every creditor and some day represent his district in Congress those who know him best are not unwilling to believe. The fall of the Delamater family—the beggary of the aged father—the crushing of the son's honorable ambition—the exile from home and friends—the suffering of innocent victims—all these illustrate the sad reverses which, in the oil-region, have "come, not single spies, but in battalions."

James Bonner, son of an Ohio clergyman and book-keeper for Noble & Delamater, lodged in the firm's new office beside the well. Seized with typhoid fever, his recovery was hopeless. The office caught fire, young Bonner's father carried him to the window, a board was placed to slide him down and he expired in a few moments. His father, overcome by smoke, was rescued with difficulty ; his mother escaped by jumping from the second story.

James Foster owned sixty acres on the west side of Oil Creek, opposite the Farrell and Caldwell tracts. The upper half, extending over the hill to Pioneer Run, he sold to the Irwin Petroleum Company of Philadelphia, whose Irwin well pumped two-hundred barrels a day. The Porter well, finished in May of 1864, flowed all summer, gradually declining from two-hundred barrels to seventy and finally pumping twenty. Other wells and a refinery paid good dividends. J. W. Sherman, of Cleveland, leased the lower end of the farm and bounced the "spring-pole" in the winter of 1861-2. His wife's money and his own played out before the second sand was penetrated. It was impossible to drill deeper "by hand-power." A horse or an engine must be had to work the tools. "Pete," a white, angular equine, was procured for one-sixteenth interest in the well. The task becoming too heavy for "Pete," another sixteenth was traded to William Avery and J. E. Steele for a small engine and boiler. Lack of means to buy coal—an expensive article, sold only for "spot cash"—caused a week's delay. The owners of the well could not muster "long green" to pay for one ton of fuel! For another sixteenth a purchaser grudgingly surrendered eighty dollars and a shot-gun! The last dollar had been expended when, on March sixteenth, 1862—just in season to celebrate St. Patrick's day—the tools punctured the third sand. A "crevice" was hit, the tools were drawn out and in five minutes everything swam in oil. The Sherman well was flowing two-thousand barrels a day! Borrowing the phrase of the parrot stripped of his feathers and blown five-hundred feet by a powder-explosion, people might well exclaim : "This beats the Old Scratch !"

To provide tankage was the first concern. Teams were dispatched for lumber and carpenters hurried to the scene. Near the well a mudhole, between two stumps, could not be avoided. In this one of the wagons stuck fast and had to be pried out. John A. Mather chanced to come along with his photographic apparatus. The men posed an instant, the horses "looked pleasant," the wagon didn't stir and he secured the artistic picture reproduced here thirty-four years after. It is an interesting souvenir of former times—times that deserve the best work of pen and pencil, camera and brush, "to hold them in everlasting remembrance."

STUCK IN A MUDHOLE NEAR THE SHERMAN WELL IN 1862.

The Sherman well "whooped it up" bravely, averaging nine-hundred barrels daily for two years and ceasing to spout in February of 1864. Pumping restored it to seventy-five barrels, which dwindled to six or eight in 1867, when fire consumed the rig and the veteran was abandoned. The product sold at prices ranging from fifty cents to thirteen dollars a barrel, the total aggregating seventeen-hundred-thousand dollars! How was that for a return? It meant one-hundred-thousand dollars for the man who traded "Pete," one-hundred-thousand for the man who invested eighty dollars and a rusty gun, one-hundred-thousand for the two men who furnished the second-hand engine, and a million —deducting the royalty—for the man who had neither cash nor credit for a load of coal!

None of the other fifty or sixty wells on the Foster farm, some of them Sherman's, was particularly noteworthy. The broad flat, the sluggish stream and the bluffs across the creek remain as in days of yore, but the wells, the shanties, the tanks, the machinery and the workmen have vanished. Sherman, hale and hearty to-day, struck a spouter in Kentucky, operated two or three years at Bradford and took up his abode at Warren. It is a treat to hear his vivid descriptions of life on Oil Creek in the infancy of developments—life crowded with transformations far surpassing the fantastic changes of "A Midsummer Night's Dream."

Among the teamsters who hauled oil from the Sherman well in its prime was "Con" O'Donnell, a fun-loving, impulsive Irishman. He saved his earnings, secured leases for himself, owned a bevy of wells at Kane City and operated in the Clarion field. Marrying a young lady of Ellicottville, N. Y., his early home, he lived some years at Foxburg and St. Petersburg. He was the rarest of practical jokers and universally esteemed. Softening of the brain afflicted him for years, death at last stilling as warm and kindly a heart as ever throbbed in a manly breast. "Con" often regaled me with his droll witticisms as we rode or drove through the Clarion district. "Peace to his ashes."

Late in the fall of 1859, "when th' frost wuz on th' punkin' an' th' bloom wuz on th' rye," David McElhenny sold the upper and lower McElhenny farms—one-hundred-and-eighty acres at the south-east corner of Cherrytree township—to Captain A. B. Funk, for fifteen-hundred dollars and one-fourth of the oil. Joining the Foster farm on the north, Oil Creek bounded the upper tract on the east and south and Pioneer Run gurgled through the western side. Oil Creek flowed through the northern and western sides of the lower half, which had the Espy farm on the east, the Boyd south and the Benninghoff north and west. McElhenny's faith in petroleum was of the mustard-seed order and he jumped at Hussey & McBride's offer of twenty-thousand dollars for the royalty. Captain Funk—he obtained the title from running steamboats lumbering on the Youghiogheny river—in February of 1860 commenced the first well on the lower McElhenny farm. All spring and summer the "spring-pole" bobbed serenely, punching the hole two-hundred-and-sixty feet, with no suspicion of oil in the first and second sands. The Captain, believing it a rank failure, would gladly have exchanged the hole "for a yellow dog." His son, A. P. Funk, bought a small locomotive-boiler and an engine and resumed work during the winter. Early in May, 1861, at four-hundred feet, a "pebble rock"—the "third sand"—tested the temper of the center-bit. Hope, the stuff that "springs eternal in the human breast," took a fresh hold. It languished as the tools bored thirty, forty, fifty feet into the "pebble" and not a drop of oil appeared. Then something happened. Flecks of foam bubbled to the top of the conductor, jets of water rushed out, oil and water succeeded and a huge pillar of pure oil soared fifty yards! The Fountain well had tapped a fountain in the rock ordained thenceforth to furnish mankind with Pennsylvania petroleum. The *first well put down to* "the third sand," and really *the first on Oil Creek that flowed* from any sand, it revealed oil-possibilities before unknown and unsuspected.

More tangible than the mythical Fountain of Youth, the Fountain well tallied three-hundred barrels a day for fifteen months. The flow ended as suddenly as it began. Paraffine clogged and strangled it to death, sealing the pores and pipes effectually. A young man "taught the young idea how to shoot" at Steam Mills, east of Titusville, where Captain Funk had lumber-mills. A visit to the Drake and Barnsdall wells, in December of 1859, determined the schoolmaster to have an oil-well of his own. Funk liked the earnest, manly youth and leased him five acres of the upper McElhenny farm. Plenty of brains, a brave heart, robust health, willing hands and thirty dollars constituted his capital. Securing two partners, "kicking down" started in the spring of 1860. Not a sign of oil could be detected at two-hundred feet, and the partners departed from the field. Summer and the teacher's humble savings were gone. He earned more money by drilling on the Allegheny river, four miles above Oil City. While thus engaged the Fountain well revolutionized the business by "flowing" from a lower rock. The ex-wielder of the birch—he had resigned the ferrule for the "spring-pole"—hastened to sink the deserted well to the depth of Funk's eye-opener. The *second* three-hundred-barrel gusher from the third sand, it rivaled the Fountain and arrived in time to help 1861 crimson the glorious Fourth !

Hon. John Fertig, of Titusville, the plucky schoolmaster of 1859–60, has been largely identified with oil ever since his initiation on the McElhenny lease.

JOHN FERTIG.   CAPT. A. B. FUNK.

The Fertig well, in which David Beatty and Michael Gorman were his partners originally, realized him a fortune. Born in Venango county, on a farm below Gas City, in 1837, he completed a course at Neilltown Academy and taught school several terms. Soon after embarking in the production of oil he formed a partnership with the late John W. Hammond, which lasted until dissolved by death twenty years later. Fertig & Hammond operated in different sections with great success, carried on a refinery and established a bank at Foxburg. Mr. Fertig was Mayor of Titusville three terms, School-Controller, State Senator and Democratic candidate for Lieutenant-Governor in 1878. He has been vice-president of the Commercial Bank from its organization in 1882 and is president of the Titusville Iron-Works. Head of the National Oil-Company, he was also chief officer of the Union Oil-Company, an association of refining companies. For three years its treasurer—1892–5—he tided the United States Pipe-Line Company over a financial crisis in 1893. As a pioneer producer—one of the few survivors connected with developments for a generation—a refiner and shipper, banker, manufacturer and business-man, John Fertig is most distinctively a representative of the oil-country. From first to last he has been admirably prudent and aggressive, conservative and enterprising in shaping a career with much to cherish and little to regret.

Frederick Crocker drilled a notable well on the McElhenny, near the Foster line, jigging the "spring-pole" in 1861 and piercing the sand at one-hundred-and-fifty feet. He pumped the well incessantly two months, getting clear water for his pains. Neighbors jeered, asked if he proposed to empty

9

the interior of the planet into the creek and advised him to import a Baptist colony. Crocker pegged away, remembering that "he laughs best who laughs last." One morning the water wore a tinge of green. The color deepened, the gas "cut loose," and a stream of oil shot upwards! The Crocker well spurted for weeks at a thousand-barrel clip and was sold for sixty-five-thousand dollars. Shutting in the flow, to prevent waste, wrought serious injury. The well disliked the treatment, the gas sought a vent elsewhere, pumping coaxed back the yield temporarily to fifty barrels and in the fall it yielded up the ghost.

Bennett & Hatch spent the summer of 1861 drilling on a lease adjoining the Fountain, striking the third sand at the same depth. On September eighteenth the well burst forth with thirty-three-hundred barrels per day! This was "confusion worse confounded," foreigners not wanting "the nasty stuff" and Americans not yet aware of its real value. The addition of three-thousand barrels a day to the supply—with big additions from other wells—knocked prices to twenty cents, to fifteen, to ten! All the coopers in Oildom could not make barrels as fast as the Empire well—appropriate name—could fill them. Bradley & Son, of Cleveland, bought a month's output for five-hundred dollars, loading one-hundred-thousand barrels into boats under their contract! The despairing owners, suffering from "an embarrassment of riches," tried to cork up the pesky thing, but the well was like Xantippe, the scolding wife of Socrates, and would not be choked off. They built a dam around it, but the oil wouldn't be dammed that way. It just gorged the pond, ran over the embankment and greased Oil Creek as no stream was ever greased before! Twenty-two-hundred barrels was the daily average in November and twelve-hundred in March. The torrent played April-fool by stopping without notice, seven months from its inception. Cleaning out and pumping restored it to six-hundred barrels, which dropped two-thirds and stopped again in 1863. An "air blower" revived it briefly, but its vitality had fled and in another year the grand Empire breathed its last.

These wells boomed the territory immensely. Derricks and engine-houses studded the McElhenny farms, which operators hustled to perforate as full of holes as a strainer. To haul machinery from the nearest railroad doubled its cost. Pumping five to twenty barrels a day, when adjacent wells flowed more hundreds spontaneously, lost its charm and most of the small fry were abandoned. Everybody wanted to get close to the third-sand spouters, although the market was glutted and crude ruinously cheap. A town—Funkville - arose on the northern end of the upper farm, sputtered a year or two, then "folded its tent like the Arabs and silently stole away." A search with a microscope would fail to unearth an atom of Funkville or the wells that created it. Fresh strikes in 1862 kept the fever raging. Davis & Wheelock's rattler daily poured out fifteen-hundred barrels. The Densmore triplets, bunched on a two-acre lease, were good for six-hundred, four-hundred and five-hundred respectively. The Olmstead, American, Canfield, Aikens, Burtis and two Hibbard wells, of the vintage of 1863, rated from two-hundred to five-hundred each. A band of less account—thirty to one-hundred barrels—assisted in holding the daily product of the McElhenny farms, from the spring of 1862 to the end of 1863, considerably above six-thousand barrels. The mockery of fate was accentuated by a dry-hole six rods from the Sherman and dozens of poor wells in the bosom of the big fellows. Disposing of his timber-lands and saw-mills in 1863, Captain Funk built a mansion and removed to Titusville. Early in 1864 he sold his wells and oil-properties and died on August second, leaving an estate of two-millions. He built schools and churches, dispensed freely to the needy and was honest to the core. Pleased with the work of a clerk, he deeded him an interest in the last well he ever drilled, which the lucky young man sold for one-hundred-thousand dollars.

Almost simultaneously with the Empire, in September of 1861, the Buckeye well, on the George P. Espy farm, east of lower McElhenney, set off at a thousand-barrel jog. It was located on a strip of level ground too narrow for tanks, which had to be erected two-hundred feet up the hill. The pressure of gas sufficed to force the oil into these tanks for a year. The production fell to eighty barrels and then, tiring of a climbing job that smacked of Sisyphus and the rolling stone, took a permanent rest. From this famous well J. T. Briggs, manager of the Briggs and the Gillettee Oil-Companies, shipped to Europe in 1862 the first American petroleum ever sent across the Atlantic. The Buckeye Belle stood about hip-high to its consort, a dozen other wells on the Espy produced mildly and Northrop Brothers operated a refinery.

Improved methods of handling and new uses for the product advanced crude to five dollars in the spring of 1864. Operations encroached upon the higher lands, exploding the notion that paying territory was confined to flats bordering the streams. Pioneer Run, an affluent of Oil Creek, bisecting the

PIONEER AS IT LOOKED IN 1864-5.

western end of the upper McElhenney and Foster farms, panned out flatteringly. Substantial wells, yielding fifteen barrels to three-hundred, lined the ravine thickly. The town of Pioneer attracted the usual throngs. David Emery, Lewis F. Emery, Frank W. Andrews and not a few leading operators resided there for a time. The Morgan House, a rude frame of one story, dished up meals at which to eat beef-hash was to beefashionable. Clark & McGowen had a feed-store, offices and warehouses abounded, tanks and derricks mixed in the mass and boats loaded oil for refineries down the creek or the Allegheny river. The characteristic oil-town has faded from sight, only the weather-beaten railroad-station and a forlorn iron tank staying. John Rhodes, the last resident, was killed in February of 1892 by a train. He lived alone in a small house beside the track, which he was crossing when the engine hit him, the noisy waters in the culvert drowning the sound of the cars. Rhodes hauled oil in the old days to Erie and Titusville, became a producer, met with reverses, attended to some wells for a company, worked a bit of garden and felt independent and happy.

Matthew Taylor, a Cleveland saloonist, whom the sequel showed to be no saloonatic, took a four-hundred-dollar flyer at Pioneer, on his first visit to Oildom. A well on the next lease elevated values and Taylor returned home in two weeks with twenty-thousand dollars, which subsequent deals quadrupled. A Titusville laborer—"a broth of a b'y wan year frum Oireland"—who stuck fifty dollars into an out-of-the-way Pioneer lot, sold his claim in a month for five-thousand. He bought a farm, sent across the water for his colleen and "they lived happily ever after." The driver of a contractor's team, assigned an interest in a drilling-well for his wages, cleaned up thirty, thousand dollars by the transaction and went to Minnesota. Could the mellowest melodrama unfold sweeter melodies?

North and west of the lower McElhenny farm, at the bend in Oil Creek, lay John Benninghoff's two big blocks of land, through which Benninghoff Run flowed southward. Pioneer Run crossed the north-east corner of the property, the greater part of which was on the hills. Five acres on Oil Creek and the slopes on Pioneer Run were first developed. Leases for a cash-bonus and liberal royalty were gobbled greedily. Up Benninghoff Run and back of the hills operations spread. For one piece of ground the owner declined tempting offers, because he would not permit his potato-patch to be trodden down! Some wells pumped and some flowed from twenty-five to three-hundred barrels a day seven days in the week. William Jenkins, the Huidekoper Oil-Company, the DeKalb Oil-Company and Edward Harkins had regular bonanzas. The Lady Herman, which Robert Herman had the politeness to name for his wife, was a genuine beauty. The first well ever cased and the first pump-station—it hoisted oil to Shaffer—were on the hillside at the mouth of Benninghoff Run. The platoon of wells in the illustration of that locality, as they appeared in 1866, includes these and a hint of the barn beside the homestead. The busy scene—pictured now for the first time—was photographed within an hour of its obliteration. The artist had not finished packing his outfit when lightning struck one of the derricks and a disastrous fire swept the hill as bare as Old Mother Hubbard's cupboard! Wealth deluged the thrifty land-holder, oil converting his broad acres into a veritable Golconda. He awoke one morning to find himself rich. He was awakened one night to find himself famous, the newspapers devoting whole pages—under "scare-heads"—to the unpretending farmer in the southern end of Cherrytree. "And thereby hangs a tale."

Suspicious of banks, Benninghoff stored his money at home. Purchasing a cheap safe, he placed it in a corner of the sitting-room and stocked it with a half-million dollars in gold and greenbacks! Cautious friends warned him to be careful, lest thieves might "break through and steal." James Saeger, of Saegertown, a handsome, popular young fellow, who sometimes played cards, heard of the treasure in the flimsy receptacle. "Jim" belonged to a respectable family and had been a merchant at Meadville. Napoleon melted silver statues of the apostles to put the precious metal in circulation and Saeger concluded to give Benninghoff's pile an airing. He spoke to George Miller of the ease with which the safe could be cracked and engaged two Baltimore burglars, McDonald and Elliott, to manage the job. Jacob Shoppert, of Saegertown, and Henry Geiger, who worked for Benninghoff and slept in the house, were enlisted. The deed, planned with extreme care not to miss fire, was fixed for a night when Joseph Benninghoff, the son, was to attend a dance.

On Thursday evening, January sixteenth, 1868, Saeger, Shoppert, McDonald and Elliott left Saegertown in a two-horse sleigh for Petroleum Centre, twenty-nine miles distant. At midnight they knocked at Benninghoff's door. Geiger answered the rap and was quickly gagged, said to be as arranged previously. John Benninghoff, his wife and daughter were bound and the experts proceeded to open the safe. The frail structure was soon ransacked. The marauders bundled up their booty, sampled Mrs. Benninghoff's pies, drank a gallon of milk and departed at their leisure, leaving the inmates of the house securely tied. Joseph returned in an hour or two and relieved the prisoners from their unpleasant predicament. An examination of the safe showed that two-hundred-and-sixty-five-thousand dollars had been taken! The bulk of this was in gold. A package of two-hundred-thousand dollars, in large bills, done up in a brown paper, the looters passed unnoticed! The alarm was given, the wires flashed the news everywhere and the press teemed with sensational reports. By noon on Friday the oil-regions had been set agog and people all over the United States were talking of "the Great Benninghoff Robbery."

Saegar and his pals drove back and stopped at Louis Warlde's hotel to divide the spoils. McDonald, Elliott and Saeger took the lion's share, Geiger and Shoppert received smaller sums and Warlde accepted thirteen-hundred dollars for his silence. The Baltimore toughs lingered in the neighborhood a week and then sought the wintry climate of Canada, Saeger staying around home. Intense excitement prevailed. Hundreds of detectives, eager to gain reputation and the reward of ten-thousand dollars, spun theories and looked wise. Ex-Chief-of-Police Hague, of Pittsburg, was especially alert. For three months the search was vain. George Miller, whom McDonald wished to put out of the road "to keep his mouth shut," in a quarrel with Saeger over a game of cards, blurted out: "I know about the Benninghoff robbery!" Saeger pacified Miller with a thousand dollars, which the latter scattered quickly. Jacob Shoppert was his boon companion and the pair spent money at a rate that caused officers to shadow them. Shoppert visited a town on the edge of Ohio and was arrested. Calling for a pen and paper, he wrote to Louis Warlde, the Saegertown hotel-keeper, reproaching him for not sending money. The jailer handed the detectives the letter, on the strength of which Warlde, who had started a brewery in Ohio, and Miller were arrested. The three were convicted and sentenced to a short term in the penitentiary. Geiger's complicity in the plot could not be proved beyond a doubt and he was acquitted. Officer Hague captured McDonald and Elliott in Toronto, but Canadian lawyers picked flaws in the papers and they could not be extradited. Escaping to Europe, they were heard of no more. Saeger, who had not been suspected until after his departure, went west and was lost sight of for many a day.

Three years later a noted cattle-king of the Texas-Colorado trail entered a saloon in Denver to treat a party of friends. The bar-tender, Gus. Peiflee, formerly of Meadville, recognized the customer as "Jim." Saeger. He telegraphed east and Chief-of-Police Rouse, of Titusville, posted off to Denver with Joseph Benninghoff. They secured extradition-papers and arrested Saeger, who coolly remarked; "You'll be a devilish sight older before you see me in Pennsylvania." Their lawyers informed them that a hundred of Saeger's cow-boys were in the city — reckless, lawless fellows, certain to kill whoever attempted to take him away. Rouse and Benninghoff dropped the matter and returned alone. Saeger is living in Texas, prosperous and respected. He is just in his dealings, a bountiful giver, and not long ago sent five-thousand dollars to the widow of George Miller. Perhaps he may yet turn up in Washington as Congressman or United-States Senator. This is the story of a robbery that attracted more attention than the first woman in bloomers.

John Benninghoff was born in Lehigh county, where his ancesters were among the first German immigrants, on Christmas Day, 1801. His father, Frederick Benninghoff, settled near New Berlin, Union county, in John's boyhood. There the son married Elizabeth Heise in 1825 and in 1828 located on a farm near Oldtown, Clearfield county. Thence he removed to Venango county, living close to Cherrytree village four years. In 1836 he bought a piece of land on the south border of Cherrytree township, near what was to become Petroleum Centre. He added to his purchase as his means permitted, until he owned about three-hundred acres, with solid buildings and modern improvements. He was in easy circumstances prior to the oil-developments that enriched him. Contrary to the general opinion, the robbery did not impoverish him, as one-half the money was untouched. His twelve children—eight boys and four girls—grew up and eight are still living. Selling his farms in Venango, he removed to Greenville, Mercer county, in the spring of 1868 and died in March, 1882. At his death he had sixty-one grandchildren and fifteen great-grandchildren. He left his family a large estate. The Benninghoff farms, so far as oil is concerned, are utterly deserted.

West and north of Benninghoff were the farms of John and R. Stevenson. On the former, extending south to Oil Creek, Reuben Painter drilled a well in 1863. The contractor reporting it dry, Painter moved the machinery and surrendered the lease. He and his brothers operated profitably in Butler and McKean counties, Reuben dying at Olean in 1892. In November of 1864 the Ocean Oil-Company of Philadelphia bought John Stevenson's lands. The Ocean well began flowing at a six-hundred-barrel pace on September first, 1865, with the Arctic a good second. Fifty others varied from fifty to two-hundred barrels. Thomas McCool built a refinery and the farm paid the company about two-thousand per cent! The principal wells on both Stevenson tracts clustered far above the flats, the derricks and buildings resembling "a city set on a hill." Major Mills, justly proud of his King of the Hills, an elegant producer, delighted to visit it with his wife and two young daughters, one of them now Mrs. John D. Archbold, of New York. Painter's supposed dry-hole, drilled seventeen feet deeper, gushed furiously, proving to be the best well in the collection! Said the Ocean manager, as he watched the oily stream ascend "higher 'n a steeple": "A million dollars wouldn't touch one side of this property!" Sinking a four-inch hole seventeen feet farther would have given Reuben Painter this splendid return two years earlier! He missed a million dollars by only seventeen feet! A Gettysburg soldier, from whose nose a rifle-ball shaved a piece of cuticle the size of a pin-head, wittily observed: "That shot came mighty near missing me!" Inverting this remark, Painter had cause to exclaim: "That million came mighty near hitting me!"

Although surrounded by farms unrivaled as oil-territory and sold to Woods & Wright of New York at a fancy price, James Boyd's seventy-five acres in Cornplanter township, south of the lower McElhenny, dodged the petroleum-artery. The sands were there, but so barren of oil that nine-tenths of the forty wells did not pay one-tenth their cost. The Boyd farm was for months the terminus of the railroad from Corry. Hotels and refineries were built and the place had a short existence, a brief interval separating its lying-in and its laying-out.

G. W. McClintock, in February of 1864, sold his two-hundred-acre farm, on the west side of Oil Creek, midway between Titusville and Oil City, to the Central Petroleum Company of New York, organized by Frederic Prentice and George H. Bissell. This notable farm embraced the site of Petroleum Centre and Wild-Cat Hollow, a circular ravine three-fourths of a mile long, in which two-hundred paying wells were drilled. Brown, Catlin & Co.'s medium well, finished in August of 1861, was the first on the McClintock tract. The company bored a multitude of wells and granted leases only to actual operators, for one-half royalty and a large bonus. For ten one-acre leases one-hundred-thousand dollars cash and one-half the oil, offered by a New York firm in 1865, were refused. The McClintock well, drilled in 1862, figured in the thousand-barrel class. The Coldwater, Meyer, Clark, Anderson, Fox, Swamp-Angel and Bluff wells made splendid records. Altogether the Central Petroleum Company and the corps of lessees harvested at least five-millions of dollars from the McClintock farm!

Aladdin's lamp was a miserly glim in the light of fortunes accruing from petroleum. The product of a flowing well in a year would buy a tract of gold-territory in California or Australia larger than the oil-producing regions. Millions of dollars changed hands every week. The Central Company staked off a half dozen streets and leased building-lots at exorbitant figures. Board-dwellings, offices, hotels, saloons and wells mingled promiscuously. It mattered nothing that discomfort was the rule. Poor fare, worse beds and the worst liquors were tolerated by the hordes of people who flocked to the land of derricks. Edward Fox, a railroad contractor who "struck the town" with eighty-thousand dollars, felicitously baptized the infant Petroleum Centre. The owners of the ground opposed a borough organization and the town traveled at a headlong go-as-you-please. Sharpers and prostitutes flourished, with no fear of human or divine law, in the metropolis of rum and debauchery. Dance-houses, beside which "Billy" McGlory's Armory-Hall and "The." Allen's Mabille in New York were Sunday-school models, nightly counted their revelers by hundreds. In one of these dens Gus Reil, the proprietor, killed poor young Tait, of Rouseville. Fast women and faster men caroused and gambled, cursed

and smoked, "burning the candle at both ends" in pursuit of—pleasure!
Frequently the orgies eclipsed Monte Carlo—minus some of the glitter—and
the Latin Quartier combined. Some readers may recall the night two "dead
game sports" tossed dice twelve hours for one-thousand dollars a throw! But
there was a rich leaven of first-class fellows. Kindred spirits, like "Sam"
Woods, Frank Ripley, Edward Fox and Col. Brady were not hard to discover.
Spades were trumps long years ago for Woods, who has taken his last trick

and sleeps in an Ohio grave. Ripley is in Duluth, Fox is "out west" and
Brady is in Harrisburg. Captain Ray and A. D. Cotton had a bank that
handled barrels of money. For two or three years "The Centre"—called that
for convenient brevity—acted as a sort of safety-valve to blow off the surplus
wickedness of the oil-regions. Then "the handwriting on the wall" manifested itself. Clarion and Butler speedily reduced the four-thousand population

to a mere remnant. The local paper died, houses were removed and the giddy Centre became "a back number." The sounds of revelry were hushed, flickering lights no longer glared over painted harlots and the streets were deserted. Bissell's empty bank-building, three dwellings, the public school, two vacant churches and the drygoods box used as a railway-station—scarcely enough to cast a shadow—are the sole survivors in the ploughed field that was once bustling, blooming, surging, foaming Petroleum Centre!

Across the creek from Petroleum Centre, on the east side of the stream, was Alexander Davidson's farm of thirty-eight acres. A portion of this triangular "speck on the map" consisted of a mud-flat, a smaller portion of rising ground and the remainder set edgewise. Dr. A. G. Egbert, a young physician who had recently hung out his shingle at Cherrytree village, in 1860 negotiated for the farm. Davidson died and a hitch in the title delayed the deal. Finally Mrs. Davidson agreed to sign the deed for twenty-six-hundred dollars and one-twelfth the oil. Charles Hyde paid the doctor this amount in 1862 for one-half his purchase and it was termed the Hyde & Egbert farm. The Hollister well, drilled in 1861, the first on the land, flowed strongly. Owing to the dearness and scarcity of barrels, the oil was let run into the creek and the well was never tested. The lessees could not afford, as their contract demanded, to barrel the half due the land-owners, because crude was selling at twenty-five cents and barrels at three-fifty to four dollars! A company of Jerseyites, in the spring of 1863, drilled the Jersey well, on the south end of the property. The Jersey—it was a Jersey Lily—flowed three-hundred barrels a day for nine months, another well draining it early in 1864. The Maple Shade, which cast the majority into the shade by its performance, touched the right spot in the third sand on August fifth, 1863. Starting at one-thousand barrels, it averaged eight-hundred for ten months, dropped to fifty the second year and held on until 1869. Fire on March second, 1864, burned the rig and twenty-eight tanks of oil, but the well kept flowing just the same, netting the owners a clear profit of fifteen-hundred-thousand dollars! "Do you notice it?" A plump million-and-a-half from a corner of the "measly patch" poor Davidson offered in 1860 for one-thousand dollars! And the Maple Shade was only one of twenty-three flowing wells on the despised thirty-eight acres!

Companies and individuals tugged and strained to get even the smallest lease Hyde & Egbert would grant. The Keystone, Gettysburg, Kepler, Eagle, Benton, Olive Branch, Laurel Hill, Bird and Potts wells, not to mention a score of minor note, helped maintain a production that paid the holders of the royalty eight-thousand dollars a day in 1864–5! E. B. Grandin and William C. Hyde, partners of Charles Hyde in a store at Hydetown, A. C. Kepler and Titus Ridgway obtained a lease of one acre on the west side of the lot, north of the wells already down, subject to *three-quarters* royalty. A bit of romance attaches to the transaction. Kepler dreamed that an Indian menaced him with bow and arrow. A young lady, considered somewhat coquettish, handed him a rifle and he fired at the dusky foe. The redskin vamoosed and a stream of oil burst forth. Visiting his brother, who superintended the farm, he recognized the scene of his dream. The lease was secured, on the biggest royalty ever offered. Kepler chose the location and bored the Coquette well. The dream was a nightmare? Wait and see.

Drilling began in the spring of 1864 and the work went merrily on. Each partner would be entitled to one-sixteenth of the oil. Hyde & Ridgway sold their interest for ten-thousand dollars a few days before the tools reached the sand. This interest Dr. M. C. Egbert, brother of the original purchaser of the farm, next bought at a large advance. He had acquired one-sixth of the property in fee and wished to own the Coquette. Grandin and Kepler declined to sell. The well was finished and did not flow! Tubed and pumped a week, gas checked its working and the sucker-rods were pulled. Immediately the oil streamed high in the air! Twelve-hundred barrels a day was the gauge at first, settling to steady business for a year at eight-hundred. A double row of tanks lined the bank, connected by pipes to load boats in bulk. Oil was "on the jump" and the first cargo of ten-thousand barrels brought ninety-thousand dollars, representing ten days' production! Three months later Grandin and Kepler sold their one-eighth for one-hundred-and-forty-five thousand dollars, quitting the Coquette with eighty-thousand apiece in their pockets. Kepler was a dreamer whom Joseph might be proud to accept as a chum.

Dr. M. C. Egbert retained his share. Riches showered upon him. His interests in the land and wells yielded him thousands of dollars a day. Once his safe contained, by tight squeezing, eighteen-hundred-thousand dollars in currency and a pile of government bonds! He built a comfortable house and

lived on the farm. He and his family traveled over Europe, met shoals of titled folks and saw all the sights. In company with John Brown, subsequently manager of a big corporation at Bradford and now a resident of Chicago, he engaged in oil-shipments on an extensive scale. To control this branch of the trade, as the Standard Oil-Company has since done by combinations of capital, was too gigantic a task for the firm and failure resulted. The brainy, courageous doctor went to California, returned to Oildom and operated in McKean county. He has secured a foothold in the newer fields and lives in Pittsburg, frank and urbane as in the palmiest days of the Hyde & Egbert farm. If

DR. M. C. EGBERT.

Dame Fortune was strangely capricious on Oil Creek, the pluck of the men with whom "the fickle jade" played whirligig was surely admirable.

Probably no parcel of ground in America of equal size ever yielded a larger return, in proportion to the expenditure, than the Hyde & Egbert tract. Six weeks' production of the Coquette or Maple Shade would drill all the wells on the property. Charles Hyde and Dr. A. G. Egbert cleared at least three-million dollars, the latter selling one-twelfth of the Coquette alone for a quarter-million cash. Profits of others interested in the land and of the lessees trebled this alluring sum. The aggregate—eight to ten millions—in silver dollars would load a freight-train or build a column twenty miles high! Fused into a lump of gold, a dozen mules might well decline the task of drawing it a mile. Done up into a bundle of five-dollar bills, Hercules couldn't budge the bulky package. A "promoter" of the Mulberry-Sellers brand wanted an owner of the farm, when the wells were at their best, to launch the whole thing into a stock-company with five-millions capital. "Bah!" responded the gentleman, "five millions—did you say five-millions? Don't waste your breath talking until you can come around with twenty-five millions!"

A native of New-York, born in 1822, Charles Hyde was fifteen when the family settled on a farm two miles south of Titusville, now occupied by the Octave Oil-Company. At twenty he engaged with his father and two brothers, W. C. and E. B. Hyde, in merchandising, lumbering and the manufacture of salts from ashes. In 1846 he assumed charge of the lumber-mills John Titus sold the firm, originating the thrifty village of Hydetown, four miles above Titusville. The Hydes frequently procured oil from the "springs" on Oil Creek, selling it for medicine as early as 1840-1. From their Hydetown store Colonel Drake obtained some tools and supplies Titusville could not furnish. Samuel Grandin, of Tidioute, in the spring of 1860 induced Charles Hyde to buy a tenth-interest in the Tidioute and Warren Oil-Company for one-thousand dollars. The company's first well, of which he heard on his way to Pittsburg with a raft, laid the foundation of Hyde's great fortune in petroleum. He organized the Hydetown Oil-Company, which leased the McClintock farm, below Rouseville, from Jonathan Watson and drilled a two-hundred-barrel well in the summer of 1860. Mr. Hyde operated on the Clapp farm, south of McClintock, and at different points on Oil Creek and the Allegheny River. His gains from the Hyde & Egbert farm approximated two-millions. Starting the Second National Bank of Titusville in 1865, he has always been its president and chief stockholder. In 1869 he removed to Plainfield, New Jersey, cultivating four-hundred acres of suburban land and maintaining an elegant home. Plain of speech and manner, sternly honest and just, devoid of pride and pretense, Charles Hyde is a man of deeds and not of words, a fine type of the pioneer oil-producers to whom the world is indebted for the petroleum-development.

Dr Albert G. Egbert, born in Mercer county in 1828, belonged to a family of eminent physicians, his grandfather, father, two uncles, three brothers and one son practicing medicine. Predicating a future for oil upon the Drake well, his good judgment displayed itself promptly. Agreeing to purchase the Davison farm, which his modest income at Cherrytree would not enable him to pay for, his sale of a half-interest to Charles Hyde provided the money to meet the entire claim. After the wonderful success of that investment the doctor located at Franklin. He carried on oil-operations, farming and coal-mining and was always active in advancing the general welfare. Elected to Congress against immense odds, he served his district most capably, attending sedulously to his official duties and doing admirable work on committees. In public and private life he was enterprising and liberal, zealous for the right and a helpful citizen. True to his convictions and professions, he never turned his back to friend or foe. To the steady, masterful purpose of men like Dr. Egbert the oil-industry owes its rapid strides and commanding position as a commercial staple. His demise on March twenty-eighth, 1896, severs another of the links that bind the eventful past and the important present of petroleum. Early operators on Oil Creek are reduced to a handful of men whose heads are white with the snows no July sun can melt.

> " He has walk'd the way of nature ;
> The setting sun, and music at the close,
> As the last taste of sweets, is sweetest last ;
> More are men's ends mark'd than their lives before.
> Cowards die many times before their deaths,
> The valiant never taste of death but once."

The rich pickings around Petroleum Centre set many on the straight cinder-path to prosperity. The three Phillips brothers—Isaac, Thomas W. and Samuel

WELLS ON THE NIAGARA TRACT, CHERRYTREE RUN.

—came from Newcastle to coin money operating a farm south of the Espy. Prolific wells on the Niagara tract, Cherrytree Run, back of the Benninghoff farm, added to their wealth. They cut a wide swath in all the Pennsylvania fields. Two of the brothers have "ascended to the hill of frankincense and to the mountain of myrrh." The third, Thomas W., is a millionaire Congressman. During the heated debates on free-silver, in 1894, he scored the hit of the session by suggesting to convert each barrel of petroleum into legal-tender for a dollar and let it go at that. Crude was selling at sixty cents, which gave the Phillips proposition a point "sharper than a serpent's tooth" or a Demosthenean philippic. Dr. Egbert offered Isaac Phillips an interest in the Davison farm in 1862. The offer was not accepted instantly, Phillips saying he would "consider it a few days." Two weeks later he was ready to close the deal, but the plum had fallen into the lap of Charles Hyde and diverted prospective millions into another channel.

George K. Anderson figured conspicuously in this latitude, his receipts for two years exceeding five-thousand dollars a day! He built a sumptuous residence at Titusville, sought political preferment and served a term in the State Senate. Holding a vast block of Pacific-Railroad stock, he was the bosom friend of the directors and trusted lieutenant of William H. Kemble, the Philadelphia magnate whose "addition, division and silence" gave him notoriety. He bought thousands of acres of land, plunged deeply into stocks and insured his life for three-hundred-and-fifteen-thousand dollars, at that time the largest risk in the country. If he sneezed or coughed the agents of the insurance-companies grew nervous and summoned a posse of doctors to consult about the case. Outside speculations swamped him at last. The stately mansion, piles of bonds and scores of farms passed under the sheriff's hammer in 1880. Plucky and unconquerable, Anderson tried his hand in the Bradford field, operating on Harrisburg Run. The result was discouraging and he entered an insurance-office in New York. Four years ago he accepted a government-berth in New Mexico. Meeting him on Broadway the week before he left New York, his buoyant spirits seemed depressed. He spoke regretfully of his approaching departure, yet hoped it might turn out advantageously. He arrived at his post, sickened and died in a few days, "a stranger in a strange land." Relatives and loved ones were far away when he went down into the starless night of the grave. No gentle wife or child or valued friend were there to smooth the pillow of the dying man, to cool the fevered brow, to catch the last whisper, to close the glassy eyes and fold the rigid hands above the lifeless breast. The oil-regions abound with pathetic experiences, but none surpassing George K. Anderson's. Wealthy beyond the dreams of avarice, the courted politician, the confidant of presidents and statesmen, a social favorite in Washington and Harrisburg, the owner of a home beautiful as Claude Melnotte pictured to Pauline, he drained the cup of sorrow and misfortune. Reverses beset him, his riches took wings, bereavements bore heavily upon him, he was glad to secure a humble clerkship and death ended the sad scene in a distant territory. Does not human life contain more tears than smiles, more pain than pleasure, more cloud than sunshine in the checkered passage from the cradle to the tomb?

Frank W. Andrews, born in Vermont and reared in Ohio, taught school in Missouri, hunted for gold at Pike's Peak and landed on Oil Creek in the winter of 1863-4. Hauling oil nine months supplied funds to operate on Cherrytree Run. He drilled four dry holes. One on the McClintock farm and three more on Pithole Creek followed. This was not a flattering start, but Andrews had lots of sand and persistence. Emerging from the Pithole excitement with limited cash and unlimited machinery, he returned to Oil Creek and operated extensively. His first well at Pioneer flowed three-hundred barrels a day. Fifty others at Shamburg, on the Benninghoff farm and Cherrytree Run brought him hundreds of thousands of dollars. He was rated at three-millions in 1870. Keeping up with the tidal wave southward, he put down two-hundred wells in the Franklin, Clarion and Butler districts. Failures of banks and manufactories in which he had a large stake shattered his fortune. With the loss of money he did not lose his manliness and self-reliance. In the Bradford region he pressed forward vigorously. Again he "plucked the flower of success" and was fast recuperating when thrown from his horse and fatally injured. Upright, unassuming and refined, Mr. Andrews merited the confidence and esteem of

HYDE & EGBERT TRACT AND McCRAY FARM IN 1870.

his fellows. A farmer in Butler county, who granted him a lease which numbers had been refused, estimated him in homely phrase: "Frank Andrews iz th' nices' kind uv man an' hes a winnin' way 'ud jes' coax a settin' hen offen her nest!"

The bluff overlooking Petroleum Centre from the east formed the western side of the McCray farm. At its base, on the Hyde & Egbert plot, were several of the finest wells in Pennsylvania, the Coquette almost touching McCray's line. Dr. M. C. Egbert leased part of the slope and drilled three wells. Other parties drilled five and the eight behaved so handsomely that the owner of the land declined an offer, in 1865, of a half-million dollars for his eighty acres. A well on top of the hill, not deep enough to hit the sand and supposed to be dry, postponed further operations five years. His friends distanced Jeremiah in their lamentations that McCray had spurned the five-hundred-thousand dollars. He may have thought of Shakespeare's " tide in the affairs of men," but he sawed wood and said nothing. Jonathan Watson, advised by a clairvoyant, in the spring of 1870 drilled a three-hundred-barrel well on the uplands of the Dalzell farm, close to the southern boundary of the McCray. The clairvoyant's astonishing guess revived interest in Petroleum Centre, which for a year or two had been on the down grade. Besieged for leases, McCray could not meet a tithe of the demand at one-thousand dollars an acre and half the oil. Derricks clustered thickly. Every well tapped the pool underlying fifteen acres, pumping as if drawing from a lake of petroleum. Within four months the daily production was three-thousand barrels. This meant nineteen-hundred barrels for the land-owner –fifteen-hundred from royalty and four-hundred from wells he had drilled—a regular income of nine-thousand dollars a day! Cipher it out—nineteen-hundred barrels at four-fifty to five dollars, with eleven-hundred barrels for the lessees—and what do you find? Fourteen-thousand dollars a day for the last quarter of 1870 and nine months of 1871, from one-sixth of a farm sold in 1850 for seventeen-hundred dollars !

James S. McCray, a farmer's son, born in 1824 on the flats below Titusville, at twenty-two set out for himself with two dollars in his pocket. Working three years in a saw-mill on the Allegheny, he saved his earnings and in 1850 was able to buy a team and take up the farm decreed to enrich him beyond his wildest fancies. He married Miss Martha G. Crooks, a willing helpmeet in adversity and wise counsellor in prosperity. His first venture in oil, a share in a two-acre lease at Rouseville, he sold to drill a well on the Blood farm, elbowing his own. From this he realized seventy-thousand dollars. For his own farm he refused a million dollars in 1871. Sharpers dogged his footsteps and endeavored to rope him into all sorts of preposterous schemes. He told me one project, which was expected to control the coal-trade of the region, bled him two-hundred-and-sixty-thousand dollars! Instead of selling his oil right along, at an average figure of nearly five dollars, he stored two-hundred-thousand barrels in iron-tanks, to await higher prices. In my presence H. I. Beers, of McClintockville, bid him five-thirty-five a barrel for the lot. McCray stuck out for five-fifty. He kept the oil for years, losing thousands of barrels by leakage and evaporation, and sold the bulk of it at one to two dollars. Had he dealt with Beers he would have been six-hundred-thousand dollars richer! Mr. McCray removed to Franklin in 1872 and died some years ago. He rests in the cemetery beside his faithful wife and only daughter. The wells on his farm drooped and withered and the famous fifteen-acre field has long been a pasture. A robust character, strong-willed and kindly, sometimes queerly contradictory and often misjudged, James S. McCray could adopt the words of King Lear : " I am a man more sinned against than sinning."

The Dalzell or Hayes farm, on which the first well—fifty barrels—was drilled in 1861, boasted the Porcupine, Rhinoceros, Ramcat, Wildcat, and a menagerie of thirty others ranging from ten barrels to three-hundred. At the north end of the farm, in the rear of the Maple Shade and Jersey wells, the Petroleum Shaft and Mining Company attempted to sink a hole seven feet by seventeen to the third sand. The shaft was dug and blasted one-hundred feet, at immense cost. The funds ran out, gas threatened to asphyxiate the workmen, the big pumps could not exhaust the water and the absurd undertaking was abandoned.

The story of the Story farm does not lack romantic ingredients. William Story owned five-hundred acres south of the G. W. McClintock farm, Oil Creek, the Dalzell and Tarr farms bounding his land on the east. He sold in 1859 to Ritchie, Hartje & Co., of Pittsburg, for thirty-thousand dollars. George H. Bissell had negotiated for the property, but Mrs. Story objected to signing the deed. Next day Bissell returned to offer the wife a sufficient inducement, but the Pittsburg agent had been there the previous evening and secured her signature to the Ritchie-Hartje deed by the promise of a silk dress! Thus a twenty-dollar gown changed the ultimate ownership of millions of dollars! The long-haired novelist, who soars into the infinite and dives into the unfathomable, may try to imagine what the addition of a new bonnet would have accomplished.

The seven Pittsburgers organized a stock company in 1860 to develop the farm. By act of Legislature this was incorporated on May first, 1861, as the Columbia Oil-Company, with a nominal capital of two-hundred-and-fifty-thousand dollars—ten-thousand shares of twenty-five dollars each. Twenty-one-thousand barrels of oil were produced in 1861 and ninety-thousand in 1862, shares selling at two to ten dollars. Foreign demand for oil improved matters. On July eighth, 1863, the first dividend of thirty per cent. was declared, followed in August and September by two of twenty-five per cent. and in October by one of fifty per cent. Four dividends, aggregating one-hundred-and-sixty per cent., were declared the first six months of 1864. The capital was increased to two-and-a-half-millions, by calling in the old stock and giving each holder of a twenty-five-dollar share five new ones of fifty dollars apiece. Four-hundred per cent. were paid on this capital in six years. The original stockholders received their money back *forty-three* times and had ten times their first stock to keep on drawing fat dividends! Suppose a person had bought one-hundred shares in 1862 at two dollars, in eight years he would have been paid one-hundred-and-seven-thousand dollars for his two hundred and have five-hundred fifty-dollar shares on hand! From a mere speck of the Story farm the Columbia Oil-Company in ten years produced oil that sold for ten-millions of dollars! Wonder not that men, dazzled by such returns, blind to the failures that littered the oily domain, clutched at the veriest phantoms in the mad craze for boundless wealth.

ANDREW CARNEGIE.

Andrew Carnegie, the colossus of the iron-trade, was a Columbia stock-holder. He started in life as a messenger-boy. William Melville and David Mc-Cargo were his associates in the telegraph service of the Pennsylvania Railroad. One day Edgar Thomson, president of the road and head of the great steel-works at Braddock, asked Mr. Pitcairn to send him a smart lad to help in his office. As the story goes, Pitcairn decided that he could best spare the canny young Scot. Melville was exceptionally quick, McCargo never neglected the smallest details of his work and Carnegie, who possessed the cautious deliberation of his race, was sent to Mr. Thomson. His shrewdness and fidelity speedily won Thomson's favor. The railroad-king died and his clever clerk eventually controlled the steel-plant ten miles east of Pittsburg. Now Andrew Carnegie bosses the steel-industry, owns the largest steel-plants in the world, manufactures massive armor-plate for war-ships—blow-holes blew holes in its reputation "once upon a time"—and has acquired forty or fifty-millions by the sweat of his workmen's brows. He has parks and castles in Scotland, spends much of his time and cash abroad, coaches with princes and nobles and lets H. C. Frick fricasee the toilers at Braddock and Homestead. The Homestead riots, precipitated by a ruffianly horde of Pinkerton thugs, aroused a storm of indignation which defeated Benjamin Harrison for the presidency and elected Grover Cleveland on the issue of tariff-reform. Mr. Carnegie writes soul-stirring magazine-articles on the duties of capital to labor and has established numerous public-libraries. He is stoutly built and exceedingly healthy. His enormous fortune may yet endow some magnificent charity. Melville keeps books in a pipe-line office at Pittsburg and McCargo is superintendent of the Allegheny-Valley Railway.

Splendidly managed throughout, the policy of the Columbia Company was to operate its lands systematically. Wells were not drilled at random over the farm, nor were leases granted to speculators. There was no effort to make a big showing of production and exhaust the territory in the shortest time possible. For twenty-five years the Story farm yielded profitably. The wells, never amazingly large, held on tenaciously. The Ladies' well produced sixty-five-thousand barrels, the Floral sixty-thousand, the Big Tank fifty-thousand, the Story Centre forty-five-thousand, the Breedtown forty-thousand, the Cherry Run fifty-five-thousand, the Titus pair one-hundred-thousand and the Perry thirty-five-thousand. The company erected machine-shops, built houses for employés, and the village of Columbia prospered. The Columbia Cornet Band, superbly appointed, its thirty members in rich uniforms, its instruments the finest and its drum-major an acrobatic revelation, could have given Gilmore's or Sousa's points in ravishing music. G. S. Bancroft superintended the wells and D. H. Boulton, now of Franklin, assisted President D. B. Stewart, of Pittsburg, in conducting affairs generally. The village has vanished, the cornet band is hushed forever, the fields are the prey of weeds and underbrush and brakemen no more call out "Columby!" A few small wells, hidden amid the hills, produce a morsel of oil, but the farm, despoiled of sixteen-million dollars of greasy treasure, would not bring one-fourth the price paid William Story for it in the fall of 1859. "So passes away earthly glory" is as true to-day as when Horace evolved the classic phrase two-thousand years ago.

On the east side of Oil Creek, opposite the southern half of the Story farm, James Tarr owned and occupied a triangular tract of two-hundred acres. He was a strong-limbed, loud-voiced, stout-hearted son of toil, farming in summer and hauling lumber in winter to support his family. Although uneducated, he had plenty of "horse sense" and native wit. His quaint speech coined words and terms that are entrenched firmly in the nomenclature of Oildom. Funny stories have been told at his expense. One of these, relating to his daughter, whom he had taken to a seminary, has appeared in hundreds of newspapers. According to the revised version, the principal of the school expressing a fear that the girl had not "capacity," the fond father, profoundly ignorant of what was meant, drew a roll of greenbacks from his pocket and exclaimed: "Damn it, that's nothing! Buy her one and here's the stuff to pay for it!" The fact that it is pure fiction may detract somewhat from the piquancy of this incident.

Tarr realized his own deficiencies from lack of schooling and spared no pains, when the golden stream flowed his way, to educate the children dwelling in the old home on the south end of the farm. His daughters were bright, good-looking, intelligent girls. Scratching the barren hills for a meager corn-crop, hunting rabbits on Sundays, rafting in the spring and fall and teaming while snow lasted barely sufficed to keep the gaunt wolf of hunger from the door of many a hardy Oil-Creek settler. To their credit be it said, most of the land-

owners whom petroleum enriched took care of their money. Rough diamonds, uncut and unpolished, they possessed intrinsic worth. James Tarr was of the number who did not lose their heads and squander their substance. The richest of them all, he bought a delightful home near Meadville, provided every comfort and convenience, spent his closing years enjoyably and died in 1871. "Put yourself in his place" and, candidly, would you have done better?

For himself, George B. Delamater and L. L. Lamb, in the summer of 1860 Orange Noble leased seven acres of the Tarr farm, at the bend in Oil Creek. Dry holes the partners "kicked down" on the Stackpole and Jones farms dampening their ardor, they let the Tarr lease lie dormant some months. Contracting with a Townville neighbor—N. S. Woodford—to juggle the "spring-pole," he cracked the first sand in June, 1861. The Crescent well—so called because the faith of the owners was increasing—tipped the beam at five-hundred barrels. The first well on the Tarr farm, it flowed an average of three-hundred barrels a day for thirteen months, quitting without notice. Cleaning it out, drilling it deeper and pumping it for weeks were of no avail. Not a drop of oil could be extracted and the Crescent was abandoned. Crude was so low during most of its existence—ten to twenty-five cents—that the well, although it produced one-hundred-and-twenty-thousand barrels, did not pay the owners a dollar of profit!

10

Drilling, royalty and tankage absorbed every nickel. Like the victories of Pyrrhus, the more such strikes a fellow achieved the sooner he would be undone!

On the evening of August first, 1861, as James Tarr sat eating his supper of fried pork and johnny-cake, Heman Janes, of Erie, entered the room. "Tarr," he said, "I'll give you sixty-thousand dollars in spot cash for your farm!" Tarr almost fell off his chair. A year before one-thousand dollars would have been big money for the whole plantation. "I mean it," continued the visitor; "if you take me up I'll close the deal right here!" Tarr "took him up" and the deal, which included a transfer of several leases, was closed quickly. Janes planked down the sixty-thousand and Tarr, within an hour, had stepped from poverty to affluence. This was the first large *cash* transaction in oil-lands on the creek and people promptly pronounced Janes a fool of the thirty-third degree. An Irishman, on trial for stealing a sheep, asked by the judge whether he was guilty or not guilty, replied: "How can I tell till I hear the ividence?" Don't endorse the Janes verdict "till you hear the ividence."

A short distance below the Crescent well William Phillips, who had leased a narrow strip the entire length of the farm, was also urging a "spring-pole" actively. Born in Westmoreland county in 1824, he passed his boyhood on a farm and earned his first money mining coal. Saving his hard-won wages, he bought the keel-boat Orphan Boy and started freighting on the Ohio and Allegheny rivers. The business proving remunerative, he drilled salt-wells at Bull Creek and Wildcat Hollow. On his last trip from Warren to Pittsburg, in September of 1859, he noticed a scum of oil in front of Thomas Downing's farm, where South Oil City now stands. The story of the Drake well was in everybody's mouth and it occurred to Phillips that he could increase his growing fortune by drilling on the Downing land. At Pittsburg he consulted Charles Lockhart, William Frew, Captain Kipp and John Vanausdall and with them formed the partnership of Phillips, Frew & Co. Returning at once, he leased from Downing, erected a pole-derrick and proceeded to bore a well on the water's edge. With no machine-shops, tools or appliances nearer than Pittsburg, a hundred-and-thirty miles off, difficulties of all kinds retarded the work nine months. Finally the job was completed and the Albion well, pumping forty barrels a day, raised a commotion.

The Albion brought Phillips to the front as an oil-operator. James Tarr readily leased him part of his farm and he began Phillips No. 1 well in the spring of 1861. The Crescent's unexpected success spurred him to greater efforts. Hurrying an engine and boiler from Pittsburg, he started his second well on the flat hugging the stream twenty rods north of the Crescent. Steam-power rushed the tools at a boom-de-ay gait. The first sand, from which meanwhile No. 1 was rivaling the Crescent's yield, had not a pinch of oil. The solid-silver lining of the petroleum-cloud assumed a plated look, but Phillips heeded it not. An expert driller, he hustled the tools and on October nineteenth, at four-hundred-and-eighty feet, pierced the shell above the third sand. At dusk he shut down for the night. The weather was clear and the moon shone brightly. Suddenly a vivid flame illumined the sky. Reuben Painter's well on the Blood farm, a mile southward, had caught fire and blazed furiously. The rare spectacle of a burning well attracted everybody for miles. Phillips and Janes were among those who hastened to the fire, returning about midnight. An hour later they were summoned from bed by a man yelling at the Ella-Yaw pitch: "The Phillips is bu'sted and runnin' down the creek!" People ran to the spot on the double-quick, past the Crescent and down the bank. Gas was settling densely upon the flats and into the creek oil was pouring lavishly. Dreading a fire, lights were extinguished on the adjoining tracts and needful precautions taken. For three or four days the flow raged unhindered, then a lull occurred and tubing was inserted. After the seed-bag swelled, a stop-cock was placed on the tubing and thenceforth it was easy to regulate the flow. When oil was wanted the stop-cock was opened and wooden troughs conveyed the stuff to boats drawn up the creek by horses, the chief mode of transportation for years. The oil was forty-four gravity and four-thousand barrels a day gushed out! In June of 1862, when

WOODFORD WELL. TARR FARM IN 1862. PHILLIPS WELL.

Phillips and Major Frew, with their wives and a party of friends, inspected the well, a careful gauge showed it was doing thirty-six-hundred-and-sixty barrels! The Phillips well held the champion-belt twenty-seven years. It produced until 1871, getting down to ten or twelve barrels and ceasing altogether the night James Tarr expired, having yielded nearly *one-million* barrels! Cargoes of the oil were sold to boatmen at five cents a barrel, thousands of barrels were wasted, tens of thousands were stored in underground tanks and much was sold at three to thirteen dollars.

N. S. Woodford, Noble & Delamater's contractor, had the foresight to lease the ground between the Crescent and the Phillips No. 2. His three-thousand barreler, finished in December, 1861, drew its grist from the Phillips crevice and interfered with the mammoth gusher. When the two became pumpers neither would give out oil unless both were worked. If one was stopped the other pumped water. Ultimately the Phillips crowd paid Woodford a half-million for his well and lease, a wad for which a man would ford even the atrocious Tarr-farm mud and complacently whistle "Ta-ra-ra." He retired to his Townville home, with six-hundred-thousand dollars to show for eighteen months' operations on Oil Creek, and never bothered any more about oil. The Woodford well repaid its enormous cost. Lockhart and Frew bought out their partners at a high price and put the Phillips-Woodford interests into a stock-company capitalized at two-million dollars. The Phillips well—one result of a keen-eyed

boatman's observing an oily scum on the Allegheny River—enriched all concerned. Had Phillips failed to see the speck of grease that September day,
who can tell how different oil-region history might have been? Happily for a
good many persons, the Orphan Boy was not one of the "Ships that Pass in the
Night." What a field Oil Creek presents for the fervid fancy of a Dumas, a
Dickens, a Wilkie Collins or a Charles Reade!

Comrades in business and good-fellowship, William Phillips and John Vanausdall removed to South Oil City, lived neighbors and died twenty years ago.
They resembled each other in appearance and temper, in charitable impulse and
kindness to the poor. Phillips drilled dozens of wells—none of them dry—aided
Oil-City enterprises and was a member of the shipping firm of Munhall & Co.
until its dissolution in 1876. He was the first man to ship oil by steamer, the
Venango taking the first load to Pittsburg, and the first to run crude in bulk
down the creek. One son, John C. Phillips, and a married daughter live at Oil
City and two sons at Freeport.

Heman Janes, of Erie, the first purchaser of the Tarr farm, from 1850 to 1861
shipped large quantities of lumber to the eastern market. Passing through
Canada in 1858, he heard oil was obtained from gum-beds in Lambton county,
south of Lake Huron, and visited the place. John Williams was dipping five
barrels a day from a hole ten feet square and twenty feet deep. The best gum-
beds spread over two-hundred acres of timbered land, which Mr. Janes bought
at nine dollars an acre, the owner selling because "the stinking oil smelled five
miles off." Leasing four-hundred acres more, in 1860 he sold a half-interest in
both tracts for fifteen-thousand dollars and retired from lumbering to devote his
attention to oil. Large wells on his Canadian lands enabled him to sell the second
half of the property in 1865 for fifty-five-thousand dollars. In February, 1861,

HEMAN JANES.

he secured a thirty-day option on the J. Buchanan farm, the site of Rouseville, and tendered the price at the stipulated time, but the
transaction fell through. In March of that
year he went to West Virginia and leased one-
thousand acres on the Kanawha River, including the famous "Burning Spring." U. E.
Everett & Co. agreed to pay fifty-thousand
dollars for one-half interest in the property,
at Parkersburg, on April twelfth. All parties
met, a certified check was laid on the table
and Attorney J. B. Blair started to draw the
papers. At that moment a boy ran past,
shouting: "Fort Sumpter's fired on!" The
gentlemen hurried out to learn the particulars.
"The cat came back," but Everett didn't. A
message told him to "hold off," and he is

holding off still. Janes stayed as long as a Northerner dared and was thankful
to sell the batch of leases for seventy-five-hundred dollars. In 1862 he sued the
owners of the Phillips well for his royalty *in barrels*. They refused to furnish
the barrels, which were scarce and expensive, and the well was shut down for
months pending the litigation. The suit was for one-hundred-and-twelve-thou-
sand dollars, up to that time the largest amount ever involved in a case before
the Venango court. Edwin M. Stanton, soon to be known as the illustrious
War-Secretary, was one of the attorneys engaged by the plaintiff, for a fee of

twenty-five-thousand dollars. A compromise was arranged for half the oil. The first oil sold after this agreement was at three dollars a barrel, taken from the first twelve-hundred-barrel tank ever seen in the region. A wooden tank of that size excited more curiosity in those days than a hundred iron ones of forty-thousand barrels in this year of grace. Janes sold back half the farm to Tarr for forty-thousand dollars and two-thirds of the remaining half to Clark & Sumner for twenty-thousand, leaving him one-sixth clear of cost, the same month he bought the tract. He first suggested casing wells to exclude the water, built the first bulk-boat decked over—six-hundred barrels—to transport oil and was identified with the first practicable pipe-line. Paying seventy-five-thousand dollars for the Blackmar farm, at Pithole, he drilled three dry holes and then got rid of that land at a snug advance. Since 1878 Mr. Janes has been interested in the Bradford field and living at Erie. A man of forceful character and executive ability, hearty, vigorous and companionable, he deserves the large measure of success that rewarded him as an important factor in petroleum-affairs. In the words of the good Scottish mother to her son: "May your lot be wi' the rich in this warld and wi' the puir in the warld to come."

The amazing output of the Phillips and Woodford wells stimulated the demand for territory to the boiling point. Men were infinitely less eager to "read their title clear to mansions in the skies" than to secure a title to a fragment of the Tarr farm. Rigs huddled on the bank and in the water, for nobody thought oil existed back in the hilly sections. Sixty yards below the Phillips spouter J. F. Crane sank a well that responded as pleasantly as "the swinging of the crane." Dinsmore Brothers, at the lower end of the farm, drilled a seven-hundred-barreler late in 1861. A zoological freak introduced the animal fad, which named the Elephant, Young Elephant, Tigress, Tiger, Lioness, Scared Cat, Anaconda and Weasel wells. Reckless speculation held the fort unchecked. The third sand was sixty feet thick, the territory was durable and three-hundred walking-beams exhibited "the poetry of motion" to the music of three-four-five-six-eight-ten-dollar oil. Mr. Janes built a commodious hotel and a town of two-thousand population flourished. James Tarr sold his entire interest in 1865, for gold equivalent to two-millions in currency, and removed to Crawford county. Another million would hardly cover his royalties. Three-million dollars ahead of the game in four years, he could afford to smile at the jibes of small-souled retailers of witless ridicule. If "money talks," three-millions ought to be pretty eloquent. The churches, stores, houses, offices, wells and tanks have "gone glimmering." Tarr Farm station appears no more on railroad time-tables. Modern maps do not reveal it. Few know and fewer care who owns the place once the apple of the oilman's eye, now a shadowy relic not worth carting off in a wheelbarrow!

Producers have enjoyed quite a reputation for "resolving," and the first meeting ever held to regulate the price of crude was at Tarr farm in 1861. The moving spirits were Mr. Janes, General James Wadsworth and Josiah Oakes, the latter a New-York capitalist. The idea was to raise five-hundred-thousand dollars and buy up the territory for ten miles along Oil Creek. Wadsworth and Oakes raised over three-hundred-thousand dollars for this purpose, when the panic arising from the war ended the scheme. A contract was also made with Erie parties to lay a four-inch wooden pipe-line from Tarr farm to Oil City. On the advice of Col. Clark, of Clark & Sumner, and Sir John Hope, the eminent London banker, it was decided to abandon the project and apply for a charter for a pipe-line. This was done in the winter of 1861-2, Hon. Morrow B. Lowry,

who represented the district in the State Senate, favoring the application. Hon.
M. C. Beebe, the local member of the Legislature, opposed it resolutely, because,
to quote his own words : "There are four-thousand teams hauling oil and my
constituents won't stand this interference." The measure failing to carry, Clark
& Hope built the Standard refinery at Pittsburg.

Resistance to the South Improvement Company welded the producers solidly in 1872. The refiners organized to force a larger margin between crude and
refined. To offset this and govern the production and sale of crude, the producers established a "union," "agencies" and "councils." In October of 1872
every well in the region was shut down for thirty days. The "spirit of seventy-
six" was abroad and individual losses were borne cheerfully for the general
good. This was the heroic period, which demonstrated the manly fiber of the
great body of oil-operators. E. E. Clapp, of President, and Captain William
Harson, of Oil City, were the chief officers of these remarkable organizations.
Suspensions of drilling in 1873-4-5 supplemented the memorable "thirty-day
shut-down." At length the "union," the "councils" and the "agencies"
wilted and dissolved. The area of productive territory widened and strong
companies became a necessity to develop it. The big fish swallowed the little
ones, hence the *personal* feature so pronounced in earlier years has been almost
eliminated. Many of the operators are members of the Producers' Association,
in which Congressman Phillips, Lewis Emery, David Kirk and T. J. Vandergrift are prime factors. Its president, Hon. J. W. Lee, practiced law at Franklin, served twice as State Senator and located at Pittsburg last year. He is a
cogent speaker, not averse to legal tilts and not backward flying his colors in
the face of the enemy.

South of the Story and Tarr farms, on both sides of Oil Creek, were John
Blood's four-hundred-and-forty acres. The owner lived in an unpainted,
weatherbeaten frame house. On five acres of the flats the Ocean Petroleum-
Company had twelve flowing wells in 1861. The Maple-Tree Company's burning well spouted twenty-five-hundred barrels for several months, declined to
three-hundred in a year and was destroyed by fire in October of 1862. The
flames devastated twenty acres, consuming ten wells and a hundred tanks of
oil, the loss aggregating a million dollars. A sheet of fire, terribly grand and
up to that date the most extensive and destructive in Oildom, wrapped the flats
and the stream. Blood Well No. 1, flowing a thousand barrels, Blood No. 2,
flowing six hundred, and five other gushers never yielded after the conflagration,
prior to which the farm was producing more oil than the balance of the region.
Brewer & Watson, Ballard & Trax, Edward Filkins, Henry Collins, Reuben
Painter, James Burrows and J. H. Duncan were pioneer operators on the tract.
Blood sold in 1863 for five-hundred-and-sixty-thousand dollars and removed to
New York. Buying a brownstone residence on Fifth avenue, he splurged
around Gotham two or three years, quit the city for the country and died long
since. The Blood farm was notably prolific, but its glory has departed.
Stripped bare of derricks, houses, wells and tanks, naught is left save the rugged
hills and sandy banks. "It is no matter, the cat will mew, the dog will have
his day."

Neighbors of John Blood, a raw-boned native and his wife, enjoyed an experience not yet forgotten in New York. Selling their farm for big money, the
couple concluded to see Manhattanville and set off in high glee, arrayed in
homespun clothes of most agonizing country-fashion. Wags on the farm advised them to go to the Astor House and insist upon having the finest room in

the caravansary. Arriving in New York, they were driven to the hotel, each carrying a bundle done up in a colored handkerchief. Their rustic appearance attracted great attention, which was increased when the man marched to the office-counter and demanded "the best in the shebang, b'gosh." The astounded clerk tried to get the unwelcome guest to go elsewhere, assuring him he must have made a mistake. The rural delegate did not propose to be bluffed by coaxing or threats. At length the representative of petroleum wanted to know "how much it would cost to buy the gol-darned ranche." In despair the clerk summoned the proprietor, who soon took in the situation. To humor the stranger he replied that one-hundred-thousand dollars would buy the place. The chap produced a pile of bills and tendered him the money on the spot! Explanations followed, a parlor and bedroom were assigned the pair and for days they were the lions of the metropolis. Hundreds of citizens and ladies called to see the innocents who had come on their "first tower" as green and unsophisticated as did Josiah Allen's Wife twenty years later.

Ambrose Rynd, an Irish woolen-factor, bought five-hundred acres from the Holland Land-Company in 1800 and built a log-cabin at the mouth of Cherrytree Run. He attained the Nestorian age of ninety-nine. His grandson, John Rynd, born in the log-cabin in 1815, owned three-hundred acres of the tract when the petroleum-wave swept Oil Creek. The Blood farm was north and the Smith east. Cherrytree and Wykle Runs rippled through the western half of the property, which Oil Creek divided nicely. Developments in 1861 were on the eastern half. Starting at five-hundred barrels, the Rynd well flowed until 1863. The Crawford "saw" the Rynd and "went it one better," lasting until June of 1864. Six fair wells were drilled on Rynd Island, a dot at the upper part of the farm. The Rynd-Farm Oil-Company of New York purchased the tract in 1864. John Rynd moved to Fayette county and died in the seventies. Hume & Crawford, Porter & Milroy, B. F. Wren, the Ozark, Favorite, Frost, Northern and a score of companies operated vigorously. The third sand thickened and improved with the elevation of the hills. Five refineries handled a thousand barrels of crude per week. A snug village bloomed on the west side, the broad flat affording an eligible site. The late John Wallace and Theodore Ladd were prominent in the later stage of operations. Cyrus D. Rynd returned in 1881 to take charge of the farm and served as postmaster six years. Companies bored three-hundred wells on Cherrytree Run and its tiny branches. Their success was not startling. Kane City, two miles north, raised Cain in mild style, the territory "wearing like leather." Farther back D. W. Kenney's wells, lively as the Kilkenny cats, stirred a current that wafted in Alemagooselum City. Its unique name, the biggest feature of the "City," was worked out by Kenney, a fun-loving genius, known far and wide as "Mayor of Alemagooselum." He and his wells and town have long been "out of sight." Kane City casts an attenuated shadow. Rynd, once plump and juicy, now lean and dessicated, resembles an orange which a boy has sucked and thrown away the rind.

Rev. William Elliott, who united in one package the fervor of Paul and the snap of Ebenezer Elliott, "the Corn-Law Rhymer," lived and preached at Rynd. He organized a Sunday-school in Kenney's parish, which a devout settler undertook to superintend. At the close of the regular exercises on the opening day, Mr. Elliott asked the pious ruralist to "say a few words." The good man, wishing to clinch the lesson—about Mary Magdalene—in the minds of the youngsters, implored them to follow the example of "Miss Magdolin." The older brood tittered at this Hibernianism, the laugh swelled into a cloudburst.

Mr. Elliott nearly swallowed his pocket-handkerchief trying to shut in his smiles and a new query was born, which had a long run. It was fired at every visitor to the settlement. Small boys hurled it at the defenceless superintendent, who resigned his job and broke up the school next Sunday. Possibly Br'er Elliott, when ushered into heaven, would not be one bit surprised to hear some white-winged cherub from Alemagooselum sing out, "Say, do you know Miss Mag Dolin?"

FISHING OUT THE PREACHER'S HORSE.

The scanty herbage on the tail of the parson's horse gave rise to endless surmises. The animal stranded in a mud-hole and keeled over on his side. Four sturdy fellows tried to fish him out. In his misguided zeal one of the rescuers, tugging at the caudal appendage, pulled so hard that half the hair peeled off, leaving the denuded nag a fitting mate for Tam O'Shanter's tailless Meg.

Two museum-curio wells on the Rynd farm illustrated practically Chaplain McCabe's "Drinking From the Same Canteen." A dozen strokes of the pump every hour caused the Agitator to flow ten or fifteen minutes. The Sunday well, its companion, loafed six days in the week while the other worked, flowing on the Sabbath when the Agitator pump rested from its labors. This sort of affinity, which cost William Phillips and Noble & Delamater a mint of money, was evinced most forcibly on the McClintock farm, west side of Oil Creek, south of Rynd. William McClintock, original owner of the two-hundred acres, dying in 1859, the widow remained on the farm with her grandson, John W. Steele, whom the couple had adopted at a tender age, upon the decease of his mother. Nearly half the farm was bottom-land, fronting the creek, on the bank of which the first wells were sunk in 1861. The Vanslyke flowed twelve-hundred barrels a day, declined slowly and in its third year pumped fourteen-thousand. The Lloyd, Eastman, Little Giant, Morrison, Hayes & Merrick, Christy, Ocean, Painter, Sterrett, Chase and sixty more each put up fifty to four-hundred barrels daily. Directly between the Vanslyke and Christy, a few rods from either, New-York parties finished the Hammond well in May, 1864. Starting to flow three-hundred barrels a day, the Hammond killed the Lloyd and Christy and reduced the Vanslyke to a ten-barrel pumper. Its triumph was short-lived. Early in June the New Yorkers, elated over its performance, bought the royalty of the well and one-third acre of ground for two-hundred thousand dollars. The end of June the tubing was drawn from the Excelsior well, on the John McClintock farm, five-hundred yards east, flooding the Hammond and all the wells in the vicinity. The damage was attributed to Vandergrift & Titus's new well a short distance down the flat, nobody imagining it came from a hole a quarter-mile off. Retubing the Excelsior quickly restored one-half the Hammond's yield, which increased as the Excelsior's lessened. An adjustment followed, but the final pulling of the tubing from the Excelsior drowned the affected wells permanently. Geologists and scientists reveled in the ethics suggested by such interference, which casing wells has obviated. The

Widow-McClintock farm produced hundreds of thousands of barrels of oil and changed hands repeatedly. For years it was owned by a man who as a boy blacked Steele's boots. In 1892 John Waites renovated a number of the old wells. Pumping some and plugging others, to shut out water, surprised and rewarded him with a yield that is bringing him a tidy fortune. The action of the stream has washed away the ground on which the Vanslyke, the Sterrett and several of the largest wells were located. "Out, out, brief candle!"

Mrs. McClintock, like thousands of women since, attempted one day in March of 1863 to hurry up the kitchen-fire with kerosene. The result was her fatal burning, death in an hour and the first funeral to the account of the treacherous oil-can. The poor woman wore coarse clothing, worked hard and secreted her wealth about the house. Her will, written soon after McClintock's exit, bequeathed everything to the adopted heir, John W. Steele, twenty years old when his grandmother met her tragic fate. At eighteen he had married Miss M. Moffett, daughter of a farmer in Sugarcreek township. He hauled oil in 1861 with hired plugs until he could buy a span of stout horses. Oil-Creek teamsters, proficient in lurid profanity, coveted his varied stock of pointed expletives. The blonde driver, of average height and slender build, pleasing in appearance and address, by no means the unlicked cub and ignorant boor he has been represented, neither smoke nor drank nor gambled, but "he could say 'damn'!" Climbing a hill with a load of oil, the end-board dropped out and five barrels of crude wabbled over the steep bank. It was exasperating and the spectators expected a special outburst. Steele "winked the other eye" and remarked placidly : "Boys, it's no use trying to do justice to this occasion." The shy youth, living frugally and not the type people would associate with unprecedented antics, was to figure in song and story and be advertised more widely than the sea-serpent or Barnum's woolly-horse. Millions who never heard of John Smith, Dr. Mary Walker or Baby McKee have heard and read and talked about the one-and-only "Coal-Oil Johnnie."

The future candidate for minstrel-gags and newspaper-space was hauling oil when a neighbor ran to tell him of Mrs. McClintock's death. He hastened home. A search of the premises disclosed two-hundred-thousand dollars the old lady had hoarded. Wm. Blackstone, appointed his guardian, restricted the minor to a reasonable allowance. The young man's conduct was irreproachable until he attained his majority. His income was enormous. Mr. Blackstone paid him three-hundred-thousand dollars in a lump and he resolved to "see some of the world." He saw it, not through smoked glass either. His escapades supplied no end of material for gossip. Many tales concerning him were exaggerations and many pure inventions. Demure, slow-going Philadelphia he colored a flaming vermilion. He gave away carriages after a single drive, kept open-house in a big hotel and squandered thousands of dollars a day. Seth Slocum was "showing him the sights" and he fell an easy victim to blacklegs and swindlers. He ordered champagne by the dozen baskets and treated theatrical companies to the costliest wine-suppers. Gay ballet girls at Fox's old play-house told spicy stories of these midnight frolics. To a negro-comedian, who sang a song that pleased him, he handed a thousand-dollar pin. He would walk the streets with bank-bills stuck in the buttonholes of his coat for Young America to grab. He courted club-men and spent cash like the Count of Monte Cristo. John Morrissey sat a night with him at cards in his Saratoga gambling-house, cleaning him out of many thousands. Leeches bled him and sharpers fleeced him mercilessly. He was a spendthrift, but he didn't light cigars with hundred-dollar bills, buy a Philadelphia hotel to give a chum or destroy money "for fun." Usually somebody benefited by his extravagances.

Occasionally his prodigality assumed a sensible phase. Twenty-eight-hundred dollars, one day's receipts from his wells and royalty, went toward the erection of the soldiers' monument—a magnificent shaft of white marble—in the Franklin park. Except Dan Rice's five-thousand memorial at Girard, Erie county, this was the first monument in the Union to the fallen heroes of the civil war. Ten, twenty or fifty dollars frequently gladdened the poor who asked for relief. He lavished fine clothes and diamonds on a minstrel-troupe, touring the country and entertaining crowds in the oil-regions. John W. Gaylord, a famous artist in burnt-cork and member of the troupe, has furnished these details :

"Yes, 'Coal-Oil Johnnie' was my particular friend in his palmiest days. I was his room-mate when he cut the shines that celebrated him as the most eccentric millionaire on earth. I was with the Skiff & Gaylord minstrels. Johnnie saw us perform in Philadelphia, got stuck on the business and bought one-third interest in the show. His first move was to get five-thousand dollars' worth of woodcuts at his own expense. They were all the way from a one-sheet to a twenty-four-sheet in size and the largest amount any concern had ever owned. The cartoon, which attracted so much attention, of 'Bring That Skiff Over Here,' was in the lot. We went on the road, did a monstrous business everywhere, turned people away and were prosperous.

"Reaching Utica, N. Y., Johnnie treated to a supper for the company, which cost one-thousand dollars. He then conceived the idea of traveling by his own train and purchased an engine, a sleeper and a baggage-car. Dates for two weeks were cancelled and we went junketing, Johnnie footing the bills. At Erie we had a five-hundred-dollar supper; and so it went. It was here that Johnnie bought his first hack. After a short ride he presented it to the driver. Our dates being cancelled, Johnnie insisted upon indemnifying us for the loss of time. He paid all salaries, estimated the probable business receipts upon the basis of packed houses and paid that also to our treasurer.

"In Chicago he gave another exhibition of his eccentric traits. He leased the Academy of Music for the season and we did a big business. Finally he proposed a benefit for Skiff & Gaylord and sent over to rent the Crosby Opera-House, then the finest in the country. The manager sent back the insolent reply : 'We won't rent our house for an infernal nigger-show.' Johnnie got warm in the collar. He went down to their office in Root & Cady's music-store.

"'What will you take for your house and sell it outright?' he asked Mr. Root.

"'I don't want to sell.'

"'I'll give you a liberal price. Money is no object.'

"Then Johnnie pulled out a roll from his valise, counted out two-hundred-thousand dollars and asked Root if that was an object. Mr. Root was thunderstruck. 'If you are that kind of a man you can have the house for the benefit free of charge.' The benefit was the biggest success ever known in minstrelsy. The receipts were forty-five-hundred dollars and more were turned away than could be given admission. Next day Johnnie hunted up one of the finest carriage-horses in the city and presented it to Mr. Root for the courtesy extended.

"Oh, Johnnie was a prince with his money. I have seen him spend as high as one-hundred-thousand dollars in one day. That was the time he hired the Continental Hotel in Philadelphia and wanted to buy the Girard House. He went to the Continental and politely said to the clerk : 'Will you please tell the proprietor that J. W. Steele wishes to see him?' 'No, sir,' said the clerk; 'the landlord is busy.' Johnnie suggested he could make it pay the clerk to accommodate the whim. The clerk became disdainful and Johnnie tossed a bell-boy a twenty-dollar gold-piece with the request. The result was an interview with the landlord. Johnnie claimed he had been ill-treated and requested the summary dismissal of the clerk. The proprietor refused and Johnnie offered to buy the hotel. The man said he could not sell, because he was not the entire owner. A bargain was made to lease it one day for eight-thousand dollars. The cash was paid over and Johnnie installed as landlord. He made me bell-boy, while Slocum officiated as clerk. The doors were thrown open and every guest in the house had his fill of wine and edibles free of cost. A huge placard was posted in front of the hotel : 'Open house to-day; everything free; all are welcome!' It was a merry lark. The whole city seemed to catch on and the house was full. When Johnnie thought he had had fun enough he turned the hostelry over to the landlord, who reinstated the odious clerk. Here was a howdedo. Johnnie was frantic with rage. He went over to the Girard and tried to buy it. He arranged with the proprietor to 'buck' the Continental by making the prices so low that everybody would come there. The Continental did mighty little business so long as the arrangement lasted.

"The day of the hotel transaction we were up on Arch street. A rain setting in, Johnnie approached a hack in front of a fashionable store and tried to engage it to carry us up to the Girard. The driver said it was impossible, as he had a party in the store. Johnnie tossed him a five-hundred-dollar bill and the hackman said he would risk it. When we arrived at the hotel Johnnie said: 'See here, Cabby, you're a likely fellow. How would you like to own that rig?' The driver thought he was joking, but Johnnie handed him two-thousand dollars. A half-hour later the delighted driver returned with the statement that the purchase had been effected. Johnnie gave him a thousand more to buy a stable and that man to-day is the wealthiest hack-owner in Philadelphia."

Steele reached the end of his string and the farm was sold in 1866. When he was flying the highest Captain J. J. Vandergrift and T. H. Williams kindly urged him to save some of his money. He thanked them for the friendly advice, said he had made a living by hauling oil and could do so again if necessary, but he couldn't rest until he had spent that fortune. He spent a million and got the "rest." Returning to Oildom "dead broke," he secured the position of baggage-master at Rouseville station. He attended to his duties punctually, was a model of domestic virtue and a most popular, obliging official. Happily his wife had saved something and the reunited couple got along swimmingly. Next he opened a meat-market at Franklin, built up a nice business, sold the shop and moved to Ashland, Nebraska. He farmed, laid up money and entered the service of the Chicago, Burlington & Quincy Railroad some years ago as baggage-master. His manly son, whom he educated splendidly, is telegraph-operator at Ashland station. The father, "steady as a clock," is industrious, reliable and deservedly esteemed. Recently a fresh crop of stories regarding him has been circulated, but he minds his own affairs and is not one whit puffed up that the latest rival of Pears and Babbitt has just brought out a brand of "Coal-Oil Johnnie Soap."

John McClintock's farm of two-hundred acres, east of Steele and south of Rynd, Chase & Alden leased in September of 1859, for one-half the oil. B. R. Alden was a naval officer, disabled from wounds received in California, and an oil-seeker at Cuba, New York. A hundred wells rendered the farm extremely productive. The Anderson, sunk in 1861 near the southeast corner, on Cherry Run, flowed constantly three years, waning gradually from two-hundred barrels to twenty. Efforts to stop the flow in 1862, when oil dropped to ten or fifteen cents, merely imbued it with fresh vigor. Anderson thought the oil-business had gone to the bow-wows and deemed himself lucky to get seven-thousand dollars in the fall for the well. It earned one-hundred-thousand dollars subsequently and then sold for sixty-thousand. The Excelsior produced fifty-thousand barrels before the interference with the Hammond destroyed both. The Wheeler, Wright & Hall, Alice Lee, Jew, Deming, Haines and Taft wells were choice specimens. William and Robert Orr's Auburn Oil-Works and the Pennechuck Refinery chucked six-hundred barrels a week into the stills. The McClintocks have migrated from Venango. Some are in heaven, some in Crawford county and some in the west. If Joseph Cooke's conundrum—"Does Death End All?"—be negatived, there ought to be a grand reunion when they meet in the New Jerusalem and talk over their experiences on Oil Creek.

Eight miles east of Titusville, at Enterprise, John L. and Foster W. Mitchell, sons of a pioneer settler of Allegheny township, were lumbering and merchandising in 1859. They had worked on the farm and learned blacksmithing from their father. The report of Col. Drake's well stirred the little hamlet. John L. Mitchell mounted a horse and rode at a John-Gilpin gallop to lease Archibald

Buchanan's big farm, on both sides of Oil Creek and Cherry Run.   The old
man agreed to his terms, a lease was executed, the rosy-cheeked mistress and
all the pupils in the log school-house who could write witnessed the signatures
and Mitchell rode back with the document in his pocket.   He also leased John
Buchanan's two-hundred acres, south of Archibald Buchanan's three-hundred,
on the same terms—one-fourth the oil for ninety-nine years.   Forming a partnership with Henry R. Rouse and Samuel Q. Brown, he "kicked down" the first
well in 1860 to the first sand.   It pumped ten barrels a day and was bought by
A. Potter, who sank it and another to the third
sand in 1861.   A three- hundred-barreler for
months, No. 1 changed hands four times,
was bought in 1865
by Gould & Stowell
and produced oil—
it pumped for fifteen
years—that sold for
two - hundred - and -
ninety-thousand dol-
lars !   This veteran
was   the   third   or
fourth producing-
well in the region.
The Curtis, usually
considered "the first
flowing - well," in
July of 1860 spouted
freely   at   two-hun-
dred feet.   It was
not tubed and sur-
face-water soon mastered the flow of oil.   The
Brawley—sixty-thousand barrels in eight months
—the Goble & Flower, the Shaft, the Sherman
and the Nausbaum were moguls of 1861-2.   Beech
& Gillett, Alfred Willoughby, Taylor & Rockwell, Shreve & Glass, Allen Wright,
Wesley Chambers—his infectious laugh could be heard five squares—and a host
of companies operated in 1861-2-3.   Franklin S. Tarbell, E. M. Hukell, E. C.
Bradley, Harmon Camp, George Long and J. T. Jones arrived later.   The territory was singularly profitable.   Mitchell & Brown erected a refinery, divided the
tracts into hundreds of acre-plots for leases and laid out the town of Buchanan
Farm.   Allen Wright, president of a local oil-company, in February of 1861
printed his letter-heads "Rouseville" and the name was adopted unanimously.

Rouseville grew swiftly and for a time was headquarters of the oil-industry.
Churches and schools arose, good people feeling that man lives not by oil
alone any more than by bread.   Dwellings extended up Cherry Run and the
slopes of Mt. Pisgah.   Wells and tanks covered the flats and there were few
drones in the busy hive.   If Satan found mischief for the idle only, he would have
starved in Rouseville.   Stores and shops multiplied.   James White fitted up an
opera-house and C. L. Stowell opened a bank.   Henry Patchen conducted the
first hotel.   N. W. Read enacted the role of "Petroleum V. Nasby, wich iz postmaster."   The receipts in 1869 exceeded twenty-five-thousand dollars.   Miss
Nettie Dickinson, afterwards in full charge of the money-order department at

Pittsburg and partner with Miss Annie Burke in a flourishing Oil-City book-store, ran the office in an efficient style Postmaster-General Wilson would have applauded. Yet moss-backed croakers in pants, left over from the Pliocene period, think the gentle sex has no business with business! The town reached high-water mark early in the seventies, the population grazing nine-thousand. Production declined, new fields attracted live operators and in 1880 the inhabitants numbered seven-hundred, twice the present figure. Rouseville will go down in history as an oil-town noted for progressiveness, intelligence, crooked streets and girls "pretty as a picture."

The Buchanan Farm Oil-Company purchased Mitchell & Brown's interest and the Buchanan Royalty Oil-Company acquired the one-fourth held by the land-owners. Both realized heavily, the Royalty Company paying its stock-holders—Arnold Plumer, William Haldeman and Dr. C. E. Cooper were principals—about a million dollars. The senior Buchanan, after receiving two or three-hundred-thousand dollars—fifty times the sum he would ever have gained farming—often denounced "th' pirates that robbed an old man, buyin' th' farm he could 'ave sold two year later fur two millyun!" The old man has been out of pirate-range twenty-five years and the Buchanan families are scattered. Most of the old-time operators have handed in their final account. Poor Fred Rockwell has mouldered into dust. Wright, Camp, Taylor, Beech, Long, Shreve, Haldeman, Hostetter, Cooper, Col. Gibson and Frank Irwin are "grav'd in the hollow ground." Death claimed "Hi" Whiting in Florida and last March stilled the cheery voice of Wesley Chambers. The earnest, pleading tones of the Rev. R. M. Brown will be heard no more this side the walls of jasper and the gates of pearl.

A millionaire by his persevering application and wise management, John L. Mitchell married Miss Hattie A. Raymond and settled at Franklin. He organized the Exchange Bank in 1871 and was its president until ill-health obliged him to resign. Amply endowed with worldly goods, fortunate in his home and family, free from business-cares, without cause to bemoan the past or fear the future, he is serenely enjoying the evening of life. Equally successful and favored, Foster W. Mitchell also located at the county-seat and built the Exchange Hotel. He operated extensively on Oil Creek and in the northern districts, developed the Shaw farm and established a bank at Rouseville, subsequently transferring it to Oil City. He was active in politics and in the producers' organizations, treasurer of the Centennial Commission and an influential force in the Oil Exchange. Enterprising, liberal and discerning, quick to plan and execute, he is a public-spirited, prominent citizen. David H. Mitchell likewise gained a fortune in oil, founded a bank and died at Titusville. The surviving brothers, now retired bankers and living quietly, rank with the most substantial financiers of Northwestern Pennsylvania. Samuel Q. Brown, their relative and associate in various undertakings, was a merchant and banker at Pleasantville. Retiring from these pursuits, he removed to Philadelphia and then to New York to oversee the financial work of the Tidewater Pipe-Line.

Born in New York in 1824, Henry R. Rouse studied law, taught school in Warren county and engaged in lumbering and storekeeping at Enterprise. He served in the legislatures of 1859-60, acquitting himself manfully. Promptly catching the inspiration of the hour, he shared with William Barnsdall and Boone Meade the honor of putting down the third oil-well in Pennsylvania. With John L. Mitchell and Samuel Q. Brown he leased the Buchanan farm and invested in oil-lands generally. Fabulous wealth began to reward his efforts.

Had he lived "he would have been a giant or a bankrupt in petroleum." Operations on the John Buchanan farm were pushed actively. Near the upper line of the farm, on the east side of Oil Creek, at the foot of the hill, Merrick & Co. drilled a well in 1861, eight rods from the Wadsworth. On April seventeenth, at the depth of three-hundred feet, gas, water and oil rushed up, fairly lifting the tools out of the hole. The evening was damp and the atmosphere surcharged with gas. People ran with shovels to dig trenches and throw up a bank to hold the oil, no tanks having been provided. Mr. Rouse and George H. Dimick, his clerk and cashier, with six others, had eaten supper and were sitting in Anthony's Hotel discussing the fall of Fort Sumter. A laborer at the Merrick well bounded into the room to say that a vein of oil had been struck and barrels were wanted. All ran to the well but Dimick, who went to send barrels. Finishing this errand, he hastened towards the well. A frightful explosion hurled him to the earth. Smouldering coals under the Wadsworth boiler had ignited the gas. In an instant the two wells, tanks and an acre of ground saturated with oil were in flames, enveloping ninety or a hundred persons. Men digging the ditch or dipping the oil wilted like leaves in a gale. Horrible shrieks rent the air. Dense volumes of black smoke ascended. Tongues of flame leaped hundreds of feet. One poor fellow, charred to the bone, died screaming with agony over his supposed arrival in hell. Victims perished scarcely a step from safety. Rouse stood near the derrick at the fatal moment. Blinded by the first flash, he stumbled forward and fell into the marshy soil. Throwing valuable papers and a wallet of money beyond the circuit of fire, he struggled to his feet, groped a dozen paces and fell again. Two men dashed into the sea of flame and dragged him forth, his flesh baked and his clothing a handful of shreds. He was carried to a shanty and gasped through five hours of excruciating torture. His wonderful self-possession never deserted him, no word or act betraying his fearful suffering. Although obliged to sip water from a spoon at every breath, he dictated a concise will, devising the bulk of his estate in trust to improve the roads and benefit the poor of Warren county. Relatives and intimate friends, his clerk and hired boy, the men who bore him from the broiling furnace and honest debtors were remembered. This dire calamity blotted out nineteen lives and disfigured thirteen men and boys permanently. The blazing oil was smothered with dirt the third day. Tubing was put in the well, which flowed ten-thousand barrels in a week and then ceased. Nothing is left to mark the scene of the sad tragedy. The Merrick, Wadsworth, Haldeman, Clark & Banks, Trundy, Comet and Imperial wells, the tanks and the dwellings have been obliterated. Dr. S. S. Christy—he was Oil City's first druggist—Allen Wright, N. F. Jones, W. B. Williams and William H. Kinter, five of the six witnesses to Rouse's remarkable will, are in eternity, Z. Martin alone remaining.

Warren's greatest benefactor, the interest of the half-million dollars Rouse bequeathed to the county has improved roads, constructed bridges and provided a poor-house at Youngsville. Rouse was distinguished for noble traits, warm impulses, strong attachments, energy and decision of character. He dispensed his bounty lavishly. It was a favorite habit to pick up needy children, furnish them with clothes and shoes and send them home with baskets of provisions. He did not forget his days of trial and poverty. His religious views were peculiar. While reverencing the Creator, he despised narrow creeds, deprecated popular notions of worship and had no dread of the hereafter. To a preacher, in the little group that watched his fading life, who desired an hour before the end to administer consolation, he replied: "My account is made up. If I am a debtor, it would be cowardly to ask for credit now. I do not care to discuss the matter." He directed that his funeral be without display, that no sermon be preached and that he be laid beside his mother at Westfield, New York. Thus lived and died Henry R. Rouse, of small stature and light frame, but dowered with rare talents and heroic soul. Perhaps at the Judgment Day, when deeds outweigh words, many a strict Pharisee may wish he could change places with the man whose memory the poor devoutly bless. As W. A. Croffut has written of James Baker in "The Mine at Calumet":

> " ' Perfess '?　He didn't perfess.　He hed
> One simple way all through—
> He merely practiced an' he sed
> That that wud hev to do.
> ' Under conviction'?　The idee!
> He never done a thing
> To be convicted fer.　Why, he
> Wuz straighter than a string."

Cherry Run, once the ripest cherry in the orchard, had a satisfactory run. A spice of romance flavored its actual realities. Not two miles up the stream William Reed, in 1863, drilled a dry-hole six-hundred feet deep. Two miles farther, in the vicinity of Plumer, a test well was sunk seven-hundred feet, with no better result. Wells near the mouth of the ravine produced very lightly. Fifty-thousand dollars would have been an extreme price for all the land from Rouseville to Plumer, the tasteful village Henry McCalmont named in honor of Arnold Plumer. In May of 1864 Taylor & Rockwell opened a fresh vein on the run. At two-hundred feet their well threw oil above the derrick and flowed sixty barrels a day regularly. Operators reversed their opinion of the territory. To the surprise of his acquaintants, who deemed him demented, Reed started another well four rods below his failure of the previous year. It was on the right bank of the run, on a five-acre patch bought from John Rynd in 1861 by Thomas Duff, who sold two acres to Robert Criswell. Reed was not over-stocked with cash and Criswell joined forces with him to sink the second well. I. N. Frazer took one-third interest. At the proper depth the outlook was gloomy. The sand appeared good, but days of pumping failed to bring oil. On July eighteenth, 1864, the well commenced flowing three-hundred barrels a day, holding out at this rate for months. Criswell realized thirty-thousand dollars from his share of the oil and then sold his one-fourth interest in the land and well for two-hundred-and-eighty-thousand to the Mingo Oil-Company. He operated in the Butler field, lived at Monterey, removed to Ohio and died near Cincinnati. One son, David S., a well-known producer, resides at Oil City; another, Robert W., is on the editorial staff of a New-York daily. Frazer sold for one-hundred-thousand dollars and next loomed up as "the discoverer of Pithole." Reed sold to Bishop & Bissell for two-hundred-thousand dollars, after pocketing seventy-five-thousand from oil. Coming to Venango county with Frederic Prentice in 1859, he drilled wells by contract, sometimes "a solid Muldoon" and sometimes "a broken Reed." He returned east—his birthplace —with the proceeds of the world-famed well bearing his name. An idea haunted him that Captain Kidd's treasure was buried at a certain part of the Atlantic coast. He boarded at a house on the shore and hunted land and sea for the hidden deposit. He would dig in the sand, sail out some distance and peer into the water. One day he went off in his skiff, a storm arose, the boat drifted away and that was the last ever seen of William Reed.

The Reed well put Cherry Run at the head of the procession. Within sixty days it enriched Reed, Criswell and Frazer nearly seven-hundred-thousand dollars. The new owners drilled three more on the same acre, getting back every cent of their purchase-money and fifty per cent. extra for good measure. In other words, the five-acre collection of rocks and stumps, with eleven producing wells and one duster, harvested two-million dollars! The Mountain well mounted high, the Phillips & Egbert was a fillip and the Wadsworth & Wynkoop rolled out oil in wads worth a wine-coop of gold-eagles. The fever to

lease or buy a spot to plant a derrick burned fiercely. The race to gorge the ravine with rigs and drilling appliances would shut out Edgar Saltus in his "Pace that Kills." Soon three-hundred wells lined the flats and lofty banks guarding the purling streamlet. Clanking tools, wheezy engines and creaking pumps assailed the ears. Smoke from a myriad soft-coal fires attacked the eyes. An endless cavalcade of wagons churned the soil into vicious batter. The activities of the Foster, McElhenny, Farrell, Davison and Tarr farms were condensed into one surging, foaming caldron, quickening the pulse-beats and sending the brain see-sawing.

Across the run the Curtin Oil-Company farmed out forty acres. The Baker well, an October biscuit, flowed one-hundred barrels a day all the winter of 1864-5 and pumped six years. Water, bane of flannel-suits and uncased oil-wells, deluged it and its neighbors. Hugh Cropsey, a New-York lawyer and last owner of the well nearest the Baker, "ran engine," saved a trifle, pulled up stakes in 1869 and tried his luck at Pleasantville. Returning to Cherry Run, he resuscitated a well on the hill and was suffocated by gas in a tank containing a few inches of fresh crude. His heirs sold me the old well, which pumped nine months without varying ten gallons in any week and repaid twice its cost. Unchangeable as the laws of the Medes and Persians, its production was the steadiest in the chronicles of grease. One Saturday evening N. P. Stone, super-intendent of the St. Nicholas Oil-Company, bought it from me at the original price. His men took charge of it at noon on Tuesday. At five o'clock the well quit forever, "too dead to skin !" Cleaning out, drilling deeper, casing, torpe-doing and weeks of pumping could not persuade it to shed another drop of oil or water. This close shave was a small by-play in a realistic drama teeming with incidents far stranger than "The Strange Adventures of Miss Brown." B. H. Hulseman, president of the St. Nicholas Oil-Company, was a wealthy leather-merchant in Philadelphia. He spent much of his time on Cherry Run, lost heavily in speculations, entered the oil-exchange and died at Oil City. Kind-hearted, sincere and unpretending, his good remembrance is a legacy to cherish lovingly.

Two-hundred yards above the Baker a half-dozen wells crowded upon a half-acre. True to its title, the Vampire sucked the life-blood from its pal and pro-duced bounteously. The Munson, owned by the first sacrifice to nitro-glycerine, sustained the credit of its environment. The Wade was the star-performer of the group. James Wade, an Ohio teamster, earned money hauling oil. Con-cluding to wade in, he secured a bantam lease and engaged Thomas Donnelly to drill a well. It surpassed the Reed, flowing four-hundred barrels a day at the start. Frank Allen, agent of a gilt-edged New-York company, rode from Oil City to see a well described to him as "livelier than chasing a greased pig at a county-fair." His exalted conceptions of petroleum befitted the representative of a company capitalized at three-millions, in which August Belmont, Russell Sage and William B. Astor were said to be stockholders. The fuming, gassing stream of oil suited him to a t. "I'll give you three-hundred-thousand dollars for it," he said to Wade, whom the offer well-nigh paralyzed. The two men went into the grocery close by, Wade signed a transfer of the well and Allen handed him a New-York draft. The happiest being in the pack, Wade packed his carpet-bag, hitched his horses to the wagon, bade the boys good-bye and drove to Oil City to get the paper cashed. He wore greasy clothes and did not wear the air of a millionaire. "Is Mr. Bennett in ?" he asked a clerk at the bank. "Naw ; what do you want ?" was the reply. "I want a draft cashed." "Oh, you do, eh? I guess I can cash it !" The clerk's haughty demeanor fell below zero upon beholding the draft. He invited Wade to be seated. Mr. Bennett, the urbane cashier, returned in a few moments. The bank hadn't half the cur-rency to meet the demand on the instant. Wade left directions to forward the money to his home in Ohio, where he and his faithful steeds landed two days later. He bought fine farms for his brothers and himself, invested two-hundred-thousand dollars in government-bonds and wisely enjoyed, amid the peaceful scenes of agricultural life, the fruits of his first and last oil-venture. Few have been as sensible, for the petroleum-coast is encrusted with financial wrecks—vast fortunes amassed only to be lost on the perilous sea of speculation. The world has heard of the *prizes* in the lottery of oil, while the *blanks*—tenfold more numerous—are glossed over by the glamour of the Sherman, Empire, Noble, Phillips, Reed and other wells, "familiar as household words."

Thomas Johnson, of Oil City, held one-eighth of the Curtin interest and

11

Patrick Johnson had a bevy of patrician wells at the summit of the tallest hill in the valley. The curtain has been rung down, the lights are out, the players have dispersed and none can hint of "Too Much Johnson." The farm of sixty acres adjacent to the Curtin and the Criswell nook Hamilton McClintock traded to Daniel Smith in 1858 for a yoke of oxen. Smith sold it in 1860 for five-hundred dollars and sank the cash in a dry-hole on Oil Creek. C. J. Cornen and Henry I. Beers, the mainstay of the Cherry Run Petroleum-Company, bought the farm in 1863 for sixty-five-hundred dollars, clearing two-millions from the investment. Cornen served as State Senator in Connecticut and died in the eighties. His sons operate in Warren county and down the Allegheny. Mr. Beers, who settled at McClintockville, for thirty years has been prominent in business and politics. The Yankee well, erratic as George Francis Train, was the first glory of the Smith tract. The Reed caused a rush for one-acre leases at four-thousand-dollars bonus and half the oil. Picking up gold-dollars at every step would have been less lucrative. The wells were stayers and Daniel Smith was not "a Daniel come to judgment" in his estimate of the farm he implored J. W. Sherman to buy for two-hundred-and-fifty dollars.

Queerly enough, the farms above the Smith were failures. Hundreds of wells clear up to Plumer never paid the expense of recording the leases. The territory was a roast for scores of stock-companies. Below Plumer a mile Bruns & Ludovici, of New York, built the Humboldt Refinery in 1862. Money was lavished on palatial quarters for the managers, enclosed grounds, cut-stone walls, a pipe-line to Tarr Farm and the largest refining capacity in America. Inconvenient location and improved methods of competitors forced the Humboldt to retire. Part of the machinery was removed, the structures crumbled and some of the dressed stone forms the foundations of the National Transit Building at Oil City. Plumer, which had a grist-mill, store, blacksmith-shop and tavern in 1840 and four-thousand population in 1866, is quiet as its briar-grown graveyard. The Brevoort Oil-Company, Murray & Fawcett and John P. Zane raked in shekels on Moody Run, which emptied into Cherry Run a half-mile south-west of the Reed well. Zane, whole-souled, resolute and manly, operated in the northern district and died at Bradford in 1894. A "forty-niner," he supported John W. Geary for Mayor of San Francisco, built street-railways and worked gold-mines in California. He wrote on finance and petroleum, hated selfishness and stood firmly on the platform laid down in the beatitudes.

Seventy-five wells were drilled on Hamilton McClintock's four-hundred acres in 1860-1. Here was Cary's "oil-spring" and expectations of big wells soared high. The best yielded from one-hundred to three-hundred barrels a day. Low prices and the war led to the abandonment of the smaller brood. A company bought the farm in 1864. McClintockville, a promising village on the flat, boasted two refineries, stores, a hotel and the customary accessories, of which the bridge over Oil Creek is the sole reminder. Near the upper boundary of the farm the Reno Railroad crossed the valley on a giddy center-trestle and timber abutments, not a splinter of which remains. General Burnside, the distinguished commander, superintended the construction of this mountain-line, designed to connect Reno and Pithole and never completed. Occasionally the dignified general would be hailed by a soldier who had served under him. It was amusing to behold a greasy pumper, driller or teamster step up, clap Burnside on the shoulder, grasp his hand and exclaim: "Hello, General! Deuced glad to see you! I was with you at Fredericksburg! Come and have a drink!"

The Clapp farm of five-hundred acres had a fair allotment of long-lived wells. George H. Bissell and Arnold Plumer bought the lower half, in the closing days of 1859, from Ralph Clapp. The Cornplanter Oil Company purchased the upper half. The Hemlock, Cuba, Cornwall—a thousand-barreler—and Cornplanter, on the latter section, were notably productive. The Williams, Stanton, McKee, Elizabeth and Star whooped it up on the Bissell–Plumer division. Much of the oil in 1862–3 was from the second sand. Four refineries flourished and the tract coined money for its owners. A mile east was the prolific Shaw farm, which put two-hundred-thousand dollars into Foster W. Mitchell's purse. Graff & Hasson's one-thousand acres, part of the land granted Cornplanter in 1796, had a multitude of medium wells that produced year after year. In 1818 the Indian chief, who loved fire-water dearly, sold his reservation to William Connely, of Franklin, and William Kinnear, of Centre county, for twenty-one-hundred-and-twenty-one dollars. Matthias Stockberger bought Connely's half in 1824 and, with Kinnear and Reuben Noyes, erected the Oil-Creek furnace, a foundry, mill, warehouses and steamboat-landing at the east side of the mouth of the stream. William and Frederick Crary acquired the business in 1825 and ran it ten years. William and Samuel Bell bought it in 1835 and shut down the furnace in 1849. The Bell heirs sold it to Graff, Hasson & Co. in 1856 for seven-thousand dollars. James Hasson located on the property with his family and farmed five years. Graff & Hasson sold three-hundred acres in 1864 to the United Petroleum Farms Association for seven-hundred-and-fifty-thousand dollars. James Halyday settled on the east side in 1803. His son James, the first white baby in the neighborhood, was born in 1809. The Bannon family came in the forties, Thomas Moran built the Moran House—it still lingers—in 1845 and died in 1857. Dr. John Nevins arrived in 1850 and in the fall of 1852 John P. Hopewell started a general store. Hiram Gordon opened the "Red Lion Inn," Samuel Thomas shod horses and three or four families occupied small habitations. And this was the place, when 1860 dawned, that was to become the petroleum-metropolis and be known wherever men have heard a word of "English as she is spoke."

Cornplanter was the handle of the humble settlement, towards which a stampede began with the first glimmer of spring. To trace the uprising of dwellings, stores, wharves and boarding-houses would be as difficult as perpetual motion. People huddled in shanties and lived on barges moored to the bank. Derricks peered up behind the houses, thronged the marshy flats, congregated on the slopes, climbed the precipitous bluffs and established a foothold on every ledge of rock. Pumping-wells and flowing-wells scented the atmosphere with gas and the smell of crude. Smoke from hundreds of engine-houses, black, sooty and defiling, discolored the grass and foliage. Mud was everywhere, deep, unlimited, universal—yellow mud from the newer territory—dark, repulsive, oily mud around the wells—sticky, tricky, spattering mud on the streets and in the yards. J. B. Reynolds, of Clarion county, and Calvin and William J. McComb, of Pittsburg, opened the first store under the new order of things in March of 1860. T. H. and William M. Williams joined the firm. They withdrew to open the Pittsburg store next door. Robson's hardware-store was farther up the main street, on the east side, which ended abruptly at Cottage Hill. William P. Baillee—he lives in Detroit—and William Janes built the first refinery, on the same street, in 1861, a year of unexampled activity. The plant, which attracted people from all parts of the country—Mr. Baillee called it a "pocket-still"—was enlarged into a refinery of five stills, with an output of two-hundred barrels of refined oil every twenty-four hours. Fire destroyed it and the firm built another on the flats near by. On the west side, at the foot of a steep cliff, Dr. S. S. Christy opened a drug-store. Houses, shops, offices, hotels and saloons hung against the side of the hill or sat loosely on heaps of earth by

MAIN STREET, EAST SIDE OF OIL CREEK, OIL CITY, IN 1861.

the creek and river. One evening a half-dozen congenial spirits met in Williams & Brother's store. J. B. Reynolds, afterwards a banker, who died several years since, thought Cornplanter ought to be discarded and a new name given the growing town. He suggested one which was heartily approved. Liquid re-freshments were ordered and the infant was appropriately baptized OIL CITY.

Peter Graff was laid to rest years ago. The venerable James Hasson sleeps in the Franklin cemetery. His son, Captain William Hasson, is an honored resident of the city that owes much to his enterprise and liberality. Capable, broad-minded and trustworthy, he has been earnest in promoting the best inter-ests of the community, the region and the state. A recent benefaction was his splendid gift of a public park—forty acres—on Cottage Hill. He was the first burgess and served with conspicuous ability in the council and the legislature. Alike as a producer, banker, citizen, municipal officer and lawgiver, Captain Hasson has shown himself "every inch a manly man."

When you talk of any better town than Oil City, of any better section than the oil-regions, of any better people than the oilmen, of any better state than Pennsylvania, "every potato winks its eye, every cabbage shakes its head, every beet grows red in the face, every onion gets stronger, every sheaf of grain is shocked, every stalk of rye strokes its beard, every hill of corn pricks up its ears, every foot of ground kicks" and every tree barks in indignant dissent.

Such was the narrow ravine, nowhere sixty rods in width, that figured so grandly as the Valley of Petroleum.

A SPLASH ON OIL CREEK.

The dark mud of Oil Creek! Unbeautiful mud,
That couldn't and wouldn't be nipped in the bud!
    Quite irreclaimable,
    Wholly untamable;
There it was, not a doubt of it,
People couldn't keep out of it;
    On all sides they found it,
    So deep none dare sound it—
    No way to get 'round it.
To their necks babies crept in it,
To their chins big men stept in it;
    Ladies—bless the sweet martyrs!
    Plung'd far over their garters;
    Girls had no exemption,
    Boys sank past redemption;
To their manes horses stall'd in it,
To their ear-tips mules sprawl'd in it!
    It couldn't be chain'd off,
    It wouldn't be drain'd off;
    It couldn't be tied up,
    It wouldn't be dried up;
    It couldn't be shut down,
    It wouldn't be cut down.
Riders gladly abroad would have shipp'd it,
Walkers gladly at home would have skipp'd it.
    Frost bak'd it,
    Heat cak'd it;
    To batter wheels churned it,
    To splashes rains turned it,
    Bad teamsters gol-durned it!
Each snow-flake and dew-drop, each shower and flood
Just seem'd to infuse it with lots of fresh blood,
    Increasing production,
    Increasing the ruction,
    Increasing the suction!
    Ev'ry flat had its fill of it,
    Ev'ry slope was a hill of it,
    Ev'ry brook was a rill of it;
    Ev'ry yard had three feet of it,
    Ev'ry road was a sheet of it;
    Ev'ry farm had a field of it,
    Ev'ry town had a yield of it.
      No use to glare at it;
      No use to swear at it;
    No use to get mad about it,
    No use to feel sad about it;
No use to sit up all night scheming
Some intricate form of blaspheming;
    No use in upbraiding—
    You *had* to go wading,
    Till wearied humanity,
    Run out of profanity,
    Found rest in insanity;
Or winged its bright way—unless dropp'd with a thud—
To the land of gold pavements and no Oil-Creek mud!

VIII.

PITHOLE AND AROUND THERE.

The Meteoric City that Dazzled Mankind—From Nothing to Sixteen-Thousand Population in Three Months—First Wells and Fabulous Prices—Noted Organizations—Shamburg, Red-Hot and Cash-Up—"Spirits" Trying Their Hand—The Pleasantville Furore—Facts Surpassing Fiction in the Wild Scramble for the Almighty Dollar.

> " A lively place in days of yore, but something ails it now."—*Wordsworth*.
> " Ah! who shall lift that wand of magic power?"—*Longfellow*.
> "Wealth flow'd from wood and stream and soil,
> The rock poured forth its amber oil,
> And, lo! a magic city rose."—*Marjorie Meade*.
> " Pithole was the most remarkable town in the oil-regions."—*Edwin C. Bell*.
> " It went up like a rocket and came down like a stick."—*Popular Phrase*.
> "We are such stuff as dreams are made of."—*Shakespeare*.

ITHOLE, "the magic city," had little in its antecedents to betoken the meteoric rise and fall of the most remarkable oil-town that ever "went up like a rocket and came down like a stick." The unpoetic name of Pithole Creek was applied to the stream which flows through Allegheny township and bounds Cornplanter for several miles on the east. It empties into the Allegheny River eight miles above Oil City and was first mentioned by Rev. Alfred Brunson, an itinerant Methodist minister, in his "Western Pioneer" in 1819. Upheavals of rock left a series of deep pits or chasms on the hills near the mouth of the stream. From the largest of these holes a current of warm air repels leaves or pieces of paper. Snow melts around the cavity, which is of unknown depth, and the air is a mephitic vapor or gas. A story is told of three hunters who, finding the snow melted on a midwinter day, determined to investigate. One of them swore it was an entrance to the infernal regions and that he intended to warm himself. He sat on the edge of the hole, dangled his feet over the side, thanked the devil for the opportune heat, inhaled the gas and tumbled back insensible. His companions dragged him away and the investigation ended summarily. Seven miles up the creek, in the northeast corner of Corn-

Ruins of the Metropolitan Hotel, Pithole, Pa.

157

planter, Rev. Walter Holmden was a pioneer settler. Choosing a tract of two-hundred acres, he built a log-house on the west bank of the creek, cleared a few acres, struggled with poverty and died in 1840. Mr. Holmden was a fervent Baptist preacher. Thomas Holmden occupied the farm after the good old man's decease, with the Copelands and Blackmers and James Rooker as neighbors. Developments had covered the farms from the Drake well to Oil City. Operators ventured up the ravines, ascended the hills and began to take chances miles from either side of Oil Creek. Successful wells on the Allegheny River broadened opinions regarding the possibilities of petroleum. Nervy men invaded the eastern portion of Cornplanter, picking up lands along Pithole Creek and its tributaries. I. N. Frazer, fresh from his triumph on Cherry Run as joint-owner of the Reed well, desired fresh laurels. He organized the United-States Oil-Company, leased part of the Holmden farm for twenty years and started a well

FRAZER WELL, ON HOLMDEN FARM, PITHOLE, IN MAY, 1865.

in the fall of 1864. The primitive derrick was reared in the woods below the Holmden home. At six-hundred feet the "sixth sand"—generally called that at Pithole—was punctured. Ten feet farther the tools proceeded, the drillers watching intently for signs of oil. On January seventh, 1865, the torrent broke loose, the well flowing six-hundred-and-fifty barrels a day and ceasing finally on November tenth. A picture of the well, showing Frazer with his back to the tree beside his horse and a group of visitors standing around, was secured in May. Kilgore & Keenan's Twin wells, good for eight-hundred barrels, were finished on January seventeenth and nineteenth. The unfathomable mud and disastrous floods of that memorable season retarded the hegira from other sections, only to intensify the excitement when it found vent. Duncan & Prather bought Holmden's land for twenty-five-thousand dollars and divided the flats and slopes into half-acre leases. The first of May witnessed a small clearing in the forest, with three oil-wells, one drilling-well and three houses as its sole evidences of human handiwork.

Ninety days later the world heard with unfeigned surprise of a "city" of sixteen-thousand inhabitants, possessing most of the conveniences and luxuries of the largest and oldest communities! Capitalists eager to invest their greenbacks thronged to the scene. Labor and produce commanded extravagant figures, every farm for miles was leased or bought at fabulous rates, money circulated like the measles and for weeks the furore surpassed the frantic ebullitions of Wall Street on Black Friday! New strikes perpetually inflated the mania. Speculators wandered far and wide in quest of the subterranean wealth that promised to outrival the golden measures of California or the silver-lodes of Nevada. The value of oil-lands was reckoned by millions. Small interests in single wells brought hundreds-of-thousands of dollars. New York, Philadelphia, Boston and Chicago measured purses in the insane strife for territory. Hosts of adventurers sought the new Oil-Dorado and the stocks of countless "petroleum companies" were scattered broadcast over Europe and America. An ambitious operator sold *seventeen*-sixteenths in one well and shares in leases were purchased ravenously. A half-acre lease on the Holmden farm realized bonuses of twenty-four-thousand dollars before a well was drilled on the property and the swarm of dealers resembled the plague of locusts in Egypt in number and persistence!

Everything favored the growth of Pithole. The close of the war had left the country flooded with paper currency and multitudes of men thrown upon their own resources. Hundreds of these flocked to the inviting "city," which presented manifold inducements to venturesome spirits, keen shysters, unscrupulous stock-jobbers, needy laborers and dishonest tricksters. The post-office speedily ranked third in Pennsylvania, Philadelphia and Pittsburg alone excelling it. Seven chain-lightning clerks assisted Postmaster S. S. Hill to handle the mail. Lines of men extending a block would await their turns for letters at the general-delivery. It was a roystering time! Hotels, theaters, saloons, drinking-dens, gambling-hells and questionable resorts were counted by the score. A fire-department was organized, a daily paper established and a mayor elected. Railways to Reno and Oleopolis were nearly completed before "the beginning of

JOHN A. MATHER.

the end" came with terrible swiftness. In November and December the wells declined materially. The laying of pipe-lines to Miller Farm and Oleopolis, through which the oil was forced to points of shipment by steam-pumps, in one week drove fifteen-hundred teams to seek work elsewhere. Destructive fires accelerated the final catastrophe. The graphic pen of Dickens would fail to give an adequate idea of this phenomenal creation, whose career was a magnified type of dozens of towns that suddenly arose and as suddenly collapsed in the oil-regions of Pennsylvania.

Pithole had many wells that yielded freely for some time. The Homestead, on the Hyner farm, finished in June of 1865, proved a gusher. On August first the Deshler started at one-hundred barrels; on August second the Grant, at four-hundred-and-fifty barrels; on August twenty-eighth the Pool, at eight-hundred barrels; on September fifth the Ogden, at one-hundred barrels, and on September fifteenth Pool & Perry's No. 47, at four-hundred barrels. The Frazer improved during the spring to eight-hundred barrels, while the Grant reached seven-hundred in September. On November twenty-second the Eureka joined the chorus at five-hundred barrels. The daily production of the Holmden farm exceeded five-thousand barrels for a limited period, with a proportionate yield of seven-dollar crude from adjacent tracts.   John A. Mather, the veteran Titusville photographer, discarded his camera to become a full-fledged oilman. He bored a well that tinctured the suburban slope of Balltown a glowing madder.   The frenzy spread.   J. W. Bonta and James A. Bates paid James Rooker two-hundred-and-eighty-thousand dollars for his hundred-acre farm, south of the Holmden.   Rooker, a hard-working tiller of the soil, lived in a kind of rookery and earned a poor subsistence by constant toil.   He stuck to the money derived from the sale of his farm, moved west and died at a goodly age.   A neighbor refused eight-hundred-thousand dollars for his barren acres.   "I don't keer ter hev my buckwheat tramped uver," he explained, "but you kin hev this farm next winter fer a million!" He kept the farm, reaped his crop and was not disturbed until death compelled him to lodge in a plot six feet by two.

Bonta & Bates did not linger for "two blades of grass to grow where one grew before."   Within two months they disposed of ninety leases for four-hundred-thousand dollars and half the oil!   They spent eighty-thousand on the Bonta House, a sumptuous hostlery.   Duncan & Prather leased building-lots at a yearly rental of one-hundred to one-thousand dollars.   First, Second and Holmden streets bristled with activity.   The Danforth House stood on a lot subleased for fourteen-thousand dollars bonus.   Sixty hotels could not accommodate the influx of guests.   Beds, sofas and chairs were luxuries for the few. "First come, first served," was the rule.   The many had to seek the shaving-pile, the hay-cock or the tender side of a plank.   Some mingled promiscuously in "field-beds"—rows of "shake-downs" on attic floors.   Besides the Bonta and Danforth, the United States, Chase, Tremont, Buckley, Lincoln, Sherman, St. James, American, Northeast, Seneca, Metropolitan, Pomeroy and fifty hotels of minor note flourished.   If palaces of sin, gorgeous bar-rooms, business-houses and places of amusement abounded, churches and schools marked the moral sentiment.   Fire wiped out the Tremont and adjoining houses in February of 1866.   Eighty buildings went up in smoke on May first and June thirteenth. Thirty wells and twenty-thousand barrels of oil went the same road in August. The best buildings were torn down, to bloom at Pleasantville or Oil City.   The disappearance of Pithole astonished the world no less than its marvelous growth. The Danforth House sold for sixteen dollars, to make firewood!   The railroads were abandoned and in 1876 only six voters remained.   A ruined tenement, a deserted church and traces of streets alone survive.   Troy or Nineveh is not more desolate.

In July of 1865 Duncan & Prather granted Henry E. Picket, George J. Sherman and Brian Philpot, of Titusville, a thirty-day option on the Holmden farm for one-million-three-hundred-thousand dollars.   Mr. Sherman arranged to sell the property in New York at sixteen-hundred-thousand!   The wells already down produced largely, seventy more were drilling and the annual ground-rents footed up sixty-thousand dollars.   The Ketcham forgeries tangled the funds of the New-Yorkers and negotiations were opened with H. H. Honore, of Chicago. After dark on the last day of the option Honore tendered the first payment— four-hundred-thousand dollars.   It was declined, on the ground that the business day expired at sundown, and litigation ensued.   A compromise resulted in the transfer of the property to Honore.   The deal involved the largest sum ever paid in the oil-regions for a single tract of land. The bubble burst so quickly that the Chicago purchaser, like Benjamin Franklin, "paid too much for the whistle." Col. A. P. Duncan commanded the Fourth Cavalry Company, the first mustered in Venango county, every member of which carried to the war a small Bible presented by Mrs. A. G. Egbert, of Franklin. Tall, erect, of military bearing and undoubted integrity, he lived at Oil City and died years ago. Duncan & Prather owned one of the two banks that handled car-loads of money in the dizziest town that ever blasted radiant hopes and shriveled portly pocket-books.

Pithole was the Mecca of a legion of operators whose history is part and parcel of the oil-development. Phillips Brothers, giants on Oil Creek, bought farms and drilled extensively. Frederic Prentice and W. W. Clark, who figured in two-thirds of the largest transactions from Petroleum Centre to Franklin, held a full hand. Frank W. Andrews, John Satterfield, J. R. Johnson, J. B. Fink, A. J. Keenan—the first burgess—D. H. Burtis, Heman Janes, "Pap" Sheak-ley, L. H. Smith and hundreds of similar calibre were on deck. John Galloway, known in every oil-district of Pennsylvania and West Virginia as a tireless hustler, did not let Pithole slip past unnoticed. He has been an operator in all the fields since his first appearance on Oil Creek in the fall of 1861. Sharing in the prosperity and adversity of the oil-regions, he has never been hoodooed or bankrupted. His word is his bond and his promise to pay has always meant one-hundred cents on the dollar. More largely interested in producing than ever, he attends to business at Pittsburg and lives at Jamestown, happy in his deserved success, in the love of his family and the esteem of countless friends. Mr. Galloway's pedestrian feats would have crowned him with olive-wreaths at the Olympic games.

JOHN GALLOWAY.

Deerfoot could hardly have kept up with him on a twenty-mile tramp to see an important well or hit a farmer for a lease before breakfast. He's a good one!

The Swordsman's Club attained the highest reputation as a social organization. One night in 1866, when Pithole was at the zenith of its fame, John Satterfield, Seth Crittenden, Alfred W. Smiley, John McDonald, George Burchill, George Gilmore, Pard B. Smith, L. H. Smith, W. H. Longwell and other congenial gentlemen met for an evening's enjoyment. The conversation turned upon clubs. Smiley jumped to his feet and moved that "we organize a club." All assented heartily and the Swordman's Club was organized there and then, with Pard B. Smith as president and George Burchill as secretary. Elegant rooms were fitted up, the famous motto of "R. C. T." was adopted and the club gave a series of most elaborate "promenade concerts and balls" in 1866-7. Invitations to these brilliant affairs were courted by the best people of Oildom. The club dissolved in 1868. Its membership included four congressmen, two ex-governors wore its badge and scores of men conspicuous in the state and nation had the honor of belonging to the Swordman's. At regular meetings "the feast of reason and the flow of soul" blended merrily with the flowing bowl. Sallies of bright wit, spontaneous and never hanging fire, were promptly on schedule time. Good fellowship prevailed and C. C. Leonard immortalized the club in his side-splitting "History of Pithole." Verily the years slip by. Long ago the ephemeral town went back to its original pasture, long ago the facetious historian went back to dust, long ago many a good clubman's sword turned into rust. Pard B. Smith runs a livery in Cleveland, Longwell is in Oil City, Smiley——he represented Clarion county twice in the Legislature—manages the pipe-line at Foxburg, L. H. Smith is in New York and others are scattered or dead. On November twenty-first, 1890, the "Pioneers of Pithole"—among them a number of Swordsmen—had a reunion and banquet at the Hotel Brunswick, Titusville. These stanzas, composed and sung by President Smith and "Alf" Smiley, were vociferously cheered :

> " 'Twas side by side, as Swordsmen true,
> In Pithole long ago,
> We met the boys on common ground
> And gave them all a show.
> In social as in business ways
> Our honor was our law,
> And when a brother lost his grip
> He on the boys could draw.

> CHORUS : "We're the boys, the same old boys,
> Who were there in sixty-five ;
> If any Swordsman comes our way
> He'll find us still alive.

> " What if grim age creeps on apace,
> Our souls will ne'er grow old ;
> We will, as in the Pithole days,
> Stand true as Swordsmen bold.
> In those old days we had our fun,
> But stood for honor true ;
> Here, warmly clasping hand-to-hand,
> Our friendship we renew."

Scarcely less noted was the organization heralded far and wide as "Pithole's Forty Thieves." Well-superintendents, controlling the interests of outside companies, were important personages.

GOVERNOR SHEAKLEY.

Distant stockholders, unable to understand the difficulties and uncertainties attending developments, blamed the superintendents for the lack of dividends. No class of men in the country discharged their duties more faithfully, yet cranky investors in wildcat stocks termed them "slick rascals," "plunderers" and "robbers." Some joker suggested that once a band of Arabian Knights—fellows who stole everything—associated as "The Forty Thieves" and that the libeled superintendents ought to organize a club. The idea captured the town and "Pithole's Forty Thieves" became at once a tangible reality. Merchants, producers, capitalists and business-men hastened to enroll themselves as members. Hon. James Sheakley, of Mercer, was elected president. Social meetings were held regularly and guying greenhorns, who supposed stealing to be the object of the organization, was a favorite pastime. The practical pranks of the "Forty" were laughed at and relished in the whole region. Nine-tenths of the members were young men, honorable in every relation of life, to whom the organization was a genuine joke. They enjoyed its notoriety and delighted to gull innocents who imagined they would purloin engines, derricks, drilling-tools, saw-mills and oil-tanks. Ten years after the band disbanded its president served in Congress and was a leading debater on the Hayes-Tilden muddle. "Pap" Sheakley—as the boys affectionately called him—was the embodiment of integrity, kindliness and hospitality. He operated in the Butler field and lived at Greenville. Bereft of his devoted wife and lovely daughters by "the fell sergeant death," he sold his desolated home and accepted from President Cleveland the governorship of Uncle Sam's remotest Territory. His administration was so satisfactory that President Harrison reappointed him. There is no squarer, truer, nobler man in the public service to-day than James Sheakley, Governor of Alaska.

Rev. S. D. Steadman, the first pastor at Pithole, a zealous Methodist—was universally respected for earnestness and piety. The "Forty Thieves" sent him one-hundred-and-fifty dollars at Christmas of 1866, with a letter commending his moral teachings, his courtesy and charity. Another minister inquired of a Swordsman what the letters of the club's motto—"R. C. T."—signified.
"Religious Counsels Treasured" was the ready response. This raised the club immensely in the divine's estimation and led to a sermon in which he extolled the jolly organization! He "took a tumble" when a deacon smilingly informed him that the letters—a fake proposed in sport—symbolized "Rum, Cards, Tobacco!"

DR. G. SHAMBURG.

Two miles east of Miller-Farm station and four north-west of Pithole, on the eighty-acre tract of Oliver Stowell, the Cherry-Run Petroleum-Company finished a well in February of 1866. It was eight-hundred feet deep, drilled through the sixth sand and pumped one-hundred barrels a day. The company operated systematically, using heavy tools, tall derricks and large casing. It was managed by Dr. G. Shamburg, a man of character and ability, who studied the strata carefully and gathered much valuable data. The second well equalled No. 1 in productiveness and longevity, both lasting for years. J. B. Fink's, a July posy of two-hundred barrels, was the third. The grand rush began in December, 1867, the Fee and Jack Brown wells, on the Atkinson farm, flowing four-hundred barrels apiece. A lively town, eligibly located in a depression of the table-lands, was properly named Shamburg, as a compliment to the genial doctor. The Tallman, Goss, Atkinson and Stowell farms whooped up the production to three-thousand barrels. Frank W. and W. C. Andrews, Lyman and Milton Stewart, John W. Irvin and F. L. Backus had bought John R. Tallman's one-hundred acres in 1865. Their first well began producing in September, 1867, and in 1868 they sold two-hundred-thousand barrels of oil for nearly eight-hundred-thousand dollars! A. H. Bronson—bright, alert, keen in business and popular in society—paid twenty-five-thousand for the Charles Clark farm, a mile north-east. His first well—three-hundred barrels—paid for the property and itself in sixty days. Operations in the Shamburg pool were almost invariably profitable and handsome fortunes were realized. A peculiarity was the presence of green and black oils, a line on the eastern part of the Cherry

Run Company's land defining them sharply. Their gravity and general properties were identical and the black color was attributed to oxide of iron in the rock. Dr. Shamburg died at Titusville and the town he founded is taking a perpetual vacation.

Carl Wageforth, a genius well known in early years as one of the owners of the Story farm, started a "town" in the woods two miles above Shamburg. The "town" collapsed, Wageforth clung to his store a season and next turned up in Texas as the founder of a German colony. He secured a claim in the Lone Star State about thrice the size of Rhode Island, settled it with thrifty immigrants from the "Faderland" and bagged a bushel of ducats. He made and lost fortunes in oil and could no more be kept from breaking out occasionally than measles or small-pox.

East of Petroleum Centre three miles, on the bank of a pellucid stream, John E. McLaughlin drilled a well in 1868 that flowed fourteen-hundred barrels. The sand was coarse, the oil dark and the magnitude of the strike a surprise equal to the answer of the dying sinner who, asked by the minister if he wasn't afraid to meet an angry God, unexpectedly replied: "Not a bit; it's the other chap I'm afraid of!" Excepting the half-dozen mastodons on Oil Creek, the McLaughlin was the biggest well in the business up to that date. Wide-awake operators struck a bee-line for leases. A town was floated in two weeks, a Pithole grocer erecting the first building and labeling the place "Cash-Up" as a gentle hint to patrons not to let their accounts get musty with age. The name fitted the town, which a twelvemonth sufficed to sponge off the slate. Small wells and dry-holes ruled the roost, even those nudging "the big 'un" missing the pay-streak. The McLaughlin—a decided freak—declined gradually and pumped seven years, having the reservoir all to itself. Located ten rods away in any direction, it would have been a duster and Cash-Up would not have existed! A hundred surrounding it did not cash-up the outlay for land and drilling.

North of Pithole the tide crossed into Allegheny township. Balltown, a meadow on C. M. Ball's farm in July, 1865, at the end of the year paraded stores, hotels, a hundred dwellings and a thousand people. Fires in 1866 scorched it and waning production did the rest. Dawson Centre, on the Sawyer tract, budded, frosted and perished. The Morey House, on the Copeland farm, was the oasis in the desert, serving meals that tickled the midriff and might cope with Delmonico's. Farms on Little Pithole Creek were riddled without swelling the yield of crude immoderately. Where are those oil-wells now? Echo murmurs "where?" In all that section of Cornplanter and Allegheny townships a derrick, an engine-house or a tank would be a novelty of the rarest breed.

Attracted by the quality of the soil and the beauty of the location—six-hundred feet above the level of Oil Creek and abundantly watered—in 1820 Abraham Lovell forsook his New York farm to settle in Allegheny township, six miles east by south of Titusville. Aaron Benedict and Austin Merrick came in 1821. John Brown, the first merchant, opened a store in 1833. A pottery, tannery, ashery, store and shops formed the nucleus of a village, organized in 1850 as the borough of Pleasantville. Three wells on the outskirts of town, bored in 1865-6, produced a trifling amount of oil. Late in the fall of 1867 Abram James, an ardent spiritualist, was driving from Pithole to Titusville with three friends. A mile south of Pleasantville his "spirit-guide" assumed control of Mr. James and humped him over the fence into a field on the William Porter farm. Powerless to resist, the subject was hurried to the northern end of the field, contorted violently, jerked through a species of "couchee-couchee dance" and pitched to the ground! He marked the spot with his finger, thrust a penny into the dirt and fell back pale and rigid. Restored to consciousness, he told his astonished companions it had been revealed to him that streams of oil lay beneath and extended several miles in a certain direction. Putting no faith in "spirits" not amenable to flasks, they listened incredulously and resumed their journey. James negotiated a lease, borrowed money—the "spirit-guide" neglected to furnish cash—and planted a derrick where he had planted the penny. On February twelfth, 1868, at eight-hundred-and-fifty deep, the Harmonial Well No. 1 pumped one-hundred-and-thirty barrels!

The usual hurly-burly followed. People who voted the James adventure a fish-story writhed and twisted to drill near the spirited Harmonial. New strikes increased the hubbub and established the sure quality of the territory. Scores of wells were sunk on the Porter, Brown, Tyrell, Beebe, Dunham and other farms for miles. Prices of supplies advanced and machine-shops in the oil-regions ran night and day to meet orders. Land sold at five-hundred to five-thousand dollars an acre, often changing hands three or four times a day. Interests in wells going down found willing purchasers. Strangers crowded Pleasantville, which trebled its population and buildings during the year. It was a second edition of Pithole, mildly subdued and divested of frothy sensationalism. If gigantic gushers did not dazzle, dry-holes did not discourage. If nobody cleared a million dollars at a clip, nobody cleared out to avoid creditors. Nobody had to loaf and trust to Providence for daily bread. Providence wasn't running a bakery for the benefit of idlers and work was plentiful at Pleasantville. The production reached three-thousand barrels in the summer of 1868, dropping to fifteen-hundred in 1870. Three banks prospered and imposing brick-blocks succeeded unsubstantial frames. Fresh pastures invited the floating mass to Clarion, Armstrong and Butler. Small wells were abandoned, machinery was shipped southward and the pretty village moved backward gracefully. Pleasantville had "marched up the hill and then marched down again."

Abram James, a man of fine intellect, nervous temperament and lofty principle, lived at Pleasantville a year. He located a dozen paying wells in other sections, under the influence of his "spirit-guide." The Harmonial was his greatest hit, bringing him wealth and distinction. His worst break—a dry-hole on the Clarion river eighteen-hundred feet deep—cost him six-thousand dollars in 1874. None questioned his absolute sincerity, although many rejected his theories of the supernatural. Whether he is still in the flesh or has become a spirit has not been manifested to his old friends in Oildom.

"Spirits" inspired four good wells at Pithole. One dry hole, a mile southeast of town, seriously depressed stock in their skill as "oil-smellers." An enthusiastic disciple of the Fox sisters, assured of "a big well," drilled two-hundred feet below the sixth sand in search of oil-bearing rock. He drilled himself into debt and Sheriff C. S. Mark—six feet high and correspondingly broad—whom nobody could mistake for an ethereal being, sold the outfit at junk-prices.

Red-Hot, in the palmy era of the Shamburg excitement a place of much sultriness, is cold enough to chill any stray visitor who knew the mushroom at its warmest stage. Windsor Brothers, of Oil City—they built the Windsor Block—drilled a well in 1869 that flowed three-hundred-and-fifty barrels. Others

followed rapidly, people flocked to the newest centre of attraction and a typical
oil-town strutted to the front. The territory lacked the staying quality, the
Butler region was about to dawn and 1871 saw Red-Hot reduced to three
houses, a half-dozen light wells and a muddy road. Lightning-rod pedlars,

RED-HOT, A TYPICAL OIL-TOWN, IN 1870.

book-agents and medical fakirs no longer disturb its calm serenity. Not a
scrap of the tropical town has been visible for two decades.

Tip-Top filled a short engagement. Operations around Shamburg and
Pleasantville directed attention to the Captain Lyle and neighboring farms,
midway between these points. "Ned" Pitcher's well, drilled in 1866 on the
Snedaker farm, east of Lyle, had started at eighty barrels and pumped twenty
for two years. Pithole was booming and nobody thought of Ned's pitcher
until 1868. Many of the wells produced fairly, but the territory soon depreciated
and the elevated town—aptly named by a poet with an eye to the eternal fitness
of things—lost its hold and glided down to nothingness. The hundred-eyed
Argus could not find a sliver that would prick a thumb or tip a top.

Eight miles north-east of Titusville, where Godfrey Hill drilled a dry-hole
in 1860 and two companies drilled six later, the Colorado district finally rewarded gritty operators. Enterprise was benefited by small wells in the vicinity. Down Pithole Creek to its junction with the Allegheny the country was
punctured. Oleopolis straggled over the slope on the river's bank, a pipe-line,
a railroad to Pithole and minor wells contributing to its support. The first well
tackled a vein of natural gas, which caught fire and consumed the rig. The
driller was alone, the owner of the well having gone into the shanty. In a
twinkling flames enveloped the astonished knight of the temper-screw, who
leaped from the derrick, clothes blazing and hair singed off, and headed for the
water. "Boss," he roared in his flight, "jump into the river and say your
prayers quick ! I've bu'sted the bung and hell's running out !"

The Pithole bubble was blown at an opportune moment to catch suckers.
Hundreds of oil-companies had come into existence in 1864, hungry for territory
and grasping at anything within rifle-shot of an actual or prospective "spouter."
The speculative tide flowed and ebbed as never before in any age or nation.
Volumes could be written of amazing transitions of fortune. Scores landed at
Pithole penniless and departed in a few months "well heeled." Others came
with "hatfuls of money" and went away empty-handed. Thousands of stockholders were bitten as badly as the sailor whom the shark nipped off by the
waist-band. It was rather refreshing in its way for "country Reubens" to do
up Wall-street sharpers at their own game. Shrewd Bostonians, New-Yorkers
and Philadelphians, magnates in business and finance, were snared as readily

as hayseeds who buy green-goods and gold-bricks. There were no flies on the smooth, glib Oily Gammon whose mouth yielded more lubricating oil than the biggest well on French Creek. His favorite prey was a pilgrim with a bursting wallet or the agent of an eastern petroleum-company. A well pouring forth three, six, eight, ten, twelve or fifteen-hundred barrels of five-dollar crude every twenty-four hours was a spectacle to fire the blood and turn the brain of the most sluggish beholder. "Such a well," he might calculate, "would make me a millionaire in one year and a Crœsus in ten." The wariest trout would nibble at bait so tempting. The schemer with property to sell had "the very thing he wanted" and would "let him in on the ground-floor." He met men who, driving mules or jigging tools six months ago, were "oil-princes" now. Here lay a tract, "the softest snap on top of the earth," only a mile from the Great Geyser, with a well "just in the sand and a splendid show." He could have it at a bargain-counter sacrifice—one-hundred-thousand dollars and half the oil. The engine had given out and the owner was about to order a new one when called home by the sudden death of his mother-in-law. Settling the old lady's estate required his entire attention, therefore he would consent to sell his oil-interests "dirt-cheap" to a responsible buyer who would push developments. The price ought to be two or three times the sum asked, but the royalty from the big wells sure to be struck would ultimately even up matters. The tale was plausible and the visitor would "look at the property." He saw real sand on the derrick-floor and everything besmeared with grease. The presence of oil was unmistakable. Drilling ten feet into the rich rock would certainly tap the jugular and—glorious thought!—perhaps outdo the Great Geyser itself. He closed the deal, telegraphed for an engine—he was dying to see that stream of oil climbing skywards—and chuckled gleefully. The keen edge of his delight might have been dulled had he known that the well was *through*, not merely *to*, the sand and absolutely guiltless of the taint of oil! He did not suspect that barrels of crude and buckets of sand from other wells had been dumped into the hole at night, that the engine had been disabled purposely and that another innocent was soon to cut his wisdom-teeth! He found out when the well "came in dry" that Justice Dogberry was not a greater ass and that the fool-killer's snickersnee was yearning for him. Possibly he might by persistent drilling find paying wells and get back part of his money, but nine times out of ten the investment was a total loss and the disgusted victim quit the scene with a new interpretation of the scriptural declaration: "I was a stranger and ye took me in."

The methods of "turning an honest penny" varied to fit the case. To "doctor" a well by dosing it with a load of oil was tame and commonplace. In three instances wells sold at fancy prices were connected by underground pipes with tanks of oil at a distance. When the parties arrived to "time the well" the secret pipe was opened. The oil ran into the tubing and pumped as though coming direct from the sand! The deception was as perfect as the oleomargarine the Pennsylvania State Board of Agriculture pronounced "dairy butter of superior quality!" "Seeing is believing" and there was the oil. They had seen it pumping a steady stream into the tank, timed it, gauged it, smelled it. The demonstration was complete and the cash would be forked over, a twenty-barrel well bringing a hundred-barrel price! A smart widow near Pithole sold her farm at treble its value because of "surface indications" she created by emptying a barrel of oil into a spring. The farm proved good territory, much to the chagrin of the widow, who roundly abused the purchasers

12

for "cheatin' a poor lone woman!" Selling stock in companies that held lands, or interests in wells to be drilled "near big gushers"—they might be eight or ten miles off—was not infrequent. On the other hand, a very slight risk often brought an immense return. Parties would pay five-hundred dollars for the refusal of a tract of land and arrange with other parties to sink a well for a small lease on the property. If the well succeeded, one acre would pay the cost of the entire farm; if it failed, the holders of the option forfeited the trifle that secured it and threw up the contract. It was risking five-hundred dollars on the chance, not always very remote, of gaining a half-million.

Sometimes the craze to invest bordered upon the ludicrous. Sixteenths and fractions of sixteenths in producing, non-producing, drilling, undrilled and never-to-be-drilled wells "went like hot cakes" at two to twenty-thousand dollars. A newcomer, in his haste to "tie onto something," shelled out one-thousand dollars for a share in a gusher that netted him two quarts of oil a day! Another cheerfully paid fifteen-thousand for the sixteenth of a flowing well which discounted the Irishman's flea—"you put your finger on the varmint and he wasn't there"—by balking that night and declining ever to start again! At a fire in 1866 water from a spring, dashed on the blaze, added fuel to the flames. An examination showed that oil was filling the spring and water-wells in the neighborhood. From the well in Mrs. Reichart's yard the wooden pump brought fifty barrels of pure oil. L. L. Hill's well and holes dug eight or ten feet had the same complaint. Excitement blew off at the top gauge. The *Record* devoted columns to the new departure. Was the oil so impatient to enrich Pitholians that, refusing to wait for the drill to provide an outlet, it burst through the rocks in its eagerness to boom the district? Patches of ground the size of a quilt sold for two, three or four-hundred dollars and rows of pits resembling open graves decorated the slope. In a week a digger discovered that a break in the pipe-line supplied the oil. The leak was repaired, the pits dried up, the water-wells resumed their normal condition and the fiasco ended ignominiously. It was a modern version of the mountain that set the country by the ears to bring forth a mouse.

In the swish and swirl of Pithole teamsters—a man with two stout horses could earn twenty dollars a day clear—drillers and pumpers played no mean part. They received high wages and spent money freely. Variety-shows, music-halls—with "pretty waiter-girls"—dance-houses, saloons, gambling-hells and dens of vice afforded unlimited opportunities to squander cash and decency and self-respect. Many a clever youth, flushed with the idea of "sowing his wild oats," sacrificed health and character on the altars of Bacchus and Venus. Many a comely maiden, yielding to the wiles of the betrayer, rounded up in the brothel and the potter's field. Many a pious mother, weeping for the wayward prodigal who was draining her life-blood, had reason to inquire: "Oh, where is my boy to-night?" Many a husband, forgetting the trusting wife and children at home, wandered from the straight path and tasted the forbidden fruit. Many a promising life was blighted, many a hopeful career blasted, many a reputation smirched and many a fond heart broken by the pitfalls and temptations of Pithole. Dollars were not the only stakes in the exciting game of life—good names, family ties, bright prospects, domestic happiness and human souls were often risked and often lost. "The half has never been told."

Mud was responsible for the funniest – to the spectators—mishap that ever convulsed a Pithole audience. A group of us stood in front of the Danforth House at the height of the miry season. Thin mud overflowed the plank-crossing and a grocer laid short pieces of scantling two or three feet apart for pedestrians to step on. A flashy sport, attired in a swell suit and a shiny beaver, was the first to take advantage of the improvised passage. Half-way across the scantling to which he was stepping moved ahead of his foot. In trying to recover his balance the sport careened to one side, his hat flew off and

he landed plump on his back, in mud and water three feet deep! He disappeared beneath the surface as completely as though dropped into the sea, his head emerging a moment later. Blinded, sputtering and gasping for breath, he was a sight for the gods and little fishes! Mouth, eyes, nose and ears were choked with the dreadful ooze. Two men went to his assistance, led him to the rear of the hotel and turned the hose on him. His clothes were ruined, his gold watch was never recovered and for weeks small boys would howl: "His name is Mud!"

AN INVOLUNTARY MUD-BATH.

John Galloway, on one of his rambles for territory, ate dinner at the humble cabin of a poor settler. A fowl, tough, aged and peculiar, was the principal dish. In two weeks the tourist was that way again. A boy of four summers played at the door, close to which the visitor sat down. A brood of small chickens approached the entrance. "Poo', ittey sings," lisped the child, "oo mus' yun away; here's 'e yasty man 'at eated up oos mammy." The good woman of the shanty had stewed the clucking-hen to feed the unexpected guest.

A maiden of uncertain age owned a farm which various operators vainly tried to lease. Hoping to steal a march on the others, one smooth talker called the second time. "I have come, Miss Blank," he began, "to make you an offer." He didn't get a chance to add "for your land." The old girl, not a gosling who would let a prize slip, jumped from her chair, clasped him about the neck and exclaimed: "Oh! Mr. Blank, this is so sudden, but I'm yours!" The astounded oilman shook her off at last and explained that he already had a wife and five children and wanted the farm only. The clinging vine wept and stormed, threatened a breach-of-promise suit and loaded her dead father's blunderbuss to be prepared for the next intruder.

Col. Gardner, "a big man any way you take him," was Chief-of-Police at Pithole. He has operated at Bradford and Warren, toyed with politics and military affairs and won the regard of troops of friends. Charles H. Duncan, of Oil City—his youthful appearance suggests Ponce de Leon's spring—served in the borough-council, of which James M. Guffey, the astute Democratic leader and successful producer, was clerk. Col. Morton arrived in August of 1865 with a carpet-bag of job-type. His first work—tickets for passage over Little Pithole Creek—the first printing ever done at Pithole, was never paid for. The town had shoals of trusty, generous fellows—"God's own white boys," Fred Wheeler dubbed them—whose manliness and enterprise and liberality were always above par.

A young divine preached a sermon at Pithole, on the duty of self-consecration, so effectively that a hearer presented him with a bundle of stock in a company operating on the Hyner farm. The preacher sold his shares for ten-thousand dollars and promptly retired from the pulpit to study law ! Rev. S. D. Steadman, while a master of sarcasm that would skewer a hypocrite on the point of irony, was particularly at home in the realm of the affections and of the ideal. In matters of the heart and soul few could with surer touch set aflow the founts of tender pathos. He met his match occasionally. Rallying a friend on his Calvinism, he said, "I believe Christians may fall from grace." "Brother Steadman," was the quick rejoinder, "you need not argue that ; the flock you're tending is convincing proof that the doctrine is true of your membership."

A good deal of fun has been poked at the Georgia railroad which had cow-catchers at the rear, to keep cattle from walking into the cars, and stopped in the woods while the conductor went a mile for milk to replenish a crying baby's nursing-bottle. On my last trip to Pithole by rail there were no other passengers. The conductor sat beside me to chat of former days and the decadence of the town at the northern end of the line. Four miles from Oleopolis fields of wild strawberries "wasted their sweetness on the desert air." In reply to my hint that the berries looked very tempting, the conductor pulled the bell-rope and stopped the train. All hands feasted on the luscious fruit until satisfied. Coleridge, who observed that "Doubtless the Almighty *could* make a finer fruit than the wild strawberry, but doubtless He never did," would have enjoyed the scene. "Don't hurry too much," the conductor called after me at Pithole, "we can start forty minutes behind time and I'll wait for you !" The rails were taken up and the road abandoned in the fall, but the strawberry-picking is as fresh as though it happened yesterday.

Long ago teamsters would start from the mines with twenty bushels of fifteen-cent coal. By the time they reached Pithole it would swell to thirty-five bushels of sixty-cent coal. With oil for back-loading the teamsters made more money then than a bond-juggler with a cinch on the United-States treasury.

A farmer's wife near Pleasantville, who had washed dishes forty years, became so tired of the monotony that, the day her husband leased the farm for oil-purposes, she smashed every piece of crockery in the house and went out on the woodpile and laughed a full hour. It was the first vacation of her married life and dish-washing women will know how to sympathize with the poor soul in her drudgery and her emancipation.

Pithole, Shamburg, Red-Hot, Tip-Top, Cash-Up, Balltown and Oleopolis have passed into history and many of their people have gone beyond the vale of this checkered pilgrimage, yet memories of these old times come back freighted with thoughts of joyous days that shall return no more forever.

PITHOLE REVISITED.

The following lines, first contributed by me to the Oil-City *Times* in 1870, went the rounds twenty-five years ago :

Not a sound was heard, not a shrill whistle's scream,
    As our footsteps through Pithole we hurried ;
Not a well was discharging an unctuous stream
    Where the hopes of the oilmen lay buried !

We walk'd the dead city till far in the night—
    Weeds growing where wheels once were turning—
While seeking to find by the struggling moonlight
    Some symptom of gas dimly burning.

No useless regret should encumber man's breast,
    Though dry-holes and Pitholes may bound him ;
So we lay like a warrior taking his rest,
    Each with his big overcoat 'round him.

Few and short were the prayers we said,
    We spoke not a sentence of sorrow,
But steadfastly gazed on the place that was dead
    And bitterly long'd for the morrow !

We thought, as we lay on our primitive bed,
    An old sand-pump reel for a pillow,
How friends, foes and strangers were heartily bled
    And ruin swept on like a billow !

Lightly we slept, for we dreamt of the scamp,
    And in fancy began to upbraid him,
Who swindled us out of our very last stamp—
    In the grave we could gladly have laid him !

We rose half an hour in advance of the sun,
    But little refreshed for retiring !
And, feeling as stiff as a son of a gun,
    Set off on a hunt for some firing.

Slowly and sadly our hard-tack went down,
    Then we wrote a brief sketch of our story
And struck a bee-line for Oil City's fair town,
    Leaving Pithole alone in its glory !

GENERAL VIEW OF BRADFORD

UP THE ALLEGHENY RIVER.

IX.

A BEE-LINE FOR THE NORTH.

"Be sure you're right, then go ahead."—*Davy Crockett.*
"Jes foller de no'th star an' yu'll cum out right, shuah."—*Uncle Remus.*
"Stay, stay thy crystal tide,
Sweet Allegheny!
I would by thee abide,
Though early friends denied—
They were with thee allied,
Sweet Allegheny.—*Marjorie Meade.*
"Better a year of Bradford than a cycle of Cathay."—*Bradford Era.*
"Don't you turn to the right, don't you turn to the left, but keep in the middle of the road.
—*Popular Melody.*

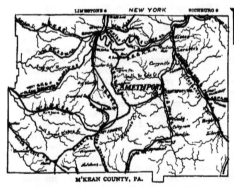

M'KEAN COUNTY, PA.

IN transforming the bleak, uninteresting Valley of Oil Creek into the rich, attractive Valley of Petroleum the course of developments was southward from the Drake well. Although some persons imagined that a pool or a strip bordering the stream would be the limit of successful operations, others entertained broader ideas and believed the petroleum sun was not doomed to rise and set on Oil Creek. The Evans well at Franklin confirmed this view. Naturally the Allegheny River was regarded with favor as the base of further experiments. Quite as naturally the town at the junction of the river and the creek was benefited. The Michigan Rock-Oil Company laid out building-lots and Oil City grew rapidly. Across the Allegheny, on the Downing and Bastian farms, William L. Lay laid out the village of Laytonia in 1863 and improved the ferriage. Phillips & Vanausdall, who struck a thirty-barrel well on the Downing farm in 1861, established a ferry above Bastian's and started the suburbs of Albion and Downington. In 1865 these were merged into Imperial City, which in 1866 was united with Laytonia and Leetown to form Venango City. In 1871 the boroughs of Venango City and Oil City were incorporated as the city of Oil City, with William M. Williams as mayor. Three passenger-bridges, one railroad bridge and an electric street-railway connect the north and south sides of the "Hub of Oildom." Beautiful homes, first-class schools and churches, spacious business-blocks, paved streets, four railroads, electric-lights, water-works, pipe-line offices, strong banks, enormous tube-works, huge refineries, bright newspapers, a paid fire-department, all the modern conveniences and twelve-thousand clever people make Oil City one of the busiest and most desirable towns in or out of Pennsylvania.

The largest of twenty-five or thirty wells drilled around Walnut Bend, six miles up the river, in 1860-65, was rated at two-hundred barrels. Four miles farther, two miles north-east of the mouth of Pithole Creek, John Henry settled on the north bank of the river in 1802. Henry's Bend perpetuates the name of this brave pioneer, who reared a large family and died in 1858. The farm opposite Henry's, at the crown of the bend, Heydrick Brothers, of French Creek township, leased in the fall of 1859. Jesse Heydrick organized the Wolverine Oil-Company, the second ever formed to drill for petroleum. Thirty shares of stock constituted its capital of ten-thousand-five-hundred dollars. The first well, one-hundred-and-sixty feet deep, pumped only ten barrels a day, giving Wolverine shares a violent chill. The second, also sunk in 1860, at three-hundred feet flowed fifteen-hundred barrels! Beside this giant the Drake well was a midget. The Allegheny had knocked out Oil Creek at a stroke, the production of the Heydrick spouter doubling that of all the others in the region put together. It was impossible to tank the oil, which was run into a piece of low ground and formed a pond through which yawl-boats were rowed fifty rods! By this means seven-hundred barrels a day could be saved. At last the tubing was drawn, which decreased the yield and rendered pumping necessary. The well flowed and pumped about one-hundred-thousand barrels, doing eighty a day in 1864-5, when the oldest producer in Venango county. It was a celebrity in its time and proved immensely profitable. In December of 1862 Jesse Heydrick went to Irvine, forty miles up the river, to float down a cargo of empty barrels. Twenty-five miles from Irvine, on the way back, the river was frozen from bank to bank. He sawed a channel a mile, ran the barrels to the well, filled them, loaded them in a flat-boat and arrived at Pittsburg on a cold Saturday before Christmas. Oil was scarce, the zero weather having prevented shipments, and he sold at thirteen dollars a barrel. A thaw set in, the market was deluged with crude and in four days the price dropped to two dollars! Stock-fluctuations had no business in the game with petroleum.

Wolverine shares climbed out of sight. Mr. Heydrick bought the whole batch, the lowest costing him four-thousand dollars and the highest fifteen-thousand. He sold part of his holdings on the basis of fifteen-hundred-thousand dollars for the well and farm of two-hundred acres, forty-three-thousand times the original value of the land! Heydrick Brothers bored seventy wells on three farms in President township, one of which cost eighteen months' labor and ten-thousand dollars in money and produced nine barrels of oil. They disposed of it, the new owner fussed with it and for five years received fifteen barrels of oil a day.

Accidents and incidents resulting from the Wolverine operations would fill a dime-novel. Jesse Heydrick, organizer of the company, went east with two or three-hundred-thousand dollars, presumably to "play Jesse" with the bulls and bears of Wall Street. He returned in a year or more destitute of cash, but loaded with entertaining tales of adventure. He told a thrilling story of his abduction from a New-York wharf and shipment to Cuba by a band of kidnappers, who stole his money and treated him harshly. He endured severe hardships and barely escaped with his life and a mine of experience. Working his way north, he resumed surveying, prepared valuable maps of the Butler field and was a standard authority on oil-matters in the district. For years he has been connected with a pipe-line in Ohio. Mr. Heydrick is cultured and social, brimful of information and interesting recitals, and not a bilious crank who thinks the world is growing worse because he lost a fortune. His brother, Hon. C. Heydrick, of Franklin, was president of the Oil-City Bank, incorporated in 1864 as a bank of issue and forced to the wall in 1866 by the failure of Culver, Penn & Co. Judge Heydrick served one year on the Supreme Bench with distinguished honor. He ranks with the foremost lawyers of Pennsylvania in ability and legal attainments. James Heydrick was a skilled surveyor and Charles W. resided at the old homestead on French Creek. Heydrick Brothers were "the Big Four" in developments that brought the Allegheny River region into the petroleum-column. It is singular that the Heydrick well, located at random thirty-six years ago, was the largest ever struck on the banks of the crooked, zig-zagged, ox-bowed stream.

Eight rods square on the Heydrick tract leased for five-thousand dollars and fifty per cent. of the oil, while the Wolverine shares attested the increasing wealth of the oil-interest and the pitch to which oil-stocks might rise. Hussey & McBride secured the Henry farm and obtained a large production in 1860-1. The Walnut Tree and Orchard wells headed the list. Warren & Brother pumped oil from Pithole to Henryville, a small town on the flats, of whose houses, hotels, stores and shipping-platforms no scrap survives. The Commercial Oil-Company bought the Culbertson farm, above Henry, and drilled extensively on Muskrat and Culbertson Runs. Patrick McCrea, the first settler on the river between Franklin and Warren, the first Allegheny ferryman north of Franklin and the first Catholic in Venango county, migrated from Virginia in 1797 to the wilds of North-western Pennsylvania. C. Curtiss purchased the McCrea tract of four-hundred acres in 1861 and stocked it in the Eagle Oil-Company of Philadelphia. Fair wells were found on the property and the town of Eagle Rock attained the dignity of three-hundred buildings. An eagle could fly away with all that is left of the town and the wells.

Farther along Robert Elliott, who removed from Franklin, owned one-thousand acres on the south side of the river and built the first mill in President township. Rev. Ralph Clapp built a blast-furnace in 1854-5, a mile from the mouth of Hemlock Creek, at the junction of which with the Allegheny a big hotel, a store and a shop are situated. Mr. Clapp gained distinction in the pulpit and in business, served in the Legislature and died in 1865. His son, Edwin E. Clapp, has a block of six-thousand acres, the biggest slice of undeveloped territory in Oildom. Productive wells have been sunk on the river-front, but he has invariably refused to sell or lease save once. To Kahle Brothers, for the sake of his father's friendship for their father, he leased two-hundred acres, on which many good wells are yielding nicely. Preferring to keep his own lands untouched until he "gets good and ready," he operated largely at Tidioute, he and his brother, John M. Clapp, acquiring great wealth. He was chairman of the Producers' Council and active in the memorable movements of 1871-3. He built for his home the President Hotel, furnishing it with

EDWIN E. CLAPP.

every comfort and luxury except the one no *bachelor* can possess. From him Macadam, Talbot and Nicholson could have learned much about road making. At his own expense he has constructed many miles of first-class roads in President, grading, ditching and leveling in a fashion to make a bicycler's mouth water. There is not a scintilla of pride or affectation in his composition. It is told that an agent of the Standard Oil-Company appointed a time to meet him "on important business." The interview lasted two minutes. "What is the business?" interrogated Clapp. "Our company authorizes me to offer you one-million dollars for your lands in President and I am prepared to pay you the money." "Anything else?" "No."

"Well, the land isn't for sale; good-morning!" Off went Clapp as coolly as though he had merely received a bid for a bushel of potatoes. Whether true or not, the story is characteristic. As a friend to swear by, a helper of the poor, a believer in fair-play, a prime joker and an inimitable weaver of comic yarns few can equal and none excel the "President of President."

Around Tionesta, the county-seat of Forest, numerous holes were punched. Thomas Mills, who operated in Ohio and missed opening the Sisterville field by a scratch, drilled in 1861-2. The late George S. Hunter—he built Tionesta's first bridge and ought to have a monument for enterprise—hunted earnestly for paying territory. Up Tionesta Creek operations extended slowly, but developments in 1882-3 atoned for the delay. Then Forest county was "the cynosure of all eyes," each week springing fresh surprises. Balltown had a crop of dry holes, followed by wells of all grades from twenty barrels to fifteen-hundred. At Henry's Mills and on the Cooper lands, north-east of Balltown and running into Warren county, spouters were decidedly in vogue. Reno No. 1 well, finished in December of 1882, flowed twenty-eight-hundred barrels! Reno No. 2, McCalmont Oil-Company's No. 1, Patterson's and the Anchor Oil-Company's No. 14 went over the fifteen-hundred mark. In the midst of these gushers Melvin, Walker & Shannon's duster indicated spotted territory, uncertain as the verdict of a petit jury. The Forest splurge held the entire oil-trade on the ragged edge for months. Every time one or more fellows took to the woods to manipulate a wildcat-well oil took a tumble. Notwithstanding the magnitude of the business, with thirty-six-million barrels of oil in stock and untold millions of dollars invested, the report from Balltown or Cooper of a new strike caused a bad break. Some owners of important wells worked them as "mysteries" to "milk the trade." Derricks were boarded tightly, armed men kept intruders from approaching too near and information was withheld or falsified until the gang of manipulators "worked the market." To offset this leading dealers employed "scouts," whose mission was to get correct news at all hazards. The duties of these trusty fellows involved great labor, night watches, incessant vigilance and sometimes personal danger. The "mystery" racket and the introduction of "scouts" were new elements in the business, necessitated by the peculiar tactics of a small clique whose methods were not always creditable. The passing of the Forest field, which declined with unprecedented rapidity, practically ended the system that had terrorized the oil-exchanges in New York, Oil City, Bradford and Pittsburg. The collapse of the Cooper pool was more unexpected than the striking of a gusher would be under any circumstances. Its influence upon oil-values was ridiculously disproportionate to its merits.

Closely allied to Balltown and Cooper in its principal features, its injurious effects and sudden depreciation, was the field that taught the Forest lesson. On May nineteenth, 1882, the oil-trade was paralyzed by the report of a big well in Cherry Grove township, Warren county, miles from previous developments. The general condition of the region was prosperous, with an advancing market and a favorable outlook. The new well—the famous "646"—struck the country like a cyclone. Nobody had heard a whisper of the finding of oil in the hole George Dimick was drilling near the border of Warren and Forest. The news that it was flowing twenty-five-hundred barrels flashed over the wires with disastrous consequences. The excitement in the oil-exchanges, as the price of certificates dropped thirty to fifty per cent. in a few moments, was indescribable. Margins and small holders were wiped out in a twinkling and the losses aggregated millions. It was a panic of the first water, far-reaching and ruinous. A plunge from one-thirty to fifty-five cents for crude meant distress and bankruptcy to thousands of producers and persons carrying oil. Men comfortably off in the morning were beggared by noon. Other wells speedily followed "646." The Murphy, the Mahoopany and scores more swelled the daily yield to thirty-thousand barrels. Five-hun-

IN THE MIDDLE FIELD.

dred wells were rushed down with the utmost celerity. Big companies bought lands at big prices and operated on a big scale. Pipe-lines were laid, iron-tanks erected and houses reared by the hundred. Cherry Grove dwarfed the richest portions of the region into insignificance. It bade fair to swamp the business, to flood the world with cheap oil, to compel the abandonment of entire districts and to crush the average operator. But if the rise of Cherry Grove was vividly picturesque, its fall was startlingly phenomenal. One dark December morning the workmen noticed that the Forest Oil-Company's largest gusher had stopped flowing. Within a week the disease had spread like an epidemic. Spouters ceased to spout and obstinately declined to pump. The yield was counted by dozens of barrels instead of thousands. In January one-fourth the wells were deserted and the machinery removed. Three-hundred wells on April first yielded hardly two-thousand barrels, three-quarters what "646" or the Murphy had done alone! The suddenness of the topple cast Oil Creek into the shade and eclipsed Pithole itself. Piles of junk represented miles of pipe-lines and acres of tanks. The Cooper fever was breaking out and, with Henry's Mills and Balltown, repeated in 1883 the hurrah of 1882. For eleven months the Forest–Warren pools fretted and fumed, producing five-million barrels of oil and having the trade by the throat. In that brief period Cherry Grove came and went, Cooper threatened and subsided, and Balltown was bowled out. Nine-tenths of the operators figured as heavy losers. Pennsylvania's production shrank from ninety-thousand barrels to sixty-thousand and a healthy reaction set in. Petroleum developments often presented remarkable peculiarities, but the strangest of all was the readiness with which speculators time and again fell a prey to the schemes of Forest–Warren jobbers, whose "picture is turned to the wall."

The professional "oil-scout" first became prominent at Cherry Grove. He was neither an Indian fighter nor a Pinkerton detective, although possessing the courage and sharpness of both. He combined a knowledge of woodcraft and human-nature with keen discernment, acute judgment and infinite patience.

S. B. HUGHES.

S. B. Hughes, J. C. Tennent, P. C. Boyle, J. C. McMullen, Frank H. Taylor, Joseph Cappeau, James Emery and J. H. Rathbun were captains in the good work of worrying and circumventing the "mystery" men. Hughes rendered service that won the confidence of his employers and brought him a competence. Never caught napping, for one special feat he was said to have received ten-thousand dollars. It was not uncommon for him and his comrades to keep their boots on a week at a stretch, to snatch a nap under a tree or on a pile of casing, to creep on all-fours inside the guard-lines and watch pale Luna wink merrily and the bright stars twinkle while reclining on the damp ground to catch the faintest sound from a mystified well. Boyle and Tennent made brilliant plays in the campaign of 1882-3. Captain J. T. Jones, failing to get correct information regarding "646," lost heavily on long oil when the Cherry Grove gusher hypnotized the market and sent Tennent from Bradford to size up the wells and the movements of those manipulating them. Michael Murphy, learning that Grace & Dimick were quietly drilling a wildcat-well on lot 646, smelled a large-sized rodent and concluded to share in the sport. For one-hundred dollars an acre and one-eighth the usufruct Horton, Crary & Co., the Sheffield tanners, sold him lot 619, north-east of 646. Murphy had cut his eye-teeth as an importer—John S. Davis was his partner—of oil-barrels, an exporter of crude and an operator at Bradford. He pushed a well on the south-west corner of his purchase and secured lands in the vicinity. Grace & Dimick held back their well a month to tie up lots and complete arrangements regarding the market. Everything was managed adroitly. The trade had not a glimmer of suspicion that a bombshell might be fired at any moment. Murphy's rig burned down on May fifteenth, he was in Washington trying to close a deed for another tract and "646" was put through the sand. On June second Murphy's No. 1, which he guarded strictly after rebuilding the rig, flowed sixteen-hundred barrels. His No. 2, finished on July third, flowed thirty-six-hundred barrels in twenty-four hours! The Mahoopany and a half-dozen others aided in the demoralization of prices. Murphy sold eighty acres of lot 619 for fifty-thousand dollars to the McCalmont Oil-Company. The Anchor Oil-Company's gusher on lot 647 caught fire, without curtailing the flow, and was burning furiously as "Jim" Tennent arrived from Bradford. The scouts had their hands full, with the "white-sand pools" and the keenest masters of "mystery wells" to demand their best licks.

Watching Murphy's dry-hole on lot 633 was Tennent's initial job. The Whale Oil-Company's duster on lot 648 next claimed the attention of the scouts. It had been drilled below the sand-level and the tools left at the bottom. On Sunday night, July ninth, 1882, Boyle, Tennent and two companions raised the tools by hand, measured the well with a steel-line and telegraphed their principals that it was dry. This report jumped the market on Monday morning from forty-nine cents to sixty. The Shannon well on the Cooper tract needed constant care and the scouts divided the labor. Tennent and Rathbun one night sought to crawl near the well. A twig snapped off and a guard fired, the ball grazing "Jim's" ear. In December Boyle and W. C. Edwards drilled Grandin No. 4 below the sand before the owners knew the rock had been reached. Its failure surprised the trade as much as the success of "646." Boyle actually posted the guards to keep intruders away and they refused to let W. W. Hague, an owner of the well, inside the line until the contractor appeared and permitted him to pass! Boyle and Tennent did fine work north of the Cooper field. At the Shultz well Tennent, in order to make a quick trip of a half-mile to the pipeline telegraph, clung to the tail of Cappeau's horse and kept up with the animal's gallop. Mercury might not have endorsed that style of locomotion, but it served the purpose and got the news to Jones ahead of everybody else. Tennent played the market skillfully, cleared twenty-five-thousand dollars on Macksburg lands and operated with tolerable success in McKean county. Nine years ago he removed to his thousand-acre prairie farm in Kansas, the land of sockless statesman and nimble grasshoppers.

Boyle, brimful of novel resources, puzzled the "mystery" chaps by his bold ingenuity and usually beat them at their own game. He squarely overmatched the field-marshals of manipulation. His fertile brain originated the plan of drilling Grandin No. 4 and other test wells. The night he went to drill the Grace through the sand he paid the ferryman at Dunham's Mills not to answer any calls until morning, thus cutting off all chance of pursuit and surprise. At the well Boyle wrote an order to deliver the well to Tennent, signing it Pickwick, and the drillers retired to bed! Somebody had been there before them and poured back the sand-pumpings. At the Patterson well Boyle devised a code of tin-horn signals that outwitted the men inside the derrick and flashed the result to Gusher City. The number of expedients continually devised was a marvel. Thanks to the energy and ability of these tireless scouts, of whose midnight exploits, wild rides, hairbreadth escapes and queer adventures many pages could be written, the effect of "mysteries" was frequently neutralized and at length the whole system of guarded wells, bull-dogs and shot-guns was eliminated.

The Forest-Warren white-sand pools marked a new era in developments, with new ideas and new methods to hoodoo speculation. Cherry Grove had wilted from twenty-five-thousand barrels in September to three-thousand in December, when Cooper Hill loomed above the horizon and Balltown appeared on deck. Shallow wells had been sunk far up Tionesta Creek in 1862-3. Near

the two dwellings, saw-mill, school-house and barn dubbed Foxburg, the stamping-ground of deer-hunters and bark-peelers, Marcus Hulings—his name is a
synonym for successful wildcatting—in 1876 drilled a well that smacked of oil.
The derrick stood ten years and globules of grease bubbled up from the depths,
a thousand feet beneath. C. A. Shultz, a piano-tuner, taking his cue from the
Hulings well, interested Frederick Morck, a Warren jeweler, and leased the
Fox estate and contiguous lands in 1881. The Blue-Jay and two Darling wells,
small producers, created a ripple which dry-holes evaporated. They were on
Warrant 2991, Howe township, known to fame as the Cooper tract, north-west
of Foxburg. The conditions of the lease required a well at the western end of
the warrant. Cherry Grove was at its zenith, crude was flirting with the fifties
and operators considered the Blue-Jay chick a lean bird. J. Mainwaring leased
one-hundred acres from Morck & Shultz and built a rig at the head of a wild

ravine, in the sunless woodland, a half-mile from
Tionesta Creek. He lost faith and the Mainwaring lease and rig passed to P. M. Shannon,
of Bradford. Born in Clarion county, Philip
Martin Shannon enlisted at fourteen, served
gallantly through the war, traveled as salesman
for a Pittsburg house and in 1870 cast his lot
with the oilmen at Parker. A pioneer at Millerstown and its burgess in 1874, he filled the
office capably and in 1876 received a big majority at the Republican primary for the legislative nomination. The county ring counted
him out. He drifted with the tide to Bullion,
removed to Bradford in 1879, was elected mayor
in 1885 and discharged his official duties with
excellent discretion. Temperate in habits and
upright in conduct, Mayor Shannon had been

P. M. SHANNON.

an observer and not a participant in the nether side of oil-region life and knew
where to draw the line. He was a favorite in society, high in Masonic circles
and efficient in securing lands for firms with which he had become connected.
Pittsburg is now his home and he manages the company that is developing the
Wyoming field. Mr. Shannon is always generous and courteous. He could
give a scout "the marble heart," lecture an offender, denounce a wrong or
decline to furnish points regarding his mystery-well in a good-natured way that
disarmed criticism. He retains his old-time geniality and prosperity has not
compelled him to buy hats three sizes larger than he wore at Parker and Millerstown "in the days of auld lang-syne."

A. B. Walker and T. J. Melvin joined Shannon in his Cooper venture. A
road was cut through the dense forest from the Fox farm-house up the steep
hill to the Mainwaring derrick. An engine and boiler were dragged to the spot
and Captain Haight contracted to drill the hole. Melvin and Walker, believing
the well a failure at eighteen-hundred feet, went to Cherry Grove on July twenty-
fifth, 1882. Shannon stayed to urge the drill a trifle farther and it struck the
sand at one o'clock next day. He drove in two pine-plugs, sent a messenger for
his partners and filled the well with water to shut in the oil. The well wouldn't
consent to be plugged and drowned. The stream broke loose at three o'clock,
hurling the tools and plugs into the Forest ozone. Shannon and Haight, standing in the derrick, narrowly escaped death as the tools crashed through the

roof and fell to the floor. More plugs, sediment and old clothes were jammed down to conceal the true inwardness of the well, news of which was expected to pulverize the market. Heavy flows following the expulsion of the tools led the owners to anticipate a big strike. Outposts were established and guards, each armed with a Winchester rifle, were changed every six hours. The wildcat-well, eight miles from a telegraph-wire, became an entrenched camp with a half-dozen wakeful scouts besieging the citadel. Vicksburg was not guarded more vigilantly. If a twig cracked or an owl hooted a shower of bullets whizzed in the direction of the noise. Through August the well was permitted to slumber, oil that forced a passage in spite of the obstructions running into pits inside "the dead-line." The trade staggered under the adverse fear of the mystery. Bradford operators formed a syndicate with the owners in lands and speculation and sold a million barrels of crude short. When everything was ready to spring the trap some of the parties went to drill out the plugs and usher in the market-crusher. "We have a jack-pot to open at our pleasure" remarked one of them, voicing the sentiment of all. None looked for anything smaller than fifteen-hundred barrels. The four drillers were discharged and two trusted lieutenants turned the temper-screw and dressed the bits. Ten plugs and a mass of dirt must be cleaned out. From a distance the scouts timed every motion of the walking-beam, gluing their eyes to field-glasses that not a symptom of a flow might slip their eager gaze, "like stout Cortez when he stared at the Pacific upon a peak in Darien." Swift horses were fastened to convenient trees, saddled and bridled for a race to the telegraph-office. A slice of bread and a can of beans served for food. For days the drilling continued. On September four-teenth the last splinter of the plugs was extracted, the sand was cut deeper and —the well didn't respond worth a cent! The faithful scouts, who had stood manfully between the trade and the manipulators, rushed the report. It was a bracer to the market. Bears who pinned their hopes to the Shannon well, the pivot upon which petroleum hinged, scrambled to cover their shorts at heavy loss. Balltown duplicated some of the Cooper experiences, mystery-wells on Porcupine Run agitating the trade in the spring of 1883. The Cherry Grove, Cooper Hill and Balltown pools yielded eight or nine-million barrels. Opera-tions extended to Sheffield and the cream was soon skimmed off. The middle field had enjoyed a very lively inning.

Two miles back of Trunkeyville, on the west side of the Allegheny, Cal-vert, Gilchrist & Risley drilled the Venture well in April, 1870, on the Tuttle farm. Fisher Brothers, of Oil City, and O. D. Harrington, of Titusville, bought the well for fifteen-thousand dollars when it touched the third sand. It was eight-hundred feet deep, flowed three-hundred barrels and started the Fagundas field. The day after it began flowing the Fishers, Adnah Neyhart, Grandin Brothers and David Bently paid one-hundred-and-twenty-thousand dollars for the Fagundas farm of one-hundred-and-sixty acres. Mrs. Fagundas, one son and one daughter died within three months of the sale. Neyhart & Grandin bought a half-interest in David Beatty's farm for ninety-thousand dollars. The Lady Burns well, on the Wilkins farm, finished in June, seconded the Venture. A daily production of three-thousand barrels and a town of twenty-five-hundred population followed quickly. A mile from Fagundas operations on the Hunter, Pearson, Guild and Berry farms brought the suburb of Gillespie into being. The territory lasted and a small yield is obtained to-day. A half-dozen houses, the Venture derrick, Andrews & Co.'s big store and the office in which whole-souled M. Compton—he's in Pittsburg with the Forest Oil-Company now—

labored as secretary of the Producers' Council, hold the fort on the site of well-nigh forgotten Fagundas. William H. Calvert, who projected the Venture well, died at Sistersville, West Virginia, on February seventeenth, 1896. He had drilled on Oil Creek and at Pithole, operated in the southern field and was negotiating for a block of lands near Sistersville when a clot of blood on the brain cut short his active life.

David Beatty had drilled on Oil Creek in 1859-60 with John Fertig. He settled on a farm in Warren county "to get away from the oil." His farm was smothered in oil by the Fagundas development. He removed to the pretty town of Warren, building an elegant home on the bank of Conewango Creek. Fortune hounded him and insisted upon heaping up his riches. John Bell drilled a fifty-barrel well eighty rods above the mansion. Wells surrounding his lot and in his yard emitted oil. Mr. Beatty resigned himself to the inevitable and lived at Warren until called to his final rest some years ago. His case resembled the heroine in Milton Nobles's Phenix, where "the villain still pursued her." The boys used to relate how a negro, the first man to die at Oil City after the advent of petroleum, was buried in a lot on the flats. Somebody wanted that precise spot next day to drill a well and the corpse was planted on the hill-side. The next week that particular location was selected for a well and the body was again exhumed. To be sure of getting out of reach of the drill the friends of the deceased boated his remains down the river to Butler county. Twelve years later the bones were disinterred—an oil-company having leased the old grave-yard—and put in the garden of the dead man's son, to be handy for any further change of base that may be required.

At East Hickory the Foster well, drilled in 1863, flowed three-hundred barrels of amber oil. Two-hundred wells were sunk in the Hickory district, which proved as tough as Old Hickory to nineteen-twentieths of the operators. Three Hickory Creeks—East Hickory and Little Hickory on the east and West Hickory—enter the river within two miles. Near the mouth of West Hickory three Scotchmen named McKinley bored a well two-hundred-and-thirty feet in 1861. They found oil and were preparing to tube the well when the war broke out and they abandoned the field. A well on the flats, drilled in 1865, flowed two-hundred barrels of lubricating oil, occasioning a furore. One farm sold for a hundred-thousand dollars and adjacent lands were snapped up eagerly.

Ninety-five years ago hardy lumbermen settled permanently in Deerfield township, Warren county, thirty miles above the mouth of Oil Creek. Twenty years later a few inhabitants, supported by the lumber trade, had collected near the junction of a small stream with the Allegheny. Bold hills, grand forests, mountain rills and the winding river, sprinkled with green islets, invested the spot with peculiar charms. Upon the creek and hamlet the poetic Indian name of Tidioute, signifying a cluster of islands, was fittingly bestowed. Samuel Grandin, who located near Pleasantville, Venango county, in 1822, removed to Tidioute in 1839. He owned large tracts of timber-lands and increased the mercantile and lumbering operations that gave him prominence and wealth. Mr. Grandin maintained a high character and died at a ripe age. His oldest son, John Livingston Grandin, returned from college in 1857 and engaged in business with his father, assuming almost entire control when the latter retired from active pursuits. News of Col. Drake's well reached the four-hundred busy residents of the lumber-center in two days. Col. Robinson, of Titusville, rehearsed the story of the wondrous event to an admiring group in Samuel Grandin's store. Young J. L. listened intently, saddled his horse and in an hour purchased thirty acres of the Campbell farm, on Gordon Run, below the village, for three-hundred dollars. An "oil-spring" on the property was the attraction. Next morning he contracted with H. H. Dennis, a man of mechanical skill, to drill a well "right in the middle of the spring." The following day a derrick—four pieces of scantling—towered twenty feet, a spring-pole was procured, the "spring" was dug to the rock, and the "tool" swung at the *first* oil-well in Warren county and among the first in Pennsylvania. Dennis hammered a drilling-tool from a bar of iron three feet long, flattening one end to cut two-and-a-half inches, the diameter of the hole. In the upper end of the drill he formed a socket, to hold an inch-bar of round iron, held by a key riveted though and lengthened as the depth required. Two or three times a day, when the "tool" was drawn out to sharpen the bit and clean the hole, the key had to be cut off at each joint! With this rude outfit drilling began the first week of September, 1859, and the last week of October the well was down one-hundred-and-thirty-four feet. Tubing would not go into the hole and it was enlarged to four inches. The discarded axle of a tram-car, used to carry lumber from Gordon Run to the river, furnished iron for the reamer. Days, weeks and months were consumed at this task. At last, when the hole had been enlarged its full depth, the reamer was let down "to make sure the job was finished." It stuck fast, never saw daylight again and the well sunk with so much labor had not one drop of oil!

Other wells in the locality fared similarly, none finding oil nearer than Dennis Run, a half-mile distant. There scores of large wells realized fortunes for their owners. In two years James Parshall was a half-million ahead. He settled at Titusville and built the Parshall House—a mammoth hotel and opera-house—which fire destroyed. The "spring" on the Campell farm is in existence and the gravel is impregnated with petroleum, supposed to percolate through fissures in the rocks from Dennis Run.

During the summer of 1860 developments extended across and down the river a mile from Tidioute. The first producing well in the district, owned by King & Ferris, of Titusville, started in the fall at three-hundred barrels and boomed the territory amazingly. It was on the W. W. Wallace lands—five-hundred acres below town—purchased in 1860 by the Tidioute & Warren Oil-Company, the third in the world. Samuel Grandin, Charles Hyde and Jonathan Watson organized it. J. L. Grandin, treasurer and manager of the company, in eight years paid the stockholders twelve-hundred-thousand dollars dividends on a capital of ten-thousand! He leased and sub-leased farms on both sides of the Allegheny, drilling some dry-holes, many medium wells and a few large ones. He shipped crude to the seaboard, built pipe-lines and iron-tanks and became head of the great firm of Grandins & Neyhart. Elijah Bishop Grandin—named from the father of C. E. Bishop, founder of the Oil-City *Derrick*—who had carried on a store at Hydetown and operated at Petroleum Centre, resumed his residence at Tidioute in 1867 and associated with his brother and brother-in-law, Adnah Neyhart, in producing, buying, storing and transporting petroleum. Mr. Neyhart and Joshua Pierce, of Philadelphia, had drilled on Cherry Run, on Dennis Run and at Triumph and engaged largely in shipping oil to the coast. Pierce & Neyhart—J. L. Grandin was their silent partner—dissolved in 1869. The firm of Grandins & Neyhart, organized in 1868, was marvelously successful. Its high standing increased confidence in the stability of financial and commercial affairs in the oil-regions. The brothers established the Grandin Bank and Neyhart, besides handling one-fourth of the crude produced in Pennsylvania,

13

opened a commission-house in New York to sell refined, under the skilled
management of John D. Archbold, now vice-president of the Standard Oil-
Company. They and the Fisher Brothers owned the Dennis Run and Triumph
pipe-lines and piped the oil from Fagundas, where they drilled a hundred prolific wells and were the largest operators. They bought properties in different
portions of the oil-fields, extended their pipe-lines to Titusville and erected
iron-tankage at Parker and Miller Farm. The death of Mr. Neyhart terminated
their connection with oil-shipments.

Owning thousands of acres in Warren and Forest counties, the Grandins
were heavily interested in developments at Cherry Grove, Balltown and Cooper.
As those sections declined they gradually withdrew from active oil-operations,
sold their pipe-lines and wound up their bank. J. L. Grandin removed to Bos-

J. L. GRANDIN.

E. B. GRANDIN.

ton and E. B. to Washington, to embark in new enterprises and enjoy, under
most favorable conditions, the fruits of their prosperous career at Tidioute.
Their business for ten years has been chiefly loaning money, farming and lumbering in the west. They purchased seventy-two-thousand acres in the Red
River Valley of Dakota—known the world over as "the Dalrymple Farm "—
sold it down to forty-thousand and in 1895 harvested from twenty-four-thousand
acres six-hundred-thousand bushels of wheat, oats and barley. On this tract
they employ hundreds of men and horses, scores of ploughs and reapers and

steam-threshers and illustrate how to farm profitably on the biggest scale. With Hunter & Cummings, of Tidioute, and J. B. White, of Kansas City, as partners, they organized the Missouri Lumber and Mining Company. The company owns two-hundred-and-forty-thousand acres of timber-land in Missouri and cut fifty-million feet of lumber last year in its vast saw-mills at Grandin, Carter county. Far-seeing, clear-headed, of unblemished repute and liberal culture, such men as J. L. and E. B. Grandin reflect honor upon humanity and deserve the success an approving conscience and the popular voice commend heartily.

Above Tidioute a number of "farmers' wells"—shallow holes sunk by hand and soon abandoned—flickered and collapsed. On the islands in the river small wells were drilled, most of which the great flood of 1865 destroyed. Opposite the town, on the Economite lands, operations began in 1860. Steam-power was used for the first time in drilling. The wells ranged from five barrels to eighty, at one-hundred-and-fifty feet. They belonged to the Economites, a German society that enforced celibacy and held property in common. About 1820 the association founded the village of Harmony, Butler county, having an exclusive colony and transacting business with outsiders through the medium of two trustees. The members wore a plain garb and were distinguished for morality, simplicity, industry and strict religious principles. Leaving Harmony, they located in the Wabash Valley, lost many adherents, returned to Pennsylvania and built the town of Economy, in Beaver county, fifteen miles below Pittsburg. They manufactured silks and wine, mined coal and accumulated millions of dollars. A loan to William Davidson, owner of eight-thousand acres in Limestone township, Warren county, obliged them to foreclose the mortgage and bid in the tract. Their notions of economy applied to the wells, which they numbered alphabetically. The first, A well, yielded ten barrels, B pumped fifty and C flowed seventy. The trustees, R. L. Baker and Jacob Henrici, erected a large boarding-house for the workmen, whose speech and manners were regulated by printed rules. Pine and oak covered the Davidson lands, which fronted several miles on the Allegheny and stretched far back into the township. Of late years the Economite society has been disintegrating, until its membership has shrunk to a dozen aged men and women. Litigation and mismanagement have frittered away much of its property. It seems odd that an organization holding "all things in common" should, by the perversity of fate, own some of the nicest oil-territory in Warren, Butler and Beaver counties. A recent strike on one of the southern farms flows sixty barrels an hour. Natural gas lighted and heated Harmony and petroleum appears bound to stick to the Economites until they have faded into oblivion.

Below the Economite tract numerous wells strove to impoverish the first sand. G. I. Stowe's, drilled in 1860, pumped eight barrels a day for six years. The Hockenburg, named from a preacher who wrote an essay on oil, averaged twelve barrels a day in 1861. The Enterprise Mining and Boring Company of New York leased fifteen rods square on the Tipton farm to sink a shaft seven feet by twelve. Bed rock was reached at thirty feet, followed by ten feet of shale, ten of gray sand, forty of slate and soap-rock and twenty of first sand. The shaft, cribbed with six-inch plank to the bottom of the first sand, tightly caulked to keep out water, was abandoned at one-hundred-and-sixty feet, a gas explosion killing the superintendent and wrecking the timbers. Of forty wells on the Tipton farm in 1860-61 not a fragment remained in 1866.

Tidioute's laurel wreath was Triumph Hill, the highest elevation in the neig'::borhood. Wells nine-hundred feet deep pierced sixty feet of oil-bearing sand, which produced steadily for years. Grandins, Fisher Brothers, M. G. Cushing, E. E. Clapp, John M. Clapp and other leading operators landed bounteous pumpers. The east side of the hill was a forest of derricks, crowded like trees in a grove. Over the summit and down the west side the sand and the

VIEW ON EAST SIDE OF TRIUMPH HILL IN 1874.

development extended. For five years Triumph was busy and prosperous, yielding hundreds-of-thousands of barrels of oil and advancing Tidioute to a town of five-thousand population. Five churches, the finest school-buildings in the county, handsome houses, brick blocks, superior hotels and large stores greeted the eye of the visitor. The Grandin Block, the first brick structure, built of the first brick made in Deerfield township, contained an elegant opera-house. Three banks, three planing-mills, two foundries and three machine-shops flourished. A dozen refineries turned out merchantable kerosene. Water-works were provided and an iron bridge spanned the river. Good order was maintained and Tidioute—still a tidy village—played second fiddle to no town in Oildom for intelligence, enterprise and all-round attractiveness.

The tidal wave effervesced at intervals clear to the Colorado district. Perched on a hill in the hemlock woods, Babylon was the rendezvous of sports,

VIEW ON WEST SIDE OF TRIUMPH HILL IN 1874.

strumpets and plug-uglies, who stole, gambled, caroused and did their best to break all the commandments at once. Could it have spoken, what tales of horror that board-house under the evergreen tree might recount! Hapless wretches were driven to desperation and fitted for the infernal regions. Lust and liquor goaded men to frenzy, resulting sometimes in homicide or suicide. In an affray one night four men were shot, one dying in an hour and another in six weeks. Hogan, who had laughed at the feeble efforts of the township-constable to suppress his resort, was arrested, tried for murder and acquitted on the plea of self-defence. The shot that killed the first victim was supposed to have been fired by "French Kate," Hogan's mistress. She had led the demimonde in Washington and led susceptible congressmen astray. Ben met her at Pithole, where he landed in the summer of 1865 and ran a variety-show that would make the vilest on the Bowery blush to the roots of its hair. He had been a prize-fighter on land, a pirate at sea, a bounty-jumper and blockade-runner and prided himself on his title of "the Wickedest Man in the World." Sentenced to death for his crimes against the government, President Lincoln pardoned him and he joined the myriad reckless spirits that sought fresh adventures in the Pennsylvania oil-fields. In a few months the Scripture legend—"Babylon has fallen"—applied to the malodorous Warren town. The tiger can "change his spots"—by moving from one spot to another—and so could Hogan. He was of medium height, square-shouldered, stout-limbed, exceedingly muscular and trained to use his fists. He fought Tom Allen at Omaha, sported at Saratoga and in 1872 ran "The Floating Palace"—a boat laden with harlots and whiskey—at Parker. The weather growing too cold and the law too hot for comfort, he opened a den and built an opera-house at Petrolia. In "Hogan's Castle" many a clever young man learned the short-cut to disgrace and perdition. Now and then a frail girl met a sad fate, but the carnival of debauchery went on without interruption. Hogan put on airs, dressed in the loudest style and would have been the burgess had not the election-board counted him out! A fearless newspaper forcing him to leave Petrolia, Hogan went east to engage in "the sawdust swindle," returned to the oil-regions in 1875, built an opera-house at Elk City, decamped from Bullion, rooted at Tarport and Bradford and departed by night for New York. Surfeited with revelry and about to start for Paris to open a joint, he heard music at a hall on Broadway and sat down to wait for the show to begin. Charles Sawyer, "the converted soak," appeared shortly, read a chapter from the Bible and told of his rescue from the gutter. Ben was deeply impressed, signed the pledge at the close of the service, agonized in his room until morning and on his knees implored forgiveness. How surprised the angels must have been at the spectacle of the prodigal in this attitude! After a fierce struggle, to quote his own words, "peace filled my soul chock-full and I felt awful happy." He claimed to be converted and set to work earnestly to learn the alphabet, that he might read the Scriptures and be an evangelist. He married "French Kate," who also professed religion, but it didn't strike in very deep and she eloped with a tough. Mr. Moody welcomed Hogan and advised him to traverse the country to offset as far as possible his former misdeeds. Amid the scenes of his grossest offenses his reception varied. High-toned Christians, who would not touch a down-trodden wretch with a ten-foot pole, turned up their delicate noses and refused to countenance "the low impostor." They forgot that he sold his jewelry and most of his clothes, lived on bread and water and endured manifold privations to become a bearer of the gospel-message. Even ministers who proclaimed that "the blood of Christ cleanses from *all* sin" doubted Hogan's salvation and showed him the cold shoulder in the chilliest orthodox fashion. He stuck manfully and for eighteen years has labored zealously in the vineyard. Judging from his struggles and triumphs, is it too much to believe that a front seat and a golden crown are reserved for the reformed pugilist, felon, robber, assassin of virtue and right bower of Old Nick? Unlike straddlers in politics and piety, who want to go to Heaven on velvet cushions and pneumatic tires,

> " He doesn't stand on one foot fust,
> An' then stand on the other,
> An' on which one he feels the wust
> He couldn't tell you nuther."

The expectation of an extension of the belt northward was not fulfilled immediately. Wells at Irvineton, on the Brokenstraw and tributary runs, failed to find the coveted fluid. Captain Dingley drilled two wells on Sell's Run, three miles east of Irvineton, in 1873, without slitting the jugular. A test well at Warren, near the mouth of Conewango Creek, bored in 1864 and burned as pumping was about to begin, had fair sand and a mite of oil. John Bell's operations in 1875 opened an amber pool up the creek that for a season crowded the hotels three deep with visitors. They bored dozens of wells, yet the production never reached one-thousand barrels and in four months the patch was cordoned by dry holes and as quiet as a cemetery. The crowds exhaled like morning dew. Warren is a pretty town of four-thousand population, its location and natural advantages offering rare inducements to people of refinement and enterprise. Its site was surveyed in 1795 and the first shipment of lumber to Pittsburg was made in 1801. Incorporated as a borough in 1832, railroad communication with Erie was secured in 1859, with Oil City in 1867 and with Bradford in 1881. Many of the private residences are models of good taste. Massive brick-blocks, solvent banks, churches, stores, high-grade schools, shaded streets and modern conveniences evidence its substantial prosperity. Hon. Thomas Struthers—he built sections of the Philadelphia & Erie and the Oil-Creek railroads and established big iron-works

CHARLES W. STONE.

—donated a splendid brick building for a library, opera-house and post-office. His grandson, who inherited his millions and died in February of 1896, was a mild edition of "Coal-Oil Johnnie" in scattering money. Lumbering, the principal industry for three generations, enriched the community. Col. Lewis F. Watson represented the district twice in Congress and left an estate of four-millions, amassed in lumber and oil. He owned most of the township bearing his name. Hon. Charles W. Stone, his successor, ranks with the foremost members of the House in ability and influence. A Massachusetts boy, he set out in life as a teacher, came to Warren to take charge of the academy, was county-superintendent, studied law and rose to eminence at the bar. He was elected Lieutenant-Governor of the State, served as Secretary of the Commonwealth and would be Governor of Pennsylvania to-day had "the foresight of the Republicans been as good as their hindsight." He has profitable oil-interests, is serving his third term in Congress and has been nominated for the fourth. Alike fortunate in his political and professional career, his social relations, his business connections and his personal friendships, Charles W. Stone holds a place in public esteem few men are privileged to attain.

At Clarendon and Stoneham hundreds of snug wells yielded three-thousand barrels a day from a regular sand that did not exhaust readily. Southward the Garfield district held on fairly and a narrow-gauge railroad was built to Farnsworth. The Wardwell pool, at Glade, four miles east of Warren, fizzed after the manner of Cherry Grove, rich in buried hopes and dissipated greenbacks. P. M. Smith and Peter Grace drilled the first well—a sixty barreler—close to the ferry in July of 1883. Dry-holes and small wells alternated with provoking uncertainty until J. A. Gartland's twelve-hundred-barrel gusher on the Clark farm, in May of 1884, inaugurated a panic in the market that sent crude down to fifty cents. The same day the Union Oil-Company finished a four-hundred barrel spouter and May ended with fifty-six wells producing and a score of dusters. June and July continued the refrain, values see sawing as reports of dry-holes or fifteen-hundred-barrel-strikes, some of them worked as "mysteries," bamboozled the trade. Wardwell's production ascended to twelve-thousand barrels and fell by the dizziest jumps to as many hundred, the porous rock draining with the speed of a lightning-calculator. Tiona developed a lasting deposit of superior oil. Kane has a tempting streak, in which Thomas B. Simpson and other Oil-City parties are interested. Gas has been found at Wilcox, Johnsonburg and Ridgway, Elk county taking a slick hand in the game. Kinzua, four miles north-east of Wardwell, revealed no particular cause why the spirit of mortal ought to be proud. Although Forest and Warren, with a slice of Elk thrown in, were demoralizing factors in 1882-3-4, their aggregate output would be only a light luncheon for the polar bear in McKean county.

The United States Land-Company, holding a quarter-million acres in McKean and adjoining counties, in 1837 sent Col. Levitt C. Little from New Hampshire to look after its interests. He located on Tuna Creek, eight miles from the southern border of New York state. The Websters arrived in 1838, journeying by canoe from Olean. Other families settled in the valley, founding the hamlet of Littleton, which in 1858 adopted the name of Bradford and became a borough in 1872, with Peter T. Kennedy as burgess. The vast forests were divided into huge blocks, such as the Bingham, Borden, Clark & Babcock, Kingsbury and Quintuple tracts. Lumber was rafted to distant points and thousands of hardy woodmen "shantied" in rough huts each winter. They beguiled the long evenings singing coarse songs, playing cards, imbibing the vintage of Kentucky or New England from a black jug and telling stories so bald the mules drooped their ears to hide their blushes. But they were openhearted, sternly honest, sticklers for fair-play, hard-working and admirable forerunners of the approaching civilization. To the sturdy blows of the rugged chopper and raftsman all classes are indebted for fuel, shelter and innumerable comforts. Like the rafts they steered to Pittsburg and the wild beasts they hunted, most of these brave fellows have drifted away never to return.

Six-hundred inhabitants dwelt peacefully at Bradford ten years after the Pithole bubble had been blown and pricked. The locomotive and track of a branch of the Erie Railroad had supplanted A. W. Newell's rude engine, which transported small loads to and from Carrollton. An ancient coach, weather-beaten and worm-eaten, sufficed for the scanty passenger-traffic and the quiet borough bade fair to stay in the old rut indefinitely. The collection of frames labeled Tarport—a suit of tar and feathers presented to a frisky denizen begot the name—snuggled on a muddy road a mile northward. Seven miles farther, at Limestone, the "spirits" directed Job Moses to buy ten-thousand acres of land. He bored a half-dozen shallow wells in 1864, getting some oil and gas.

Jonathan Watson skirmished two miles east of Limestone, finding slight tinges of greasiness. A mile south-west of Moses the Crosby well was dry. Another mile south the Olmsted well, on the Crooks farm, struck a vein of oil at nine-hundred feet and flowed twenty barrels on July fourteenth, 1875. The sand was poor and dry-holes south and west augured ill for the territory. Frederick

FREDERICK CROCKER.

Crocker drilled a duster early in 1875 on the Kingsbury lands, east side of Tuna Creek. He had grit and experience and leased an angular piece of ground formed by a bend of the creek for his second venture. It was part of the Watkins farm, a mile above Tarport. A half-mile south-west, on the Hinchey farm, the Foster Oil-Company had sunk a twenty-barrel well in 1872, which somehow passed unnoticed. On September twenty-sixth, 1875, from a shale and slate at nine-hundred feet, the Crocker well flowed one-hundred-and-seventy barrels. This opened the gay ball which was to transmute the Tuna Valley from its arcadian simplicity to the intense bustle of the grandest petroleum-region the world has ever known. The valley soon echoed and re-echoed the music of the tool-dresser and rig-builder, the click of the drill and the vigorous profanity of the imported teamster. Frederick Crocker, who drilled on Oil Creek in 1860 and devised the valve which kept the Empire well alive, had won another victory and the great Bradford field was born. He lived at Titusville fifteen years, erected the home afterwards occupied by Dr. W. B. Roberts, sold his Bradford property, operated in the Washington district and died at Idlewild on February twenty-second, 1895. Mr. Crocker possessed real genius, decision and the qualities which "from the nettle danger pluck the flower success." Active to the close of his long and useful eighty-three years, he met death calmly and was laid to rest in the cemetery at Titusville.

Scarcely had the Crocker well tanked its initial spurt ere "the fun grew fast and furious." Rigs multiplied like rabbits in Australia. Train-loads of lively delegates from every nook and cranny of Oildom crowded the streets, overran the hotels and taxed the commissary of the village to the utmost. Town-lots sold at New-York prices and buildings spread into the fields. At B. C. Mitchell's Bradford House, headquarters of the oil-fraternity, operators and land-holders met and drillers "off tour" solaced their craving for "the good things of this life" playing billiards and practising at the hotel-bar. Hundreds of big contracts were closed in the second-story room where Lewis Emery, "Judge" Johnson, Dr. Book and the advance-guard of the invading hosts assembled. Main street blazed at night with the light of dram-shops and the gaieties incidental to a full-fledged frontier-town. Noisy bands appealed to lovers of varieties to patronize barnlike-theatres, strains of syren music floated from beer-gardens, dance-halls of dubious complexion were thronged and gambling-dens ran unmolested. The free-and-easy air of the community, too intent chasing oil and cash to bother about morality, captivated the ordinary stranger and gained "Bad Bradford" notoriety as a combination of Pit-hole and Petroleum Centre, with a dash of Sodom and Pandemonium, condensed into a single package. In February of 1879 a city-charter was granted and James Broder was elected mayor. Radical reforms were not instituted with undue haste, to jar the sensitive feelings of the incongruous masses gathered from far and near. Their accommodating nature at last adapted itself to a new state of affairs and accepted gracefully the restrictions imposed for the general welfare. Checked temporarily by the Bullion spasm in 1876–7, the influx redoubled as the lower country waned. Fires merely consumed frame-structures to hasten the advent of costly brick-blocks. Ten churches, schools, five banks, stores, hotels, three newspapers, street-cars, miles of residences and fifteen-thousand of the liveliest people on earth attested the permanency of Bradford's boom. Narrow-gauge railroads circled the hills, traversed spider-web trestles and brought tribute to the city from the outlying districts. The area of oil-territory seemed interminable. It reached in every direction, until from sixteen-thousand mouths seventy-five thousand acres poured their liquid treasure. The daily production waltzed to one-hundred-thousand barrels! Iron-tanks were built by the thousand to store the surplus crude. Two, three or four-thousand-barrel gushers were lacking, but wells that yielded twenty-five to two-hundred littered the slopes and valleys. The field was a marvel, a phenomenon, a revelation. Bradford passed the mushroom stage safely and was not snuffed out when developments receded and the floaters wandered south in quest of fresh excitement. To-day it is a thriving railroad and manufacturing centre, the home of ten-thousand intelligent, independent, go-ahead citizens, proud of its past, pleased with its present and confident of its future.

To trace operations minutely would be an endless task. Crocker sold a half-interest in his well and drilled on an adjacent farm. Gillespie, Buchanan & Kelly came from Fagundas in 1874 and sank the two Fagundas wells—twenty and twenty-five barrels—a half-mile west of Crocker, in the fall and winter. Butts No. 1, a short distance north, actually flowed sixty barrels in November of 1874. Jackson & Walker's No. 1, on the Kennedy farm, north edge of town, on July seventeenth, 1875, flowed twenty barrels at eleven-hundred feet. The dark, pebbly sand, the best tapped in McKean up to that date, encouraged the belief of better strata down the Tuna. On December first, two months after Crocker's strike, the yield of the Bradford district was two-hundred-and-ten barrels. The Crocker was doing fifty, the Olmsted twenty-five, the Butts fifteen, the Jackson & Walker twenty and all others from one to six apiece. The oil, dark-colored and forty-five gravity, was loaded on Erie cars direct from the wells, most of which were beside the tracks. The Union Company finished the first pipe-line and pumped oil to Olean the last week of November. Prentice, Barbour & Co. were laying a line through the district and 1875 closed with everything ripe for the millenium these glimmerings foreshadowed.

Lewis Emery, richly dowered with Oil-Creek experience and the get-up-and-get quality that forges to the front, was an early arrival at Bradford. He secured the Quintuple tract of five-thousand acres and drilled a test well on the Tibbets farm, three miles south of town. Its success confirmed his judgment of the territory and began the wonderful Quintuple development. The Quintuple rained staying wells on the lucky, plucky graduate from Pioneer, quickly placing him in the millionaire-class. He built blocks and refineries, opened an immense hardware-store, constructed pipe-lines, established a daily-paper, served two terms in the Senate and opposed the Standard "tooth and toe-nail." Thoroughly earnest, he champions a cause with unflinching tenacity. He owns a big ranche in Dakota, big lumber-tracts and saw-mills in Kentucky, a big oil-production and a big share in the United-States Pipe-Line. He has traveled

over Europe, inspected the Russian oil-fields and gathered in his private
museum the rarest collection of curiosities and objects of interest in the state.
Senator Emery is a staunch friend, a fighter who "doesn't know when he is
whipped," liberal, progressive, fluent in conversation and firm in his convictions.

Hon. David Kirk sticks faithfully to Emery in his hard-sledding to array
petroleumites against the Standard. He manages the McCalmont Oil-Company, which operated briskly in the Forest pools, at Bradford and Richburg.
Mr. Kirk is a rattling speaker, positive in his sentiments and frank in expressing his views. He extols Pennsylvania petroleum, backs the outside pipe-lines
and is an influential leader of the Producers' Association.

　　Dr. W. P. Book, who started at Plumer, ran big hotels at Parker and Millerstown and punched a hole in the Butler field occasionally, leased nine-hundred acres below Bradford in the summer of 1875. He bored two-hundred wells, sold the whole bundle to Captain J. T. Jones and went to Washington Territory with eight-hundred-thousand dollars to engage in lumbering and banking. Captain Jones landed on Oil Creek after the war, in which he was a brave soldier, and drilled thirteen dry-holes at Rouseville! Repulses of this stripe would wear out most men, but the Captain had enlisted for the campaign and proposed to stand by his guns to the last. His fourteenth attempt—a hundred-barreler on the Shaw farm—recouped former losses and inaugurated thirty years of remarkable prosperity. Fortune smiled upon him in the Clarion field. Pipe-lines, oil-wells, dealings in the exchanges, whatever he touched turned into gold. Not handicapped by timid partners, he paddled his own canoe and became the largest individual operator in the northern region. Acquiring tracts that proved to be the heart of the Sistersville field, he is credited with rejecting an offer last year of five-million dollars for his West-Virginia and Pennsylvania properties! From thirteen wells, good only for post-holes if they could be dug up and retailed by the foot, to five-millions in cash was a pretty stretch onward and upward. He preferred staying in the harness to the obscurity of a mere coupon-clipper. He lives at Buffalo, controls his business, enjoys his money, remembers his old friends and does not put on airs because of marching very near the head of the oleaginous procession.

Theodore Barnsdall has never lagged behind since he entered the arena in 1860. He operated on Oil Creek and has been a factor in every important district. Marcus Hulings, reasoning that a paying belt intersected it diagonally, secured the Clark & Babcock tract of six-thousand acres on Foster Brook, north-east of Bradford. Hundreds of fine wells verified his theory and added a half-million to his bank-account. Sitting beside me on a train one day in 1878, Mr. Hulings refused three-hundred-thousand dollars, offered by Marcus Brownson, for his interest in the property. He projected the narrow-gauge railroad from Bradford to Olean and a bevy of oil-towns—Gillmor, Derrick City, Red Rock and Bell's Camp—budded and bloomed along the route. Frederic Prentice built pipe-lines and tanks, leased a half-township, started thirty wells in a week on the Melvin farm and organized the Producers' Consolidated Land and Petroleum Company, big in name, in quality and capital. The American Oil-Company's big operations wafted the late W. A. Pullman a million and the presidency of the Seaboard Bank in New York, filled Joseph Seep's stocking and saddled a hundred-thousand dollars on James Amm. The Hazelwood Oil-Company, guided by Bateman Goe's prudent hand, drilled five-

FREDERICK BODEN.

hundred wells and counted its gains in columns of six figures. Frederick Boden—true-blue, clear-grit, sixteen ounces to the pound—forsook Corry to extract a stream of wealth from the Borden lands, six miles east of Tarport. Prompt, square and manly, he merited the good-luck that rewarded him in Pennsylvania and followed him to Ohio, where for two years he has been operating extensively. Boden's wells boosted the territory east and north. From its junction with the Tuna at Tarport—Kendall is the post-office—to its source off in the hills, Kendall Creek steamed and smoked. Tarport expanded to the proportions of a borough. Two narrow-gauge roads linked Bradford and Eldred, Sawyer City, Rew City, Coleville, Rixford and Duke Centre—oil-towns in all the term implies—keeping the rails from rusting. Other narrow-gauges diverged to Warren, Mt. Jewett and Smethport. The Erie extended its branch south and the Rochester & Pittsburg crossed the Kinzua gorge over the highest railway viaduct—three-hundred feet—in this nation of tall projects and tall achievements.

Twenty-nine years ago a stout-hearted, strong-limbed, wiry youth, fresh from the Emerald Isle, asked a man at Petroleum Centre for a job. Given a pick and shovel, he graded a tank-bottom deftly and swiftly. He dug, pulled tubing, drove team and earned money doing all sorts of chores. Reared in poverty, he knew the value of a dollar and saved his pennies. To him Oildom, with its "oil-princes"—George K. Anderson, Jonathan Watson, Dr. M. C. Egbert, David Yanney, Sam Woods, Joel Sherman and the Phillips Brothers were in their glory—was a golden dream. He learned to "run engine," dress tools, twist the temper-screw and handle drilling and pumping-wells expertly. Although neither a prohibitionist nor a prude, he never permitted mountain-dew, giddy divinities in petticoats or the prevailing follies to get the better of him in his inordinate desire for riches. Drop by drop for three years his frugal store increased and he migrated to Parker early in the seventies. Such was the young man who "struck his gait" in the northern end of Armstrong county, who was to outshine the men he may have envied on Oil Creek, to scoop the biggest prize in the petroleum-lottery and weave a halo of glittering romance around the name of John McKeown.

Working an interest in an oil-well, he hit a paying streak and joined the pioneers who had sinister designs on Butler county, proverbial for "buckwheat-batter" and "soap-mines." At Lawrenceburg, a suburb of Parker, he boarded with a comely widow, the mother of two bouncing kids and owner of a little cash. He married the landlady and five boys blessed the union of loyal hearts. His wife's money aided him to develop the Widow Nolan farm, east of the coal-bank near Millerstown. Regardless of Weller's advice to "beware of vid-ders," he wedded one and from another obtained the lease of a farm on which his first well produced one-hundred-and-fifty barrels a day for a year, a fortune in itself. This was the beginning of McKeown's giant strides. In partnership with William Morrisey, a stalwart fellow-countryman—dead for years—he drilled at Greece City, Modoc and on the Cross-Belt. He held interests with Parker & Thompson and James Goldsboro, played a lone hand at Martinsburg, invested in the Karns Pipe-Line and avoided speculation. He agreed with Thomas Hayes, of Fairview, in 1876, to operate in the Bradford field. Hayes went ahead to grab a few tracts at Rixford, McKeown remaining to dispose of his Butler properties. He sold every well and every inch of land at top figures. No slave ever worked harder or longer hours than he had done to gain a firm footing. No task was too difficult, no fatigue too severe, no undertaking too hazardous to be met and overcome. Avarice steeled his heart and hardened his muscles. Wrapped in a rubber-coat and wearing the slouch-hat everybody recognized, he would ride his powerful bay-horse knee-deep in mud or snow at all hours of the night. It was his ambition to be the leading oil-operator of the world. While putting money into Baltimore blocks, bank-stocks and western ranches, he always retained enough to gobble a slice of seductive oil-territory.

Plunging into the northern field "horse, foot and dragoons," he bought out Hayes, who returned to Fairview with a snug nest-egg, and captured a huge chunk of the Bingham lands. Robert Simpson, agent of the Bingham estate, fancied the bold, resolute son of Erin and let him pick what he wished from the fifty-thousand acres under his care. McKeown selected many juicy tracts, on which he drilled up a large production, sold portions at excessive prices and cleared at least a million dollars in two or three years! As Bradford declined he turned his gaze towards the Washington district, bought a thousand acres of land and at the height of the excitement had ten-thousand barrels of oil a day! His object had been attained and John McKeown was the largest oil-producer in the universe.

Down in Washington, as in Butler and McKean, he attended personally to his wells, hired the workmen, negotiated for all materials and managed the smallest details. He removed his family to the county-seat and lived in a plain, matter-of-fact way. It had been his intention to erect a forty-thousand-dollar house and reside at Jamestown, N. Y. Ground was purchased and the foundation laid. The local papers spoke of the acquisition he would be to the town, one suggesting to haul him into politics and municipal improvements, and McKeown resented the notoriety by pulling up stakes and locating at Washington. It often amused me to hear him denounce the papers for calling him rich. He was more at home in a derrick than in a drawing-room. The din of the tools boring for petroleum was sweeter to his ears than "Lohengrin" or "The Blue Danube." Watching oil streaming from his wells delighted his eye more than a Corot or a Meissonier in a gilt frame. For claw-hammer coats, tooth-pick shoes and vulgar show he had no earthly use. Democratic in his habits and speech, he heard the poor man as patiently as the banker or the schemer with a "soft snap." Clothes counted for nothing in his judgment of people. He enjoyed the hunt for riches more than the possession. In no sense a liberal man, sometimes he thawed out to friends who got on the sunny side of his frosty nature and wrote checks for church or charity. Hard work was his diversion, his chief happiness. His wells and lands and income grew to dimensions it would have strained the nerves and brains of a half-dozen men to supervise. He had mortgaged his robust constitution by constant exposure and the foreclosure could not always be postponed. Repeated warnings were unheeded and the strong man broke down just when he most needed the vitality his lavish drafts exhausted. Eminent physicians hurried from Pittsburg and Philadelphia to his relief, but the paper had gone to protest and on Sunday forenoon, February eighth, 1891, at the age of fifty-three, John McKeown passed into eternity. Father Hendrich administered the last rites to the dying man. He sank into a comatose state and his death was painless. The remains were interred in the Catholic cemetery at Lawrenceville, in presence of a great multitude that assembled to witness the curtain fall on the most eventful life in the oil-regions.

Estimates of McKeown's wealth ranged from three-millions to ten. A guess midway would probably be near the mark. When asked by Dunn or Bradstreet how he should be rated, his invariable answer was: "I pay cash for all I get." O. D. Bleakley, of Franklin, was appointed guardian of the sons and Hon. J. W. Lee is Mrs. McKeown's legal adviser. The oldest boy has married, has received his share of the estate and is spending it freely. A younger son was drowned in a pond at the school to which his mother sent the bright lad. Once McKeown, desiring to have Dr. Agnew's candid opinion at the lowest cost, put on his poorest garb and secured a rigid examination upon his promise to pay the great Philadelphia practitioner ten dollars "as soon as he could earn the money." He thanked the doctor, returned in a business-suit, told of the ruse he had adopted and cemented the acquaintance with a check for one-hundred dollars. In Baltimore he posed as a hayseed at a forced sale of property the mortgagors calculated to bid in at a fraction of its value. He deposited a million dollars in a city-bank and appeared at the sale in the old suit and slouched hat he had packed in his satchel for the occasion. Stylish bidders at first ignored the seedy fellow whose winks to the auctioneer elevated the price ten-thousand dollars a wink. One of them hinted to the stranger that he might be bidding beyond his limit. "I guess not," replied John, "I pay cash for what I get." The property was knocked down to him for about six-hundred-thousand dollars. He requested the attorney to tele-phone to the bank whether his check for the amount would be honored. "Good for a million!" was the response. Now his triumphs and his spoils have shrunk to the little measure of the grave!

> " Through the weary night on his couch he lay
> With the life-tide ebbing fast away.
> When the tide goes out from the sea-girt lands
> It bears strange freight from the gleaming sands :
> The white-winged ships, which long may wait
> For the foaming wave and the wind that's late;
> The treasures cast on a rock-bound shore
> From stranded ships that shall sail no more,
> And hopes that follow the shining seas—
> Oh! the ocean wide shall win all these.
> But saddest of all that drift to the sea
> Is the human soul to eternity,
> Floating away from a silent shore,
> Like a fated ship, to return no more."

The Bradford Oil-Company—J. T. Jones, Wesley Chambers, L. G. Peck and L. F. Freeman were the principal stockholders—owned a good share of the land on which Greater Bradford was built and ten-thousand acres in the northern field. The company drilled three-hundred wells in McKean and Alle-gany, realized fifty-thousand dollars from city-lots and its stock rose to two-thousand dollars a share. In 1881 Captain Jones bought out his copartners. The Enterprise Transit Company, managed by John Brown, achieved reputa-tion and currency. The McCalmont Oil-Company—organized during the Bul-lion phantom by David Kirk, I. E. Dean, Tack Brothers and F. A. Dilworth—humped itself in the middle and northern fields, sometimes paying three-hun-dred-thousand dollars a year in dividends. Kirk & Dilworth founded Great Belt City, in Butler county, cutting up a farm and selling hundreds of lots. "Farmer" Dean, manager of the company, operated in the lower fields, lived two years at Richburg, toured the country to preach the gospel according to the Greenbackers and won laurels on the rostrum. Frank Tack—frank and trustworthy—was vice-president of the New-York Oil-Exchange and his brother is dead. The Emery Oil-Company, the Quintuple, Mitchell & Jones, Whitney & Wheeler, Melvin and Fuller, George H. Vanvleck, George V. Forman, John L. McKinney & Co., Isaac Willets and Peter T. Kennedy were shining lights in the McKean-Allegany firmament. Kennedy owned the saw-mill when Brad-ford was a lumber-camp and his estate—he died at fifty—inventoried eleven-hundred-thousand dollars. Hundreds of small operators left Bradford happy as men should be with as much money as their wives could spend ; other hundreds dumped their well-winnings into the insatiable maw of speculation.

The Bradford field was young when Col. John J. Carter, of Titusville, paid sixty-thousand dollars for the Whipple farm, on Kendall Creek. Friends shook their heads over the purchase, up to that time the largest by a private individual in the district, but the farm produced fifteen-hundred-thousand barrels of oil and demonstrated the wisdom of the deed. Other properties were developed by this indefatigable worker, until his production was among the largest in the northern region and he could have sold at a price to number him with the millionaires. Unani-

COL. JOHN J. CARTER.

mously chosen President of the Bradford, Bordell & Kinzua Railroad Company, he completed the line in ninety days from the issue of the charter and in eighteen months returned the stockholders eighty per cent. in dividends. Its superior management and heavy earnings attracted favorable attention to this road both at home and abroad. President Carter displayed signal ability in handling the property, which his intelligent methods saved to its owners, while every other narrow-gauge in the system fell into the clutches of receivers or sold as junk to meet court-charges for costly litigation.

All "Old-Timers" remember the "Gentlemen's Furnishing-House of John J. Carter," the finest establishment of the kind west of New York. Young Carter, with a splendid military record, located at Titusville in the summer of 1865, immediately after being mustered out of the service, and engaged in mercantile pursuits ten years. Like other progressive men, he took interests in the wild-cat ventures that made Pithole, Shamburg, Petroleum Centre and Pleasantville famous. From large holdings in Venango, Clarion and Forest he reaped a rich harvest. One tract of four-thousand acres in Forest, purchased in 1886 and two-thirds of it yet undrilled, he expects to hand down to his children as a proof of their father's business-foresight. He scanned the petroleum-horizon around Pittsburg carefully and retained his investments in the middle and upper fields. Taylorstown and McDonald, with their rivers of oil, burst forth with the fury of a flood and disappeared. Sistersville, in West Virginia, had given the trade a taste of its hidden treasures from a few scattered wells. Much salt-water, little oil and deep drilling discouraged operators. How to produce oil at a profit, with such quantities of water to be pumped out, was the problem. Col. Carter visited the scene, comprehended the situation, devised his plans and bought huge blocks of the choicest territory before the oil-trade thought Sistersville worth noticing. This bold stroke added to the value of every well and lease in West Virginia, inspired the faltering with courage and rewarded him magnificently. Advancing prices rendered the princely yield of his scores of wells immensely profitable. Purchases based on fifty-cent oil—the trade had small faith in the outcome—he sold on the basis of dollar-fifty oil. Col. Carter is in the prime of vigorous manhood, ready to explore new fields and surmount new obstacles. He occupies a beautiful home, has a superb library, is a thorough scholar and a convincing speaker. His recent argument before the Ohio Legislature, in opposition to the proposed iniquitous tax on crude-petroleum, was a masterpiece of effective, pungent, unanswerable logic. None who admire a brave, manly, generous character will say that his success is undeserved

The Bradford field extended to the north-east part of McKean and into Allegany county, New York. In 1867 an adventurous operator put down a well in Independence township, forty-five miles north-east of Bradford. In 1880 O. P. Taylor—dead now—began operations in Scio township, drilling a half-dozen wells, two of which had a little oil. On May thirtieth, 1881, one was finished in a ravine close to the quiet village of Richburg, Wirt township. It started at thirty barrels, causing much excitement, which Samuel Boyle's three-hundred-barreler in July sent up to fever-heat. The latter flowed a dark oil, heavier than the usual product, but this made no difference in the scramble for leases. Eighteen months sufficed to define the Allegany field, which was confined to seven-thousand acres. Twenty-nine-hundred wells were bored and the maximum yield of the district was nineteen-thousand barrels. Richburg and Bolivar, both old villages, quadrupled their size in three months. Narrow-gauge railroads soon connected the new field with Olean, Friendship and Bradford. The territory was shallow in comparison with parts of McKean, where eighteen-hundred feet was not an uncommon depth for wells. Timber and water were abundant, good roads presented a pleasing contrast to the unfathomable mud of Clarion and Butler and the country was decidedly attractive. Efforts to find an outlet to the belt failed in every instance. The climax had been reached and a gradual decline set in. Allegany was the northern limit of remunerative developments in the United States, which the next turn of the wheel once more diverted southward. The McCalmont Oil-Company and Phillips Brothers were leaders in the Richburg field. The country had been settled by Seventh-day Baptists, whose "Sunday was on Saturday." Not to offend these devout people by discriminating in favor of Sunday, operators "whipped the devil around the stump" by drilling and pumping their wells seven days a week!

Canada has oil-fields of considerable importance. The largest and oldest is in Enniskillen township, Lambton county, a dozen miles from Port Sarnia, at the foot of Lake Huron. Black Creek, a small tributary of the Detroit river, flows through this township and for many years its waters had been coated with a greasy liquid the Indians sold as a specific for countless diseases. The precious commodity was of a brown color, exceedingly odorous, unpleasant to the taste and burned with great intensity. In 1860 several wells were started, the projectors believing the floating oil indicated valuable deposits within easy reach of the surface. James Williams, who had previously garnered the stuff in pits, finished the first well that yielded oil in paying quantity. Others followed in close succession, but months passed without the sensation of a genuine spouter. Late in the summer of the same year that operations commenced, John Shaw, a poor laborer, managed to get a desirable lease on the bank of the creek. He built a cheap rig, provided a spring-pole and "kicked down" a well, toiling all alone at his weary task until money and credit and courage were exhausted. Ragged, hungry and barefooted, one forenoon he was refused boots and provisions by the village-merchant, nor would the blacksmith sharpen his drills without cash down. Reduced to the verge of despair, he went back to his derrick with a heavy heart, ate a hard crust for dinner and decided to leave for the United States next morning if no signs of oil were discovered that afternoon. He let down the tools and resumed his painful task. Twenty minutes later a rush of gas drove the tools high in the air, followed the next instant by a column of oil that rose a hundred feet! The roar could be heard a mile and the startled populace rushed from the neighboring hamlet to see the unexpected marvel. Canada boasted its first flowing-well and the tidings flew like wild-fire. Before dark hundreds of excited spectators visited the spot. For days the oil gushed unchecked, filling a natural basin an acre in extent, then emptying into the creek and discoloring the waters as far down as Lake St. Clair. None knew how to regulate its output and bring the flow under control. Thus it remained a week, when a delegate from Pennsylvania showed the owner how to put in a seed-bag and save the product. The first attempt succeeded and thenceforth the oil was cared for properly. Opinions differ as to the actual production of this novel strike, although the best judges placed it at five-thousand barrels a day for two or three weeks! The stream flowed incessantly the full size of the hole, a strong pressure of gas forcing it out with wonderful speed. The well produced generously four months, when it "stopped for keeps." Persons who visited the well at its best will recall the surroundings. A pond of oil large enough for a respectable regatta lay between it and Black Creek, whose greasy banks for miles bore traces of the lavish inundation of crude. The locality was at once interesting and high-flavored and a conspicuous feature was Shaw himself. Radiant in a fresh suit of store-clothes, he moved about with the complacency incident to a green ruralist who has "struck ile."

One of the persons earliest on the ground after the well began to flow was the storekeeper who had refused the proprietor a pair of boots that morning. With the cringing servility of a petty retailer he hurried to embrace Shaw, coupling this outbreak of affection with the assurance that everything in the shop was at his service. It is gratifying to note that Shaw had the spirit to rebuke this puppyism. Bringing his ample foot into violent contact with the dealer's most vital part, he accompanied a heavy kick with an emphatic command to go to the place Heber Newton and Pentecost have ruled out. Shaw was entirely uneducated and fell a ready prey to sharpers on the watch for easy victims. Cargoes of oil shipped to England brought small returns and his sudden wealth slipped away in short order. Ere long the envied possessor of the big well was obliged to begin life anew. For a few years he struggled along as an itinerant photographer, traveling with a "car" and earning a precarious subsistence taking "tin-types." Death closed the scene in 1872, the luckless pioneer expiring at Petrolea in absolute want. Thus sadly ended another illustration of the adverse fortune which frequently overtakes men whose energy and grit confer benefits upon mankind that surely entitle them to a better fate.

As might be imagined, Shaw's venture gave rise to operations of great magnitude. Hosts flocked to the scene in quest of lands and developments began on an extensive scale. Among others a rig was built and a well drilled without delay as close to the Shaw as it was possible to place the timbers. The sand

14

was soon reached by the aid of steam-power and once more the oil poured forth enormously, the new strike proving little inferior to its neighbor. It was named the Bradley, in honor of the principal owner, E. C. Bradley, afterwards a leading operator in Pennsylvania, president of the Empire Gas-Company and still a resident of Oildom. The yield continued large for a number of months, then ceased entirely and both wells were abandoned. Of the hundreds in the vicinity a good percentage paid nicely, but none rivalled the initial spouters. The influx of restless spirits led to an "oil-town," which for a brief space presented a picture of activity rarely surpassed. Oil Springs, as the mushroom city was fittingly termed, flourished amazingly. The excessive waste of oil filled every ditch and well, rendering the water unfit for use and compelling the citizens to quench their thirst with artificial drinks. The bulk of the oil was conveyed to Mandaumin, Wyoming or Port Sarnia, over roads of horrible badness, giving employment to an army of teamsters. A sort of "mud canal" was formed, through which the horses dragged small loads on a species of flatboats, while the drivers walked along the "tow-path" on either side. The mud had the consistency of thin batter and was seldom under three feet deep. To those who have never seen this unique system of navigation the most graphic description would fail to convey an adequate idea of its peculiar features. Unlike the Pennsylvania oil-fields, the petroleum-districts of Canada are low and swampy, a circumstance that added greatly to the difficulty of moving the greasy staple during the wet season. Ultimately roads were cut through the soft morasses and railways were constructed, although not before Oil Springs had seen its best days and begun a rapid descent on the down grade. Salt-water quickly put a stop to many wells, the production declined rapidly and the town was depopulated. Operations extended towards the north-west, where Petrolea, which is yet a flourishing place, was established in 1864. Bothwell, twenty-six miles south of Oil Springs, had a short career and light production. Canadian operators were slower than the Yankees of the period and the tireless push of the Americans who crowded to the front at the beginning of the developments around Oil Springs was a revelation to the quiet plodders of Enniskillen and adjacent townships. The leading refineries are at London, fifty miles east of Wyoming and one of the most attractive cities in the Dominion.

The Tidioute belt, varying in narrowness from a few rods to a half-mile, was one of the most satisfactory ever discovered. When lessees fully occupied the flats Captain A. J. Thompson drilled a two-hundred-barrel well on the point, at the junction of Dingley and Dennis Runs. Quickly the summit was scaled and amid drilling wells, pumping wells, oil-tanks and engine-houses the town of Triumph was created. Triumph Hill turned out as much money to the acre as any spot in Oildom. The sand was the thickest—often ninety to one-hundred-and-ten feet—and the purest the oil-region afforded. Some of the wells pumped twenty years. Salt-water was too plentiful for comfort, but half-acre plots were grabbed at one-half royalty and five-hundred dollars bonus. Wells jammed so closely that a man could walk from Triumph to New London and Babylon on the steam-boxes connecting them. Percy Shaw—he built the Shaw House—had a "royal flush" on Dennis Run that netted two-hundred-thousand dollars. From an investment of fifteen-thousand dollars E. E. and J. M. Clapp cleared a half-million.

"Spirits" located the first well at Stoneham and Cornen Brothers' gasser at Clarendon furnished the key that unlocked Cherry Grove. Gas was piped from the Cornen well to Warren and Jamestown. Walter Horton was the moving spirit in the Sheffield field, holding interests in the Darling and Blue Jay wells and owning forty-thousand acres of land in Forest county. McGrew Brothers, of Pittsburg, spent many thousands seeking a pool at Garland. Grandin & Kelly's operations below Balltown exploded the theory that oil would not be found on the south side of Tionesta Creek. Cherry Grove was at its apex when, in July of 1884, with Farnsworth and Garfield boiling over, two wells on the Thomas farm, a mile south-east of Richburg, flowed six-hundred barrels apiece. They were among the largest in the Allegany district, but a three-line mention in the Bradford *Era* was all the notice given the pair.

To the owner of a tract near "646," who offered to sell it for fifty-thousand dollars, a Bradford operator replied: "I would take it at your figure if I thought my check would be paid, but I'll take it at forty-five-thousand whether the check is paid or not!" The check was not accepted.

John Shaw, whose gusher brought the "gum-beds" of Enniskillen into the petroleum-column, narrowly escaped anticipating Drake three years. Shaw removed from Massachusetts to Canada in 1838 and was regarded as a visionary schemer. In 1856 he sought to interest his neighbors in a plan to *drill a well through the rock* in search of the reservoir that supplied Bear Creek with a thick scum of oil. They hooted at the idea and proposed to send Shaw to the asylum. This tabooed the subject and postponed the advent of petroleum until 1859.

Tack Brothers drilled a dry-hole twenty-six-hundred feet in Millstone township, Elk county. Grandin & Kelly drilled four-thousand feet in Forest county and got lots of geological information, but no oil.

Get off the train at Trunkeyville—a station-house and water-tank—and climb up the hill towards Fagundas. After walking through the woods a mile an opening appears. A man is plowing. The soil looks too poor to raise grasshoppers, yet that man during the oil-excitement refused an offer of sixty-thousand dollars for this farm. His principal reason was that he feared a suitable house into which to move his family could not be obtained! On a little farther a pair of old bull-wheels, lying unused, tells that the once productive Fagundas pool has been reached. A short distance ahead on an eminence is a church. This is South Fagundas. No sound save the crowing of a chanticleer from a distant farm-yard breaks the silence. The merry voices heard in the seventies are no longer audible, the drill and pump are not at work, the dwellings, stores and hotels have disappeared. The deserted church stands alone. A few landmarks linger at Fagundas proper. There is one store and no place where the weary traveler can quench his thirst. The nearest resemblance to a drinking-place is a boy leaning over a barrel drinking rain-water while another lad holds him by the feet. Fagundas is certainly "dry." The stranger is always taken to the Venture well. Its appearance differs little from that of hundreds of other abandoned wells. The conductor and the casing have not been removed. Robert W. Pimm, who built the rig, still lives at Fagundas. He will be remembered by many, for he is a jovial fellow and was "one of the boys." The McQuade—the biggest in the field—the Bird and the Red Walking-beam were noted wells. If Dr. Stillson were to hunt up the office where he extracted teeth "without pain" he would find the building used as a poultry-house. Men went to Fagundas poor and departed with sufficient wealth to live in luxury the rest of their lives; others went wealthy and lost everything in a vain search for the greasy fluid. Passing through what was known as Gillespie and traversing three miles of a lonely section, covered with scrub-oak and small pine, Triumph is reached. It is not the Triumph oil-men knew twenty-five years ago, when it had four-thousand population, four good hotels, two drug-stores, four hardware-stores, a half-dozen groceries and many other places of business. No other oil-field ever held so many derricks upon the same area. The Clapp farm has a production of twelve barrels per day. Traces of the town are almost completely blotted out. The pilgrim traveling over the hill would never suspect that a rousing oil-town occupied the farm on which an industrious Swede has a crop of oats. Along Babylon hill, once dotted with derricks thickly as trees in the forest, nothing remains to indicate the spot where stood the ephemeral town.

Five townships six miles square—Independence, Willing, Alma, Bolivar and Genesee, with Andover, Wellsville, Scio, Wirt and Clarksville north—form the southern border of Allegany county, New York. The first well bored for oil in the county—the Honeyoe—was the Wellsville & Alma Oil-Company's duster in Independence township, drilled eighteen-hundred feet in September, 1877. Gas at five-hundred feet caught fire and burned the rig and signs of oil were found at one-thousand feet. The second was O. P. Taylor's Pikeville well, Alma township, finished in November, 1878. Taylor, the father of the Allegany field, decided to try north of Alma and in July of 1879 completed the Triangle No. 1, in Scio township, the first in Allegany to produce oil. It

originated the Wellsville excitement and first diverted public attention from Bradford. Triangle No. 2, drilled early in 1880, pumped twelve barrels a day. S. S. Longabaugh, of Duke Centre, sank a dry-hole, the second well in Scio, three miles north-east of Triangle No. 1. Operations followed rapidly. Richburg No. 1, Wirt township, in which Taylor enlisted three associates, responded at a sixty-barrel gait in May of 1881 to a huge charge of glycerine. Samuel Boyle, who had struck the first big well at Sawyer City, completed the second well at Richburg in June, manipulated it as a "mystery" and torpedoed it on July thirteenth. It flowed three-hundred barrels of blue-black oil, forty-two gravity, from fifty feet of porous sand and slate. Tay-

O. P. TAYLOR.

lor's exertions and perseverance showed indomitable will, bravery and pluck. He was a Virginian by birth, a Confederate soldier and a cigar-manufacturer at Wellsville. It is related that while drilling his first Triangle well the tools needed repairs and he had not money to send them to Bradford. His Wellsville acquaintances seemed amazingly "short" when he attempted a loan. His wife had sold her watch to procure food and she gave him the cash. The tools were fixed, the well was completed and it started Taylor on the road to the fortune he and his helpmeet richly earned. The pioneer died in the fall of 1883. The record of his adventures, trials and tribulations in opening a new oil-district would fill a volume. He was prepared for the message: "Child of Earth, thy labors and sorrows are done."

The bee-line to the north was squarely "on the belt."

OILY OOZINGS.

Kerosene is often the last scene.

The ladies—God bless them!—are nothing if not consistent—at times. It used to be a fad with Bradford wives to keep a stuffed owl in the parlor for ornament and a stuffed club in the hall for the night-owl's benefit.

> The Oil-Creek girls are the dandy girls,
> For their kiss is most intense;
> They've got a grip like a rotary-pump
> That will lift you over the fence.

The steel of a rimmer was lost in a drilling well on Cherry Run. After fishing for it for a long time the well-owner, becoming discouraged, offered a man one-thousand dollars to take it out. He broomed the end of a tough block, ran it down the well attached to the tools and in ten minutes had the steel out.

> The woman who eagerly seized the oil-can
> And to pour kerosene in the cook-stove began,
> So that people for miles to quench the fire ran,
> While she soar'd aloft like a flash in the pan,
> Didn't know it was loaded.

At a drilling well near Rouseville the tools were lowered on Monday morning and, after running a full screw, were drawn minus the bit, with the stem-box greatly enlarged. After fishing several days for it the drillers were greatly surprised to find the lost bit standing in the slack-tub. The tools had been lowered in the darkness with no bit on.

> An Oil-City tramp on the pavement drear
> Saw something that seem'd to shine;
> He pick'd it up and gave a big cheer—
> 'Twas a nickel bright, the price of a beer—
> And shouted "The world is mine!"

"Breathe through the nostrils" is good advice. People should breathe through the nose and not use it so much for talking and singing through. Yet every rule has exceptions. A pair of mules hauled oil at Petroleum Centre in the flush times of the excitement. The mud was practically bottomless. A visitor was overheard telling a friend that the bodies of the mules sank out of sight and that they were breathing through their ears, which alone projected above the ooze. Petroleum Centre and many more departed oil-towns suggest the old jingle:

> "There was an old woman lived under a hill,
> If she hadn't moved she'd be there still;
> But she moved!"

About St. Valentine's Day in 1866, when the burning of the Tremont House led to the discovery of oil in springs and wells, was a hilarious time at Pithole. Every cellar was fairly flooded with grease. People pumped it from common pumps, dipped it from streams, tasted it in tea, inhaled it from coffee-pots and were afraid to carry lights at night lest the very air should cause explosion and other unhappiness. It became a serious question what to drink. The whiskey could not be watered—there was no water. Dirty shirts could not be washed—the very rain was crude oil. Dirt fastened upon the damask cheeks of Pithole damsels and found an abiding-place in the whiskers of every bronzed fortune-hunter. Water commanded an enormous price and intoxicating beverages were cheap, since they could scarcely be taken in the raw. The editor of the *Record*, a strict temperance man, was obliged to travel fourteen miles every morning by stone-boat to get his glass of water. Stocks of oil-companies were the only thing in the community thoroughly watered. Tramps, hobos, wandering vagrants and unwashed disbelievers that "cleanliness is next to Godliness" pronounced Pithole a terrestrial paradise. They were willing to reverse Muhlenburg's sentiment and "live alway" in that kind of dry territory.

X.

ON THE SOUTHERN TRAIL.

Down the Allegheny—Reno, Scrubgrass, Bullion—Clarion District—St. Petersburg, Antwerp, Edenburg—Parker to Greece City—Butler's Rich Pastures—The Cross-Belt—Petrolia, Karns, Millerstown—Thorn-Creek Geysers—McDonald Mammoths—Invasion of Washington—West Virginia Plays the Deuce—General Gleanings.

> "I'm comin' from de Souf, Susanna doant yo cry."—*Negro Melody.*
> "Let us battle for elbow-room."—*James Parish Steele.*
> "We must take the current when it serves, or lose our ventures."—*Shakespeare.*
> "Peter Oleum came down like a wolf on the fold."—*Byron Parodied.*
>   "Liberal as noontide speeds the ambient ray
>   And fills each crevice in the world with day."—*Lytton.*
> "How soon our new-born-light attains to full-aged noon."—*Francis Quarles.*
> "What lavish wealth men give for trifles light and small."—*W. S. Hawkins.*
> "Who, grown familiar with the sky, will grope henceforward among groundlings?"
>   —*Robert Browning.*

ST. GEORGE'S WELLS and RESIDENCE, A.V.R.R.

OUTH and west of Oil Creek for many miles the petroleum-star shed its effulgent luster. Down the Allegheny adventurous operators groped their way patiently, until Clarion, Armstrong, Butler, Washington and West Virginia unlocked their splendid storehouses at the bidding of the drill. Aladdin's wondrous lamp, Stalacta's wand or Ali Babi's magic sesame was not so grand a talisman as the tools which from the bowels of the earth brought forth illimitable spoil. No need of fables to varnish the tales of struggles and triumphs, of disappointments and successes, of weary toil and rich reward that have marked the oil-development from the Drake well to the latest strike in Tyler county. Men who go miles in advance of developments to seek new oil-fields run big chances of failure. They understand the risk and appreciate the cold fact that heavy loss may be entailed. But "the game is worth the powder" in their estimation and impossibility is not the sort of ability they swear by. "Our doubts are traitors and make us lose the good we oft might win" is a maxim oil-operators have weighed carefully. The man who has faith to attempt something is a man of power, whether he hails from Hong Kong or Boston, Johannesburg or Oil City. The man who will not improve his opportunity, whether seeking salvation or petroleum, is a sure loser. His stamina is as fragile as a fifty-cent shirt and will wear out quicker than religion that is used for a cloak only. Muttering long prayers without working to answer them is not the way to angle for souls, or fish, or oil-wells. It demands nerve and vim and enterprise to stick thousands of dollars in a hole ten, twenty, fifty or "a hundred miles from anywhere," in hope of opening a fresh vein of petroleum. Luckily men possessing these qualities have not been lacking since the first well on Oil Creek sent forth the feeble squirt that has grown to a mighty river. Hence prolific territory, far from being scarce, has sometimes been too plentiful for the financial health of the average producer, who found it hard to cipher out a profit selling dollar-crude at forty cents. As old fields exhausted new ones were explored in every direction, those south of the original strike presenting a very respectable figure in the oil-panorama. If "eternal vigilance is the price of liberty," eternal hustling is the price of oil-

operations. Maria Seidenkovitch, a fervid Russian anarchist, who would rather hit the Czar with a bomb than hit a thousand-barrel well, has written:

"There is no standing still! Even as I pause
  The steep path shifts and I slip back apace;
  Movement was safety; by the journey's laws
  No help is given, no safe abiding-place;
  No idling in the pathway, hard and slow—
  I must go forward or must backward go!"

Down the Allegheny three miles, on a gentle slope facing bold hills across the river, is the remnant of Reno, once a busy, attractive town. It was named from Gen. Jesse L. Reno, who rose to higher rank than any other of the heroes Venango "contributed to the death-roll of patriotism." He spent his boyhood at Franklin, was gradu-

GEN. JESSE L. RENO.

ated from West Point in the class with George B. McClellan and "Stonewall" Jackson, served in the Mexican war, was promoted to Major-General and fell at the battle of South Mountain in 1862. The Reno Oil-Company, organized in 1865 as the Reno Oil and Land Company, owns the village-site and twelve-hundred acres of adjacent farms. The company and the town owed their creation to the master-mind of Hon. C. V. Culver, to whose rare faculty for developing grand enterprises the oil-regions offered an inviting field. Visiting Venango county early in the sixties, a canvass of the district convinced him that the oil-industry, then an infant beginning to creep, must attain giant proportions. To meet the need of increased facilities for business, he conceived the idea of a system of banks at convenient points and opened the first at Franklin in 1861. Others were established at Oil City, Titusville and suitable trade-centres until the combination embraced twenty banks and banking-houses, headed by the great office of Culver, Penn & Co. in New York. All enjoyed large patronage and were converted into corporate banks. The speculative mania, unequaled in the history of the world, that swept over the oil-regions in 1864-5, deluged the banks with applications for temporary loans to be used in purchasing lands and oil-interests. Philadelphia alone had nine-hundred stock-companies. New York was a close second and over seven-hundred-million dollars were capitalized—on paper—for petroleum speculations! The production of oil was a new and unprecedented business, subject to no known laws and constantly overturning theories that set limits to its expansion. There was no telling where flowing-wells, spouting thousands of dollars daily without expense to the owners, might be encountered. Stories of sudden fortunes, by the discovery of oil on lands otherwise valueless, pressed the button and the glut of paper-currency did the rest.

Mr. Culver directed the management and employment of fifteen-million dollars in the spring of 1865! People literally begged him to handle their money, elected him to Congress and insisted that he invest their cash and bonds. The Reno Oil-Company included men of the highest personal and commercial standing. Preliminary tests satisfied the officers of the company that the block of land at Reno was valuable territory. They decided to operate it, to improve the town and build a railroad to Pithole, in order to command the trade of Oil Creek, Cherry Run and "the Magic City." Oil City opposed the railroad strenuously, refusing a right-of-way and compelling the choice of a circuitous route, with difficult grades to climb and ugly ravines to span. At length a consolidation of competing interests was arranged, to be formally ratified on March twenty-ninth, 1866. Meanwhile rumors affecting the credit of the Culver banks were circulated. Disastrous floods, the close of the war and the amazing collapse of Pithole had checked speculation and impaired confidence in oil-values. Responsible parties wished to stock the Reno Company at five-million dollars and Mr. Culver was in Washington completing the railroad-negotiations which, in one week, would give him control of nearly a million. A run on his banks was started, the strain could not be borne and on March twenty-seventh, 1866, the failure of Culver, Penn & Co. was announced. The assets at cost largely exceeded the liabilities of four-million dollars, but the natural result of the suspension was to discredit everything with which the firm had been identified. The railroad-consolidation, confessedly advantageous to all concerned, was not confirmed and Reno stock was withheld from the market. While the creditors generally co-operated to protect the assets and adjust matters fairly, a few defeated measures looking to a safe deliverance. These short-sighted individuals sacrificed properties, instituted harassing prosecutions and precipitated a crisis that involved tremendous losses. Many a man standing on his brother's neck claims to be looking up far into the sky watching for the Lord to come!

The fabric reared with infinite pains toppled, pulling down others in its fall. The Reno, Oil-Creek & Pithole Railroad, within a mile of completion, crumbled into ruin. The architect of the splendid plans that ten days of grace would have carried to fruition displayed his manly fiber in the dark days of adversity and he has been amply vindicated. Instead of yielding to despair and "letting things take their course," he strove to realize for the creditors every dollar that could be saved from the wreck. Animated by a lofty motive, for thirty years Mr. Culver has labored tirelessly to discharge the debts of the partnership. No spirit could be braver, no life more unselfish, no line of action more steadfastly devoted to a worthy object. He had bought property and sought to enhance its value, but he had never gambled in stocks, never dealt in shares on the mere hazard of a rise or gone outside the business—except to help customers whose necessities appealed to his sympathy—with which he was intimately connected. Driven to the wall by stress of circumstances and general distrust, he has actually paid off all the small claims and multitudes of large ones against his banks. How many men, with no legal obligation to enforce their payment, would toil for a generation to meet such demands? Thistles do not bear figs and banana-vendors are not the only persons who should be judged by their fruits. It is a good thing to achieve success and better still to deserve it. Gauged by the standard of high resolve, earnest purpose and persistent endeavor—by what he has tried to do and not by what may have been said of him—Charles Vernon Culver can afford to accept the verdict of his peers and of the Omniscient Judge, who "discerns the thoughts and intents of the heart."

> " I will go on then, though the limbs may tire,
>   And though the path be doubtful and unseen;
> Better with the last effort to expire
>   Than lose the toil and struggle that have been,
> And have the morning strength, the upward strain,
> The distance conquered in the end made vain."

Reorganized in the interest of Culver, Penn & Co.'s creditors, the Reno Company developed its property methodically. No. 18 well, finished in May of 1870, pumped two-hundred barrels and caused a flutter of excitement. Fifty others, drilled in 1870-1, were so satisfactory that the stockholders might have shouted "Keno!" The company declined to lease and very few dry-holes were put down on the tract. Gas supplied fuel and the sand, coarse and

pebbly, produced oil of superior gravity at five to six-hundred feet. Reno grew, a spacious hotel was built, stores prospered, two railroads had stations and derricks dotted the banks of the Allegheny. The company's business was conducted admirably, it reaped liberal profits and operated in Forest county. Its affairs are in excellent shape and it has a neat production to-day. Mr. Culver and Hon. Galusha A. Grow have been its presidents and Hon. J. H. Osmer is now the chief officer. Mr. Osmer is a leader of the Venango bar and has lived at Franklin thirty-one years. His thorough knowledge of law, sturdy independence, scorn of pettifogging and skill as a pleader gained him an immense practice. He has been retained in nearly all the most important cases before the court for twenty-five years and appears frequently in the State and the United-States Supreme Courts. He is a logical reasoner and brilliant orator, convincing juries and audiences by his incisive arguments. He served in Congress with distinguished credit. His two sons have adopted the legal profession and are associated with their father. A man of positive individuality and sterling character, a friend in cloud and sunshine, a deep thinker and entertaining talker is James H. Osmer.

J. H. OSMER.

Cranberry township, a regular petroleum-huckleberry, duplicated the Reno pool at Milton, with a vigorous offshoot at Bredinsburg and nibbles lying around loose. Below Franklin the second-sand sandwich and Bully-Hill successes were special features. A mile up East Sandy Creek—it separates Cranberry and Rockland—was Gas City, on a toploftical hill twelve miles south of Oil City. A well sunk in 1864 had heaps of gas, which caught fire and burned seven years. E. E. Wightman and Patrick Canning drilled five good wells in 1871 and Gas City came into being. Vendergrift & Forman constructed a pipeline and telegraph to Oil City. Gas fired the boilers, lighted the streets, heated the dwellings and great quantities wasted. The pressure could be run up to three-hundred pounds and utilized to run engines in place of steam, were it not for the fine grit with the gas, which wore out the cylinders. Wells that supplied fuel to pump themselves seemed very similar to mills that furnished their own motive-power and grist for the hoppers. A cow that gave milk and provided food for herself by the process could not be slicker. Gas City vaporized a year or two and flickered out. The last jet has been extinguished and not a glimmer of gas or symptom of wells has been visible for many years.

Fifteen of the first sixteen wells at Foster gladdened the owners by yielding bountifully. To drill, to tube, to pump, to get done-up with a dry-hole, "aye, there's the rub" that tests a fellow's mettle and changes blithe hope to bleak despair. Foster wells were not of that complexion. They lined the steep cliff that resembles an Alpine farm tilted on end to drain off, the derricks standing like sentries on the watch that nobody walked away with the romantic landscape. Lovers of the sterner moods of nature would revel in the rugged scenery, which discounts the overpraised Hudson and must have fostered sublime emotions in the impassive redmen. Indian-God Rock, inscribed with untranslatable hieroglyphics, presumably tells what "Lo" thought of the surroundings. Six miles south of the huge rock, which somebody proposed to boat to Franklin and set in the park as an interesting memento of the aborigines, was "the burning well." For years the gas blazed, illuminating the hills and keeping a plot of grass constantly fresh and green. The flood in 1865 overflowed the hole, but the gas burned just as though water were its native element. It was the fad for sleighing parties to visit the well, dance on the sward when snow lay a yard deep ten rods away and hold outdoor picnics in January and February. This practically realized the fancy of the boy who wished winter would come in summer, that he might coast on the Fourth of July in shirt-sleeves and linen-pants. Here and there in the interior of Rockland township morsels of oil have been unearthed and small wells are pumping to-day.

C. D. Angell leased blocks of land from Foster to Scrubgrass in 1870-71 and jabbed them with holes that confirmed his "belt theory." His first well—a hundred-barreler—on Belle Island, a few rods below the station, opened the Scrubgrass field. On the Rockland side of the river the McMillan and 99 wells headed a list of remunerative producers. Back a quarter-mile the territory was tricky, wells that showed for big strikes sometimes proving of little account. A town toddled into existence. Gregory—the genial host joined the heavenly host long ago—had a hotel at which trains stopped for meals. James Kennerdell ran a general store and the post-office. The town was busy and had nothing scrubby except the name. The wells retired from business, the depot burned down, the people vanished and Kennerdell Station was established a half-mile north. Wilson Cross continued his store at the old stand until his death in March, 1896. Within a year paying wells have been drilled near the station and two miles southward. On the opposite bank Major W. T.

Baum, of Franklin, has a half-dozen along the base of the hill that net him a princely return. A couple of miles north-west, in Victory township, Conway Brothers, of Philadelphia, recently drilled a well forty-two hundred feet. The last sixty feet were sand with a flavor of oil, the deepest sand and petroleum recorded up to the present time. Careful records of the strata and temperature were taken. Once a thermometer slipped from Mr. Conway's hand and tumbled to the bottom of the well, the greatest drop of the mercury in any age or clime.

Sixty farmers combined in the fall of 1859 to drill the first well in Scrubgrass township, on the Rhodabarger tract. They rushed it like sixty six-hundred feet, declined to pay more assessments, kicked over the dashboard and spilled the whole combination. The first productive well was Aaron Kepler's, drilled on the Russell farm in 1863, and John Crawford's farm had the largest of the early ventures. On the Witherup farm, at the mouth of Scrubgrass Creek, paying wells were drilled in 1867. Considerable skirmishing was done at intervals without startling results. The first drilling in Clinton township was on the Kennerdell property, two miles west of the Allegheny, the Big-Bend Oil-Company sinking a dry-hole in 1864-5. Jonathan Watson bored two in 1871, finding traces of oil in a thin layer of sand. The Kennerdell block of nine-hundred acres figured as the scene of milling operations from the beginning of the century. David Phipps — the Phipps families are still among the most prominent in Venango county—built a grist-mill on the property in 1812, a saw-mill and a woolen-factory, operated an iron-furnace a mile up the creek and founded a natty village. Fire destroyed his factory and Richard Kennerdel bought the place in 1853. He built a woolen-mill that attained national celebrity, farmed extensively, conducted a large store and for thirty years was a leading business-man. A handsome fortune, derived from manufacturing and oil-wells on his lands, and the respect of all classes rewarded the enterprise, sagacity and hospitality of this progressive citizen. The factory he reared has been dismantled, the pretty little settlement amid the romantic hills of Clinton is deserted and the man to whom both owed their development rests from his labors. Mr. Kennerdell possessed boundless energy, decision and the masterly qualities that surmount obstacles, build up a community and round out a manly character. Cornen Brothers have a production on the Kennerdell tract, which they purchased in 1892. During the Bullion furore a bridge was built at Scrub-grass and a railroad to Kennerdell was constructed. Ice carried off the bridge and the faithful old ferry holds the fort as in the days of John A. Canan and George McCullough.

Phillips Brothers, who had operated largely on Oil Creek and in Butler county, leased thousands of acres in Clinton and drilled a number of dry-holes. Believing a rich pool existed in that latitude, they were not deterred by reverses that would have stampeded operators of less experience. On August ninth, 1876, John Taylor and Robert Cundle finished a two-hundred-barrel spouter on the George W. Gealy farm, two miles north of Kennerdell. They sold to Phillips Brothers, who were drilling on adjacent farms. The new strike opened the Bullion field, toward which the current turned forthwith. H. L. Taylor and John Satterfield, the biggest operators in Butler, visited the Gealy well and offered a half-million dollars for the Phillips interests in Clinton. A hundred oilmen stood watching the flow that August morning. The parties consulted briefly and Isaac Phillips invited me to walk with him a few rods. He said: "Taylor & Satterfield wish to take our property at five-hundred-thousand dollars. This is a good deal of money, but we have declined it. We think there will be a million in this field for us if we develop it ourselves." They carried out this programme and the estimate was approximated closely.

The Sutton, Simcox Taylor, Henderson, Davis, Gealy, Newton and Berringer farms were operated rapidly. Tack Brothers paid ten-thousand dollars to Taylor for thirty acres and Porter Phipps leased fifteen acres, which he sold to Emerson & Brownson, whose first well started at seven-hundred barrels. Phillips Brothers' No. 3 well, on the Gealy farm, was a four-hundred-barreler. In January, 1877, Frank Nesbit's No. 2, Henderson farm, flowed five-hundred barrels, and in February the Galloway began at two-hundred. The McCalmont Oil-Company's Big Medicine, on the Newton farm, tipped the beam at one-thousand barrels on June seventh. Mitchell & Lee's Big Injun flowed three-thousand barrels on June eighteenth, the biggest yield in the district.

Ten rods away a galaxy of Franklinites drilled the driest kind of a dry-hole. In August the McCalmont No. 31 and the Phillips No. 7 gauged a plump thousand apiece. These were the largest wells and they exhausted speedily. The oil from the Gealy No. 1 was hauled to Scrubgrass until connections could be laid to the United Pipe-Lines. The Bullion field, in which a few skeleton-wells produce a few barrels daily, extended seven miles in length and three-eighths of a mile in width. Like the business-end of a healthy wasp, "it was little, but—oh, my!" It swerved the tide from Bradford and ruled the petroleum-roost eighteen months. Summit City on the Simcox farm, Berringer City on the Berringer farm, and Dean City on the McCalmont farm, flourished during the excitement. The first house at Summit was built on December eighth, 1876. In June of 1877 the town boasted two-hundred buildings and fifteen-hundred population. Abram Myers, the last resident, left in April of 1889. All three towns have "faded into nothingness" and of the five-hundred wells producing at the summit of Bullion's short-lived prosperity not a dozen survive. Westward a new strip was opened last year the wells on several farms yielding their owners a pleasant income.

Major St. George—the kindly old man sleeps in the Franklin cemetery—had a bunch of wells and lived in a small house close to the Allegheny-Valley track, near the siding in Rockland township that bears his name. At Rockland Station a stone chimney, a landmark for many years, marked the early

VIEW ON RITCHEY RUN.

abode of Hon. Elisha W. Davis, who operated at Franklin, was speaker of the House of Representatives two terms and spent the closing years of his active life in Philadelphia. Emlenton, the lively town at the southeastern corner of Venango county, was a thriving place prior to the oil-development. Wells in the vicinity were generally small, Ritchey Run having some of the best. This romantic stream, south of the town, borders Clarion county for a mile or two from its mouth. John Kerr, a squatter, cleared a portion of the forest and was drowned in the river, slipping off a flat rock two miles below his bit of land. The site of Emlenton was surveyed for Joseph B. Fox and Andrew McCaslin. Fox, a rich Quaker, was the pioneer settler

and founded the town of Foxburg, four miles south of Emlenton, the intervening territory forming part of his estate. McCaslin owned the land above the Valley Hotel and the public-school. He was elected sheriff in 1832 and built an iron-furnace. As a compliment to Mrs. Fox—Miss Hannah Emlen—he named the hamlet Emlenton. Doctor James Growe built the third house in the settlement. The covered wooden-bridge, usually supposed to have been brought over in the Mayflower, withstood floods and ice-gorges until April of 1883. John Keating, who had the second store, built a furnace near St. Petersburg and held a thousand acres of land. Oil-producers were well represented in the growing town, which has been the home of Marcus Hulings, L. E. Mallory, D. D. Moriarty, M. C. Treat and R. W. Porterfield. James Bennett, a leader in business, built the opera-house and the flour-mills and headed the company that built the Emlenton & Shippenville Railroad, which ran to Edenburg at the height of the Clarion development. Emlenton is supplied with natural-gas and noted for good schools, good hotels and get-up-and-get citizens.

Dr. A. W. Crawford, who served three terms in the Legislature, was appointed consul to Antwerp by President Lincoln, in 1861. At the time he reached Antwerp a cheap illuminant was unknown on the continent. Gas was used in the cities, but the people of Antwerp depended mainly upon rape-seed oil. Only wealthy people could afford it and the poorer folks went to bed in the dark. From Antwerp to Brussels the country was shrouded in gloom at night. Not a light could be seen outside the towns, in the most populous section on earth. A few gallons of American refined had appeared in Antwerp previous to Dr. Crawford's arrival. It was regarded as an object of curiosity. A leading firm inquired about this new American product and Dr. Crawford was the man who could give the information. He was from the very part of the country where the new illuminant was produced. The upshot of the matter was that Dr. Crawford put the firm in communication with American shippers, which led to an order of forty barrels by Aug. Schmitz & Son, Antwerp dealers. The article had tremendous prejudice to overcome, but the exporters succeeded in finally disposing of their stock. It yielded them a net return of forty francs. The oil won its way and from the humble beginning of forty barrels in 1861, the following year witnessing a demand for fifteen-hundred-thousand gallons. By 1863 it had come largely into use and since that time it has become a staple article of commerce. Dr. Crawford served as consul at Antwerp until 1866, when he returned home and began a successful career as an oil-producer. It was fortunate that Col. Drake chanced upon the shallowest spot in the oil-regions where petroleum has ever been found, when he located the first well, and equally lucky that a practical oilman represented the United States at Antwerp in 1861. Had Drake chanced upon a dry-hole and some other man been consul at Antwerp, oil-developments might have been retarded for years.

> "Oft what seems a trifle,
> A mere nothing in itself, in some nice situations
> Turns the scale of Fate and rules important actions."

Fertig & Hammond drilled medium wells on the Fox estate of twelve-hundred acres, near Foxburg Station, in 1870-71. They established a bank and operations in the neighborhood were pressed actively by the Fox heirs and producers from the upper districts. Foxburg was the jumping-off point for pilgrims to the Clarion field, which Galey No. 1 well, on Grass Flats, inaugurated in August, 1871. Others on the Flats, ranging from thirty to eighty barrels, boomed Foxburg and speedily advanced St. Petersburg, three miles inland, from a sleepy village of thirty houses to a busy town of three-thousand population. In September of 1871 Marcus Hulings, whose great specialty was opening new fields, finished a hundred-barrel well on the Ashbaugh farm, a mile beyond St. Petersburg. The town of Antwerp was one result. The first building, erected in the spring of 1872, in sixty days had the company of four groceries, three hotels, innumerable saloons, telegraph-office, school-house and two-hundred dwellings. Its general style was summed up by the victim of a poker-game in the expressive words: "If you want to get a smell of brimstone before supper go to Antwerp!" Fire in 1873 wiped it off the face of the planet.

Charles H. Cramer, now proprietor of a hotel in Pittsburg, left the Butler field to drill the Antwerp well, in which he had a quarter-interest. James M. Lambing, for whom he had been drilling, jokingly remarked: "When you return 'broke' from the wildcat well on the Ashbaugh farm I will have another job for you." It illustrates the ups and downs of the oil-business in the seventies to note that, when the well was completed, Lambing had met with financial reverses and Cramer was in a position to give out jobs on his own hook. Victor Gretter was one of the spectators of the oil flowing over the derrick. The waste suggested to him the idea of the oil-saver, which he patented. This strike reduced the price of crude a dollar a barrel. Antwerp would have been more important but for its nearness to St. Petersburg, which disastrous fires in 1872–3 could not prevent from ranking with the best towns of Oildom. Stages from Foxburg were crowded until the narrow-gauge railroad furnished improved facilities for travel. Schools, churches, hotels, newspapers, two banks and an opera-house flourished. The Pickwick Club was a famous social organization. The Collner, Shoup, Vensel, Palmer and Ashbaugh farms and Grass Flats produced three-thousand barrels a day. Oil was five to six dollars and business strode ahead like the wearer of the Seven-League Boots. Now the erstwhile busy town is back to its pristine quietude and the farms that produced oil have resumed the production of corn and grass.

A jolly Dutchman near St. Petersburg, who married his second wife soon after the funeral of the first, was visited with a two-hours' serenade in token of disapproval. He expostulated pathetically thus : "I say, poys, you ought to be ashamed of myself to be making all dish noise ven der vas a funeral here purty soon not long ago." This dispersed the party more effectually than a bull-dog and a revolver could have done.

A girl just returned to St. Petersburg from a Boston high-school said, upon seeing the new fire-engine at work : "Who would evah have dweamed such a vewy diminutive looking apawatus would hold so much wattah !"

"Where are you going?" said mirth-loving Con. O'Donnell to an elderly man in a white cravat whom he overtook on the outskirts of Antwerp and proposed to invite to ride in his buggy. "I am going to heaven, my son. I have been on my way for eighteen years." "Well, good-bye, old fellow ! If you have been traveling toward heaven for eighteen years and got no nearer than Antwerp, I will take another route."

The course of operations extended past Keating Furnace, up and beyond Turkey Run, a dozen miles from the mouth of the Clarion River. Good wells on the Ritts and Neeley farms originated Richmond, a small place that fizzled out in a year. The Irwin well, a mile farther, flowed three-hundred barrels in September of 1872. The gas took fire and burned three men to death. The entire ravine and contiguous slopes proved desirable territory, although the streak rarely exceeded a mile in breadth. Turkey City, in a nice expanse to the east of the famous Slicker farm, for months was second only to St. Petersburg as a frontier town. It had four stages to Foxburg, a post-office, daily mail-service and two passable hotels. George Washington, who took a hack at a cherry-tree, might have preferred walking to the drive over the rough, cut-up roads that led to and from Turkey City. The wells averaged eleven-hundred feet, with excellent sand and loads of gas for fuel. Richard Owen and Alan Cochran, of Rouseville, opened a jack-pot on the Johnson farm, above town. Wells lasted for years and this nook of the Clarion district could match pennies with any other in the business of producing oil.

Captain John Kissinger, a pioneer settler, died in 1880 at the age of eighty-five. He was the father of thirty-four children, nine of whom perished by his dwelling taking fire during the absence of the parents from home. His second wife, who survived him ten years, weighed three-hundred pounds.

Northward two miles was Dogtown, beautifully situated in the midst of a rich agricultural section. The descendants of the first settlers retain the characteristics of their German ancestors. Frugal, honest and industrious, they live comfortably in their narrow sphere and save their gains. The Delo farm, another mile north, was for a time the limit of developments. True to his instincts as a discoverer of new territory, Marcus Hulings went six miles north-east of St. Petersburg, leased B. Delo's farm and drilled a forty-barrel well in the spring of 1872. Enormous quantities of gas were found in the second sand. The oil was piped to Oil City. A half-mile east, on the Hummell farm, Salem township, Lee & Plumer struck a hundred-barreler in July of 1872. The Hummell farm had been occupied for sixty years by a venerable Teuton, whose rustic son of fifty-five summers described himself as "the pishness man ov the firm." The new well, twelve-hundred feet deep, had twenty-eight feet of nice sand and considerable gas. Its success bore fruit speedily in the shape of a "town" dubbed Pickwick by Plumer, who belonged to the redoubtable Pickwick Club at St. Petersburg. A quarter-mile ahead, on a three-cornered plot, Triangle City bloomed. The first building was a hotel and the second a hardware store, owned by Lavens & Evans. Charles Lavens operated largely in the Clarion region and in the northern field, lived at Franklin several years and removed to Bradford. He is president of the Bradford Commercial Bank and a tip-top fellow at all times and under all circumstances. Evans may claim recognition as the author, in the muddled days of shut-downs and suspensions in 1872, of the world-famed platform of the Grass-Flats producers : "Resolved that we don't care a damn !" The three tailors of Tooley street, who issued a manifesto as "We, the people of England," were outclassed by Evans and his friends. News of their action was flashed to every "council" and "union" in the oil-country, with more stimulating effect than a whole broadside of formal declarations. Triangle, Pickwick and Paris City have passed to the realm of forgetfulness.

Major Henry Wetter, the embodiment of honor and energy, was the largest operator in the district until swamped by the low price of oil. Death overtook him while struggling against heavy odds to recuperate his health and fortune. How sad it is that the flower must die before the fruit can bloom !

Marcus Hulings, a leader in the world of petroleum, was born near Philipsburg, Clarion county, and began his career as a producer in 1860. For some years he had been a contractor and builder and he turned his practical knowledge of mechanics to good account. His earliest oil-venture was a well on the Allegheny River above Oil City, for which he refused sixty-thousand dollars. To be nearer the producing-fields, he removed to Emlenton and resided there a number of years. The Hulings family had been identified with Venango county from the first settlement, one of them establishing a ferry at Franklin a century ago. Prior to that date the family owned and lived on what is now Duncan's Island, at the junction of the Susquehanna and Juniata Rivers, fifteen miles north-west of Harrisburg. Marcus was a pathfinder in Forest county and opened the Clarion region. He leased Clark & Babcock's six-thousand acres in McKean county and drilled hundreds of paying wells. Deciding to locate at Oil City, he built an elegant home on the South Side and bought a delightful place in Crawford county for a summer residence. His liberality, enterprise and energy seemed inexhaustible. He donated a magnificent hall to Allegheny College, Meadville, aided churches and schools, relieved the poor and was active in political affairs. Besides his vast oil-interests he had mines in Arizona and California, mills on the Pacific coast and huge lumbertracts in West Virginia. Self-poised and self-reliant, daring yet prudent, brave and trustworthy, he was one of the grandest representatives of the petroleumindustry. Neither puffed up by prosperity nor unduly cast down by adversity, he met obstacles resolutely and accepted results manfully. My last talk with him was at Pittsburg, where he told of his endeavor to organize a company to

15

develop silver-claims in Mexico. He had grown older and weaker, but the earnestness of youth was still his possession. His eyes sparkled and his face lightened as he shook my hand at parting and said : " You will hear from me soon. If this company can be organized I would not exchange my Mexican properties for the wealth of the Astors !"

He died in a few weeks, his dream unfulfilled. Losses in the west had reduced his fortune without impairing his splendid courage, hope and patience. He united the endurance of a soldier with the skill of a commander. Marcus Hulings deserved to enjoy a winter of old age as green as spring, as full of blossoms as summer, as generous as autumn. His son, Hon. Willis J. Hulings,

served in the Legislature three terms. He introduced the bills prohibiting railroad-discriminations and was a strong debater on the floor. Senator Quay favored him for State Treasurer and attempted to stampede the convention which nominated William Livsey. This was the beginning of the differences between Quay and the combine which culminated in the rout of the latter and the triumph of the Beaver statesman in 1895-6. Mr. Hulings lives at Oil City, has a beautiful home and is colonel of the Sixteenth Regiment of the National Guards. He practiced law in 1877-81, then devoted his attention to oil-operations, to mining and lumbering, in which he is at present actively engaged.

John Lee drilled his first well on the Hoover farm, near Franklin, in 1860, and he is operating to-day in Clinton and Rockland townships. He has had his share of storm and sunshine, from dusters at Nickelville to a slice of the Big Injun at Bullion, in the shifting panorama of oil-developments for thirty-six years, but his fortitude and manliness never flinched. He is no sour dyspeptic, whose conduct depends upon what he eats for breakfast and who cannot believe the world is O. K. if he drills a dry-hole occasionally.

Frederick C. Plumer and John Lee, partners in the Clarion and Butler fields, were successful operators. Their wells on the Hummell farm netted handsome returns. By a piece of clever strategy they secured the Diviner tract, drilled a well that extended the territory two miles south of Millerstown and sold out for ninety-thousand dollars. Plumer quit with a competence, purchased his former hardware-store at Newcastle, took a flyer in the Bullion district and died at Franklin, his birthplace and boyhood home, in 1879. "Fred" was a thorough man of affairs, prompt, courteous, affable and popular. His long sickness was borne cheerfully and he faced the end—he died at thirty-one—without repining. His wife and daughter have joined him in the land of deathless reunions.

> " Over the river !
> Sailing on waters where lotuses smile.
> Passing by many a tropical isle,
> Sighting savannas there mile upon mile,
>   Over the river !
> Music forever and beauty for aye,
> Sunlight unending—the sunlight and day.
> Never a farewell to weep on the way,
>   Over the river!"

East, north and west the area of prolific territory widened. Wells on the Young farm started a jaunty development at Jefferson Furnace. Once the scene of activity in iron-manufacture, the old furnace had been neglected for three decades. Oil awakened the spot from its Rip-Van-Winkle slumber. A narrow-gauge railroad crossed Beaver Creek on a dizzy trestle, which afforded an enticing view of derricks, streams, hills, dales, cleared farms and wooded slopes. The wells have pumped out, the railroad has been switched off and the stout furnace stands again in its solitary dignity. James M. Guffey, J. T. Jones, Wesley Chambers and other live operators kept branching out until Beaver City, Mongtown, Mertina, Edenburg, Knox, Elk City, Fern City and Jerusalem, with Cogley as a supplement, were the centers of a production that aggregated ten-thousand barrels a day. The St. Lawrence well, on the Bowers farm, a mile north of Edenburg, was finished in June of 1872 and directed attention to Elk township. For two years it pumped sixty-

BEAVER CREEK AT JEFFERSON FURNACE.

nine barrels a day, six days each week, the owners shutting it down on Sunday. Previously Captain Hasson, of Oil City, and R. Richardson, then of Tarr Farm and now of Franklin, had drilled in the vicinity. Ten dusters north of the Bowers farm augured poorly for the St. Lawrence. It disappointed the prophets of evil by striking a capital sand and producing with a regularity surpassed only by one well on Cherry Run. It was not "a lovely toy, most fiercely sought, that lost its charm by being caught."

The St. Lawrence jumped the northern end of the Clarion district to the front. Hundreds of wells ushered in new towns. Knox, on the Bowers farm, attained a post-office, a hardware store and a dozen dwellings, its proximity to Edenburg preventing larger growth. The cross-roads collection of five-houses and a store known as Edenburg progressed immensely. John Mendenhall and J. I. Best's farm-houses, 'Squire Kribbs's country-store and justice-mill, a blacksmith-shop and three dwellings constituted the place at the date of the St. Lawrence advent. The nearest hotel—the Berlin House—was three miles northward. In six months the quiet village became a busy, hustling, prosperous town of twenty-five hundred population. It had fine hotels, fine stores, banks and people whom a destructive fire—it eliminated two-thirds of the buildings in one night—could not "send to the bench." When the flames had been subdued, a crowd of sufferers gathered at two o'clock in the morning, sang "Home, Sweet Home," and at seven were clearing away the embers to rebuild. Narrow-gauge railroads were built and the folks didn't scare at the cars. Elk City flung its antlers to the breeze two miles east. Isaac N. Patterson—he is president of the Franklin Savings Bank and a big operator in Indiana—had a creamy patch on the Kaiser farm. Jerusalem's first arrival—Guffey's wells created it—was a Clarion delegate with a tent and a cargo of liquids. He dealt the drink over a rough board, improvised as a counter, so briskly that his receipts in two days footed up seven-hundred dollars. He had no license, an officer got on the trail and the vendor decamped. He is now advance-agent of a popular show, wears diamonds the size of walnuts and tells hosts of oil-region stories. The Clarion field was not inflamed by enormous gushers, but the wells averaged nicely and possessed the cardinal virtue of enduring year after year. It is Old Sol, steady and persevering, and not the flashing meteor, "a moment here, then gone forever," that lights and heats the earth and is the fellow to bank upon.

An Edenburg mother fed her year-old baby on sliced cucumbers and milk, and then desired the prayers of the church "because the Lord took away her darling." "How is the baby?" anxiously inquired one lady of another at Beaver City. "Oh, baby died last week, I thank you," was the equivocal reply.

Some of the oilmen were liberally endowed with the devotional sentiment. When the news of a blazing tank of oil at Mertina reached Edenburg, a jolly operator telegraphed the fact to Oil City, with the addendum: "Everything has gone hellward." A half-hour later came his second dispatch: "The oil is blazing, with big flames going heavenward." Such a happy blending of the infernal with the celestial is seldom witnessed in ordinary business.

The behavior of some people in a crisis is a wonderful puzzle, sometimes funnier than a pig-circus. At the St. Petersburg fire, which sent half the town up in smoke, an old woman rescued from the Adams House, with a bag of money containing four-hundred dollars, was indignant that her fifty-cent spectacles had been left to burn. A male guest stormed over the loss of his satchel, which a servant had carried into the street, and threatened a suit for damages. The satchel was found and opened. It had a pair of dirty socks, two dirty collars, a comb and a toothbrush! The man with presence of mind to throw his mother-in-law from the fourth-story window and carry a feather-pillow down stairs was not on hand. St. Petersburg had no four-story buildings.

John Kiley and "Ed." Callaghan headed a circle of jolly jokers at Triangle

City and Edenburg. Hatching practical sells was their meat and drink. One evening they employed a stranger to personate a constable from Clarion and arrest a pipe-line clerk for the paternity of a bogus offspring. In vain the astonished victim protested his innocence, although he acknowledged knowing the alleged mother of the alleged kid. The minion of the law turned a deaf ear to his prayers for release, but consented to let him go until morning upon paying a five-dollar note. The poor fellow thought of an everlasting flight from Oildom and was leaving the room to pack up his satchel when the "constable" appeared with a supply of fluids. The joke was explained and the crowd liquidated at the expense of the subject of their pleasantry. Kiley was an oil-man and operated in the northern fields. Callaghan slung lightning in the telegraph-office. He married at Edenburg and went to Chicago. His wife procured a divorce and married a well-known Harrisburger.

A letter from his feminine sweetness, advising him to hurry up if he wished her not to marry his rival, so flustrated an Edenburg druggist that he imbibed a full tumbler of Jersey lightning. An irresistible longing to lie down seized him and he stretched himself for a nap on a lounge in a room back of the store. John Kiley discovered the sleeping beauty, spread a sheet over him and prepared for a little sport. He let down the blinds, hung a piece of crape on the door and rushed out to announce that "Jim" was dead. People flocked to learn the particulars. Entering the drug-store a placard met their gaze: "Walk lightly, not to disturb the corpse!" They were next taken to the door of the rear apartment, to see a pair of boots protruding from beneath a sheet. Nobody was permitted to touch the body, on a plea that it must await the coroner, but the friends were invited to drink to the memory of the deceased pill-dispenser and suggest the best time for his funeral. Thus matters continued two hours, when the "corpse" wakened up, kicked off the sheet and walked out! His friends at first refused to recognize him, declaring the apparition was a ghost, but finally consented to renew the acquaintance upon condition that he "set 'em up" for the thirsty multitude.

A Clarion operator, having to spend Sunday in New York, strayed into a fashionable church and was shown to a swell seat. Shortly after a gentleman walked down the aisle, glared at the stranger, drew a pencil from his pocket, wrote a moment and handed him a slip of paper inscribed, "This is my pew." The unabashed Clarionite didn't bluff a little bit. He wrote and handed back the paper: "It's a darned nice pew. How much rent do you ante up for it?" The New-Yorker saw the joke, sat down quietly and when the service closed shook hands with the intruder and asked him to dinner. The acquaintance begun so oddly ripened into a poker-game next evening, at which the oilman won enough from the city clubman to pay ten years' pew-rent. At parting he remarked: "Who's in the wrong pew now?" Then he whistled softly: "Let me off at Buffalo!"

Clarion's products were not confined to prize pumpkins, mammoth corn and oil-wells. The staunch county supplied the tallest member of the National Guard, in the person of Thomas Near, twenty-one years old, six feet eleven in altitudinous measurement and about twice the thickness of a fence-rail. The Clarion company was mustered in at Meadville. General Latta's look of astonishment as he suryeyed the latitude and longitude of the new recruit was exceedingly comical. He rushed to Governor Hartranft and whispered, "Where in the name of Goliath did you pick up that young Anak?" At the next annual review Near stood at the end of the Clarion column. A staff-officer, noticing a man towering a foot above his comrades, spurred his horse across the field and yelled : "Get down off that stump you blankety-blank son of a gun !" The tall boy did not "get down" and the enraged officer did not discover how it was until within a rod of the line. His chagrin rivaled that of Moses Primrose with the shagreen spectacles. Poor Near, long in inches and short in years, was not long for this world and died in youthful manhood.

Counselled by "spirits," Abram James selected a block of land on Blyson Run, twenty miles up the Clarion River, as the location of a rich petroleum-field. His luck at Pleasantville induced numbers to believe him an infallible oil-smeller. The test-well that was to deluge Blyson with crude was bored eighteen-hundred feet. It had no sand or oil and the tools were stuck in the hole ! The "spirits" couldn't have missed the mark more widely if they had directed James to mine for gold in a snow-bank.

Hon. James M. Guffey, one of Pennsylvania's most popular and successful citizens, began his career as a producer in the Clarion district. Born and reared on a Westmoreland farm, his business aptitude early manifested itself. In youth he went south to fill a position under the superintendent of the Louis-

JAMES M. GUFFEY.

ville & Nashville Railroad. The practical training was put to good use by the earnest young Pennsylvanian. Its opportunities for dash and energy to gain rich rewards attracted him to the oil - region. Profiting by what he learned from the experiences of others, in Venango county—a careful observer, he did not have to scorch himself to find out that fire is hot—he located at St. Petersburg in 1872. Clarion was budding into prominence as a prospective oil-field. Handling well-machinery as agent of the Gibbs & Sterrett Manufacturing Company brought him into close relations with operators and operations in the new territory. He improved his advantages, leased lands, secured interests in promising farms, drilled wells and soon stepped to the front as a first-class producer. Fortune smiled upon the plucky Westmorelander, whose tireless push and fearless courage cool judgment and sound discretion tempered admirably While always ready to accept the risks incident to producing oil and developing untried sections, he was not a reckless plunger, going ahead blindly and not counting the cost. He decided promptly, moved forward resolutely and took nobody's dust. Those who endeavored to keep up with him had to "ride the horse of Pacolet" and travel fast. He invested in pipe-lines and local enterprises, helped every deserving cause, stood by his friends and his convictions, believed in progress and acted strictly on the square. Not one dollar of his splendid winnings came to him in a manner for which he needs blush, or apologize or be ashamed to look any man on earth straight in the face. He did not get his money at the expense of his conscience, of his self-respect, of his generous instincts or of his fellow-men. Of how many millionaires, in this age of shoddy and chicanery, of jobbery and corruption, of low trickery and inordinate desire for wealth, can this be said?

Mr. Guffey is an ardent Democrat, but sensible voters of all classes wished him to represent them in Congress and gave him a superb send off in the oil-portion of the Clarion district. Unfortunately the fossils in the back-townships prevented his nomination. The uncompromising foe of ring rule, boss-domination and machine-crookedness, he is a leader of the best elements of his party and not a noisy ward-politician. His voice is potent in Democratic councils and his name is familiar in every corner of the producing-regions. His oil-operations have reached to Butler, Forest, Warren, McKean and Allegheny counties. He furnished the cash that unlocked the Kinzua pool and extended the Bradford field miles up Foster Brook. In company with John Galey, Michael Murphy and Edward Jennings, he drilled the renowned Matthews well and owned the juiciest slice of the phenomenal McDonald field. He started developments in Kansas, putting down scores of wells, erecting a refinery and giving the state of Mary Ellen Lease a product drouths cannot blight nor grasshoppers devour. He was largely instrumental in developing the natural-gas fields of Western Pennsylvania, Ohio and Indiana, heading the companies that piped it into Pittsburg, Johnstown, Wheeling, Indianapolis and hundreds of small towns. He owns thousands of acres of the famous gas-coal lands of his native county, vast coal-tracts in West Virginia and valuable reality in Pittsburg. He lives in a handsome house at East Liberty, brightened by a devoted wife and four children, and dispenses a bountiful hospitality. Quick to mature and execute his plans, he dispatches business with great celerity, keeping in touch constantly with the details of his manifold enterprises. He is the soul of honor in his dealings, liberal in his benefactions and always approachable. His charm of manner, kindness of heart, keen intuition and rare geniality draw men to him and inspire their confidence and regard. He is a striking personality, his lithe frame, alert movements, flowing hair, luxuriant mustache, rolling collar, streaming tie, frock-coat and broad-brimmed hat suggesting General Custer. When at last the vital fires burn low, when his brave heart beats weak and slow, when the evening shadows lengthen and he enters the deepening dusk at the ending of many happy years, James M. Guffey will have lived a life worth living for its worth to himself, to his family, to the community and to the race.

> " The grass is softer to his tread
>   For rest it yields unnumber'd feet ;
> Sweeter to him the wild rose red
>   Because it makes the whole world sweet."

Thomas McConnell, Smith K. Campbell, W. D. Robinson and Col. J. B. Finlay, of Kittanning, in 1860 purchased two acres of land on the west bank of the Allegheny, ninety rods above Tom's Run, from Elisha Robinson. Organizing the Foxburg Oil-Company of sixteen shares, they drilled a well four-hundred-and-sixty feet. An obstruction delayed work a few days, the war broke out and the well was abandoned. The same parties paid Robinson five-thousand dollars in 1865 for one-hundred acres and sold thirty to Philadelphia capitalists. The latter formed the Clarion and Allegheny-River Oil-Company and sunk a well which struck oil on October tenth, the first produced in the upper end of Armstrong county and the beginning of the Parker development.

Venango was drooping and operators sought the southern trail. The Robin-son farm was not perforated as quickly as "you could say Jack Robinson," the owners choosing not to cut it into small leases, but other tracts were seized eagerly. Drilled deeper, the original Robinson well was utterly dry! Had it been finished in 1860-1 the territory might have been condemned and the Parker field never heard of!

John Galey's hundred-barrel well, drilled in 1869 on the island above Parker, relieved the monotony of commonplace strikes—twenty to fifty barrels —on the Robinson and adjacent farms and elevated the district to the top rung of the ladder. Parker's Landing—a ferry and a dozen houses—named from a pioneer settler, ambled merrily to the head of the procession. The center of operations that stretched into Butler county and demonstrated the existence of three greasy streaks, Parker speedily became a red-hot town of three-thousand inhabitants. Hotels, stores, offices, banks and houses crowded the strip of land at the base of the steep cliff, surged over the hill, absorbed the suburbs of Lawrenceburg and Farrentown and proudly wore the title of "Parker City." Hosts of capital fellows made life a perpetual whirl of business and jollity. Operators of every class and condition, men of eminent ability, indomitable hustlers, speculators, gamblers and adventurers thronged the streets. It was the vim and spice and vigor of Oil City, Rouseville, Petroleum Centre and Pit-hole done up in a single package. A hundred of the liveliest laddies that ever capered about a "bull-ring" traded jokes and stories and oil-certificates at the Oil-Exchange. Two fires obliterated nine-tenths of the town, which was never wholly rebuilt. Developments tended southward for years and the sun of Parker set finally when Bradford's rose in the northern sky. The bridge and a few buildings have held on, but the banks have wound up their accounts, the multitudes have dispersed, the residence-section of the cliff is a waste and the glory of Parker a tradition. As the ghost of Hamlet's father observed con-cerning the bicycle academy, where beginners on wheels were plentiful: "What a falling off was there!"

Galey leased lands, sunk wells and sold to Phillips Brothers for a million dollars. He played a strong hand in Butler and Allegheny and removed to Pittsburg, his present headquarters. He possessed nerve, energy and endur-ance and, like the country-boy applying for a job, "wuz jam'd full ov day's work." He would lend a hand to tube his wells, lay pipes, move a boiler or twist the tools. There wasn't a lazy bone in his anatomy. Rain, mud, storm or darkness had no terrors for the bold rider, who bestrode a raw-boned horse and "took Time by the forelock." A young lady from New York, whose father was interested with Galey in a tract of oil-land, accompanied him on one of his visits to Millerstown. She had heard a great deal about her father's partner and the producers, whom she imagined to be clothed in broadcloth and diamonds. When the stage from Brady drew up at the Central Hotel a gor-geous chap was standing on the platform. He sported a stunning suit, a huge gold-chain, a diamond-pin and polished boots, the whole outfit got up regard-less of expense. "Oh, papa, I see a producer! That must be Mr. Galey," exclaimed the girl as this prototype of the dude met her gaze. The father glanced at the object, recognized him as a neighboring bar-tender and spoiled his daughter's fanciful notion by the curt rejoinder: "That blamed fool is a gin-slinger!" Butler had long been a sort of by-word for poverty and mean-ness, the settlers going by the nickname of "Buckwheats." This was an unjust imputation, as the simple people were kind, honest and industrious, in these respects presenting a decided contrast to some of the new elements in the wake of the petroleum-development. The New-York visitor drove out in the afternoon to meet his business-associate. A mile below the Diviner farm a man on horseback was seen approaching. Mud covered the panting steed and his rider. The young lady, anxious to show how much she knew about the country, hazarded another guess. "Oh! papa," she said earnestly, "I'm sure that's a Buckwheat!" The father chuckled, next moment greeted the rider warmly and introduced him to his astonished daughter as "My partner, Mr. Galey!" A hearty laugh followed the father's version of the day's incidents.

John McKeown drilled on the Farren hill and the slopes bordering the north bank of Bear Creek. Glory Hole popped up on B. B. Campbell's Bear-Creek farm. Campbell—bluff, whole-souled "Ben"—is a Pittsburg capitalist, big in body and mind, outspoken and independent. "The Campbells are coming" could not have found a better herald. He produced largely, bought stacks of farms, refined and piped oil and was an important factor in the Armstrong–Butler development. At the Ursa Major well, the first on the farm, large casing and heavy tools were first used, with gratifying results. "Charley" Cramer juggled the temper-screw and laughed at the chaps who solemnly predicted the joints would not stand the strain and the engine would not jerk the tools out of the hole. The tool-dresser on Cramer's "tower"— drilling went on night and day, each "tower" lasting twelve hours and the men changing at noon and midnight—was A. M. Lambing, now the learned and zealous parish-priest at Brad-dock. The well, completed in

JAMES M. LAMBING.

June of 1871 and good for a hundred barrels, was owned by James M. Lambing, to whom more than any other man the world is indebted for the extension of the Butler field.

Born in Armstrong county, in 1861 young Lambing concluded to invest some time and labor—his sole capital—in a well at the mouth of Tubb's Run, two miles above Tionesta. A dry-hole was the poor reward of his efforts. Enlisting in the Eighty-third Regiment, he received disabling injuries, was discharged honorably, returned to Forest county in 1863, superintended the Denver Petroleum-Company, dealt in real estate and in 1866 commenced operating at Tidioute. A vein of bad luck in 1867 exhausting his last dollar, he sold his gold-watch and chain to pay the wages of his drillers. Facing the future bravely, he worked by the day, contracted to bore wells at Pleasantville, Church Run, Shamburg and Red Hot and bore up cheerfully during three years of adversity. In the winter of 1869 he traded an engine for an interest in a well at Parker that smelled of oil. For another interest he drilled the Wilt & Crawford well and secured leases on Tom's Run. His Pharos, Gipsy Queen and Lady Mary wells enabled him to strike out boldly. In company with his brother — John A. Lambing — C. D. Angell and B. B. Campbell, he ventured beyond the prescribed limits to the Campbell, Morrison and Gibson farms. He "wildcatted" farther south, at times with varying success, pointing the way to Modoc and Millerstown. Reverses beset him temporarily, but hope and courage and integrity remained and he recovered the lost ground. Charitable, enterprising and sincere, no truer, squarer, manlier man than James M. Lambing ever marched in the grand cavalcade of Pennsylvania oil-producers. He and John A. retired from the business years ago to engage in other pursuits. James M. settled at Corry and served so capably as mayor that the citizens wanted to elect him for life. His noble, womanly wife, a real helpmeet always, makes his hospitable home an earthly paradise. He has an office in Pittsburg and customers for his Ajax machinery wherever oil is produced. "Who can blot his name with any just reproach?"

Well-known operators figured in the vicinity of Bear Creek. Joseph Overy drilled rows of good wells, pushed south and founded the town embalmed as St. Joe in compliment to its progenitor. Marcus Brownson— he was active in Venango and McKean and died at Titusville—had a walkover on the Walker

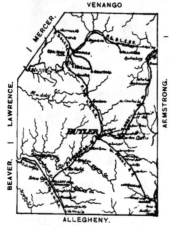

farm, a mile in advance. On Donnelly's eleven-hundred acres, offered in 1868 for six-thousand-dollars, scores of medium wells yielded from 1871 to 1878. S. D. Karns drained the Morrison farm and John McKeown hit the "sucker-rod belt"—so called from its extreme narrowness—near Martinsburg. Ralph Brothers tickled the sand on the Sheakley farm. Up the stream operations jogged and Argyle City sprouted on the hillside. Two miles ahead, upon the line dividing the Jameson and Blaney farms, Dimick, Nesbit & Co. finished a wildcat well on April seventeenth, 1892. This was the noted Fanny Jane— gallantly named in honor of a pretty girl—which pumped one-hundred barrels and gave birth to Petrolia, seven miles south by west of Parker. George H. Dimick, examining lands in Fairview township, Butler county, decided that a natural basin at the junction of South Bear Creek and Dougherty Run was oil-territory. Fifty men were raising a barn on the Campbell farm, overlooking this basin. Proceeding to the spot, he proposed to drill a test well if the owners of the soil would lease enough land to warrant the undertaking. Terms were agreed upon which secured twenty acres of the Blaney farm, sixteen of the Jameson, ten of the W. A. Wilson, ten of the James Wilson and ten of the Graham, at one-eighth royalty. The nearest producing wells at that date were three miles north. The Fanny Jane stirred the blood of the oil-clans. The moving mass began to arrive in May and by July two-thousand people had their home at Petrolia.

A charter was obtained and Mr. Dimick was chosen burgess at the first borough-election, in February of 1873. The town expanded like the turnip Longfellow said "grew and it grew and it grew all it was able." Hotels, stores, shops and offices lined the valley and dwellings crowned the hills. A narrowgauge railroad from Parker was built in 1874, extended to Karns City and Millerstown and ultimately to Butler. Fisher Brothers paid sixty-thousand

dollars for the Blaney farm and wells multiplied in all directions. A dog-fight or a street-scrap would gather hundreds of spectators. The Argyle Savings Bank handled hundreds-of-thousands of dollars daily. Ben Hogan erected a big opera-house and May Marshall was the Cora Pearl of the frail sisterhood. R. W. Cram ran the post-office and news-room. "Steve" Harley wafted newsy items to the newspapers. Dr. Frank H. Johnston, now of Franklin, was the first physician. Kindred spirits met at "Sam" McBride's drug-store and

Peter Christie's Central Hotel. Poor "Sam," "Dave" Mosier, H. L. Mc-Cance and S. S. Avery are in their graves and others have wandered nobody knows whither. Petrolia continued the metropolis four years and then dropped out of the game. Some straggling houses and left-over derricks alone remain of the gayest, sprightliest, hottest, busiest town that bloomed and withered in old Butler.

George H. Dimick, the son of a Wisconsin farmer and sire of Petrolia, is liberally stocked with the never-say-die qualities of the breezy Westerner. At nineteen he taught a Milwaukee school, landed on Oil Creek in 1860 and was appointed superintendent of the two Buchanan farms by Rouse & Mitchell. He drilled on his own account in the spring of 1861, aided in settling the Rouse estate, enrolled as a private in "Scott's Nine-Hundred" and came out a captain at the close of the war. In May of 1865 he bent his footsteps towards Pithole, sold lands for the United States Petroleum-Company and drilled eleven dry-holes on the McKinney farm! Interests in the Poole, Grant, Eureka and Burchill spouters offset these losses and added thousands of dollars a week to his wealth. Staying at Pithole too long, values had shrunk to such a degree that he was virtually penniless at his departure from the "Magic City" in 1867. A whaling voyage of fifteen months in the Arctic seas and a sojourn at his boyhood home improved his health and he returned in time to share in the Pleasantville excitement. He located at Parker's Landing in 1871 as partner of McKinney & Nesbit in the sale of oil-well supplies. He operated in the Parker field, at St. Petersburg, Petrolia, Greece City and Slippery Rock. Disposing of his properties in these localities, he and Captain Peter Grace drilled the wildcat well that opened Cherry Grove and paralyzed the market in 1882. He had been active at Bradford and the middle field felt the influence of his shrewd movements. He has kept abreast of developments in the southern districts, sometimes getting several lengths ahead. He is now interested in West Virginia and Kentucky. Those who know his quick perception, his executive ability and his intense love for opening new fields would not wonder to hear of his striking a gusher at Oshkosh or Kamtschatka. Mr. Dimick is a man of active temperament, high character and sturdy industry, a genuine pathfinder and tireless explorer.

An Erie boy of fifteen when he left his father's house for the oil-region in 1862, George H. Nesbit first fired a still in a Titusville refinery and in 1863 engaged with Dinsmore Brothers at Tarr Farm. He built a small refinery at Shaffer, sold it in 1864 and in the spring of 1865 drilled wells for himself on Benninghoff and Cherry-Tree Runs. He spent two years at Pithole, gaining a fortune and remaining until the collapse swallowed the bulk of his profits. He operated at Pioneer in 1867 and a year later at Pleasantville. He and George H. Dimick prospected in 1869 for oil-belts and fresh territory, located rich leases on Hickory Creek and established the line of the Venture well at Fagundas. In 1870 Nesbit moved to Parker and, in company with John L. McKinney, sold oil-well machinery and oil-lands. McKinney & Nesbit drilled along Bear Creek, especially on the Black and Dutchess farms, prospering greatly. The firm ranked with the most enterprising and realized large returns from wells at St. Petersburg and Parker. Dimick & Nesbit, with Mr. McKinney as their associate, opened the Petrolia field in 1872. William Lardin, the contractor of the Fanny Jane, bought McKinney's interest in the well and leases. The three partners were right in the swim, their first six wells at Petrolia yielding them a thousand barrels a day. Nesbit bought the Patton farm, below town, in 1872 for twenty-thousand dollars, selling five-eighths. Five third-sand wells ranged from thirty to one-hundred barrels and oil ruled at three to five dollars. The fourth-sand was found in 1873, and in January of 1874 Nesbit & Lardin struck a thousand-barrel gusher on the Patton. The farm paid enormously and Nesbit became an "oil-prince." He developed hundreds of acres and displayed masterly tact. His check was good for a half-million any day and his luck was so remarkable that, had he fallen into the river, probably he would not have been wet. He paid the highest wages and met his bills at sight. He entered the oil-exchange at Parker, for a time was a high-roller and ended a bankrupt! The desk on which he wrote his bold, round signature on checks aggregating many hundred-thousand dollars was stored away among shocks of corn and sheaves of oats in the weather-stained barn on the Patton farm. J. N. Ireland bought the tract for seven-thousand dollars. Nesbit drifted about aimlessly, heard from occasionally at Macksburg and fetching up at last in Cincinnati. His prestige was gone, his star had waned and he never "caught on" again. He was no sluggard in business, no dullard in society, no niggard with money, no laggard in the petroleum-column. Surely the oil-region has furnished its full allotment of sad romances from real life.

> " Time, with a face like a mystery,
> And hands as busy as hands can be,
> Sits at the loom with its warp outspread,
> To catch in its meshes each glancing thread.
> Click, click! there's a thread of love wove in!
> Click, click! and another of wrong and sin !
> What a checkered thing this web will be
> When we see it unrolled in eternity!"

James E. Brown, to whom Nesbit sold one-quarter of the Patton farm, made his mark upon the industries of the state. A carpenter's son, he started a store on the site of Kittanning, saved money, purchased lands and at his death in 1880 left his family four-millions. He manufactured iron at various furnaces and owned a big block of stock in the rolling-mills at East Brady. Samuel J. Tilden was a stockholder in the works, which employed sixteen hundred men, turned out the first T-rails west of the Alleghenies and tottered to their fall in 1874. Mr. Brown cleared eight-hundred-thousand dollars in 1872 by the advance in iron. He owned oil-farms in Butler county, took stock in the Parker Bridge, the Parker & Karns City Railroad and the Karns Pipe-Line Company and conducted a bank at Kittanning. His granddaughter, Miss Findley, who inherited half his wealth, married Lord Linton, a British baronet. The aged banker—he stuck it out to eighty-two—knew how to pile up money.

Stephen Duncan Karns, who had a railroad and a town named in his honor, was a picturesque figure in the Armstrong–Butler district. With his two uncles he operated the first West-Virginia well, at the mouth of Burning-Spring Run, in 1860. His experience at his father's Tarentum salt-wells enabled him to run engine, to sharpen tools and clean out an old salt-well to be tested for oil. The well pumped forty barrels a day during the winter of 1860-1. Fort Sumter was bombarded, several Kanawha operators were killed and young Karns escaped by night in a canoe. He enlisted, served three years, led his company at Antietam and Chancellorsville and in 1866 leased one acre at Parker's Landing from Fullerton Parker. His first well, starting at one barrel a day, by months of pumping was increased to twelve barrels and earned him twenty-thousand dollars. From the Miles Oil-Company of New York he leased a farm and an abandoned well a mile below Parker. He drilled the well through the sand and it produced twenty-five barrels a day. This settled the question of oil south of Parker. "Dunc," as he was usually called by his friends, leased the Farren farm, drilled on Bear Creek, secured the famous Stonehouse farm of three-hundred acres and in 1872 enjoyed an income of five-thousand dollars a day! A mile south of Petrolia, on the McClymonds farm, Cooper Brothers were about to give up their first well as a hopeless duster. Karns thought the hole not deep enough, bought the property, resumed drilling and in two days the well was flowing one-hundred barrels! The town of Karns City blossomed into a community of twenty-five-hundred people, with three big hotels, stores, offices and dwellings galore. It fell a prey to the flames eventually. The McClymonds, Riddle and J. B. Campbell farms doubled "Dunc's" big income for many moons. He had the second well at Greece City and for a year or more was the largest producer in the oil-region. He built a pipe-line from Karns City to Harrisburg to fight the United Lines, held fifty-five-thousand dollars' stock in the Parker Bridge and controlled the Parker & Karns-City Railroad and the Exchange Bank.

Near Freeport, on the Allegheny River, thirty miles above Pittsburg, he lassoed a great farm and erected a fifty-thousand-dollar mansion. Fourteen race-horses fed in his palatial stables. Guests might bathe in champagne and the generous host spent money royally. A good strike or a point gained meant a general jollification. He played billiards skillfully, handled cards expertly and wagered heavily on anything that hit his fancy. He and his wife were in Paris during the siege. Upon his return from Europe he built the Fredericksburg & Orange Railroad, in Virginia. The glut of crude from Butler wells dropped the price in 1874 to forty cents. Losses of different kinds cramped Karns and the man worth three-millions in 1872–3 was obliged to surrender his stocks and lands and wells and begin anew! James E. Brown secured Glen-Karns, the beautiful home below Freeport. In 1880 Karns induced E. O. Emerson, the wealthy Titusville producer, to start a cattle-ranch in Western Colorado. For six years he superintended the herds on the immense plains, joining the round-ups, sleeping on the ground with the boys, roping and branding cattle and accumulating a stock of health and muscle which he thinks will carry him to the hundred-year mark. Emerson had bought from Karns the Riddle farm for eleven-thousand dollars. He deepened one well—supposed dry—to the fourth sand. It flowed six-hundred barrels and Emerson sold the tract in sixty days for ninety-thousand dollars. Karns returned from the west, practiced law a short while in Philadelphia and for some years has managed a Populist paper at Pittsburg. He ran against John Dalzell for Congress and walked at the head of the parade when General Coxey's "Army of the Commonweal" marched through the Smoky City. He enjoyed making money more than handling it, was honorable in his dealings, intensely active, comprehensive in his views and positive in his opinions. His "yes" or "no" was given promptly. "Dunc" is of slender build and nervous temperamont, easy in his manners, frank in his utterances and not scared by spooks in politics or trade. He had his share of light and shade, struggle and triumph, defeat and victory, incident and adventure in his pilgrimage.

> " How chances mock,
> . And changes fill the cup of alteration
> With divers liquors!"

Richard Jennings, over whose head the grass and flowers are growing, and his brother-in-law, the late Jacob L. Meldren, did much to develop the territory east of Petrolia. Coming from England to Armstrong county a half-century ago, they located at what is now Queenstown. Meldren bought the farm at the head of Armstrong Run on which the noted Armstrong well was struck in 1870. It opened "the Cross-Belt," an abnormal strip running nearly at right angles to the main lines and remarkable for mammoth gushers. This unprecedented "belt" upset the theories of geologists and operators. The first and only one of its kind, it resembled the mule that "had no pride of ancestry and no hope of posterity." Mr. Jennings drilled on many farms and gathered a large fortune. He was a man of character and ability, with a priceless reputation for integrity and truthfulness. Once he sent his foreman, Daniel Evans, to secure the Dougherty farm, on the southern edge of Petrolia, owned by two maiden sisters. The foreman knocked at the door, engaged board for a week, was engaged to the elder sister before the week expired and had the pleasure of reaping a harvest of greenbacks from the property in due course. It is satisfactory to find such enterprise abundantly recompensed. Not so lucky was a gay and festive operator with an ancient maiden who owned a tempting patch of land near Millerstown. He exhausted every art to get a lease, in desperation finally hinting at matrimony. The indignant lady exploded like a ton of dynamite, seizing a broom and compelling the bold visitor to beat an ungraceful retreat through the window, minus his hat and gloves! Evans leased part of the farm to his former employer, who finished the Dougherty spouter on November twenty-second, 1873. It flowed twenty-seven-hundred barrels a day from the fourth sand, loading Jennings with greenbacks and sending the speculative trade into convulsions. A patriotic citizen, devoted parent and genuine philanthropist, Richard Jennings was sincerely respected and his death was deeply mourned. His sons inherited their father's sagacity and manly principle. They have operated in the McDonald field and are prominent in banking and business at Pittsburg.

The "Cross-Belt" crossed the petroleum-horizon in dead earnest in March of 1874. Taylor & Satterfield's Boss well, on the James Parker farm, two miles east of Petrolia, flowed three-thousand barrels a day! William Hartley—General Harrison Allen defeated him for Auditor-General in 1872—organized the Stump Island Oil-Company and drilled from the mouth of the Clarion River six miles south, in 1866-7. He and John Galey owned the Island-King well at Parker's Landing and a hundred others, some of which crept well down into Armstrong county. Richard Jennings and Jacob L. Meldren had punched holes on Armstrong Run and around Queenstown, but the spouter in the Parker-farm ravine was the fellow that touched the spot and hypnotized the trade. A solid stream of oil poured into the tank as if butted through the pipe by a hundred hydraulic-rams. The billowy mass of fluid heaved and foamed and boiled and tried its level best to climb over the wooden walls and unload the roof. David S. Criswell, of Oil City, had an interest in the gusher, and Criswell City—a shop, a lunch-room and five or six dwellings—was imprinted on Heydrick & Stevenson's map. Stages between Petrolia and Brady halted at the bantling town for the convenience of pilgrims to the shrine of the Boss—a "boss" representing innumerable "bar'ls." Wells were hurried down at a spanking gait, to divy up the oily freshet. "The best-laid schemes o' mice an' men gang aft a-gley" and the uncertainty of fourth-sand wells was forcibly illustrated. Jennings had dry-holes on the Steele and Bedford farms, the latter ten rods north-west of the mastodon. Taylor & Satterfield's No. 2, thirty rods west, was a small affair. Dusters and light pumpers studded the road from Criswell to Petrolia, with the Hazelwood Oil-Company's two-hundred-barreler a trifle north to tantalize believers in a straight "belt." Lines and belts and theories and former experiences amounted to little or nothing. The only safe method was to "go it blind" and bear with exemplary resignation whatever might turn up, be it a big gusher or a measly duster.

The Boss weakened to eleven-hundred barrels in July and to a humble pumper by the end of the year. Forty rods east, on the Crawford farm, Hunter & Cummings plucked a September pippin. Their Lady Hunter, sixteen-hundred feet deep and flowing twenty-five-hundred barrels, was a trophy to enrapture any hunter coming from the chase. The Boss and the Lady Hunter were the lord and lady of the manor, none of the others approaching them in

importance. Hunter & Cummings laid a pipe-line to East Brady, to load their oil on the Allegheny-Valley Railroad. The railroad company refused to furnish cars, urging a variety of pretexts to disguise the unfair discrimination. The owners of the oil had a Roland for the Oliver of the officials. They quietly gauged their output and let it run upon the ground, notifying the company to pay for the oil. A new light dawned upon the railroaders, who discovered they had to deal with men who knew their rights and dared maintain them. Crawling off their high stool, they footed the bill, apologized meekly and thenceforth took precious care Hunter & Cummings should not have reason to complain of a car-famine. Simon Legree was not the only braggart whom good men have been obliged to knock down to inspire with decent respect for fair-play.

Hunter & Cummings stayed in the business, opening the "Pontius Pool," east of Millerstown, and sinking many wells at Herman Station, where they still have a snug production. They operated on the lands of the Brady's Bend Iron-Company, putting down the wells on the hills opposite East Brady and a number in the Bradford region. They own the Tidioute Savings Bank and large tracts in North Dakota—the scene of their "bonanza farming"—and are interested with the Grandins in the great lumber-mills at Grandin, Missouri, the largest in the south-west. In connection with these mills they are building railroads to develop their two-hundred-thousand acres of timber-lands and establish experimental farms. Both members of the firm are the architects of their own fortunes, public-spirited, generous and eminently deserving of the liberal measure of success that has attended their labors during the twenty-three years of their association as partners.

Jahu Hunter was born on a farm two miles above Tidioute in 1830. From seventeen to twenty-seven he lumbered and farmed, in 1857 engaged in merchandising and in 1861 sold his store and embarked in oil. He operated moderately five years, increasing his interests largely in 1866 and forming a partnership with H. H. Cummings in 1873, which continues yet. Mr. Hunter married Miss Margaret R. Magee in 1860 and one son, L. L. Hunter, survives to aid in managing his extensive business-enterprises. He occupies a delightful home at Tidioute, is president of the Savings Bank and of the chair-factory, a Mason of the thirty-second degree and a leader in all progressive movements. He has lands in various states and has prospered in manifold undertakings. He served as school-director fifteen years, contributing time and money freely in behalf of education. He believes in bettering humanity, in relieving distress, in befriending the poor, in helping the struggling and in building up the community. Retired from active work, the evening of Jahu Hunter's useful life is serene and unclouded. As the shadows lengthen he can review the past with calm content and await the future without apprehension.

Captain H. H. Cummings removed from Illinois, his birthplace in 1840, to Ohio and was graduated from Oberlin College at twenty-two. Enlisting in July, 1862, he shared the privations and achievements of the Army of the Cumberland until mustered out in June, 1865. Three months later he visited the oli-region and in January of 1866 located at Tidioute in charge of Day & Co.'s refinery. Becoming a partner, he refined and exported oil seven years and was interested in wells at Tidioute and Fagundas. The firm dissolving in 1873, he joined hands with Jahu Hunter and operated extensively in the lower country. Hunter & Cummings stood in the front rank as representative producers. Captain Cummings is president of the Missouri Mining and Lumbering Company, which has a paid-up capital of five-hundred-thousand dollars and saws forty-million feet of lumber a year. L. L. Hunter is secretary, E. B. Grandin is treasurer and Hon. J. B. White, formerly a member of the Legislature from Warren county, is general manager. As Commander of the Grand Army of the Republic in Pennsylvania, Judge Darte succeeding him this year, Captain Cummings is favorably known to veterans over the entire state. He is a man of fine attainments, broad views and noble traits—a man who sizes up to a high ideal, who can be trusted and whose friendship "does not shrink in the wash."

Taylor & Satterfield began operations in the lower fields in 1870, secured much of the finest territory in Butler and became one of the wealthiest firms in the oil-region. Harvesters rather than sowers, their usual policy was to buy lands tested by one or more wells and avoid the risk of wildcatting. In this

16

way they acquired productive farms in every part of the district, which yielded thousands of barrels a day when fully developed. Their transactions footed up many millions yearly. They established banks at Petrolia and Millerstown, employed an army of drillers and pumpers and clerks and were always ready for a big purchase that promised fat returns. In company with Vandergrift & Forman, John Pitcairn and Fisher Brothers, they built the Fairview Pipe Line from Argyle to Brady, the nucleus of the magnificent National-Transit system of oil-transportation. Captain J. J. Vandergrift, George V. Forman and John Pitcairn were associated with them in their gigantic producing-operations, which in 1879 extended to the Bradford field and grew to such magnitude that the Union Oil-Company was formed in 1881, with five-millions capital. The Union was almost uniformly successful, owning big wells and paying big dividends. In 1883 it paid Forman a million dollars for his separate holdings in Allegany county, up to that date the largest individual sale in the region. All its properties were sold to the Forest Oil-Company and the Union was dissolved, Taylor retiring and Satterfield continuing to assist in the management some months.

Hascal L. Taylor was first known in Oildom as a member of the firm of Taylor & Day, Fredonia, N. Y., whose "buckboards" had a tremendous sale in Venango, Clarion, Armstrong and Butler. He lived at Petrolia several years, having charge of the office of Taylor & Satterfield and general oversight of the Argyle Savings Bank. After his retirement from the oil-business with an ample fortune he lived at Buffalo, speculated in real-estate and purchased miles of Florida lands. He died last year, as he was arranging to erect a fifteen-story office-block in Buffalo. Mr. Taylor was of medium height and stout build, energetic, resourceful and notable in the busy world of petroleum. His only son, Emory G., clerked in the bank at Petrolia, engaged in manufacturing at Williamsport a year or two and removed to Buffalo before his father's death. He and his sister inherited the estate.

John Satterfield, a man of heart and brain, imposing in stature, frank in speech and square in his dealings, was a Mercer boy. He served four years in a regiment organized at Greenville and opened a grocery at Pithole in 1865, with James A. Waugh as partner. Selling the remnants of the grocery in 1867, he superintended wells at Tarr Farm three years and went to Parker in 1870. His work in the Butler field increased his excellent reputation for honesty and enterprise. He married Miss Matilda Martin, of Allentown, lived four years at Millerstown, removed to Titusville and built an elegant house on Delaware avenue, Buffalo. When the Union Oil-Company's accounts were closed, the books balanced and the assets transferred to the Forest he engaged in banking. He was vice-president of the Third National Bank of Buffalo and president of the Fidelity Trust Company, whose new bank-building is the boast of the Bison City. George V. Forman and Thomas L. McFarland joined him in the Fidelity. Mr. McFarland, formerly cashier of the bank at Petrolia and secretary of the Union Company, is exceedingly affable, capable and popular. Failing health induced Mr. Satterfield to go on a trip designed to include France, the Mediterranean Sea and the warmer countries of the east. With his brother-in-law, Dr. T. J. Martin, he reached Paris, took seriously ill and died on April sixth, 1894, in his fifty-fourth year. Besides his wife, who was on the ocean hastening to his bedside when the end came, he left one son and one daughter. Dr. Martin cremated the body, pursuant to the wish of the deceased, and brought the ashes home for interment. Charitable and unostentatious, upright and active, all men liked and trusted "Jack" Satterfield, whom old friends miss sadly and remember tenderly.

> The sinless land some of his friends have enter'd long ago,
> Some others stay a little while to struggle here below;
> But, be the conflict short or long, life's battle will be won
> And lovingly he'll welcome us when earthly toil is done.
> Nor will our joy be less sincere—we'll slap him on the back,
> Clasp his brave hand and warmly say: "We're glad to see you, Jack!"

The Forest Oil-Company, into which the Union was merged, reckons its capital by millions, numbers its wells by thousands and is at the head of producing companies. Its operations cover five states. The company has hundreds of wells and farms in Pennsylvania, operates extensively in Ohio, is developing large interests in Kansas and seems certain to place Kentucky and Tennessee high up in the petroleum-galaxy. From its inception as a Limited Company the management has been progressive and efficient. To meet the increasing demands of new sections the original company was closed out and the present one incorporated, with Captain Vandergrift as president and W. J. Young as vice-president and general manager.

W. J. YOUNG.

Mr. Young, who was also elected treasurer in 1890, was peculiarly fitted for his responsible duties by long experience and executive ability. Born and educated in Pittsburg, he entered the employ of a leather-merchant in 1856, spent six years in the establishment and in 1862 went to Oil City to take charge of the forwarding and storage business of John and William Hanna. The Hannas owned the steamboat Allegheny Belle No. 4 and Hanna's wharf, the site of the National-Transit machine-shops in the Third Ward. Captain John Hanna dying, John Burgess & Co. bought the firm's storage interests and admitted Young as a partner. Burgess & Co. sold to Fisher Brothers, who used the wharf and yard for shipping and appointed Mr. Young their financial agent. How capably he filled the place every operator on Oil Creek can attest. He and John J. Fisher, under the name of Young & Co., bought and shipped crude-oil in bulk-barges. His relations with the Fishers ceased in 1872 with his appointment as book-keeper of the Oil-City Savings Bank. Elected cashier of the Oil-City Trust Company in 1874, he was afterwards vice-president and president, holding the latter office until 1891. John Pitcairn retiring from the firm of Vandergrift, Pitcairn & Co., he purchased an interest in the business. The firm of Vandergrift, Young & Co. was organized and sold its property to the Forest Oil-Company, of which Mr. Young was one of the incorporators and chairman. The business of the Forest necessitated his removal to Pittsburg in 1889. He is president of the Washington Oil-Company and the Taylorstown Natural-Gas Company and has his offices in the Vandergrift building, on Fourth avenue. During his twenty-seven years' residence in Oil City he was active in promoting the welfare of the community. In 1866 he married Miss Morrow, sister-in-law and adopted daughter of Captain Vandergrift. Two daughters, one the wife of Lieutenant P. E. Pierce, West Point, N. Y., and the other a young lady residing with her parents, blessed the happy union. The hospitable home at Oil City was a

delightful center of moral and social influence. Mr. Young represented the
First Ward nine years in Common and Select Councils and was school-director
six years. He furthered every good cause and was a helpful, honored citizen.
Now at the meridian of life, his judgment matured and his acute perceptions
quickened, young in heart and earnest in spirit, a wider sphere enlarges his
opportunities. Of W. J. Young, true and tried, faithful and competent, a loyal
friend and prudent counsellor, it can never be said: "Thou art weighed in the
balance and found wanting."

Fairview, charmingly located two miles south-west of Petrolia, was on one
side of the greased streak. James M. Lambing's gas-well a mile west lighted
and heated the town, but vapor-fuel and pretty scenery could not offset the
lack of oil and the dog-in-the-manger policy of greedy land-holders. Portly
Major Adams—under the sod for years—built a spacious hotel, which Wil-
liam Lecky, Isaac Reineman, William Fleming and kindred spirits patronized.
A mile-and-a-half east of Fairview and as far south of Petrolia, on a branch of
Bear Creek, the Cooper well originated Karns City in June of 1872. S. D.
Karns laid down eight-thousand dollars for the supposed dry-hole on the
McClymonds farm, drilled forty feet and struck a hundred-barreler. Cooper
Brothers finished the second well—it flowed two-hundred barrels for months—
on the Saturday preceding "the thirty-day shut-down." Tabor & Thompson
and Captain Grace had moguls on the Riddle and Story farms. Big-hearted,
open-handed "Tommy" Thompson—a whiter man ne'er drew breath—operated profitably in Butler and McKean and was active in the movements that
made 1872-3 memorable to oil-producers. The biggest well in the bunch was
A. J. Salisbury's five-hundred-barrel spouter on the J. B. Campbell farm, in
January of 1873. Salisbury conducted the favorite Empire House, which perished in the noon-day blaze that extinguished two-thirds of Karns City in
December of 1874. One day he bought a wagon-load of potatoes from a verdant native, who dumped the tubers into the cellar and was given a check for
the purchase. He gazed at the check long and earnestly, finally breaking out:
"Vot for you gives me dose paper?" Salisbury explained that it was payment
for the murphies. "Mein Gott!" ejaculated the ruralist, "you dinks me von
tam fool to take dot papers for mein potatoes?" The proprietor strove to
enlighten the farmer, telling him to step across the street to the bank and get
his money. "I see nein monish there," replied the innocent, looking at John
Shirley's hardware-store, part of which a bank occupied. Discussing finance
with the rustic would be useless, so "Jack" sent the hotel-clerk for the cash
and counted it out in crisp documents bearing the serpentine autograph of
General Spinner.

Vandergrift & Forman paid ninety-thousand dollars for the McCafferty
farm, a mile south-west of Karns City. Mr. Forman closed the deal, going to
the house with a lawyer and a New-York draft. The honest granger, not familiar with bank-drafts, would not receive anything except actual greenbacks.
The parties journeyed to the county-seat to convert the draft into legal-tenders,
which the seller of the property carried home. William McCafferty was a
thrifty tiller of the soil and cultivated his farm thoroughly. He bought a home
at Greenville, near John Benninghoff's, put his money in Government bonds
and died in 1880. Half the farm was fine territory and repaid its cost several
times. One-twentieth of the price in 1873 would be good value to-day for the
broad acres. For John Blaney's farm, adjoining the McCafferty, Melville,
Payne & Fleming put up fifteen-thousand dollars, bored a well and sold out to

Vandergrift & Forman at fifty-thousand. The Rob Roy well, on the McCly-monds farm, produced forty-thousand barrels of fourth-sand oil, while a dry hole was sunk thirty yards away. Colonel Woodward, Mattison & McDonald, Tack & Moorhead and John Markham owned wells good for thirty to eight-hundred barrels. A cloud of dry-holes encompassed the May Marshall, on the Wallace farm. Haysville, on the Thomas Hays farm, had a brief run, a harvest of small strikes and dusters nipping it off prematurely. The epitaph of the Philadelphia baby would about fit :

> " Died when young and full of promise,
> Our own little darling Thomas;
> We can't have things here to please us—
> He has gone to dwell with Jesus."

Branching off a mile south of Karns City, on January thirty-first, 1873, the first well—one-hundred and fifty barrels—was finished on the Moore & Hepler farm of three-hundred acres. Another in February strengthened "the belt theory," belief in which induced C. D. Angell, John L. McKinney, Phillips Brothers and O. K. Warren to form a company and test the trac t. Their faith was recompensed "an hundred fold" by an array of dandy wells and the un-folding of Angelica. Operaters were feeling their way steadfastly. Two miles south-east of Angelica, on the Simon Barnhart farm, Messimer & Backus's wild-cat—also a February plant—pumped eight barrels a day. Shreve & Kingsley's, on the Stewart farm, a mile north-east, found good sand and flowed one-hun-dred-and-forty barrels, in April, 1873. The fickle tide turned in that direction and Millerstown, a dingy, pokey hamlet on a side-elevation in Donegal town-ship, a half-mile south-east of the Shreve-spouter, was on everybody's lips. Some persons and some communities have greatness thrust upon them and Millerstown was of this brood. The natives awakened one April morning to find their settlement invaded by the irrepressible oilmen.

For sixty years the quiet hamlet of Barnhart's Mills—a co lony of Barnharts settled in Donegal when the nineteenth century was in its te ens—stuck con-tentedly in the old rut, "the world unknowing, by the world unknown." It consisted chiefly of log-houses, looking sufficiently antiquated to have been imported in William Penn's good ship Welcome. A church, a school, a black-smith-shop, a grocery, a general store and a tavern had existed from time immemorial. A grist-mill ground wheat and the name of Barnhart's Mills was adopted by the post-office authorities. It yielded to Millerstown and finally to Chicora. The two-hundred villagers went to bed at dark and breakfasted by candle-light in winter. A birth, a marriage or a funeral aroused profound interest. At last news of oil "from Parker down" was heard occasionally. Petrolia arose and the Millerites shivered with apprehension. Was the petro-leum-wave to submerge their peaceful homes? The Shreve well answered the query affirmatively and the invasion was not delayed. Crowds came, proper-ties changed hands, old houses were razed and by July the ancient borough was disguised as a modern oil-town. Dr. Book built a grand hotel, Taylor & Satterfield established a bank, the United and Relief Pipe-Lines opened offices, the best firms were represented and " on to Millerstown " was the shibboleth of the hour. McFarland & Co.'s seventy-barrel well on the Thorn farm, a mile north-east of town, the third in the district, fed the oily flame. Dr. James, on R. Barnhart's lands, finished the fourth, an eighty-barreler, in June, a half-mile west of the Shreve & Kingsley, which Clark & Timblin bought for twenty-thousand dollars. Wyatt, Fertig & Hammond's mammoth flowed one-thousand barrels a day! Col. Wyatt was a real Virginian, chivalric, educated and high-strung. Hon. John Fertig was a pioneer on Oil Creek and had operated at Foxburg with John W. Hammond. The Wyatt spouted for months.

McKeown & Morissey drilled rib-ticklers on the Nolan farm. Warden & Frew, F. Prentice, Taylor & Satterfield, Captain Grace, John Preston, Cook & Goldsboro, Samuel P. Boyer, C. D. Angell and multitudes more scored big hits. McKinney Brothers & Galey secured the Hemphill and Frederick farms, on which they drilled scores of splendid wells. James M. Lambing had a chunk near the Wyatt, with Col. Brady next door. Lee & Plumer, fresh from their triumphs in Clarion, leased the Diviner farm, two miles south-west of Millerstown, for two-hundred dollars an acre bonus and one-eighth royalty. Their first well flowed fifteen-hundred barrels and they sold to Taylor & Satterfield for ninety-thousand dollars after its production paid the bonus and the drilling. Henry Greene drilled on the Johnson farm, two miles straight south of the village, and P. M. Shannon's, on the Boyle, was the lion of the eastern belt. A dry strip divided the field into two productive lines. P. H. Burchfield opened the Gillespie farm and Joseph Overy touched the Mead, four miles south of Millerstown, for a two-hundred-barreler that installed St. Joe. Dr. Hunter, of Pittsburg, monkeyed a well on the Gillespie for many weeks, inaugurating the odious "mystery" racket. Millerstown was a peach of the most approved pattern, holding its own bravely until Bradford overwhelmed the southern region. A narrow-gauge railroad connected it with Parker in 1876. Fire in 1875 swept away the central portion of the town and blotted out seven lives. Oil has receded, the operators have departed and the town is once more a placid country village.

The Barnhart and Hemphill farms yielded McKinney Brothers a lavish return, the wells averaging fifty to three-hundred barrels month after month. The two brothers, John L. and J. C., were not amateurs in oil-matters. Sons of a well-to-do lumberman and farmer in Warren county, they learned business-methods in boyhood and were fitted by habit and education to manage important enterprises. Their connection with petroleum dated back to the sixties, in the oldest districts. The knowledge stored up on Oil Creek and around Franklin and at Pleasantville was of immense benefit in the lower fields. Organizing the firm of McKinney Brothers in 1890, to operate at Parker, they kept pace with the trend of developments southward. Millerstown impressed them favorably and they paid seventy-thousand dollars for the Barnhart and two Hemphill farms, two-hundred-and-seventy acres in the heart of the richest territory. John Galey purchased an interest in the properties, which the partners developed judiciously. J. C. McKinney and Galey resided at Millerstown to oversee the numerous details of their extensive operations. In 1877, H. L. Taylor, John Satterfield, John Pitcairn and the brothers formed the partnership known as John L. McKinney & Co. It was controlled and managed by the McKinneys, until the sale of its interests to the Standard Oil-Company. John L. and J. C. McKinney sold their Ohio lands and wells in 1889 and their Pennsylvania oil-properties in 1890, since which period they have been associated with the Standard in one of its great producing branches, the South Penn Oil-Company. Noah S. Clark is president of the South Penn, with headquarters at Oil City and Pittsburg. This company has thousands of wells in Pennsylvania, Ohio and West Virginia. The wise policy that has made the Standard the world's foremost corporation has nowhere been manifested more effectively than in the formation of such companies as the Forest and the South Penn. Letting sellers of production share in the ownership and management of properties united in one grand system secures the advantages of concerted action, unlimited capital, identity of interest and combined experience. Thus men of the highest skill join hands for the good of all, using the latest appliances, buying at wholesale for cash, producing oil at the smallest cost and giving the public the fruits of systematic coöperation. In this free country "the poor man's back-yard opens into all out-doors" and many producers, like John McKeown, Captain Jones "The." Barnsdall and Michael Murphy have been conspicuously successful going it alone. Sometimes a growl is heard about monopoly, centralization and the octave of similar phrases, just as folks grumble at the weather, the heat and cold and think they could run the universe much better than its Creator does it.

> "Oh, many a wicked smile they smole,
> And many a wink they wunk;
> 'And, oh, it is an awful thing
> To think the thoughts they thunk."

Hon. John L. McKinney's talent for business displayed itself in youth. "The boy's the father to the man" and at sixteen he assumed charge of his father's accounts, superintending the sale of lumber and farm-products three years. At nineteen, in the fall of 1861, he drilled his first well, a dry-hole south of Franklin. Two leases on Oil Creek fared better and in the spring he pur-

JOHN L. MCKINNEY.   J. C. MCKINNEY.

chased one-third of a drilling well and lease on the John McClintock farm, near Rouseville. The well was spring-poled three-hundred feet, horse-power put it to four-hundred and an engine to five-hundred, at which depth it flowed six-hundred barrels, lasting two years, lessening slowly and producing enough oil to enrich the owners. Young McKinney worked his turn, "kicking the pole" all summer and visiting his home in Warren county when steam was substituted for human and equine muscle. During his absence the sand was prodded, the golden stream responded and his partner sold out for a round sum, taking no

note of his share ! He heard of the strike and found the purchasers in full possession upon his return. His contract had not been recorded, one day remained
to file it with the register and he saved his claim by a few hours ! He bought
interests on Cherry Run that profited him two-hundred thousand dollars, in 1864
leased large tracts in Greene county and in 1865 removed to Philadelphia. He
operated on Benninghoff Run in 1866, the crash of 1867 swept away his gains
and he began again "at the top of the ground." With his younger brother,
J. C. McKinney, he drilled at Pleasantville in 1868 and the next year located at
Parker's Landing, operating constantly and managing an agency for the sale of
Gibbs & Sterrett machinery. Success crowded upon him in 1871 and in 1873
McKinney Brothers & Galey were the leaders in the Millerstown field. Mrs.
McKinney, a beautiful and accomplished woman, died in 1894. Mr. McKinney
built an elegant home at Titusville and he has been an influential citizen of
"the Queen City of Oildom" for twenty years. He is president of the Commercial Bank and a heavy stockholder in local industries. He has resisted
pressing demands for his services in public office, preferring the private station,
yet participating actively in politics. John L. McKinney is earnest and manly
everywhere, steadfast in his friendships, true to his professions, liberal and
honorable always.

J. C. McKinney engaged with an engineer-corps of the Pennsylvania Railroad Company in 1861, at the age of seventeen, to survey lines southward
from Garland, on the Philadelphia & Erie Road. The survey ending at Franklin in 1863, he left the corps and started a lumber-yard at Oil City. His father
was a lumberman at Pittsfield, Warren county, and the youth of nineteen knew
every branch of the business thoroughly. He opened a yard at Franklin in
1864, resided there a number of years and in 1868 married Miss Agnes E.
Moore. His first well, drilled at Foster in 1865, produced moderately. In
company with C. D. Angell, he drilled on Scrubgrass Island—Mr. Angell
changed the name to Belle Island for his daughter Belle—in 1866 and at Pleasantville in 1868 with his brother, John L. Operating for heavy-oil at Franklin
in 1869-70, he sold his wells to Egbert, Mackey & Tafft and settled at Parker's
Landing in 1870. The firm's operations in Butler county requiring his personal
attention, he built a house and resided at Millerstown several years. There he
worked zealously, purchasing blocks of land and drilling a legion of prolific
wells. Upon the subsidence of the Butler field he removed to Titusville, buying and remodeling the Windsor mansion, which he made one of the finest
residences in the oil-region. He assists in managing the South Penn Oil-Company, to which McKinney Brothers disposed of their interests in Ohio and
Pennsylvania. In the flush of healthful vigor, wealthy and respected, he enjoys
"the good the gods provide." He keeps fast horses, handles the ribbons
skillfully, can guide a big enterprise or an untamed bicycle deftly, is companionable and utterly devoid of affectation. To the McKinneys, men of positive
character and strict integrity, the Roman eulogy applies : "A pair of noble
brothers."

"Plumer's Ride to Diviner" discounted Sheridan's Ride to Winchester in
the estimation of Millerstown hustlers. Various operators longed and prayed
for the Diviner farm of two-hundred acres, two miles south of Millerstown,
which "Ed" Bennett's three-hundred barrel well on the Boyle farm rendered
very desirable. The old, childless couple owning it declined to lease or sell,
not wishing to move out of the old house. Lee & Plumer were on the anxious
seat with the rest of the fraternity. Plumer overheard a big operator tell his

foreman one morning to offer three-hundred dollars an acre for the farm. "Fred" lost not a moment. Ordering his two-twenty horse to be saddled instantly, he galloped to the Diviner domicile in hot haste and said : "I'll give you two-hundred dollars an acre and one eighth the oil for your land and let you stay in the house !" The aged pair consulted a moment, accepted the offer and signed an agreement to transfer the property in three days. The ink was not dry when the foreman rode up, but "Fred" met him in the yard with a smile that expressed the gospel-hymn : "Too late, too late, ye cannot enter in !" The first well repaid the whole outlay in thirty days, when Taylor & Satterfield paid ninety-thousand dollars for Lee & Plumer's holdings, a snug sum to rake in from a two-mile horseback-ride. With a fine sense of appreciation the well was labeled "Plumer's Ride to Diviner," a board nailed to the walking-beam bearing the protracted title in artistic capitals.

The Millerstown fire ended seven human lives, four of them at Dr. Book's Central Hotel. A. G. Oliver, of Kane City, was roasted in the room occupied by me the previous night. Norah Canty, a waitress, descended the stairs, returned for her trunk and was burned to a cinder. Nellie McCarthy jumped from a high window to the street, fracturing both legs and sustaining injuries that crippled her permanently. In loss of life the fire ranked next to the dreadful tragedy of the burning-well at Rouseville.

P. M. Shannon, first burgess of Millerstown, had a fashion of saluting intimate friends with the query : "Where are we now?" Possibly this was the origin of the popular phrase, "Where are we at?" A zealous officer arrested a drunken loafer one afternoon. The fellow struggled to get free and the officer halted a wagon to haul the obstreperous drunk to the lock-up. The prisoner was laid on his back in the wagon and his captor tried to hold him down. A crowd gathered and the burgess got aboard to assist the peeler. He was holding the feet of the law-breaker, with his back to the end-board, at the instant the wheels struck a plank-crossing. The shock keeled Shannon backwards over the end-board into the deep, vicious mud ! The spectators thought of shedding tears at the sad plight of their chief magistrate, who sank at full length nearly out of sight. As he raised his head a ragged urchin bawled out : "Where are we now?" The laugh that ensued was a risible earthquake and

"WHERE ARE WE NOW?"

thenceforth the expression had unlimited circulation in the lower districts.

The Millerstown field produced ten-thousand barrels a day at its prime and the temptation to enlarge the productive area even St. Anthony, had he been an oil-operator, would have found it hard to resist. A half-mile west, at the Brick Church, J. A. Irons punched a hole and started a hardware-store that hatched out Irons City. St. Joe, where two-hundred lots were sold in thirty days and a beer-jerker's tent was the first business-stand, was the outcome of good wells on the Now, Meade, Boyd, Neff and Graham farms, four miles south. Three miles farther dry-holes blasted the budding hopes of Jeffersonville. Three miles south-west of Millerstown, on an elevated site, Buena Vista bade fair to knock the persimmons. The territory exhausted too speedily for comfort, other points lured the floaters, hotels and stores stood empty and a fire sent three-fourths of the neat little town up in smoke. Two miles west the Hope Oil-Company's Troutman well, reported on March twenty-second, 1873, "the biggest strike since 'sixty-five," flowed twelve-hundred barrels. The tools hung in the hole seven months, by which time the well had produced ninety-six-thousand barrels. The gusher was on the Troutman farm, a patch of rocks and stunted trees tenanted by a Frenchman. I. E. Dean, Lecky & Reineman, Captain Grace, Captain Boyer, the Reno Oil-Company and others jostled neck and neck in the race to drain the Ralston, Harper, Starr, Jenkins and Troutman lands. The result was a series of spouters that aggregated nine-thousand barrels a day. Phillips Brothers paid eighty-thousand dollars for the Starr farm and trebled their money in a year. William K. Vandergrift's Black-hawk was a five-hundred barreler and dozens more swelled the production and the excitement. The day before Husselton & Thompson's seven-hundred bar-reler, on the Gruber farm, struck the sand the boiler exploded. Two men were standing on a tank discussing politics. They saw a ton of iron heading directly towards them, concluded to postpone the argument and leaped from the tank as the flying mass tore off half the roof. The Ralston farm evoluted the embryo town of Batesville, named for the late Joseph Bates, of Oil City, and Modoc planted its wigwams on the Starr and Sutton.

Modoc stood at the top of the class for mud. The man who found a gold-dollar in a can of tomatoes and denounced the grocer for selling adulterated goods would have had no reason to grumble at the mud around Modoc. It was pure, unmixed and unstinted. The voyager who, in the spring or fall of 1873, accomplished the trip from Troutman to the frontier wells without ex-hausting his stock of profanity earned a free-pass to the happy hunting-grounds. Twenty balloon-structures were erected by May first and a red-headed dis-penser of stimulants answered to the title of "Captain Jack." Modoc was not a Tammany offshoot, but the government had an Indian war on hand and red-skinned epithets prevailed. The town soon boasted three stores, four hotels, liveries and five-hundred people. By and by the spouters wilted badly, degenerating into pumpers. On a cold, rainy night in the autumn of 1874 fire started in Max Elasser's clothing-store and one-half the town was absent at dawn next morning. Biting wind and drenching showers added to the sadness of the dismal scene. Women and children, weeping and homeless, crouched in the fields until daylight and shelter arrived. That was the last chapter in the history of Modoc. The American Hotel and a few houses escaped the flames, but the destroyed buildings were not replaced. It would puzzle a tourist now to find an atom of Modoc or the wells that vegetated about the Troutman whale.

Two miles south of Modoc the McClelland farm made a bold effort to out-shine the Troutman. Phillips Brothers owned the biggest wells, luscious fel-lows that salt-water killed off prematurely. They paid forty-five-thousand dol-lars for the Stahl & Benedict No. 1 well. The farmer leased the tract to George Nesbit and John Preston. Nesbit placed timbers for a rig on the ground and entrenched a force of men behind a fence. Preston's troops scaled the fence, dislodged the enemy, carried the timbers off the premises, built a rig and drilled a well. Such disputes were liable to occur from the ignorance or knav-ery of the natives, some of whom leased the same land to several parties. In one of these struggles for possession Obadiah Haymaker was shot dead at Murraysville, near Pittsburg. Milton Weston, a Chicago millionaire, who hired and armed the attacking party, was sent to the ·penitentiary for manslaughter. Haymaker was pleasant, sociable and worthy of a better fate.

David Morrison leased ten acres of the Jamison farm, three miles below Modoc and seven south of Petrolia, at one-fiftieth royalty. The property was situated on Connoquinessing Creek, a tributary of ·Beaver River, in the bosom of a rugged country. On August twenty-fourth, 1872, the tools pricked the sand, gas burst forth and oil flowed furiously. The gas sought the boiler-fire and the entire concern was speedily in a blaze. Unlike many others in the oil-region, the Morrison well suffered no injury from the fire. It flowed three-hundred barrels a day for a month and in October was sold to Taylor & Satter-field for thirty-eight-thousand dollars. They cleaned out the hole, which mud had clogged, restoring the yield to two-hundred barrels. S. D. Karns completed the Dogleg well, the second in the field, on Christmas day, and the third early in January, the two wells flowing seven-hundred barrels. John Preston's No. 1, a half-mile northward, flowed two-hundred barrels on January twelfth. Preston was a strong-limbed, black-haired, courageous operator, who cut his eye-teeth in the upper fields. He augmented his pile at Parker, Millerstown and Greece City, landing at last in Washington county. He was not averse to a hand at cards or a gamble in production. His word was never broken and he vied with John McKeown and John Galey in untiring energy. A truer, livelier, braver lot of men than the Butler oil-operators never stepped on God's green carpet. A mean tyrant might as well try to climb into heaven on a greased pole as to keep them at the bottom of the heap.

The first new building on the Jamison farm, a frame drug-store, was erected on September tenth. Eight-hundred people inhabited Greece City by the end of December. Drinking dens drove a thriving trade and three hotels could not stow away the crowds. J. H. Collins fed five-hundred a day. Theodore Huselton established a bank and Rev. Mr. Thorne a newspaper. A post-office was opened at New-Year. Two pipe-lines conveyed oil to Butler and Brady, two telegraph-offices rushed messages, a church blossomed in the spring and a branch of the West Penn Railroad was proposed. Greece City combined the muddiness and activity of Shaffer and Funkville with the ambition of Reno. Fifty wells were drilling in February and the surrounding farms were not permitted to "linger longer, Lucy," than was necessary to haul machinery and set the walking-beam sawing the atmosphere. Joseph Post—a jolly Rousevillean, who weighed two-hundred pounds, operated at Bradford and retired to a farm in Ohio—tested the Whitmire farm, two miles south. An extensive water-well was the best the farm had to offer and Boydstown, built in expectation of the oil that never came, scampered off. The third sand was only twelve to fifteen feet thick and the wells declined with unprecedented suddenness. The bottom seemed to drop out of the territory in a twinkling. The town wilted like a paper-collar in the dog-days. Houses were torn down or deserted and rigs carted to Millerstown. In December fire licked up three-fourths of what removals had spared, summarily ending Greece City at the fragile age of thir-teen months. "The isles of Greece, where burning Sappho loved and sung," may have been pretty slick, but the oil of Greece City would have burned out Sappho in one round.

"The meanest man I ever saw," a Butler judge remarked to a company of friends at Collins's Hotel, "has never appeared in my court as a defendant and it is lucky for him. As a matter of course he was a newspaper man—a rascal of a reporter for the Greece City *Review*, printed right in this town, and there he stands! One day he was playing seven-up with a young lady and guess what he did? He told her that whenever she had the jack of trumps it was a sure sign her lover was thinking of her. Then he watched her and whenever she blushed and looked pleased he would lead a high card and catch her jack. A man who would do that would steal a hot stove or write a libellous joke about me." The judge was a rare joker and the young man whom he apostrophised for fun didn't know a jack from a load of hay.

Parker, Martinsburg, Argyle, Petrolia, the "Cross Belt," Karns City, Angelica, Millerstown, St. Joe, Buena Vista, Modoc and Greece City had passed in review. The "belt" extended fifteen miles and the Butler field acknowledged no rival. The great Bradford district was about to distance all competitors and leave the southern region hopelessly behind, yet operators did not desist from their efforts to discover an outlet below Greece City and St. Joe. Two miles west of the county-seat Phillips Brothers stumbled upon the Baldridge pool, which produced largely. The old town of Butler, settled at the beginning of the century and not remarkable for enterprise until the oilmen shoved it forward, was dry territory. Eastward pools of minor note were revealed. William K. Vandergrift, whose three-hundred-barrel well on the Pontius farm ushered in Buena Vista's short-lived reign, drilled at Saxonburg. Along the West-Penn Railroad fair wells encouraged the quest. David Kirk entered Great Belt City in the race and the country was punctured like a bicycle-tire tripping over a road strewn with tacks pointing skyward and loaded for mischief. South of St. Joe gas blew off and Spang & Chalfant laid a line from above Freeport to pipe the stuff into their rolling-mills at Pittsburg. The search proceeded without big surprises, Bradford monopolizing public interest and Butler jogging on quietly at the rear. But the old field had plenty of ginger and was merely recovering some of the breath expended in producing forty-million barrels of crude. "I smell a rat," felicitously observed Sir Boyle Roche, "and see him floating in the air." The free play of the drill could hardly fail to ferret out something with the smell of petroleum in the soap-mine county, beyond the cut-off at Greece City and Baldridge. Bradford was sliding down the mountain it had ascended and Butler furnished the answer to the conundrum of where to look for the next fertile spot.

Col. S. P. Armstrong, who experienced a siege of hard luck in the upper latitudes, in 1884 leased a portion of the Marshall farm, on Thorn Creek, six miles south-west of the town of Butler. Operators had been skirmishing around the southern rim of the basin, looking for an annex to the Baldridge pool. Andrew Shidemantle was drilling near the mouth of the creek, on the north bank of which Johnson & Co.'s well, finished in May, found plenty of sand and salt-water and a taste of oil. More than once Armstrong was pressed for funds to pay the workmen drilling the well he began on the little stream and he sold an interest to Boyd & Semple. A vein of oil was met on June twenty-seventh, gas ignited the rig and for a week the well burned fiercely. The flames were subdued finally, the well pumped and flowed one-hundred-and-fifty barrels a day and No. 2 was started fifty rods north-east. Meanwhile Phillips Brothers set the tools dancing on the Bartlett farm, adjoining the Marshall on the north. They hit the sand on August twenty-ninth and the well flowed five-hundred barrels next day. Drilling ten feet deeper jagged a veritable reservoir of petroleum, the well flowing forty-two-hundred barrels on

September fifteenth! At last Phillips & Vanausdall's spouter on Oil Creek had been eclipsed. The trade was "shaken clear out of its boots." Glowing promises of a healthy advance in prices were frost-bitten. Scouts had been hovering around and their reckoning was utterly at fault. Brokers knew not which way to turn. Crude staggered into the ditch and speculators on the wrong side of the market went down like the Louisiana Tigers at Gettysburg. The bull-element thought the geyser "a scratch," quite sure not to be duplicated, and all hands awaited impatiently the completion of Hezekiah Christie's venture on a twenty-five acre plot hugging the Phillips lease. The one redeeming feature of the situation was that nobody had the temerity to remark, "I told you so!"

A telegraph office was rigged up near the Phillips well in an abandoned carriage, one-third mile from the Christie. About it the sharp-eyed scouts thronged night and day. On October eleventh the Christie was known to be

TELEGRAPH-OFFICE IN CARRIAGE, GROUP OF SCOUTS AND PHILLIPS WELL, THORN CREEK.

nearing the critical point. Excitement was at fever-heat among the group of anxious watchers. In the afternoon some knowing-one reported that the tools were twenty-seven feet in the sand, with no show of oil. The scouts went to condole with Christie, who was sitting in the boiler-house, over his supposed dry-hole. One elderly scout, whose rotundity made him "the observed of all observers," was especially warm in his expressions of sympathy. "That's all right, Ben," said Christie, "but before night you'll be making for the telegraph-office to sell your oil at a gait that will make a euchre-game on your coat-tails an easy matter." When the scout had gone he walked into the derrick and asked his driller how far he was in the sand. "Only twenty-two feet and we are sure to strike oil before three o'clock. Those scouts don't know what they're talking about." Christie went back to the boiler-house and waited. It was an interesting scene. About the old buggy were the self-confident scouts, many of whom had already wired their principals that the well was dry. The intervals between the strokes of the drill appeared to be hours. At length the well began to gas. Then came a low, rumbling sound and those about the carriage saw a cloud-burst of oil envelop the derrick. The Christie well was in and the biggest gusher the oil-country had ever known! The first day it did over five-thousand barrels, seven-thousand for several days after torpedoing, and for a month poured out a sea of oil. Christie refused one-hundred-thousand dollars for his monster, which cast the Cherry Grove gushers completely into the shade. Phillips No. 3, four-hundred barrels, Conners No. 1, thirty-seven-hundred, and Phillips No. 2, twenty-five hundred, were added to the string on October eighteenth, nineteenth and twenty-first. Crude tumbled, the bears pranced wildly and everybody wondered if Thorn Creek had further surprises up its sleeve. Bret Harte's "heathen Chinee" with five aces was less of an enigma.

All this time Colonel Armstrong, who borrowed money to build his first derrick and buy his first boiler, was pegging away at his second well. The sand was bored through into the slate beneath and the contractor pronounced the well a failure. The scouts agreed with him unanimously and declared the contractor a level-headed gentleman. The owner, who looked for something nicer than salt-water and forty-five feet of ungreased sand, did not lose every vestige of hope. He decided to try the persuasive powers of a torpedo. At noon on October twenty-seventh sixty quarts of nitro-glycerine were lowered into the hole. The usual low rumbling responded, but the expected flow did not follow immediately. One of the scouts laughingly offered Armstrong a cigar for the well, which the whole party declared "no good." They broke for the telegraph-office in the buggy to wire that the well was a duster. Prices stiffened and the bulls breathed more freely.

The scouts changed their minds and their messages very speedily. The rumbling increased until its roar resembled a small Niagara. A sheet of salt-water shot out of the hole over the derrick, followed by a shower of slate, stones and dirt. A moment later, with a preliminary cough to clear its passage, the oil came with a mighty rush. A giant stream spurted sixty feet above the tall derrick, dug drains in the ground and saturated everything within a radius of five-hundred feet! The Jumbo of oil-wells had been struck. Thousands of barrels of oil were wasted before the cap could be adjusted on the casing. Tanks had been provided and a half-dozen pipes were needed to carry the enormous mass of fluid. It was an inspiring sight to stand on top of the tank and watch the tossing, heaving, foaming deluge. The first twenty-four hours Armstrong No. 2 flowed *eight-thousand-eight-hundred barrels!* It dropped to six-thousand by November first, to six-hundred by December first and next morning stopped alrogether, having produced eighty-nine-thousand barrels in thirty-seven days! Armstrong then divided his lease into five-acre patches, sold them at fifteen-hundred dollars bonus and half the oil and quit Thorn Creek in the spring a half-million ahead.

Fisher Brothers were the largest operators in the field. From the Marshall farm of three-hundred acres—worth ten dollars an acre for farming—they took four-hundred-thousand dollars' worth of oil. Their biggest well flowed forty-two-hundred barrels on November fifteenth, when the total output of the field was sixteen-thousand, its highest notch. Miller & Yeagle's spouter put forty-five-hundred into the tank, sending out a stream that filled a five-inch pipe. Thomas B. Simpson, of Oil City, joined with Thomas W. Phillips in leasing the Kennedy farm and drilling a three-thousand-barrel well. They sold to the Associated Producers in December for eighty-thousand dollars and Simpson, a sensible man every day in the year, presented his wife with a Christmas check for his half of the money. McBride and Campbell, two young drillers who had gone to Thorn Creek in search of work, went a mile in advance of developments and bored a well that did five-thousand barrels a day. They sold out to

ARMSTRONG WELL

the Associated Producers for ninety-thousand and six months later their big well was classed among the small pumpers. Campbell saved what he had made, but success did not sit well on McBride's shoulders. After lighting his cigars awhile with five-dollar bills he touched bottom and went back to the drill. Hell's Half Acre, a crumb of land owned by the Bredin heirs, emptied sixty-thousand barrels into the tanks of the Associated Producers. The little truck-farm of John Mangel put ninety-thousand into its owner's pocket, although a losing venture to the operators, producing barely a hundred-thousand barrels. For the half-acre on which the red school-house was built, ten rods from the first Phillips well, the directors were offered fifty-thousand dollars. This would have endowed every school in the township, but legal obstacles prevented the sale and the district was the loser. By May the production declined to seven-thousand barrels and to one-thousand by the end of 1885. The sudden rise of the field made a score of fortunes and its sudden collapse ruined as many more. Thorn Creek, like reform measures in the Legislature, had a brilliant opening and an inglorious close.

The Thorn-Creek white-sanders encouraged wildcatting to an extraordinary degree. In hope of extending the pool or disclosing a fresh one, "men drilled who never drilled before, and those who always drilled but drilled the more." Johnson & Co., Campbell & McBride, Fisher Brothers and Shidemantle's dusters on the southern end of the gusher-farms condemned the territory in that direction. Painter Brothers developed a small pool at Riebold Station. Craig & Cappeau, who struck the initial spouter at Kane, and the Fisher Oil-Company failed to open up a field in Middlesex township. Some oil was found at Zelienople and gas at numerous points in raking over Butler county. The country south-west of Butler, into West Virginia and Ohio, was overrun by oil-prospectors, intent upon tying up lands and seeing that no lurking puddle of petroleum should escape. Test wells crossed the lines into Allegheny and Beaver counties and Shoustown, Shannopin, Mt. Nebo, Coraopolis, Undercliff and Economy figured in the newspapers as oil-centers of more or less consequence. Members of the old guard, fortified with a stack of blues at their elbow to meet any contingency, shared in these proceedings. Brundred & Marston drilled on Pine Creek, at the lower end of Armstrong county, in the seventies, a Pittsburg company repeating the dose in 1886. At New Bethlehem they bored two-thousand feet, finding seven-hundred feet of red-rock. This rock varies from one to three-hundred feet on Oil Creek and geologists assert is six-thousand feet thick at Harrisburg, diminishing as it approaches the Alleghenies. The late W. J. Brundred, agent at Oil City of the Empire Line until its absorption by the Pennsylvania Railroad, was a skilled oil-operator, practical in his ideas and prompt in his methods. His son, B. F. Brundred, is president of the Imperial Refining Company and a prosperous resident of Oil City. Joseph H. Marston died in California, whither he had gone hoping to improve his health, in 1880. He was an artist at Franklin in the opening years of developments and removed to Oil City. He owned the Petroleum House and was exceptionally genial, enterprising and popular.

> "Through many a year
> We shall remember, with a sad delight,
> The friends forever gone from mortal sight."

Pittsburg assumed the airs of a petroleum-metropolis. Natural-gas in the suburbs and east of the city changed its sooty blackness to a delicate clearness that enabled people to see the sky. Oilmen made it their headquarters and built houses at East Liberty and Allegheny. To-day more representative producers can be seen in Pittsburg than in Oil City, Titusville or Bradford. Within a hundred yards of the National-Transit offices one can find Captain Vandergrift, T. J. Vandergrift, J. M. Guffey, John Galey, Frank Queen, W. J. Young, P. M. Shannon, Frederick Hayes, Dr. M. C. Egbert, A. J. Gartland, Edward Jennings, Captain Grace, S. D. Karns, William Fleming, C. D. Greenlee, James M. Lambing, John Galloway, John J. Fisher, Henry Fisher, Frederick Fisher, J. A. Buchanan, J. N. Pew, Michael Murphy, James Patterson and other veterans in the business. These are some of the men who had the grit to open new fields, to risk their cash in pioneer-experiments, to cheapen transportation and to make kerosene "the poor man's light." They are not youngsters any more, but their hearts have not grown old, their heads have not swelled and the microbe of selfishness has not soured their kindly impulses. They are of the royal stamp that would rather tramp the cross-ties with honor than ride in a sixteen-wheeled Pullman dishonestly.

Gas east and oil west was the rule at Pittsburg. Wildwood was the chief sensation in 1889–90. This was the pet of T. J. Vandergrift, head and front of the Woodland Oil-Company, which has harvested bushels of money from the field. "Op" Vandergrift is not an apprentice in petroleum. He added to his reputation in the middle-field leading the opposition to the mystery-dodge.

Napoleon or Grant was not a finer tactician. His clever plans were executed without a hitch or a Waterloo. He neither lost his temper nor wasted his powder. The man who "fights the devil with fire" is apt to run short of ammunition, but Vandergrift knew the ropes, kept his own counsel, was "cool as a cucumber" and won in an easy canter. He is obliging, social, manfully independent and a zealous worker in the Producers' Association. It is narrated that he went to New York three years ago to close a big deal for Ohio territory he had been asked to sell. He named the price and was told a sub-boss at Oil City must pass upon the matter. "Gentlemen," he said, " I am not going to Oil City on any such errand. I came prepared to transfer the property and, if you want it, I shall

T. J. VANDERGRIFT.

be in the city until noon to-morrow to receive the money !" The cash—three-hundred-thousand dollars—was paid at eleven o'clock. Mr. Vandergrift has interests in Pennsylvania, Ohio, West-Virginia and Kentucky. He knows a good horse, a good story, a good lease or a good fellow at sight and a wildcat-well does not frighten him off the track. His home is at Jamestown and his office at Pittsburg.

Thirty-three wells at Wildwood realized Greenlee & Forst not far from a quarter-million dollars. Five in "the hundred-foot" field west of Butler repaid their cost and brought them fifty-thousand dollars from the South-Penn Oil-Company. The two lucky operators next leased and purchased eight-hundred acres at Oakdale, Noblestown and McDonald, in Allegheny and Washington counties, fifteen to twenty miles west of Pittsburg. The Crofton third-sand pool was opened in February of 1888, the Groveton & Young hundred-foot in the winter of 1889–90 and the Chartiers third-sand field in the spring of 1890. South-west of these, on the J. J. McCurdy farm, five miles north-east of Oakdale, Patterson & Jones drilled into the fifth sand on October seventeenth, 1890. The well flowed nine-hundred barrels a day for four months, six months later averaged two-hundred and by the end of 1891 had yielded a hundred-and-fifty thousand. Others on the same and adjacent tracts started at fifty to twenty-five-hundred barrels, Patterson & Jones alone deriving four-thousand barrels a day from thirteen wells. In the summer of 1890 the Royal Gas-Company drilled two wells on the McDonald estate, two miles west of McDonald Station and ten south-west of McCurdy, finding a show of oil in the so-called "Gordon

17

sand." On the farm of Edward McDonald, west side of the borough, the company struck oil and gas in the same rock the latter part of September. The well stood idle two months, was bored through the fifth sand in November, torpedoed on December twentieth and filled three tanks of oil in ten days. The tools were run down to clean it out, stuck fast and the pioneer venture of the McDonald region ended its career simultaneously with the ending of 1890. Thorn Creek had been a wonder and Wildwood a dandy, yet both combined were to be dwarfed and all records smashed by the greatest white-sand pool and the biggest gushers in America.

Geologists solemnly averred in 1883 that "the general boundaries of the oil-region of Pennsylvania are now well established," "we can have no reasonable expectation that any new and extensive field will be found" and "there are not any grounds for anticipating the discovery of new fields which will add enough to the declining products of the old to enable the output to keep pace with the consumption." Notwithstanding these learned opinions, Thorn Creek had the effrontery to "be found" in 1884, Wildwood in 1889 and the monarch of the tribe in 1891. The men who want people to discard Genesis for their interpretation of the rocks were as wide of the mark as the dudish Nimrod who couldn't hit a barn-door at thirty yards. He paralyzed his friends by announcing : "Wal, I hit the bullseye to-day the vehwy fiwst shot !" Congratulations were pouring in when he added : "Yaas, and the bweastly fawmeh made me pay twenty-five dollahs fawh the bull I didn't see when I fiwed, doncherknow !" A raw recruit instructed the architect of his uniform to sew in an iron-plate "to protect the most vital part." The facetious tailor, instead of fixing the plate in the breast of the coat, planted it in the seat of the young fellow's breeches. The enemy worsting his side in a skirmish, the retreating youth tried to climb over a stone-wall. A soldier rushed to transfix him with his bayonet, which landed on the iron-plate with the force of a battering-ram. The shock hurled the climber safely into the field, tilted his assailant backward and broke off the point of the cold steel ! The happy hero picked himself up and exclaimed fervently : "That tailor knew a devilish sight better'n me what's my most vital part !" Operators who paid no heed to scientific disquisitions, but went on opening new fields each season, believed the drill was the one infallible test of petroleum's most vital part.

In May of 1891 the Royal Gas-Company finished two wells on the Robb and Sauters tracts, south of town, across the railroad-track. The Robb proved a twenty-barreler and the Sauters flowed one-hundred-and-sixty barrels a day from the fifth sand. They attracted the notice of the oilmen, who had not taken much stock in the existence of paying territory at McDonald. Three miles north-east the Matthews well, also a May-flower, produced thirty barrels a day from the Gordon rock. On July first it was drilled into the fifth sand, increasing the output to eight-hundred barrels a day for two months. Further probing the first week in September increased it to *eleven-thousand barrels!* Scouts gauged it at seven-hundred barrels an hour for three hours after the agitation ceased ! It yielded four-hundred-thousand barrels of oil in four months and was properly styled Matthews the Great. The owners were James M. Guffey, John Galey, Edward Jennings and Michael Murphy. They built acres of tanks and kept ten or a dozen sets of tools constantly at work. Mr. Guffey, a prime mover in every field from Richburg to West Virginia, was largely interested in the Oakdale Oil-Company's eighteen-hundred acres. With Galey, Jennings and Murphy he owned the Sturgeon, Bell and Herron farms, the first six wells on which yielded twenty-eight-thousand barrels a day! The mastodon oil-field of the world had been ushered in by men whose sagacious boldness and good judgment Bradford, Warren, Venango, Clarion and Butler had witnessed repeatedly.

C. D. Greenlee and Barney Forst, who joined forces west of Butler and at Wildwood, in August of 1891 leased James Mevey's two-hundred-and-fifty acres, a short distance north-east of McDonald. Greenlee and John W. Weeks, a surveyor who had mapped out the district and predicted it would be the "richest field in Pennsylvania," selected a gentle slope beside a light growth of timber for the first well on the Mevey farm. The rig was hurried up and the tools were hurried down. On Saturday, September twenty-sixth, the fifth sand was cracked and oil gushed at the rate of one-hundred-and-forty barrels an hour.

GREENLEE & FORST WELL, M^cDONALD.

The well was stirred a trifle on Monday, September twenty-eighth, with startling effect. It put *fifteen-thousand-six-hundred barrels* of oil into the tanks in twenty-four hours! The Armstrong and the Matthews had to surrender their laurels, for Greenlee & Forst owned the largest oil-well ever struck on this continent. On Sunday, October fourth, after slight agitation by the tools, the mammoth poured out seven-hundred-and-fifty-barrels an hour for four hours, a record that may, perhaps, stand until Gabriel's horn proclaims the wind-up of oil-geysers and all terrestrial things. The well has yielded several-hundred-thousand barrels and is still pumping fifty. Greenlee & Forst's production for a time exceeded twenty-thousand barrels a day and they could have taken two or three-million dollars for their properties. The partners did not pile on the agony because of their good-luck. They kept their office at Pittsburg and

Greenlee continued to live at Butler. He is a typical manager in the field, bubbling over with push and vim. Forst had a clothing-store at Millerstown in its busy days, waltzed around the bull-ring in the Bradford oil-exchange and returned southward to scoop the capital prize in the petroleum-lottery.

Scurrying for territory in the Jumbo-field set in with the vigor of a thousand football-rushes. McDonald tourists, eager to view the wondrous spouters and hungry for any morsel of land that could be picked up, packed the Panhandle trains. Rigs were reared on town-lots, in gardens and yards. Gaslights glared, streams of oil flowed and the liveliest scenes of Oil Creek were revived and emphasized. By November first two-hundred wells were drilling and sixty rigs building. Fifty-four October strikes swelled the daily production at the close of the month to eighty-thousand barrels ! What Bradford had taken years to accomplish McDonald achieved in ninety days ! Greenlee & Forst had thirty wells drilling and three-hundred-thousand barrels of iron-tankage. Guffey, Galey & Jennings were on deck with fifteen or twenty. The Fisher Oil-Company, owning one-fourth the Oakdale's big tract and the McMichael farm, had sixteen wells reaching for the jugular, from which the Sturgeon and Baldwin spouters were drawing ten-thousand barrels a day. William Guckert—he started at Foster and was active at Edenburg, Parker, Millerstown, Bradford and Thorn Creek—and John A. Steele had two producing largely and eight going down on the Mevey farm. J. G. Haymaker, a pioneer from Allegany county, N. Y., to Allegheny county, Pa., and Thomas Leggett owned one gusher, nine drilling wells and five-hundred acres of leases. Haymaker began at Pithole, drilled in Venango and Clarion, was prominent in Butler and in 1878 optioned blocks of land on Meek's Creek that developed good territory and the thriving town of Haymaker, the forerunner of the Allegheny field. He boosted Saxonburg and Legionville and his brother, Obadiah Haymaker, opened the Murraysville gas-field and was shot dead defending his property against an attack by Weston's minions. Veterans from every quarter flocked in and new faces were to be counted by hundreds at Oakdale, Noblestown and McDonald. The National-Transit Company laid a host of lines to keep the tanks from overflowing and Mellon Brothers operated an independent pipe-line. Handling such an avalanche of oil was not child's play and it would have been utterly impossible in the era of wagons and flat-boats on Oil Creek.

McDonald territory, if unparalleled in richness, in some respects tallied with portions of Oil Creek and the fourth-sand division of Butler. Occasionally a dry-hole varied the monotony of the reports and ruffled the plumage of disappointed seekers for gushers. Even the Mevey farm trotted out dusters forty rods from Greenlee & Forst's record-breaker. The "belt" was not continuous from McCurdy and dry-holes shortened it southward and narrowed it westward, but a field so prolific required little room to build up an overwhelming production. An engine may exert the force of a thousand horses and the yield of the Greenlee & Forst or the Matthews in sixty days exceeded that of a hundred average wells in a twelvemonth. The remotest likelihood of running against such a snap was terribly fascinating to operators who had battled in the older sections. They were not the men to let the chance slip and stay away from McDonald. Hence the field was defined quickly and the line of march resumed towards the southward, into Washington county and West-Virginia.

Wrinkles, gray-hairs and sometimes oil-wells come to him who has patience to wait. Just as 1884 was expiring, the discovery of oil in a well on the Gantz lot, a few rods from the Chartiers-Railroad depot, electrified the ancient borough of Washington, midway between Pittsburg and Wheeling. The whole town gathered to see the grease spout above the derrick. Hundreds of oilmen hurried to pick up leases and jerk the tools. For six weeks a veil of mystery shrouded the well, which was then announced to be of small account. Eight others had been started, but the territory was deep, the rock was often hard and the excited populace had to wait six months for the answer to the drill.

Traveling over Washington county in 1880, Frederick Crocker noticed its strong geological resemblance to the upper oil-fields, which he knew intimately. The locality was directly on a line from the northern districts to points south that had produced oil. He organized the Niagara Oil-Company and sent agents to secure leases. Remembering the collapse of Washington companies in 1860-1, when wells on Dunkard Creek attracted folks to Greene county, farmers held back their lands until public-meetings and a house-to-house canvass satisfied them the Niagara meant business. Blocks were leased in the northern tier of townships and in 1882 a test well was drilled on the McGuigan farm. An immense flow of gas was encountered at twenty-two-hundred feet and not a drop of oil. Not disheartened, the company went west three miles and sank a well on the Buchanan farm, forty-two-hundred feet. Possibly the hole contained oil, but it was plugged and the drillers proceeded to bore thirty-six-hundred feet on the Rush farm, four miles south. Jumping eleven miles north-east, they obtained gas, salt-water and feeble spurts of oil from a well on the Scott farm. About this stage of the proceedings the People's Light and Heat Company was organized to supply Washington with natural-gas. From three wells plenty of gas for the purpose was derived. A rival company drilled a well on the Gantz lot, adjacent to the town, which at twenty-one-hundred feet struck the vein of oil that threw the county-seat into spasms on the last day of 1884.

The fever broke out afresh in July of 1885, by a report that the Thayer well, on the Farley farm, a mile south-west in advance of developments, had "come in." This well, located in an oatfield in a deep ravine, was worked as a mystery. Armed guards constantly kept watch and scouts reclining on the hill-top contented themselves with an unsatisfactory peep through a field-glass. One night a shock of oats approached within sixty feet of the derrick. The guard fired and the propelling power immediately took to its heels and ran. Another night, while a crowd of disinterested parties jangled with the guards, scouts gained entrance to the derrick from the rear, but discovered no oil. Previous to this a scout had paid a midnight visit to the well, eluded the guards, boldly climbed to the top of the derrick and with chalk marked the crown-pulley. With the aid of their glasses the vigilant watchers on the hill-top counted the revolutions and calculated the length of cable needed to reach the bottom of the well. A bolder move was to crawl under the floor of the derrick. This was successfully accomplished by several daring fellows, one of whom was caught in the act. He weighed two-hundred-and-forty pounds and his frantic struggles for a comfortable resting-place led to his discovery. A handful of cigars and a long pull at his pocket-flask purchased his freedom. The well was a failure. R. H. Thayer drilled four more good ones, one a gusher that netted him three-thousand dollars a day for months. Other operators crowded in and were rewarded with dusters of the most approved type.

The despondency following the failure of Thayer's No. 1 was dispelled on August twenty-second. The People's Light and Heat Company's well, on the Gordon farm, pierced a new sand two-hundred-and-sixty feet below the Gantz formation, and oil commenced to scale the derrick. Again the petroleum-fever raged. An owner of the well, at church on Sunday morning, suddenly awakened from his slumbers and horrified pastor and congregation by yelling: "By George! There she spouts!" The day previous he had seen the well flow and religious thoughts had been temporarily replaced by dreams of a fortune. This well's best day's record was one-hundred-and-sixty barrels. Test wells for the new Gordon sand were sunk in all directions and the Washington field had made a substantial beginning. The effect on the inhabitants was marked. The price of wool no longer formed the staple of conversation, the new indus-try entirely superseding it. Real-estate values shot skyward and the borough population strode from five-thousand to seventy-five-hundred. The sturdy Scotch-Presbyterians would not tolerate dance-houses, gambling-hells and dens of vice in a town that for twenty years had not permitted the sale of liquor. Time works wonders. Washington county, which fomented the Whisky In-surrection, was transformed into a prohibition stronghold. The festive citizen intent upon a lark had to journey to Pittsburg or Wheeling for his jag.

Col. E. H. Dyer, whom the Gantz well allured to the new district, leased the Calvin Smith farm, three miles north-east, and started the drill. He had twenty years' experience and very little cash. His funds giving out, he offered the well and lease for five-hundred dollars. Willets & Young agreed to finish the well for two-thirds interest. They pounded the rock, drilled through the fifth sand and hit "the fifty-foot" nearer China. In January of 1886 the well—Dyer No. 1—flowed four-hundred barrels a day. Expecting gas or a dry-hole, from the absence of oil in the customary sand, the owners had not erected tanks and the stream wasted for several days. Dyer sold his remaining one-third to Joseph W. Craig, a well-known operator in the Oil-City and Pittsburg oil-exchanges, for seventy-five-thousand dollars. He organized the Mascot Oil-Company, located the McGahey in another section of the field and pocketed two-hundred-thousand dollars for his year's work in Washington county. The Smith proved to be the creamiest farm in the field, returning Willets, Young and Craig six-hundred-thousand dollars. Calvin Smith was a hired man in 1876, working by the month on the farm he bought in 1883, paying a small amount and arranging to string out the balance in fifteen annual instalments. His one-eighth royalty fattened his bank-account in eighteen months to six figures, an achievement creditable to the scion of the multudinous Smith-family.

From the sinking of the Dyer well drilling went on recklessly. Everybody felt confident of a great future for Washington territory. Isaac Willets, brother of an owner of the Smith tract, paid sixty-thousand dollars for the adjoining farm—the Munce—and spent two-hundred-thousand in wells that cleared him a plump half-million. John McKeown the same day bought the farm of the Munce heirs, directly north of their uncle's, and drilled wells that yielded him five-thousand dollars a day. He removed to Washington and died there. His widow erected a sixty-thousand-dollar monument over his grave, something that would never have happened if John, plain, hard-headed and unpretentious, could have expressed his sentiments. Thayer No. 2, on the Clark farm, adjoin-ing the Gordon, startled the fraternity in May of 1886 by flowing two-thousand barrels a day from the Gordon sand. It was the biggest spouter in the heap. Lightning struck the tank and burned the gusher, the blazing oil shooting flames a hundred feet towards the blue canopy. At night the brilliant light illumined the country for miles, travelers pronouncing it equal to Mt. Vesuvius in active eruption. The burning oil ran to Gordon No. 1, on lower ground, setting it off also. In a week the Thayer blaze was doused and the stream of crude turned into the tanks of No. 1. Next night a tool-dresser, carrying a lantern on his way to "midnight tower," set fire to the gas which hung around the tanks. The flames once more shot above the tree-tops, the tool-dresser saved his life only by rolling into the creek, but the derrick was saved and no damage resulted to the well.

Captain J. J. Vandergrift leased the Barre farm, south of the Smith, and drilled a series of gushers that added materially to his great wealth. Disposing of the Barre, he developed the Taylorstown pool and reaped a fortune. T. J. Vandergrift leased the McManis farm, six miles south of Washington, and located the first Taylorstown well. Taylorstown is still on duty and W. J. Young manages the company that acquired the Vandergrift interests. South of the Barre farm James Stewart, vendor of a cure-all salve, owned a shanty and three acres of land worth four-hundred dollars. He leased to Joseph M. Craig for one-fourth royalty. The one well drilled on the lot spouted two-thousand barrels a day for weeks. It is now pumping fairly. This was salve for Stewart and liniment for Craig, whose Washington winnings exceed a half-million. "Mammy" Miller, an aged colored woman, lived on a small lot next to Stewart and leased it at one-fourth royalty to a couple of local merchants. They drilled a thousand-barrel well and "Mammy" became the most courted negress in Pennsylvania. The Union Oil-Company took four-hundred-thousand dollars from the Davis farm. Patrick Galligan, the contractor of the Smith well, leased the Taylor farm and grew rich. Pew & Emerson, who have made millions by natural-gas operations, leased the Manifold farm, west of the Smith. The first well paid them twenty-thousand dollars a month and subsequent strikes manifolded this a number of times. Pew & Emerson have risen by their energy and shrewdness and can occupy a front pew in the congregation of petroleumites.

Samuel Fergus, once county-treasurer and a man of broad mould, struck a geyser in the Fergus annex to the main pool. He drilled on his twenty-four acres solely to accommodate Robert Greene, pumper for Davis Brothers. Greene had much faith and no money, but he advised Fergus to exercise the tools at a particular spot. Fergus might have kept the whole hog and not merely a pork-chop. He sold three-eighths and carried one-eighth for Greene, who refused twenty-thousand dollars for it the day the well began flowing two-thousand barrels. "Bob" Greene, like Artemas Ward's kangaroo, was "a amoosin' cuss!" Called to Bradford soon after the gusher was struck, he met an old acquaintance at the station. His friend invited Bob into the smoker to enjoy a good cigar. He declined and in language more expressive than elegant said: "I've been a ridin' in smokers all my life. Now I'm goin' to turn a new leaf. I'm goin' to take a gentleman's car to Pittsburg and from there to Bradford I'm goin' to have a Pullman, if it takes a hull day's production." Bob took his first ride in a Pullman accordingly. The first venture induced Fergus to punch his patch full of holes and do a turn at wildcatting. His stalwart luck fired the hearts of many young farmers to imitate him, in some instances successfully. Washington has not yet gone out of the oil-business. The Cecil pool kept the trade guessing this year, but its gushers lacked endurance and the field no longer terrorizes the weakest lambkin in the speculative fold.

Greene county experienced its first baptism of petroleum in 1861-2-3, when many wells were drilled on Dunkard Creek. The general result was unsatisfactory. The idea of boring two-thousand feet for oil had not been conceived and the shallow holes did not reach the principal strata. Of fourth sand, fifth sand, Gordon rock, fifty-foot rock, Trenton rock, Berea grit, corn-meal rock, Big-Injun sand and others of the deep-down brand operators on Dunkard Creek never dreamed. Some oil was detected and more blocks of land were tied up in 1864-5. The credulous natives actually believed their county would soon be shedding oil from every hill and hollow, garden and pasture-field. The holders of the tracts—lessees for speculation only—drilled a trifle, sold interests to any suckers wanting to bite and the promised developments fizzled. E. M. Hukill, who started in 1868 at Rouseville, leased twenty-thousand acres in 1885 and located a well on D. L. Donley's farm, one-third mile south-east of the modest hamlet of Mt. Morris. Morris Run empties into Dunkard Creek near the village. The tools were swung on March second, 1886. Fishing-jobs, hard rock and varied hindrances impeded the work. On October twenty-first oil spouted, two flows occurred next day and a tank was constructed. Salt-water bothered it and the well—twenty-two-hundred feet—was not worth the pains taken for months to work it as a mystery. Hukill drilled a couple of dusters and the Gregg well at Willowtree was also a dry-hole at twenty-three hundred feet. Craig & Cappeau and James M. Guffey & Co. swept over the south-western section in an expensive search for crude. From the northern limit of McKean to the southern border of Greene county Pennsylvania had been ransacked. The Keystone players—Venango, Warren, Forest, Elk, McKean, Clarion, Armstrong, Butler, Allegheny, Beaver and Washington—put up a stiff game and the region across the Ohio was to have its innings.

C. H. Shattuck had the first well in West Virginia drilled for oil. He came from Michigan in the fall of 1859, secured land in Wirt county and bored one-hundred feet by the tedious spring-pole process. The well was on the bank of the Hughes river, from which the natives skimmed off a greasy fluid to use for rheumatism and bruises. It was dry and Shattuck settled at Parkersburg, his present abode. At Burning Springs a "disagreeable fluid" flooded a salt-well, which the owner quit in disgust. General Samuel Karns, of Pennsylvania, and his nephew, S. D. Karns, rigged it up in 1860 and pumped considerable oil. The shallow territory was operated extensively. Ford & Hanlon bored on Oil-Spring Run, Ritchie county, in 1861-2, finding heavy oil in paying quantities. W. H. Moore started the phenomenal eruption at Volcano in 1863, by drilling the first well, which produced eight-thousand barrels of lubricating oil. Sheafer & Steen's, the second well, was a good second and the Cornfield pumped seven-thousand barrels of thirty-five-gravity oil in six months. William C. Stiles and the Oil-Run Petroleum Company punched scores of wells. Volcano perched on the lubricating pedestal for years, but it is now extinct. E. L. Gale—he built the railroad freight-houses at Aspinwall and Panama and owned the site of Joliet and half the land on which Milwaukee thrives—in 1854 purchased two-thousand acres of bush twenty-five miles from Parkersburg. In 1866 the celebrated Shaw well, the first of any note on his tract, flowed one-hundred barrels of twenty-six-degree oil. Gale sent samples to the Paris Exposition in 1867 and received the only gold-medal awarded for natural oils. The Shaw well kicked up a fuss, leases brought large bonuses, excitement ran high and the "Gale Oil Field" was king of the hour. Land-grabbers annoyed Gale, who declined a million dollars for his property. He routed the herd and died at an advanced age, leaving his heirs ample means to weather the severest financial gale. The war had driven northern operators from the field and heavy-oil developments cleared the coast for the next act on the program.

Charles B. Traverneir, in the spring of 1883, on Rock Run, put down the first deep well in West Virginia. It encountered a strong flow of oil at twenty-one-hundred feet and yielded for eleven years. Volcano and Parkersburg had retired and light-oil territory was the object of the ambitious wildcatter. At Eureka, situated in a plain contiguous to the Ohio river, Brown & Rose struck the third sand in April, 1886, at thirteen-hundred feet. The well flowed seven-hundred barrels of forty-four-gravity oil, similar to the Macksburg variety and equal to the Pennsylvania article for refining. The derrick burned, with the tools at the bottom of the well, and the yield decreased to three-hundred barrels in May. Oilmen pronounced Eureka the coming oil-town and farmers asked ridiculous prices for their lands. Bradford parties leased numerous tracts and bounced the drill merrily. The third sand in West Virginia was found in what are known as "oil breaks," at irregular depths and sometimes cropping out upon the surface. Eureka is still a center of activity. The surrounding country resembles the Washington district in appearance and fertility of the soil. In 1891 Thomas Mills, who operated at Tionesta in 1862 and at Macksburg in 1883-4, leased a bundle of lands near Sistersville and sank a well sixteen-hundred feet. A glut of salt-water induced him to sell out cheap. The first important results were obtained on the Ohio side of the Ohio river, where many wells were bored. The Polecat well, drilled in 1890, daily pumped fifty barrels of oil and two-thousand of salt-water, bringing Sistersville forward a peg. Eight wells produced a thousand barrels of green oil per day in May of 1892. Operating was costly and only wealthy individuals or companies could afford to take the risks of opening such a field. Captain J. T. Jones, J. M. Guffey, Murphy & Jennings, the Carter Oil-Company, the Devonian Oil-Company, the Forest Oil-Company and the South-Penn have reduced the business to an exact science and secured a large production. Sistersville, named from the two Welles sisters, who once owned the site of the town, has been a magnet to petroleumites for two years. Gushers worthy of Butler or Allegheny have been let loose in Tyler, Wood, Monroe and Doddridge counties. The Big Moses, on Indian Creek, is a first-class gasser. Morgantown, Mannington and Sistersville are as familiar names as McDonald, Millerstown or Parker. Pipelines handle the product and old-timers from Bradford, Warren and Petrolia are seen at every turn. West-Virginia is on top for the moment, with the tendency southward and operators eagerly seeking more petroleum-worlds to conquer in Kentucky and Tennessee.

She was a radiant Sistersville girl. She descended the stairs quietly and laid her hand on the knob of the door, hoping to steal out stealthily in the gray dawn. Her father stood in the porch and she was discovered. "My daughter," said the white-haired old gentleman, "what is that—what are those you have on?" She hung her head and turned the door-knob uneasily back and forth between her fingers, but did not answer. "Did you not promise me," the old man went on, "that if I bought you a bicycle you would not wear—that is, you would ride in skirts?" She stepped impulsively toward him and paused. "Yes, father," she said, "I did and I meant it. But I didn't know these then. The more I saw of them the better I liked them. They improve on acquaintance, father. They grow on one ——" "My daughter," he interrupted, "Eve's garments grew on her!" And so it has been with the West-Virginia oil-field—it grows on one and the more he sees of it the better he likes it.

Butler, the county-seat of Butler county, was laid out in 1802 by the Cunninghams, two brothers from Lancaster, who repose in the old cemetery. The surveyor was David Dougall, who lived seventy-five years alone, in a shanty near the court-house, dying at ninety-eight. He owned a row of tumble-down frames on the public-square, eye-sores to the community, but would not sell lest his poor tenants might suffer by a change of proprietors! His memory of local events was marvelous. He walked from Detroit through the forest to Butler, following an Indian trail, and remembered when Pittsburg had only three brick-buildings. He was agent of the McCandless family and once consented to spend a night at the mansion of his friends in Pittsburg. To do honor to the occasion he wore trousers made of striped bed-ticking. Fearing fire, he would not sleep up-stairs and a bed was provided in the parlor. About midnight an alarm sounded. Dougall jumped up, grabbed his shoes and hat and walked home—thirty-three miles—before breakfast. He was an eccentric bachelor and had his coffin ready for years. It was constructed of oak, grown on one of his farms, which he willed to a friend upon condition that the legatee buried him at the foot of a particular tree and kept a night-watchman at his grave one year. He was the last of his race and the last survivor of the bold pioneers to whom Butler owed its settlement.

The Cecil pool, in Washington county, furnishes its oil from the fifty-foot sand. One well, finished last April, on a village lot, flowed thirty-three hundred barrels in twenty-four hours. The biggest strike at Legionville, Beaver county, was Haymaker's seven-hundred barreler. The Shoustown or Shannopin field, also in Beaver, sixteen miles south-east of Pittsburg, is owned principally by James Amm & Co. Tyler county, the heart of the West-Virginia region, was a backwoods district, two generations behind the age and traveling at an ice-wagon gait, until it caught "the glow of the light to come." Its beginning was small, but men who sneer at small things merely show that they have sat on a tack and been worsted in the fray. It has taken grit and perseverance to bring a hundred-thousand barrels of oil a day from the bowels of the earth in Pennsylvania, Ohio, West-Virginia and Indiana. The man who has not a liberal stock of these qualities should steep himself in brine before engaging in oil-operations. He will only hit the nail on the thumb and be as badly fooled as the chump who deems he has a cinch on heaven because he never stole sheep. Petroleum is all right and a long way from its ninth inning. The alarmist who thinks it is playing out would have awakened Noah with the cry of "Fire!"

The southern trail, with its magnificent Butler output, its Allegheny geysers, its sixteen-thousand barrels a day in Washington and its wonderful strikes in West-Virginia, was big enough to fill the bill and lap over all the edges.

WHERE IGNORANCE IS BLISS.

The first building at Triangle bore in bold letters and bad spelling a sign labeled "Tryangle Hotel."

"A Black Justice of the Peace" ran the off-color legend, painted by an artist not up in punctuation, on the weather-beaten sign of 'Squire Black, at Shippenville.

An honest Dutchman near Turkey City declined to lease his farm at one-fourth royalty, insisting upon *one-eighth* as the very lowest he would accept. He did not discover that one-eighth was not twice one-fourth until he received his first instalment of oil, when he fired off the simple expletive, "Kreutz-millionendonnerwetter!"

A farmer rather shy on grammar, who represented Butler county in the Legislature at the outset of developments around Petrolia, "brought down the house" and a unanimous appropriation by his maiden speech : "Feller citizens, if we'uns up to Butler county wuz yu'uns down to Harrisburg we'uns would give yu'uns what we'uns is after !"

General Reed, of Erie, the largest vessel-owner on the lakes, represented his district in Congress and desired a second term. The Democrats nominated Judge Thompson and Clarion county was the pivot upon which the election turned. The contest waxed furious. Near its close the two candidates brought up at a big meeting in the wilds of Clarion to debate. Lumbermen and furnacemen were out in force. Reed led off and on the homestretch told the people how he loved them and their county. He had built the fastest craft on the lakes and named the vessel Clarion. As the craft sailed from Buffalo to Erie, and from Cleveland to Detroit, and from Saginaw to Mackinaw, to Oconomowoc and Manitowoc, Oshkosh, Milwaukee and Chicago, in every port she folded her white wings and told of the county that honored him with a seat in Congress. The people were untutored in nautical affairs and listened with rapt attention. As the General closed his speech the enthusiasm was unbounded. Things looked blue for Judge Thompson. After a few moments required to get the audience out of the seventh heaven of rapture, he stepped to the front of the platform, leaned over it, motioned to the crowd to come up close and said: "Citizens of Clarion, what General Reed has told you is true. He has built a brig and a grand one. But where do you suppose he painted the proud name of Clarion?" Turning to General Reed, he said, "Stand up here, sir, and tell these honest people where you had the painter put the name of Clarion. You never thought the truth would reach back here. I shall tell these people the truth and I challenge you to deny one word of it. Yes, fellow-citizens, he painted the proud name of Clarion under the stern of the brig—under her stern, gentlemen!" The indignation of the people found vent in groans and curses. General Reed sat stunned and speechless. No excuses would be accepted and the vote of proud Clarion made Judge Thompson a Congressman.

PARKER OIL EXCHANGE IN 1874.

Top Row— J. D. Essex, — Harris. Nelson Cochran. Unknown. L. W. Weston. Hugh McKelvy. J. M'Donald. Unknown.
Warren Gray. C. Seldon. Milo Marsden. Lemuel Young. Unknown. James Green. Dr. Thorn.
Middle Row— E. Seldon. Col. Sellers. W. A. Pullman. Chas. Archbold. Harry Parker. James McCutcheon. Unknown.
U. J. Greer, Fullerton Parker, Col. Brady, Charles Heath. H. W. Batchelor. — Gephardt. Shep. Moorhead.
Full. Parker, Jr. W. C. Henry, Sam. Morrow. John Barton. R. Moorhead.
Thos. McLaughlin. Joseph Seep. Harry Marlin. Jas. Garrett. Walter Fleming,
Lower Row— Capt. J. T. Chalfant. Weston Howland. Chas. Riddell. Kem Offley. H. Boen. Chas. W. Hall. Chas. J. Frazee.
Thos. Mc'onnell. James Lowe. Richard Conn. Richard Conn. Ken. Kerr.

XI.

FROM THE WELL TO THE LAMP.

TRANSPORTING CRUDE-OIL BY WAGONS AND BOATS—UNFATHOMABLE MUD AND
SWEARING TEAMSTERS—POND FRESHETS—ESTABLISHMENT OF PIPE-LINES—
NATIONAL-TRANSIT COMPANY AND SOME OF ITS OFFICERS—SPECULATION IN
CERTIFICATES—EXCHANGES AT PROMINENT POINTS—THE PRODUCT THAT
ILLUMINES THE WORLD AT VARIOUS STAGES OF PROGRESS.

"My kingdom for a horse to haul my oil."—*Richard III. Revised.*
"We'll all dip oil, and we'll all dip oil,
We'll dip, dip, dip, and we'll all dip oil."—*Pond-Freshet Song.*
" Lines of truth run through the world of thought as pipe-lines to the sea."—*Mrs. C. A. Babcock.*
" These be piping times."—*Popular Saw.*
"Seneca predicted another hemisphere, but Columbus presented it."—*Collins.*
" Nature begets Merit and Fortune brings it into play."—*La Rochefoucauld.*
" The wise and active conquer difficulties by daring to attempt them."—*Rowe.*
"Perfection is attained by slow degrees."—*Voltaire.*
" One little bull on oil was I,
Bought a lot when the stuff was high,
Sold when low and it pump'd me dry.
One little bull on oil."—*Oil-City Blizzard.*
"It is just as dangerous to speculate in kerosene as to kindle the fire with it."—*Boston Herald.*

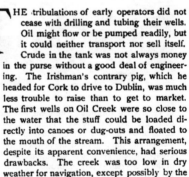

J. N. WHEELER.

THE tribulations of early operators did not cease with drilling and tubing their wells. Oil might flow or be pumped readily, but it could neither transport nor sell itself. Crude in the tank was not always money in the purse without a good deal of engineering. The Irishman's contrary pig, which headed for Cork to drive to Dublin, was much less trouble to raise than to get to market. The first wells on Oil Creek were so close to the water that the stuff could be loaded directly into canoes or dug-outs and floated to the mouth of the stream. This arrangement, despite its apparent convenience, had serious drawbacks. The creek was too low in dry weather for navigation, except possibly by the Mississippi craft that slipped along easily on the morning dew. To overcome this difficulty recourse was had to artificial methods when the production increased sufficiently to introduce flat-boats, which dispensed with barrels and freighted the oil in bulk. The system of pond-freshets was adopted. A dam at the saw-mill near the Drake well stored the fluid until the time agreed upon to

259

open the gates and let the imprisoned waters escape. Rev. A. L. Dubbs was appointed superintendent and shippers were assessed for the use of the water stored in the pond. Usually two-hundred to eight-hundred boats—boats of all shapes and sizes, from square-keeled barges, divided into compartments by cross-partitions, to slim-pointed guipers—were pulled up the stream by horses once or twice a week to be filled at the wells and await the rushing waters. Expert rivermen, accustomed to dodging snags and rocks in inland streams, managed the fleet. These skilled pilots assumed the responsibility of delivering the oil to the larger boats at Oil City, for conveyance to Pittsburg, at one-hundred to two-hundred dollars per trip.

At the appointed moment the flood-gates were opened and the water rushed forth, increasing the depth of the creek two or three feet. The boatmen stood by their lines, to cast loose when the current was precisely right. Sound judgment was required. The loaded boat, if let go too soon, ran the risk of grounding in the first shallow-place, to be battered into kindling-wood by those coming after. Such accidents occurred frequently, resulting in a general jam and loss

POND FRESHET AT OIL CITY, MOUTH OF OIL CREEK.

of vessels and cargoes. The scene was more exciting than a three-ringed circus. Property and life were imperiled, boats were ground to fragments, thousands of barrels of oil were spilled and the tangle seemed inextricable. Men, women and children lined the banks of the stream for miles, intently watching the spectacle. Persons of all nationalities, kindreds and conditions vociferated in their diversified jargon, producing a confusion of tongues that outbabeled Babel three to one. Men of wealth and refinement, bespattered and besmeared with crude—their trousers tucked into boots reaching above the knee, and most likely wearing at the same time a nobby necktie—might be seen boarding the boats with the agility of a cat and the courage of warriors, shouting, managing, directing and leading in the perilous work of safe exit. Sunday creeds were forgotten and the third commandment, constantly snapped in twain, gave emphasis to the crashing hulks and barrels. A pillar of the Presbyterian church, seeing his barge unmanned, ran screaming at the top of his voice : "Where in sheol is Parker?" This so amused his good brethren that they used it as a by-word for months.

The cry of "Pond Freshet" would bring the entire population of Oil City to witness the arrival of the boats. Sometimes the tidal wave would force them on a sand-bar in the Allegheny, smashing and crushing them like egg-shells. Oil from overturned or demolished boats belonged to whoever chose to dip it up. More than one solid citizen got his start on fortune's road by dipping oil in this way. If the voyage ended safely the oil was transferred from the guipers—fifty barrels each—and small boats to larger ones for shipment to Pittsburg. William Phillips, joint-owner of the biggest well on Oil Creek, was the first man to take a cargo of crude in bulk to the Smoky City. The pond-freshet was a great institution in its day, with romantic features that would enrapture an artist and tickle lovers of sensation to the fifth rib. One night the lantern of a careless workman set fire to the oil in one of the boats. Others caught and were cut loose to drift down the river, floating up against a pier and burning the bridge at Franklin. Running the "rapids" on the St. Lawrence river or the "Long Sault" on the Ottawa was not half so thrilling and hair-raising as a fleet of oil-boats in a crush at the mouth of Oil Creek.

The fleet of creek and river-boats engaged in this novel traffic numbered two-thousand craft. The "guiper," scow-shaped and holding twenty-five to fifty barrels, was the smallest. The "French Creekers" held ten to twelve-hundred barrels and were arranged to carry oil in bulk or barrels. At first the crude was run into open boats, which a slight motion of the water would sometimes capsize and spill the cargo into the stream. When prices ruled low oil was shipped in bulk ; when high, shippers used barrels to lessen the danger of loss. Thousands of empty barrels, lashed together like logs in a raft, were floated from Olean. The rate from the more distant wells to Oil City was one-dollar a barrel. From Oil City to Pittsburg it varied from twenty-five cents to three dollars, according to the weather, the stage of water or the activity of the demand. Each pond-freshet cost two or three-hundred dollars, paid to the mill-owners for storing the water and the use of their dams. Twice a week—Wednesday and Saturday—was the average at the busy season. The flood of petroleum from flowing-wells in 1862 exceeded the facilities for storing, transporting, refining and burning the oil, which dropped to ten cents a barrel during the summer. Thousands of barrels ran into Oil Creek. Pittsburg was the chief market for crude, which was transferred at Oil City to the larger boats. The steamer-fleet of tow-boats—it exceeded twenty—brought the empties back to Oil City. The "Echo," Captain Ezekiel Gordon; the "Allegheny Belle No. 4," Captain John Hanna; the "Leclaire," Captain Kelly; the "Ida Rees," Captain Rees, and the "Venango" were favorite passenger-steamers. The trip from Pittsburg—one-hundred-and-thirty-three miles—generally required thirty to thirty-six hours. Mattresses on the cabin-floor served as beds for thirty or forty male passengers, who did not undress and rose early that the tables might be set for breakfast. The same tables were utilized between meals and in the evening for poker-games. The busiest man on the boat was the bar-tender and the clerk was the most important. He carried letters and money for leading oil-shippers. It was not uncommon for Alfred Russell, of the "Echo," John Thompson, of the "Belle No. 4," and Ruse Russ, of the "Venango," to walk into Hanna's or Abrams's warehouse-office with large packages of money for John J. Fisher, William Lecky, John Mawhinney, William Thompson and others who bought oil. No receipts were given or taken and, notwithstanding the apparent looseness in doing business, no package was ever lost or stolen. The boats usually landed at the lower part of the eddy to put off passengers wishing to stop at the Moran and Parker Hotels. At Hanna & Co.'s and Abrams & Co.'s landing, where the northern approach of the suspension bridge now is, they put off the remaining passengers, freight and empty oil-barrels. Many a Christian-looking man was heard to swear as he left the gang-plank of the boat and struck the mud, tough and greasy and deep. He would soon tumble to the situation, roll up his trousers and "pull for the shore."

Horses and mules dragged the empty boats up Oil Creek, a terrible task in cold weather. Slush or ice and floating oil shaved the hair off the poor animals as if done with a razor. The treatment of the patient creatures—thousands were literally murdered—was frightful and few survived. For them the plea of inability availed nothing. They were worked until they dropped dead. The finest mule, ears very long, coat shiny, tail vehement, eye mischievous, heels vigorous and bray distinct and melodious, quickly succumbed to the freezing water and harsh usage. As a single trip realized more than would buy another the brutal driver scarcely felt the financial loss. A story is told of a boat-man who started in the morning for the wells to bring down a load of oil. Returning in the evening, he learned that he had been drafted into the army. Before retiring to bed he had hired a substitute for one-thousand dollars, the proceeds of his journey of eleven miles and back. William Haldeman hauled a man over the coals for beating his exhausted horse, told him to buy another and handed him five-hundred dollars for eight horses to haul a boat to the gushers at Funkville.

Pond-freshets were holidays in Oil City sufficiently memorable to go glid-ing down the ages with the biggest kind of chalk-mark. Young and old flocked to see the boats slip into the Allegheny, lodge on the gravel-bar, strike the pier of the bridge or anchor in Moran's Eddy. Hundreds of boatmen, drillers, pumpers and operators would be on board. Once the river had only a foot of water at Scrubgrass Ripple and large boats could not get to or from Pittsburg. A ship-carpenter came from New York to Titusville and spent his last dollar in lumber for six boxes sixteen feet square and twelve inches deep. He covered them with inch-boards and divided them into small compartments, to prevent the oil from running from one end to the other and swamping the vessel. This principle was applied to oil-boats thereafter and extended to bulk-barges and bulk-steamships. The ingenious carpenter floated his strange arks down to the Blood farm and bargained with Henry Balliott to fill them on credit. He performed the voyage safely, returned in due course, paid Balliott, built more boxes and went home in four months with a snug fortune. His ship had come in. Railroads and pipe-lines have relegated pond-freshets, oil-boats and Alle-gheny steamers to the rear, but they were interesting features of the petroleum-development in early days and should not be utterly forgotten.

To haul oil from inland wells to shipping-points required thousands of horses. This service originated the wagon-train of the oil-country, which at its best consisted of six-thousand two-horse teams and wagons. No such trans-port-service was ever before seen outside of an army on a march. General M. H. Avery, a renowned cavalry-commander during the war, organized a regular army-train at Pithole. Travelers in the oil-regions seldom lost sight of these endless trains of wagons bearing their greasy freight to the nearest rail-road or shipping-point. Five to seven barrels—a barrel of oil weighed three-hundred-and-sixty pounds—taxed the strength of the stoutest teams. The mud was practically bottomless. Horses sank to their breasts and wagons far above their axles. Oil dripping from innumerable barrels mixed with the dirt to keep the mass a perpetual paste, which destroyed the capillary glands and the hair of the animals. Many horses and mules had not a hair below the eyes. A long caravan of these hairless beasts gave a spectral aspect to the landscape. History records none other such roads. Houses within a quarter-mile of the roadside were plastered with mud to the eaves. Many a horse fell into the batter and was left to smother. If a wagon broke the load was dumped into the mud-canal, or set on the bank to be taken by whoever thought it worth the labor of stealing. Teamsters would pull down fences and drive through fields whenever possible, until the valley of Oil Creek was an unfathomable quagmire. Think of the bone and sinew expended in moving a thousand barrels of oil six or eight miles under such conditions. Two-thirds of the work had to be done in the fall and winter, when the elements spared no effort to increase the discomfort and difficulty of navigation by boat or wagon. To haul oil a half-dozen miles cost three to five dollars a barrel at certain periods of the year. Thousands of barrels were drawn to Shaw's Landing, near Meadville, and thousands to Garland Station and Union City, on the Philadelphia & Erie Railroad. The hauling of a few hundred barrels not infrequently consumed so much time that the shipper, in the rapid fluctuations of the market, would not realize enough to pay the wagon-freight. A buyer once paid ten-thousand dollars for one-thousand barrels at Clapp farm, above Oil City, and four-thousand for teaming it to Franklin, to be shipped by the Atlantic & Great Western Railroad to New York. Even after a plank-road had been built from Titusville to Pithole, cutting down the teaming one-half or more, the cost of laying down a barrel of crude in New York was excessive. In January of 1866 it figured as follows :

Government tax	$1 00
Barrel	3 25
Teaming from Pithole to Titusville	1 25
Freight from Titusville to New York	3 65
Cooperage and platform expenses	1 00
Leakage	25
Total	$10 40

The Oil Creek teamster, rubber-booted to the waist and flannel-shirted to the chin, was a picturesque character. He was skilled in profanity and the savage use of the whip. A week's earnings—ten, twenty and thirty dollars a day—he would spend in revelry on Saturday night. Careless of the present and heedless of the future, he took life as it came and wasted no time worrying over consequences. If one horse died he bought another. He regulated his charges by the depth and consistency of the mud and the wear and tear of morality and live-stock. Eventually he followed the flat-boat and barge and guiper to oblivion, railroads and pipe-lines supplanting him as a carrier of oil. Some of the best operators in the region adopted teaming temporarily, to get a start. They saved their money for interests in leases or drilling-wells and not a few went to the front as successful producers. The free-and-easy, devil-may-care teamster of yore, brimful of oil and tobacco and not averse to whiskey, is a tradition, remembered only by men whose polls are frosting with silver threads that do not stop at sixteen to one.

Wharves, warehouses and landings crowded Oil City from the mouth of Oil Creek to the Moran House. Barrels filled the warehouse-yards, awaiting

18

their turn to be hauled or boated to the wells, filled with crude and returned for shipment. Loaded and empty boats were coming and going continually. Firms and individuals shipped thousands of barrels daily, employing a regiment of men and stacks of cash. William M. Lecky, still a respected citizen of Oil City, hustled for R D. Cochran & Co., whose "Tiber" was a favorite tow-boat. Parker & Thompson, Fisher Brothers, Mawhinney Brothers and John Munhall & Co. were strong concerns. Their agents scoured the producing farms to buy oil at the wells and arrange for its delivery. Prices fluctuated enormously: Crude bought in September of 1862 at thirty cents a barrel sold in December at eleven dollars. John B. Smithman, Munhall's buyer, walked up the creek one morning to buy what he could at three dollars. A dispatch at Rouseville told him to pay four, if necessary to secure what the firm desired. At Tarr Farm another message quoted five dollars. By the time he reached Petroleum Centre the price had reached six dollars and his last purchases that afternoon were at seven-fifty. Business was done on honor and every agreement was fulfilled to the letter, whether the price rose or fell. Lecky, Thomas B. Simpson, W. J. Young and Isaac M. Sowers—he was the second mayor of Oil City—clerked in these shipping-offices, which proved admirable training-schools for ambitious youths. William Porterfield and T. Preston Miller tramped over Oil Creek and Cherry Run for the Fishers. Col. A. J. Greenfield, Bradley & Whiting and I. S. Gibson bought at Rouseville and R. Richardson at Tarr Farm. "Pres" Miller, "Hi" Whiting and "Ike" Gibson—square, manly and honorable—are treading the golden-streets. John Mawhinney—big in soul and body, true to the core and upright in every fiber—has voyaged to the haven of rest. William Parker is president of the Oil City Savings Bank and Thompson returned east years ago. John Munhall settled near Philadelphia and William Haldeman removed to Cleveland. The iron-horse and the pipe-line revolutionized the methods of handling crude and retired the shippers, most of whom have shipped across the sea of time into the ocean of eternity.

Fisher Brothers have a long and enviable record as shippers and producers of oil, "staying the distance" and keeping the pole in the hottest race. Men have come and men have retreated in the mad whirl of speculation and wild rush for the bottom of the sand, but they have gone on steadily for a generation and are to-day abreast of the situation. Whether a district etched its name on the Rainbow of Fame or mocked the dreams of the oil-seeker, they did not lose their heads or their credit. John J. Fisher went to Oil City in 1862 and Fisher Brothers began shipping oil by the river to Pittsburg in 1863, succeeding John Burgess & Co. The three brothers divided their forces, to give each department personal supervision, John J. managing the buying and shipping at Oil City and Frederick and Henry receiving and disposing of the cargoes at Pittsburg. Competent men bought crude at the wells and handled it in the yards and on the boats. The firm owned a fleet of bulk-boats and tow-boats and acres of barrels. Each barrel was branded with a huge F on either head. The "Big F"—widely known as Oil Creek or the Drake well—was the trade-mark of fair play and spot cash. When railroads were built the Fishers discarded boats and used more barrels than before. When wooden-tanks—a car held two—were introduced they adopted them and let the barrels slide. When pipe-lines were laid they purchased certificate-oil and continued to be large shippers until seaboard lines suspended the older systems of freighting crude by water or rail, in barrels or in tanks. From the beginning to the end of the shipping-trade Fisher Brothers were in the van.

Next devoting their attention entirely to the production of oil and gas, with the Grandins and Adnah Neyhart they invested heavily at Fagundas and laid the first pipe-line at Tidioute. They operated below Franklin and were pioneers at Petrolia. Organizing the Fisher Oil-Company, they drilled in all the Butler pools and held large interests at McDonald and Washington. At present they are operating in Pennsylvania, Ohio and West Virginia, the Fisher ranking with the foremost companies in extent and solidity. The brothers have their headquarters in the Germania Building, Pittsburg, and juicy wells in a dozen counties. Time has dealt kindly with all three, as well as with Daniel Fisher, ex-mayor of Oil City. They have loads of experience and capital and too much energy to think of adjusting their halo for retirement from active work. True men in all the relations of life, Fisher Brothers worthily represent

the splendid industry they have had no mean part in making the greatest and grandest of any age or nation. To natural shrewdness and the quick perception that comes from contact with the activities of the world they joined business-ability that would have proved successful in whatever career they undertook to map out.

The first suggestion of improvement in transportation was made in 1860, at Parkersburg, W. Va., by General Karns to C. L. Wheeler, now of Bradford. An old salt-well Karns had resurrected at Burning Springs pumped oil freely and he conceived the plan of a six-inch line of pipe to Parkersburg to run the product by gravity. The war interfered and the project was not carried out. At a meeting at Tarr Farm, in November of 1861, Heman Janes broached the idea of laying a line of four-inch *wooden-pipes* to Oil City, to obviate the risk, expense and uncertainty of transporting oil by boats or wagons. He proposed to bury the pipe in a trench along the bank of the creek and let the oil gravitate to its destination. A contract for the entire work was drawn with James Reed, of Erie. Col. Clark, of Clark & Sumner, grasped the vast possibilities the method might involve and advised applying to the Legislature for a general pipe-line charter. Reed's contract was not signed and a bill was introduced in 1862 to authorize the construction of a pipe-line from Oil Creek to Kittanning. The opposition of four-thousand teamsters engaged in hauling oil defeated the bill and the first effort to organize a pipe-line company.

J. L. Hutchings, a Jersey genius, came to the oil-country in the spring of 1862 with a rotary-pump he had patented. To show its adaptation to the oil-business he laid a string of tubing from Tarr Farm to the Humboldt Refinery, below Plumer. He set his pump working and sent a stream of crude over the hills to the refinery. The pipe was of poor quality, the joints leaked and a good deal of oil fell by the wayside, yet the experiment showed that the idea was feasible. Although eminent engineers declared friction would be fatal, the result proved that distance and grade were not insurmountable. Eminent engineers had declared the locomotive would not run on smooth rails and that a cow on the track would disrupt George Stephenson's whole system of travel, hence their dictum regarding pipe-lines had little weight. Dr. Dionysius Lardner nearly burst a flue laughing at the absurdity of a vessel without sails crossing the ocean and wrote a treatise to demonstrate its impossibility, but the saucy Sirius steamed over the herring-pond all the same. The rotary-pump at Tarr Farm confounded the scientists who worshipped theory and believed friction would knock out steam and pipe and American ingenuity and keep oil-operators forever subject to mud and pond-freshets. The two-inch line to the Humboldt Refinery planted the seed that was to become a great tree. Nobody saw this more plainly than the teamsters, who proceeded to tear up the pipe and warn producers to quit monkeying with new-fangled methods of transportation. That settled the first pipe-line and left the rampant teamsters, modern imitators of "Demetrius the silversmith," the upper dog in the fight.

Hutchings—the boys called him "Hutch"—had pump and pipe-line on the brain and would not be suppressed. He put down a line in 1863 from the big Sherman well to the terminus of the railroad at Miller Farm. The pipes were cast-iron, connected by lead-sockets and laid in a shallow ditch. The jarring of the pump loosened the joints and three-fourths of the oil started at the well failed to reach the tanks, two miles north. The teamsters were not in business solely for their health and they tore up the line to be sure it would not cut off any of their revenue. Hutchings persisted in his endeavors until debts overwhelmed him and he died penniless and disappointed. The ill-starred inventor, who lived a trifle ahead of the times, deserves a bronze statue on a shaft of imperishable granite.

The Legislature granted a pipe-line charter in 1864 to the Western Transportation Company, which laid a line from the Noble & Delemater well to Shaffer. The cast-iron pipe, five inches in diameter, was laid on a regular grade in the mode of a water-pipe. The lead points leaked like a fifty-cent umbrella, just as the Hutchings line had done, and the attempt to improve transportation was abandoned.

Samuel Van Syckle, a Jerseyite of inventive bent, arrived at Titusville in the fall of 1864. The problem of oil-transportation, rendered especially important by the opening of the Pithole field, soon engrossed his attention. In August of 1865 he completed a two-inch line from Pithole to Miller Farm. Mr. Wood and Henry Ohlen, of New York, held an interest and the First National

Bank of Titusville loaned the money to forward the project. J. N. Wheeler screwed the first joints together. Two pump-stations, a mile west of Pithole and at Cherry Run, at first helped force the oil through the pipe, which was buried two feet under ground "to be out of the way of the farmer's plow." Eight-hundred barrels a day could be run and the frantic teamsters talked of resorting to violence to cripple so formidable a rival. The pipeage was one dollar a barrel, at which rate the Pithole and Miller Farm Pipe-Line ought to have been a bonanza. Van Syckle traded heavily in oil and commanded plenty of capital. A. W. Smiley managed the line and bought oil for Van Syckle, who conducted this branch of business in his son's name. Smiley's largest transaction was a purchase of one-hundred-thousand barrels, at five dollars a barrel, from the United-States Petroleum Company, in one lot. Young Van Syckle spent money as the whim struck him. If Smiley refused his demand for a hundred or a thousand dollars, the fly youth would refuse to sign drafts and threaten to stop the whole concern. There was nothing to do in such cases but imitate Colonel Scott's coon and "come down." The Culver failure in May of 1866 compelled the First National Bank to press its claim against the line, which passed into the hands of Jonathan Watson. J. T. Briggs and George S. Stewart operated it for the bank and Watson until William H. Abbott and Henry Harley purchased the entire equipment.

Reverses beset Van Syckle, who induced George S. and Milton Stewart to erect a big refinery at Titusville to test his pet theory of "continuous distillation." Failure, tedious litigation and heavy loss resulted. Van Syckle's mind teemed with new schemes and new devices for refining. He possessed the rare faculty of finding friends willing to listen to his plans and back him with cash. Some of his ideas were valuable and they are in use to-day. Mismanagement swamped the enterprises he created and Van Syckle finally removed to Buffalo, where his checkered life closed peacefully on March second, 1894. While often unsuccessful financially, earnest men like Samuel Van Syckle benefit mankind. The oil-business is much better for the fertile brain and perseverance of the man whose pipe-line was the first to deliver oil to a railroad. His example stimulated other men combining keen perception and executive ability, who could sift the wheat from the chaff and discard the useless and impracticable.

In the fall of 1865 Henry Harley began a pipe-line from Benninghoff Run to Shaffer, the terminus of the Oil-Creek Railroad. Teamsters cut the pipes, burned the tanks and retarded the work seriously. An armed patrol arrested twenty of the ring-leaders, dispersed the mob and quelled the riot. The line—two-inch tubing of extra weight—handled oil expeditiously, a pump at Benninghoff forcing six to eight-hundred barrels a day into the tanks at Shaffer. The system was a public improvement, personal interest had to yield and four-hundred teams left the region the week Harley's line pumped its first oil. Abbott and Harley owned an interest in the Pithole line and secured control by purchasing Jonathan Watson's claim, to run it in connection with the Benninghoff line. They organized the firm of Abbott & Harley and operated both lines several months. At Miller Farm they constructed iron-tanks and loading-racks, which enabled two men to load a train of oil-cars in a few hours. Avery & Hedden laid a line from Shamburg to Miller Farm, establishing a station on the highest point of the Tallman farm and running the oil to the railroad by gravity. Abbott & Harley supplemented this with a branch from the Pithole line at the crossing of Cherry Run. Crude was a good price, operators pros-

pered and Miller Farm became a busy place. Railroads extended to the
region and pipe-lines pumped oil directly from the wells to the cars or refineries.
In the fall of 1867 Abbott & Harley acquired control of the Western Transportation Company, the only one empowered by the Legislature to pipe oil to
railway-stations. Under its charter they combined the Western and their own
two lines as the Allegheny Transportation Company. The first board of directors, elected in January of 1869, consisted of Henry Harley, president; W. H.
Abbott, secretary; Jay Gould, J. P. Harley and Joshua Douglass. T. W. Lar-
sen was appointed treasurer and William Warmcastle—genial, capable "Billy"
Warmcastle—general superintendent. Jay Gould purchased a majority of the
stock in 1868 and appointed Mr. Harley general oil-agent of the Atlantic &

W. H ABBOTT.   PIPE-LINE AT MILLER FARM IN 1800   HENRY HARLEY.

Great Western and Erie Railroads. In 1871 the Commonwealth Oil and Pipe
Company was organized in the interest of the Oil-Creek Railroad. Harley
contrived to effect a combination and reorganize the Allegheny and the Commonwealth as the Pennsylvania Transportation Company, with a capital of
nearly two-million dollars and five-hundred miles of pipes to Tidioute, Triumph,
Irvineton, Oil City, Shamburg, Pleasantville and Titusville, centering at Miller
Farm. Among the stock-holders were Jay Gould, Thomas A. Scott, William
H. Kemble, Mrs. James Fisk and George K. Anderson. The new enterprise
absorbed a swarm of small lines and was considered the acme of pipe-line
achievement.

William Hawkins Abbott was a Connecticut boy, an Ohio merchant at
twenty-five and a visitor to the Drake well in February of 1860. He remained

two days, paid ten-thousand dollars for three one-eighth interests in farms below the town and two days after William Barnsdall struck a fifty-barrel well on
one of the properties. He located at Titusville, established a market for crude
in New York, shipped extensively and in the fall of 1860, with James Parker
and William Barnsdall as partners, began the erection of the first complete
refinery in the oil-region. To convey the boilers and stills from Oil City,
whither they were shipped from Pittsburg by water, was a task greater than the
labors of Hercules. The first car-load of coal ever seen in Titusville Mr.
Abbott laid down in the fall of 1862. He opened a coal-yard and superintended
the refinery. Oil fluctuated at a rate calculated to make refiners bald-headed.
In January of 1861 Abbott paid ten dollars a barrel for crude and one-twenty-
five in March. In October of 1862 Howe & Nyce stored five-hundred barrels
of crude on the first railroad-platform at Titusville, selling it to Abbott at two-
sixty a barrel, packages included. In January of 1863 Abbott sold the oil from
the same platform for fourteen dollars and in March the same lot—it had never
been moved—brought eight dollars. Thirty days later Abbott bought it again
at three dollars a barrel and refined it. He was interested in the Noble well,
bought a large share in the Pithole and Miller Farm Pipe-Line and in 1866
formed a partnership with Henry Harley. He contributed largely to the Titusville and Pithole plank-road and all local enterprises likely to benefit the community. His generosity was comprehensive and discerning. He donated a
chapel to the Episcopal congregation, projected the Union & Titusville Railroad and was a most exemplary, public-spirited citizen. To give bountifully
was his delight. He bore financial disaster heroically and labored incessantly
to save others from loss. At seventy-two he is patient and helpful to those
about him, his daily life illustrating his real worth and illumining the pathway
of his declining years.

Born in Ohio in 1839 and graduated from the Rensselaer Polytechnic Institute as a civil-engineer in 1858, Henry Harley supervised the construction of the
Hoosac Tunnel until the war and settled at Pittsburg in 1862 as active partner
of Richardson, Harley & Co. The firm had a large petroleum commission-
house and Harley removed to Philadelphia in 1863 to manage its principal
branch. He purchased large tracts in West Virginia which did not meet his
expectations, withdrew from the commission-firm and in the latter part of 1865
built his first pipe-line. He was the confidential friend of Jay Gould and James
Fisk, whose support placed him in a position to organize the Pennsylvania
Transportation Company. For years Harley swam on the topmost wave and
was a high-roller of the loftiest stripe. Henry Villard was not more magnetic.
He told good stories, dealt out good cigars, knew champagne from seltzer and
had no trace of the miser in his intercourse with the world. He lived at Titusville in regal style and made "the grand tour of Europe" in 1872. He was on
intimate terms with railroad magnates, big politicians and Napoleons of finance.
The Pipe-Line Company got into deep waters, prosecutions and legal entangle-
ments crippled it and Henry Harley tumbled with the fabric his genius had
reared. He drifted to New York, was a familiar figure around Chautauqua
several seasons and died in 1892. His widow lives in New York and his
brother George, a popular member of the Oil-City Oil-Exchange, died last year.

In November of 1865 the Oil City & Pithole Railroad Company began a
railroad between the two towns, pushing the work with such energy that the
first train from Pithole to Oil City was run on March tenth, 1866. Vandergrift
& Forman equipped the Star Tank-Line to carry oil in tank-cars and laid the

Star Pipe-Line from West Pithole to Pithole to connect with the railroad. An unequivocal success from the start, this pipe-line has been regarded as the real beginning of the present system of oil-transportation. The lower oil-country enlarged the field for pipe-line stations. Lines multiplied in Venango, Clarion, Armstrong and Butler. Some of these were controlled by Vandergrift & Forman, who brought the business to a high standard of perfection. Each district had one or more lines running to the nearest railroad. The Pennsylvania Transportation Company secured a charter in 1875 to construct a line to the seaboard. Nothing was done except to build more lines in the oil-region. The number grew continually. Clarion had a half-dozen, the Antwerp heading the list. Parker had a brood of small-fry and Butler was net-worked. It was the fashion to talk of trunk-lines, call public meetings, subscribe for stock and—let the project die. Dr. Hostetter, the Pittsburg millionaire of "Bitters" fame, built the Conduit Line from Millerstown to the city of smoke and soot. The Karns, the Relief and others ran to Harrisville. Every fellow wanted a finger in the pipe-line pot-pie. A war of competition arose, rates were cut, business was done at heavy loss and the weaker concerns went to the wall. The companies issued certificates or receipts, instead of paying cash for crude received by their lines. When the producer ran oil into the storage-tanks of some companies he was not certain the certificates given him in return would have any value next day. He must either use the lines or leave the oil in the ground. The necessity of combining the badly-managed competitive companies into a solid organization was urgent. The Union Pipe-Line Company acquired a number of lines and operated its system in connection with the Empire Line. Under the act of 1874 Vandergrift & Forman organized the United Pipe-Lines, into which numerous local lines were merged. The first grand step had been taken in the direction of settling the question of oil-transportation for all time.

The advantages of the consolidation quickly commended the new order of things to the public. The United Lines erected hundreds of iron-tanks for storage and connected with every producing-well. Needless pipes and pumps and stations were removed to be utilized as required. The best appliances were adopted, improving the service and diminishing its cost. Uniform rates were established and every detail was systematized. Captain Vandergrift, president of the United Lines, was ably assisted in each department. Daniel O'Day, a potent force in pipe-line affairs, developed the system to an exact science. He learned the shipping-business from the very rudiments in the great Empire Line. His thorough knowledge, industry and practical talent were of incalculable value to the United Lines. He possessed in full measure the qualities adapted especially to the expansion and improvement of the giant enterprise. He had the skill to plan wisely and the ability to execute promptly. His sagacity and experience foresaw the magnificent future of the system and he laid the foundations of the United Lines broad and deep. To-day Daniel O'Day is a master-spirit of the pipe-line world, a millionaire and vice-president of the National Transit Company, which transports nine-tenths of the oil produced in the United States. He has risen by personal desert, without favoritism or partiality. His elevation has not subtracted one whit from the manly character that gained him innumerable friends in the oil-region.

Edward Hopkins, first manager of the United Pipe-Lines, was an efficient officer and died young. John R. Campbell has been treasurer from the incorporation of the lines in 1877. Born in Massachusetts and graduated from Rev. Samuel Aaron's celebrated school at Norristown, he served his apprenticeship

in the Baldwin Locomotive Works and manufactured printing-inks in Philadelphia, with William L. and Charles H. Lay as partners. In March of 1865
he visited the oil-region and in August removed to Oil City. He acquired oil-
interests, published the *Register* and was treasurer for the receiver of the Oil
City & Pithole Railroad Company. In 1867 he became book-keeper for Vandergrift & Lay, afterwards for Captain Vandergrift and later for Vandergrift &
Forman, who appointed him treasurer of their pipe-lines in 1868. He retained
the position in the United Lines and he is still treasurer of that division of the
National Transit Company. To Mr. Campbell is largely due the accurate and
comprehensive system of pipe-line accounts now universally adopted. He
aided in devising negotiable oil-certificates, reliable as government bonds and

convertible into cash at any moment. He enjoys to the fullest extent the confidence and esteem of his associates and is treasurer of a dozen large corporations. He was president term after term of the Ivy Club, one of the finest
social organizations in Pennsylvania, and a liberal promoter of important enterprises. His abiding faith in Oil City he manifests by investing in manufactures
and furthering public improvements. Active, helpful and popular in business,
in society and in the church, no eulogy could add to the high estimation in
which John R. Campbell is held wherever known.

The enormous production of the Bradford field, the increased distances and
the construction of lines to the sea presented new and difficult problems. A
natural increase in size led to a demand for pipe of better quality, for heavier
fittings and improved machinery. The largest line prior to Bradford's advent
was a four-inch pipe from the Butler field to Pittsburg, in 1875. Excepting this

and three-inch lines to Raymilton and Oil City, none of the main lines exceeded twelve miles in length. Many were gravity-lines and others used small tubing and light pumps. The greater quantities and longer distances in the northern district—the oil also congealed at a higher temperature and was harder to handle than the product of the lower fields—required greater power, larger pipes and increased facilities. The first six-inch line was laid from Tarport to Carrollton in the spring of 1879. Two four-inch lines had preceded it and a four-inch line from Tarport to Kane was completed the same season, five six-inch lines following later. The first long-distance line, a five-inch pipe from Hilliards—near Petrolia—to Cleveland, was completed in the summer of 1879. Trunk-lines to the eastern coast were begun in 1879-80. The trunk-line to Philadelphia starts at Colegrove, McKean county, and extends two-hundred-and-thirty-five miles—six-inch pipe—with a five-inch branch of sixty-six miles from Millway to Baltimore. Starting at Olean, two six-inch lines were paralleled to Saddle River, N. J. They separated there, one connecting with the refineries at Bayonne and the other going under the North and East Rivers to Hunter's Point, on Long Island. The New-York line is double under the Hudson—one pipe inside another, with tight-fitting sleeve-joints. The ends of the jacket-pipe were separated twelve inches to permit the enclosed pipe to be screwed home. The sleeve was then pushed over the gap and the space between the pipes filled with melted lead. The line is held in place by two sets of heavy chains, parallel with and about twenty feet from the pipe, one on each side. At intervals of three-hundred feet a guide-chain connects the pipe with the lateral chains and beyond each of these connections an anchor, weighing over a ton, keeps the whole in place. The completion of this part of the line was an engineering triumph not much inferior to the laying of Cyrus W. Field's Atlantic Cable.

The United Pipe-Lines Association moved forward steadily, avoiding the pitfalls that had wrecked other systems. It bought or combined the Oil-City, Antwerp, Union, Karns, Grant, Conduit, Relief, Pennsylvania, Clarion and McKean divisions of the American-Transfer, Prentice, Olean, Union Oil-Company's at Clarendon, McCalmont at Cherry Grove and smaller lines, covering the oil-region from Allegany to Butler. The United owned three-thousand miles of lines, thirty-five-million barrels of iron-tankage and one-hundred and-eighteen local pump-stations. Even these extraordinary resources were strained by the overflowing demand. Bradford was the Oliver Twist of the region, continually crying for "More!" Ohio and West Virginia entered the race and required facilities for handling an amazing amount of oil. To meet any contingency and secure the advantages of consolidation in the states producing oil the National-Transit Company increased its capital to thirty-two-million dollars. The company held the original charter granted to the Pennsylvania Company under the act of 1870. In 1880 it absorbed the American-Transfer Company, an extensive concern. On April first, 1884, it acquired the plant and business of the United Lines, thus ranking with the most powerful corporations in the land.

Men entirely familiar with the minutest details of oil-transportation and storage guided the National Transit. Captain Vandergrift was influential in the management until his retirement from active duty in 1892. President C. A. Griscom was succeeded by Benjamin Brewster and he by H. H. Rogers, the present official head of the company. John Bushnell was secretary, Daniel O'Day general manager, and James R. Snow general superintendent. Skillful, practical and keenly alive to the necessities of the oil-region, they were not kid-gloved idlers whose chief aim was to draw fat salaries. Mr. Rogers made his mark on Oil Creek in pioneer times as a forceful, intelligent, progressive business-man. He had brains, earnestness, integrity and industry and rose by positive merit to the presidency of the greatest transportation-company of the age. He is a first-class citizen, a liberal patron of education and an apostle of good roads. He endows schools and colleges, abounds in kindly deeds and does not forget his experiences in Oildom. Daniel O'Day—clever and capable, "whom not to know is to argue one's self unknown"—who has not heard of the plucky, invincible vice-president of the National Transit Company? Everybody admires the genial, resolute son of Erin whose clear head, willing hands, strong individuality and sterling qualities have raised him to a position Grover Cleveland might covet. James R. Snow invented a pump so perfect that oil would fairly flow up hill for a chance to pass through the machine. From their Broadway offices Rogers, O'Day and Snow direct by telephone and telegraph the movements of regiments of employés in Pennsylvania, Ohio, West Virginia and Indiana. They are in direct communication with every office of the company, every purchasing-agency, every pump-station on the trunk-lines and every oil-producing section of four states. No army Napoleon, Wellington or Grant commanded was better officered, better disciplined, better equipped and better managed than the grand army of National-Transit pipe-men. If "poets are born, not made," what shall be said of the wide-awake solvers of the problem of rapid transit for oil—the pipe-liners who, combining the maximum of efficiency with the minimum of cost, have placed a great staple within reach of the lowliest dwellers beneath the Stars and Stripes? Candidly, is "the best in the shop" too good for them?

No man has contributed more to the development of the oil-industry, alike as a producer, refiner and transporter, than Captain J. J. Vandergrift. His active connection with petroleum goes back to pioneer operations, widening and expanding constantly. By his energy, perseverance, uprightness and masterly traits of character he attained prominence in all branches of the oil-business. His wonderful success was not due to any caprice of fortune, but to stability of purpose, patient application and honorable methods. Vigor and decision supplemented the keen foresight that discovered the amazing possibilities of petroleum as an article of universal utility. He believed in the future of oil and shaped his course in accordance with the broadest ideas. Allied with George V. Forman, clear-headed, quick to plan and execute, the firm took a leading part in producing and carrying oil. Vandergrift & Forman constructed the Star Pipe-Line and equipped trains of tank-cars to convey crude from Pithole to Oil City. They drilled hosts of wells in Butler county and built the Fairview Pipe-Line, which finally crystallized with numerous others into the United Pipe-Lines Association and the gigantic National-Transit Company. The firm of H. L. Taylor & Co., of which they were members, originated the Union Oil-Company. Vandergrift & Forman, Vandergrift, Pitcairn & Co. and Vandergrift, Young & Co. consolidated as the Forest Oil-Company, which holds the foremost place in the production of oil. Mr. Forman operated in Allegany and McKean, developing large tracts of territory on the Bingham and Barse lands. He resided at Olean and established the finest stock-farm in the Empire State. Removing to Buffalo to engage in banking, he organized the Fidelity Trust-Company and erected for its use a palatial structure in the heart of the city. Under his presidency the Fidelity is a power in the world of finance. Shrewd, prompt and far-seeing, George V. Forman is richly dowered with the qualities of business-leadership. His influence in the oil-country was not limited to one corner or district or locality. He has enjoyed the pleasure of making money and the greater pleasure of giving liberally. He is "a man who thinks it out, then goes and does it."

Born at Pittsburg in 1827, at fifteen Jacob Jay Vandergrift chose the pathway that naturally opened before him and entered the steamboat-service, then

the chief medium of intercommunication between his native city and the west. In ten years he rose from cabin-boy to captain. He introduced the method of towing coal-barges that has since been employed in the river-traffic. The innovation attracted wide attention and gave a great impetus to mining in the Pittsburg coal-fields. Captain Vandergrift was steamboating on the Ohio when the war broke out and owned the staunch Red Fox, which the government chartered and lost near Cairo. He transported oil down the Allegheny, was concerned in West-Virginia wells—the Confederates destroyed them—and removed to Oil City in 1863 to oversee his shipping-business, with Daniel Bushnell as his first partner in producing oil. He organized the firms out of which grew the Union, the Forest, the Washington Oil-Company and the United Oil and Gas Trust. He was president of the Forest and the Washington and a leading promoter of the Anchor Oil-Company. The success of these great companies was owing largely to his peculiar ability as an organizer and manager of important enterprises. Other individuals and corporations produced oil profitably, but to Vandergrift & Forman the marvelous advance in modes of transportation is mainly attributable. They piped and railroaded oil from Pithole, extended their lines through the different fields, devised many improvements, perfected the methods of handling the product and developed the system that has eliminated jaded horses, wooden-barrels, mud-scows, slow freights and the thousand inconveniences of early transportation. Captain Vandergrift's sturdy integrity and wise forethought planned the open, clear-cut manner in which his pipe-lines conducted business. Throughout their entire existence he was president of the United Pipe-Lines and of the United Division of the National-Transit after the consolidation in 1884. Their splendid record is an unqualified tribute to his business-skill and rare sagacity. He found the region hampered by an expensive, tedious method of moving oil and left it a transportation-system that serves the industry as no other on earth is served. He substituted the steam-pump for the wearied mule, the iron-artery for the roads of bottomless mire and the huge cistern of boiler-plate for the portable tank of wooden staves that leaked at every pore. To Oil City he was a munificent benefactor. He projected the Imperial Refinery, with a capacity of fifteen-thousand barrels a week, by the sale of which he became a stockholder and officer of the Standard Oil-Company. He aided in establishing the Boiler-Works, the Barrel-Works, river-bridges, manufactories, churches and public improvements. He paid his workmen the highest wages, befriended the humble toiler and assisted every worthy object. The poor blessed his beneficent hand and all classes revered the modest citizen whose unostentatious deeds of kindness no party, race, color or creed could for one moment restrict.

Very naturally, one thus interested in a special product and its industries must be identified with its finance. Captain Vandergrift founded the Oil-City Trust-Company, one of the leading banking institutions of the state, and was prominent in organizing the Oil-Exchange, the Seaboard-National Bank of New York and the Argyle Savings-Bank at Petrolia. Removing to Pittsburg in 1881, he founded the Keystone Bank and the Pittsburg Trust-Company—nine-hundred-thousand dollars paid-up capital and four-millions deposits—and was unanimously elected president of both. He provided spacious quarters for the Oil-Exchange and established it on a sound basis. He erected the massive Vandergrift Building on Fourth avenue, in which the National-Transit Company, the Forest, the South-Penn, the Pennsylvania, the Woodland and other oil-companies are commodiously housed. The owner occupies a suite of offices on the second floor and the Pittsburg Trust-Company has its bank on the ground floor of the granite structure. He also erected the Conestoga Building, which has seven-hundred elegant offices, and the Imperial Power-Building, with factory-construction and the latest electric-motors throughout. In 1882 he organized the Pennsylvania Tube-Works—eight-hundred-thousand dollars capital—to manufacture all kinds of wrought-iron pipe. The output was so excellent that the capital was increased to two-millions and the plant doubled. The works turn out pipe from one-eighth inch to twenty-eight inches, the smallest and largest sizes in the world. The Apollo Steel-Company, which he also capitalized in 1885 at three-hundred-thousand dollars, has likewise trebled its plant and enlarged its capital to two-millions. The Penn Fuel-Company, the Bridgewater Gas Company, the Natural-Gas Company of West Virginia, the Chartiers Natural-Gas Company, the United Oil and Gas Trust, the Toledo Natural-Gas Company, the Fort-Pitt Natural-Gas Company and a number more were incorporated by Captain Vandergrift. They represent many millions of capital and have performed inestimable service in developing the fuel that proved a veritable philosopher's stone to the iron-industries of Western-Pennsylvania. As in petroleum, from the days of spring-poles and bulk-barges and pond-freshets down through all the changes of the most remarkable industrial development the world has ever seen, so Captain Vandergrift has been a pioneer, a guide and a leader in natural-gas. His hand has never been off the helm, nor has he ever grudged an atom of the energy bestowed upon the cherished pursuits of his busy life.

Forty miles north-east of Pittsburg, on a beautiful bend of the Kiskiminetas River, the new town of Vandergrift has been laid out, under the direction of Frederick Law Olmsted. It is located on a plot one mile square, two miles below Apollo, the gentle slope overlooking the valley and the river for leagues. Its residents will have within easy reach of simple thrift what luxurious people enjoy in large cities at great expense. They will have clean air and water and breathing-room, green leaves and flowers and grass, paved streets and sewers and electricity, parks and walks and drives, shade-trees and lawns and pleasant homes, for Vandergrift will be the model town of Pennsylvania. The company is paying sixty-thousand dollars a month at Apollo in wages and the big works at Vandergrift will employ thrice as many men. At first the bulk of the town will be the habitations of those employed by and associated with the company. After a little others will note its advantages and desire to share them. Provision will be made this year for an immediate population of several thousand, with the means of living comfortably, families owning their homes and controlling their own pursuits. The town is not to be a fad, a hobby, or a visionary Utopia, but a good place for men to live in, for the founder to use his money, for the world to look at and learn from. These banks and business-blocks, pipe-lines and refineries, mills and factories and the town that bears his name are enduring monuments to the enterprise and wisdom of a man who recognizes the responsibilities of wealth in his investments, in his works of philanthropy and in his gifts to the children of misfortune.

Captain Vandergrift's home in Allegheny City is a center of good cheer and genial hospitality. The host is the same kindly, companionable gentleman by his own hearth, in his office or on the street. He casts the lead of memory into the stream of the past and talks entertainingly of the old days on the Ohio, the Allegheny and Oil Creek. He is never too much engaged to welcome a comrade of his early years. He has not lost touch with men or the spirit of sympathy with the struggling and unsuccessful. His trials and vicissitudes, equally with his triumphs and successes, have strengthened his moral fiber, his manly courage and his nobility of character. Doubtful plans and purposes have had no place in his policy. Strict honesty and fairness have governed his conduct and respected the rights and privileges of his fellows. He has been quick to discover and reward talent, to grasp the details and possibilities of business and to mature plans for any emergency. Money has not shriveled his soul and .narrowed him to the prayer of selfishness: "Give *me* this day *my* daily bread." He prefers straightforwardness to a pedigree running back to the Mayflower. He realizes that golden opportunities for good are not travel- ing by a time-table and that men will not journey this way again to repair omissions and rectify mistakes. He knows that he who does right will be right and feel right. He does not lay aside his sense of justice, his love of fair-play, his earnest convictions and his desire to benefit mankind with his Sunday clothes. He believes that principle which is not exercised every day will not keep sweet a week. The story of J. J. Vandergrift's life and labor is told wherever the flame of natural-gas glows in the white heat of a furnace or the gleam of an oil-lamp brightens a happy home.

> Somehow we all feel sure, boys, that when the game is o'er—
> When the last inning's play'd, boys, this side the other shore—
> We'll hear the Umpire say boys: "The Captain's made a score."

Few persons have any conception of the labor and capital involved in stor- ing and transporting petroleum. Only those familiar with the early methods can appreciate fully the convenience and economy of the pipe-line system. It puts the producer in direct communication with the carrier and a market at all seasons, regardless of high or low water, rain or storm, mud or dust. The tanks at his wells are connected with the pipe-line by one or more of the two- inch feeders that spider-web the producing-country. Small pumps force the crude, when the location of the well prevents running it by gravity, from these tanks into a receiving-tank of the line, whence it can be piped into the trunk- lines or a storage-tank as desired. The producer who wishes his oil run noti- fies the nearest office or agent of the company—usually this requires about two minutes by wire—a gauger measures the feet and inches of fluid in the tank, opens the stop-cock, turns the stream into the line and, presto, change! the job is done. The gauger measures the oil left at the bottom of the tank, gives the producer a receipt for the difference between the two gauges and reports the result to the central station of that section of the field. There tables of the measurements of every tank in the locality are at hand, properly labeled and numbered. The right table shows at a glance the amount of oil in barrels corresponding to the feet and inches the gauger reports having run and the producer is credited accordingly, just like a depositor in a bank. These reports are summed up at a certain hour and the company learns precisely how much oil has been received each day. By a similar process the shipments are re- corded and the exact quantity in the custody of the company is known at the close of the day's business. Runs and shipments are published daily and a monthly synopsis is posted, in compliance with the laws of Pennsylvania. The producer can leave his oil in the line, subject to a slight charge for storage after thirty days, or sell it immediately. He can take certificates or acceptances of one thousand barrels each, payable on demand in crude-oil at any shipping- point in the oil-region. These certificates, good as gold and negotiable as certi- fied checks, the holder can use as collateral to borrow money, sell at sight or stow away if he looks for an advance in prices. It is not Hobson's choice with him. In an hour from the time of notifying the office his oil may be run, the amount figured up, the sale made and the currency in the owner's pocket. He has not tugged and perspired loading it in wagons or on cars, worn out his patience and his team and his profanity driving it through an ocean of mud, or risked the chances of a jam and a wreck ferrying it on the bosom of a pond- freshet. Nor has he put up one penny for the service of the pipe-line, which collects twenty cents a barrel when the oil is delivered to the purchaser. The company is not a holder of oil on its own account, except what it necessarily keeps to offset evaporation and sediment, acting merely as a common-carrier between the producer and the refiner. The system is the perfection of simplicity, accuracy and cheapness.

Pipe-lines are the natural outgrowth of the petroleum-business, which could no more get along without them than could the commerce of the world without railroads and steamships. The movement of a thousand barrels of crude in early times was a task of great magnitude, costly, time-consuming and perplexing. Sometimes barrels were not to be had, the water was too shallow for boating or the mud too deep for teaming. Often a big well wasted half its product and gorged transportation, harassing the soul and depleting the purse of the luckless owner. Fancy attempting to handle a hundred-thousand barrels a day with the primitive appliances! Whew! You might as well try to cart off Niagara in kegs. Butler and McKean rushed wells by the hundred every week, swelling the production extravagantly. The supply was enormously in excess of the demand. Operators wouldn't stop drilling and the surplus oil had to be cared for in some way. The United Lines and the National-Transit Company spent millions of dollars to provide adequate facilities. Not only was the vast output to be taken from the wells, but a large percentage must be stored. To pipe a hundred-and-forty-thousand barrels a day was a grand achievement, even without the burden of husbanding much of the stuff for weeks, months and years. A wilderness of iron-tanks—thirty to forty thousand barrels each—went up at Olean, Oil City, Raymilton, Parker and distributing points. Stocks increased and tanks multiplied until forty-million barrels were piled up! Think of the mountains of pipe, the acres of iron-plates, the legions of workmen and the stacks of cash all this required. Six pipes were laid to New York and the Tidewater Company built a six-inch line to New Jersey. The trunk-lines of the National-Transit alone are five-thousand miles in length, besides which the Tidewater and the United-States pipe oil eastward. Fifty-thousand barrels of crude a day flow through these underground arteries to the refineries at Hunter's Point, Bayonne and Philadelphia. Other thousands are piped to Baltimore, Buffalo, Cleveland, Pittsburg and refineries in the oil-region. The pipe used in transporting crude would girdle the earth twice and leave a long string for extra-measure. Truly "these be piping times."

McDonald gushers poured out their floods, but the National-Transit and Mellon Lines were on deck with pumps and pipes that snatched the contents of the tanks and whirled them to the sea. John McKeown's leviathan at Washington electrified the neighborhood by starting at three-hundred barrels an hour, with only three small tanks to hold the product. It filled the first in forty minutes. Superintendent Glenn Braden set up a pump in thirty minutes more that would empty the tank in a half-hour. All night it was nip and tuck between the spouter and the pump, big Goliath and puny David. The pump won, the oil was safe in the line and not a drop spilled! West-Virginia's geysers burst forth and the Southern Trunk-Line—three-hundred miles of eight-inch and six-inch pipe—linked Morgantown to Philadelphia. Lima tried to drown Ohio in crude and an eight-inch line quietly dumped the deluge into Chicago. Part of it fired the half-mile row of boilers at the Columbian Exposition, with not a cinder, a speck of ashes or a whiff of smoke to dim the lustrous flame of fuel-oil. Indiana, the home of some pretty big statesmen, some pretty big oil-territory and "the Hoosier Schoolmaster," had a surfeit of crude.

which the pipe-lines bore to the huge refinery at Whiting, to Cleveland and the Windy City. Thus the development of new fields, remote from railroads, has been rendered possible.

Trunk-lines require pipe of extra weight, manufactured expressly for the purpose from wrought-iron, lap-welded, cut into lengths of eighteen feet and tested to a pressure of two-thousand pounds to the square inch. Pumping-stations, supplied with powerful machinery, are located at suitable points, generally twenty-five to thirty miles apart. The stations on the National-Transit trunk-lines usually comprise a boiler-house forty feet square, built of brick and roofed with corrugated iron, lighted by electricity and containing seven or eight tubular boilers of eighty to one-hundred horse-power. For greater safety from fire the immense pumps are in a separate brick-building. The largest pumps are triple-expansion crank and fly-wheel engines, the invention of John S. Klein, superintendent of the company's machine-shops at Oil City. Each of these giants can force twenty-five-thousand barrels of oil a day through three six-inch pipes from one station to the next. A low-duty engine is run when the main-pump is stopped for repairs or any cause. At each station two or more storage-tanks—thirty to thirty-five thousand barrels apiece— are provided. One receives the oil from the preceding station while the pump is emptying the other into the receiver at the station beyond. The movement is incessant. Night and day, never tiring and never resting, the iron-arteries throb and pulsate with the greasy liquid that rushes swiftly a yard beneath the surface, duplicate machinery obviating the necessity of delay or interruption. Five or six boilers are fired at once and two are held in reserve, in case of accident. Loops are laid around some of the stations, that a pump may send the oil two or three times the average distance and the total disability of a station not blockade the line. When lofty hills are surmounted the pressure on the pump reaches twelve to fifteen-hundred pounds. Independent telegraph-lines connect the stations with one another and the main-offices. The engineers handle the key and click messages expertly. The lines are patrolled regularly to detect leaks, although the system of checking from tank to tank makes it impossible for a serious break to pass unnoticed. To clear the incrustations of paraffine, especially in cold weather, a scraper or "go-devil" is sent through the pipes. The best of these instruments—a spindle with a ball-and-socket-joint near its center to follow the bends of the pipe, fitted with steel-blades set radially and kept in position by three arms in front and rear—was devised by Mr. Klein. Oblique vanes, put in motion by the running oil, rotate the spindle and the blades scrape the pipe as the "go-devil" is propelled forward. A catch-box is placed at the end of each division and the queer traveler can be closely timed. The great battery of boilers, the huge engine-pumps—one on the Lima-Chicago line weighs a hundred tons—the electric-plants and the intricate maze of steam-pipes and water-pipes suggest the machinery of an ocean-steamship.

If the railroad is "the missionary of punctuality," as Robert Burdette concisely expresses it, surely the pipe-line is the messenger of efficiency. With wondrous speed and unfailing certainty it conveys crude-oil from the wells to the refineries in or out of the region, climbing hills, descending ravines, fathoming rivers and traversing plains and forests. Methods of refining have kept pace with progress in transportation. The smoky, dangerous, inconvenient kettle-still of the pioneer on Oil Creek has given place to the mammoth refinery of to-day, with its labor-saving appliances, its hundreds of skilled employés and

19

its improved processes. Instead of the ill-smelling, sputtering, explosive mixture of earlier years, the world now receives the water-white kerosene that burns as steadily and safely as a wax-taper. Seventy tank-vessels carry it over the seas to Europe, Asia and Africa. It is delivered at your house in neat cans, or the grocer will sell it by the pint, quart, gallon or barrel. The light is pure as heaven's own sunshine, grateful to the eye and beautifying to the home. No other substance approaches petroleum in the number and utility of its products. Long years of patient research and experiment have extracted from it one-hundred-and-fifty articles of value in art, science, mechanics and domestic economy. It supplies healing-salves, ointments, cosmetics, soaps, dainty toilet-accessories and—oh, girly Vassar girls—chewing-gum! Refuse tar and scum are converted into lamp-black and coarse lubricants. Scarcely a particle of it goes to waste. Noxious gases and poisonous acids no longer pollute the air and the streams around refineries, offending human nostrils and killing helpless fish. The amazing vastness of its development is equalled only by the marvelous variety of petroleum's commercial uses.

At every stage of its journey from the hole in the ground to the abode of the purchaser of kerosene, oil is handled with a view to the best results. The pipe-line relieves the producer from worry and fatigue and a large outlay, furnishing him prompt service and a cash market at his own door every business-day in the year. It enables the refiner to fill the consumer's lamp at a trifling margin above the price of crude. For seventy cents a barrel—less than half it cost formerly to haul it a mile—the line collects oil from the wells, pumps it into the trunk-lines and delivers it in New York. Contrast this charge with the four, five, eight or ten dollars exacted in the days of boats and wagons, barrels and tank-cars and endeavor to figure the saving to the public wrought by the pipe-lines, to say nothing of greater convenience and expedition. The existing transportation-system may be a monopoly, but the country is hungry for more monopolies of the same sort. If it be monopoly to bring order out of chaos, to build one strong enterprise from a dozen weaklings, to consolidate into a grand corporation a score of feeble lines and reduce freight-rates seventy-five to ninety-five per cent., the National-Transit Company is the rankest monopoly of the century. It practices the kind of monopoly that converts a row of tottering shanties into a stately business-block. It is guilty of furnishing storage solid as the Rock of Gibraltar to the men who drilled oil down to forty cents a barrel and tiding them over the period of excessive production. This is the brand of monopoly that keeps industry alive, that supplies foreign nations with an American product and benefits humanity. If Van Syckle, Abbott and Harley were plucky and courageous in braving the wrath of four-thousand teamsters, how much more brain and brawn, muscle and money, dollars and sense were needed to lay trunk-lines that sent ten-thousand tank-cars to the junk-pile and diminished the revenues of railroads millions of dollars annually! The owners of these lines have grown rich, as they ought to do, because for every dollar of their winnings they have saved producers and consumers of petroleum ten.

Pipe-line certificates afforded an excellent medium for speculation. The commodity they represented was subject to fluctuations of five to fifty per cent., which made it particularly fascinating to speculators in stocks. Oil-exchanges were established at Oil City, Titusville, Parker, Bradford, Pittsburg, New York, Philadelphia and elsewhere. In a single year the clearances exceeded eleven-billion barrels. Bulls and bears reveled in excitement and brokers had customers from every quarter of the country. The forerunner of these institutions was "the Curbstone Exchange" at Oil City in 1870. The bulk of the buying and selling was done in front of Lockhart, Frew & Co.'s office, Centre street, near the railroad track. Producers, dealers and spectators would congregate on the sidewalk, discuss the situation, tell stories and buy or sell oil. The group in the illustration includes a number of well-known citizens. Most of them have left Oil City and not a few have gone from earth. Acquaintances will recognize Dr. Knox, John Mawhinney, James Mawhinney, John D. Archbold, Dr. Baldwin, A. H. Bronson, P. H. Judd, L. D. Kellogg, A. E. Fay, George Porter, Edward Higbee, William M. Williams, John W. Austin, J. M. Butters, Joseph Bates, George W. Parker, William H. Porterfield, Charles W. Frazer, Edward Simmons, Samuel H. Lamberton, James H. Magee, Isaac Lloyd and William

OIL CITY "CURBSTONE EXCHANGE" IN 1870.

Elliott. Charles Lockhart and William Frew were pioneer refiners at Pittsburg and heavy buyers of crude at Oil City. William G. Warden entered into partnership with them and established the great Atlantic Refinery at Point Breeze. In 1874 the refineries controlled by Warden, Frew & Co. consolidated with the Standard Oil-Company of Ohio, forming the nucleus of the Standard Oil-Trust. Mr. Warden built the Gladstone, the first large apartment-house in Philadelphia, and died in April of 1895. He married a daughter of Daniel Bushnell and was one of the most enterprising and charitable citizens of Pennsylvania. His surviving contemporaries are old in reminiscences of Oil Creek and the days when pipe-lines and oil-certificates were unguessed probabilities.

Trades were made in offices, at wells, on streets, anywhere and everywhere. Purchasers for Pittsburg, Baltimore and Philadelphia refiners started brokerage in 1868, on a commission of ten cents a barrel from buyers and five from sellers. The Farmers' Railroad, completed to Oil City in 1867, brought so many operators to town that a car was assigned them, in which they bought and sold "spot," "regular" and "future oil." There were no certificates, no written obligations, no margins to bind a bargain, but everything was done on honor and no man's word was broken. "Spot oil" was to be moved and paid for at once, "regular" allowed the buyer ten days to put the oil on the cars and "future" was taken as agreed upon mutually. Large lots frequently changed hands in this passenger-car, really the first oil-exchange. The business increased, an exchange on wheels had manifest disadvantages and in December of 1869 it was decided to effect a permanent organization. Officers were elected and a room was rented on Centre street. It removed to the Sands Block in 1871, to the Opera-House Block in January of 1872 and to a temporary shed next the Empire-Line office in the fall, when South-Improvement complications dissolved the organization. For about fifteen months hotels, streets, or offices sufficed for accommodations. In February of 1864 the exchange was reorganized, with George V. Forman as president, and occupied quarters in the Collins House four years. Gradually rules were adopted and methods introduced that brought about the system afterwards in vogue. In April of 1878 the formal opening of the splendid Oil-Exchange Building took place. The structure contained offices, committee-rooms, telegraph-lines, reading-rooms and all conveniences for its four-hundred members. H. L. Foster, now of Chicago, was president term after term. The late H. L. McCance, secretary for years, was a first-class artist, with a skill for caricature worthy of Thomas Nast. Some of the most striking cartoons pertaining to oil were the work of his ready pencil. F. W. Mitchell & Co. inaugurated the advancing of money on certificates, their bank's transactions in this line ranging from one to four-million dollars a day. The application of the clearing-house system in 1882 simplified the routine and facilitated deliveries. The volume of business was immense, the clearances often amounting to ten or fifteen-million barrels a day. Only the New York and the San Francisco stock-exchanges surpassed it. If speculation were piety, everybody who inhaled the air of Oil City would have been saved and the devil might have put up his shutters. During rapid fluctuations the galleries would be packed with men and women who had "taken a flyer" and watched the antics of the bulls and bears intently. Fortunes were gained and lost. Many a "lamb" was shorn and many a "duck" lamed. It was a raging fever, a delirium of excitement, compressing years of ordinary anxiety and haste into a week. Now the exchange is deserted and speculative trade in oil is dead. Part of the big building is a clothier's store and offices are rented for sleeping-apartments. Myer Lowentritt, Stewart Simpson, "Eddie" Selden, Samuel Justus and a half-dozen others are seen occasionally, but days pass without a solitary transaction, the surging crowds have vanished and activity is a dream of bygone years.

Parker had a lively oil-exchange when the Armstrong and Butler fields were at their height. The most prominent men in speculative trade lived in the town or were represented in the exchange. Thomas B. Simpson was a large operator. George Darr was agent of Daniel Goettel, who once engineered the greatest bull-movement in the history of oil and was supposed to have "cornered" the market. Charles Ball and Henry Loomis earned sixty-thousand dollars brokerage a year and died within a month of each other. Trade slackened and expired. The boys shifted to Bradford and Pittsburg and a constable sold the building to satisfy Mrs. W. H. Spain's claim for ground-rent! The five-thousand-dollar library and the costly pictures, dust-covered and neglected, sold for a trifle and went to South Oil City. A jollier, bigger-hearted crowd of fellows than the members of the Parker Exchange never played a practical joke or helped a poor sufferer out of "a deuce of a fix."

The Bradford Oil-Exchange started on January first, 1883, with five-hundred members and a forty-thousand-dollar building. Five-hundred others, with Hon. David Kirk as president, organized the Producers' Petroleum-Exchange and erected a spacious brick-block, occupying it on January second, 1884. Both exchanges whooped it up briskly, both have subsided and the buildings are stores and offices. Titusville's handsome exchange, on the site of the American Hotel, has gone the same road. Captain Vandergrift built the Pittsburg Oil-Exchange, the finest of them all, fitting it up superbly. A bank and offices have succeeded the festive dealers in crude. From the Mining-Stock Exchange, the Miscellaneous Security Board and several more of similar types the New-York Consolidated Stock and Petroleum Exchange developed a huge concern, with twenty-four-hundred members and a lordly building—erected in 1887—on Broadway and Exchange Place. The membership was the largest in the country, with the exception of the Produce Exchange, and the business in oil at times exceeded the transactions of the Stock Exchange. Seats sold as high as three-thousand dollars. Charles G. Wilson has been president since the organization of the Petroleum and Stock Board, which absorbed the National Petroleum Exchange—L. H. Smith was its president—and in 1885 adopted the elongated name that has burdened it eleven years. Oil is not mentioned once a week, because the stocks have declined to a skeleton and the certificates represent scarcely a half-million barrels. Phila-

delphia had an exchange of lesser degree and a score of oil-region towns sharpened their appetite for speculation by establishing branch-concerns and bucket-shops. The almost entire disappearance of the speculative trade is not the least remarkable feature of the petroleum-development.

Since the elimination of exchanges producers generally sell their oil in the shape of credit-balances. For their convenience the Standard Oil-Company has established purchasing-agencies throughout the region. The quantity of crude to the credit of the seller on the pipe-line books is ascertained from the National-Transit office, a check is given and all the trouble the producer has is to draw his money from the bank. It is

JOSEPH SEEP.

handier than a pocket in a shirt, easier than rolling off a log in a mill-pond, and the happy "victim of monopoly" goes on his way rejoicing after the manner of Philip's converted eunuch. If he reside at a distance, be sojourning at Squedunk or in London, traveling with the Czar or showing the Prince of Wales a good time, a message to the agency will deliver his oil to Harry Lewis and the cash to his own order in a twinkling. The whole chain of purchasing-agencies is managed by Joseph Seep, whose headquarters are at Oil City. The Standard has the knack of selecting A-1 men for responsible positions—men who are not misfits, square pegs in round holes or small potatoes in the hill. Among the capable thousands who represent the great corporation none is better adapted to his important place than the head of the purchasing-agencies. He has the tact, the experience, the knowledge of human nature and the strength of character the position demands. For twenty-five years he has purchased crude for the company, up Oil Creek, at Oil City and down the Allegheny. You may not belong to his church or his party, you may differ from him on silver and woman-suffrage, you may even call the Standard an "octopus"—Col. J. A. Vera first did this at a meeting near St. Petersburg in 1874—and wish to turn its picture to the wall, but you like "Joe" Seep for his candor, his manliness, his admirable blending of suavity and firmness. He hails from the succulent blue-grass of Kentucky, combines Southern ease and Northern vigor, lives at Titusville and enjoys his wealth. It would strain Chicago's convention-hall to hold his legions of friends. His heart and his purse are alike generous. He produces oil, buys oil, ships oil and "pays the freight" on three-fourths of the oil handled in Oildom. He and George Lewis and Harry Lewis—"match 'em if you can"—have bought enough oil to fill a sea on which the navies of the world might race and leave room for the Yale crew that crew too soon. Seep and the Lewises are the gilt-edged stripe of men who don't drop banana-skins on the sidewalk to trip up a neighbor or squirm with envy because somebody else has a streak of good-luck. When Seep's last shipment has been made, the account is closed and the Recording Angel's ledger shows his big credit-balance, St. Peter will "throw the gates wide open," bid him welcome and never think of springing the old gag: "Not for Joseph, not for Joe !"

Sudden shifts in the market brought queer experiences in the days of wild oil-speculation, enriching some dabblers and impoverishing others. Stories of gains and losses were printed in newspapers, repeated in Europe and exaggerated at home and abroad. A bull-clique at Bradford, acting upon "tips from the inside," dropped four-hundred-thousand dollars in six months. An Oil-City producer cleared three-hundred-thousand one spring, loaded for a further rise and was bankrupted by the frightful collapse Cherry Grove ushered in. A Warren minister risked three-thousand dollars, the savings of his lifetime, which vanished in a style that must have taught him not to lay up treasures on earth. A Pittsburg cashier margined his own and his grandmother's hundred-thousand dollars. The money went into the whirlpool and the old lady went to the poor-house. A young Warrenite put up five-hundred dollars to margin a block of certificates, kept doubling as the price advanced and quit fifty-thousand ahead. He looked about for a chance to invest, but the craze had seized him and he hazarded his pile in oil. Cherry Grove swept away his fortune in a day. A Bradford hotel-keeper's first plunge netted him a hundred dollars one forenoon. He thought that beat attending bar and haunted the Producers' Exchange persistently. He mortgaged his property in hope of calling the turn, but the sheriff raked in the pot and the poor landlord was glad to drive a beer-wagon. Such instances could be multiplied indefinitely. Hundreds of producers lost in the maelstrom all the earnings of their wells, while the small losers would be like the crowd John beheld in his vision on Patmos, "a great company whom no man can number." Wages of drivers, pumpers, drillers, laborers and servant-girls were swallowed in the quicksands of the treacherous sea.

Of course there were many winners and many happy strokes of fortune. In 1876 Peter Swenk, of Ithaca, N. Y., purchased through a Parker broker ten-thousand barrels at two dollars and left orders to buy five-thousand more should the market break to one-seventy-five. Returning home, he was taken violently ill and the market suddenly fell forty cents, five cents below his margins. The day was stormy and Swenk could obtain no reports except from Oil City, where the break was eight cents greater than at Parker. The storm saved Swenk, although he did not know it for months, by crippling the wires and shutting off communication between Oil City and Parker the last hour of business. Concluding the margins were exhausted and the broker had sold the oil to save himself, Swenk went west to start anew. Weeks after his departure his Ithaca friends received urgent telegrams from the broker at Parker. They forwarded the messages, which informed him that, as the market stood, he was worth nineteen-thousand dollars and would be wise to sell. Swenk wired to close the whole matter and started for Parker. The market jumped another peg just before the order to sell arrived and Swenk received twenty-two-thousand dollars profits. He paid the broker double commission, returned home and bought a splendid farm. The faithful broker who managed this singular deal is now virtually a pauper at Bradford and a slave of rum. Last time we met he staggered up to me, his eyes bleared and his clothing in tatters, pressed my hand and said : "Gimme ten cents ; I'm dying for a drink !"

A big spurt in April of 1895 temporarily revived interest in oil-speculations. Again the exchange at Oil City was thronged. Exciting scenes of former years were renewed as the price climbed ten cents a clip. It was refreshing after the long stagnation to see the pool once more stirred to its depths. From one-ten on April fourth the price strode to two-eighty on April seventeenth. Certificates were scarce and credit-balances were snapped up eagerly. A few big winnings resulted, then the reaction set in, the spasm subsided and matters resumed their customary quietude. Connected with this phenomenal episode the papers in May told this breezy tale of "Bailey's Jag Investment :"

"C. J. Bailey, of Parkersburg, drew seventy-five-hundred dollars out of the Commercial Bank of Wheeling as the earnings of a three-hundred-dollar investment, made involuntarily and unknowingly. Bailey is a traveling salesman. A little less than a month ago he made a trip through the West-Virginia oil-fields. At Sistersville he got in with a crowd of oil-men, with the result that next day he had a big head, a very poor recollection of what had happened and was three-hundred dollars short, according to his memorandum-book. He wisely decided that the less publicity he gave his loss the better it would be and kept still. On Friday he was coming to Wheeling on the Ohio River Railroad, when a stranger approached him with :

"'You are J. C. Bailey, I believe.'"

"'Yes,' replied Bailey.

"'Well, you will find seven-thousand-five-hundred dollars to your credit in the Commercial Bank at Wheeling,' replied the stranger. 'I put it there day before yesterday and was about to advertise for you.'"

"Bunco was the first thought of Bailey ; but as the stranger did not ask for any show of money and talked all right, he asked for an explanation. It turned out that the stranger was one of the men with whom Bailey had been out in Sistersville. He was also secretary and treasurer of an oil-company, which had struck a rich well in the back-country pool two weeks before. Bailey, while irresponsible, had put three-hundred dollars into the company's capital-stock, on the advice of his friends. Meantime the well had been drilled, coming in a gusher of three-thousand barrels a day, one-tenth of which belonged to Bailey on his three-hundred-dollar investment. Bailey came to Wheeling, went to the bank and found the money awaiting him. He drew five-thousand dollars to send to his wife. Bailey's good fortune is not over yet, for the well is a good producer and the company holds large leases, on which several more good wells are sure to be drilled."

What of the brokers and speculators? They are scattered like chaff. A thousand have "gone and left no sign." President Foster, of the Oil City Exchange, an accomplished musician, traveler and orator, is a Chicagoan. John Mawhinney, John S. Rich—the fire at Rouseville's burning-well nearly destroyed his sight—H. L. McCance, George Cornwall, Wesley Chambers, Dr. Cooper,

A. D. Cotton, T. B. Porteous, Isaac Reineman, I. S. Gibson, Charles J. Fraser, W. K. Vandergrift, B. W. Vandergrift, B. F. Hulseman, Charles Haines, Michael Geary, Patrick Tiernan, "Shep" Moorhead, Melville, McCutcheon, Fullerton Parker, George Harley, Marcus Brownson and a host of other familiar figures will nevermore be seen in any earthly exchange. "Jimmy" Lowe—he was a telegrapher at first—Arthur Lewis, M. K. Bettis, George Thumm, I. M. Sowers and a dozen more drifted to Chicago. "Dick" Conn, "Sam" Blakeley, Wade Hampton, "Rod" Collins, Major Evans, Col. Preston and Charles W. Owston are residents of New York. "Tom" McLaughlin buys oil for the Standard at Lima. "Ajax" Kline is dissecting the Tennessee field for the Forest Oil-Company. "Cal" Payne is Oil-City manager of the Standard's gas-interests. "Tom" Blackwell is in Seep's purchasing-agency. John J. Fisher is flourishing at Pittsburg. "Charley" Goodwin holds the fort at Kane. Daniel Goettel and W. S. McMullan are running a large lumber-plant in Missouri. O. C. Sherman is a Baptist preacher and Jacob Goettel fills a Methodist pulpit. Frank Ripley and "Fin" Frisbee are heavy-weights in Duluth real-estate. C. P. Stevenson, the leading Bradford broker, dwells at his ease on a plantation in North Carolina. B. F. Blackmarr lives at Meadville and "Billy" Nicholas is a citizen of Minneapolis. Some are in California, some in Alaska, some in Florida, some in Europe and two or three in India. Go whither you may, it will be a cold day if you don't stumble across somebody who belonged to an oil-exchange or had a cousin whose husband's brother-in-law knew a man who was acquainted with another man who once saw a man who met an oil-broker. It is sad to think how the capital fellows who juggled certificates at Oil City, Parker and Bradford have thinned out and the pall of obliteration has been spread over the exchanges.

> "So fallen! So lost! the light withdrawn
> Which once they wore,
> The glory of their past has gone
> Forevermore!"

A pretty girl might as well expect to escape admiring glances as petroleum to escape a fire occasionally. "Uncle Billy" Smith's lantern ignited the first tank at the Drake well and a long procession has followed in its smoky trail. The lantern-fiend has been a prolific cause of oil-conflagrations, boiling-over refinery-stills have not been slack in this particular, the cigarette with a fool at one end and a spark at the other has done something in the same line, but lightning is the champion tank-destroyer. The result of an electric-bolt and a tank of inflammable oil engaging in a debate may be imagined. At first tanks were covered loosely with boards or wooden roofs. The gas formed a vapor which attracted lightning and kept up a large production of fires each season. One vicious stroke cremated sixty tanks of oil at the Atlantic Refinery in 1883. In July and August of 1880, a quarter-million barrels of McKean crude went up by the lightning-route. On June eleventh, 1880, a flash collided with the first *steel-tank* on which lightning had ever experimented and set the oil blazing. The tank was on a hill-side three-hundred feet from the west bank of Oil Creek, at Titusville. Several houses and the Acme Refinery, located between it and the stream, were consumed. While the burning oil flowed down the hill a sheet of solid flame covered ten acres. Bursting tanks, exploding stills and burning oils were an unpleasant promonition of the red-hot hereafter prepared for the wicked. The fire raged three days with the fury of the furnace heated seven-fold to give Shadrack, Meshach and Abed-nego a roast. The Titusville Battery checked it somewhat by cannonading the tanks with solid shot, which made

holes that let the oil run into the creek. This plan was tried successfully in
Butler and McKean. The old log-house that sheltered the generations of Campbells on the site of Petrolia met its fate by the firing of Taylor & Satterfield's
twenty-thousand-barrel tank on the hill above, which fell a prey to lightning.
Three tanks opposite the mouth of Bear Creek, below Parker, stood together
and burned together, the one singed by Jupiter's shaft setting off its mates. The
scene at night was of the grandest, multitudes gathering to watch the huge
waves of flame and dense clouds of smoke. As the oil burned down—just as
it would consume in a lamp—the tank-plates would collapse and the blazing
crude would overflow. Thousands of barrels would pour into the Allegheny,
covering the water for a mile with flame and painting a picture beside which a

FIRST STEEL OIL-TANK STRUCK BY LIGHTNING, AT TITUSVILLE, JUNE 11, 1880.

volcanic eruption resembled the pyrotechnics of a lucifer-match. Many tanks
were burned prior to the use of close iron-roofs, which confine the gas and do
not offer special inducements to "the artillery of heaven" to score a hit. Of
late years such fires have been rarities. All oil in the pipe-line to which the
burned tank belonged was assessed to meet the amount lost. This was known
as General Average, as unwelcome in oil as General Apathy in politics,
General Depression in business, General Dislike in society or General Weyler
in Cuba.

George B. Harris, a pioneer refiner, died at Franklin in January of 1892,
aged sixty years. A member of the firm of Sims & Co., he built the first or
second refinery in Venango county, near the lower end of Franklin. He pros-

pered for years, but reverses swept away his fortune and he was poor when death closed the scene.

A party of young men from New England started a refinery on Oil Creek in the sixties. Their industry, correct habits and attention to business attracted favorable notice. Mr. Trefts, of machinery fame, one day observed to a friend : " You mark my words ; some day these young men will be rich and their names shall be a power in the land. I know it will be so from their industry and good habits." This assertion was prophetic. The young man at the head of that modest firm of young men was H. H. Rogers, now president of the National-Transit Company. Speaking of his election as supervisor of streets and highways at Fair Haven, a New-York paper indulged in this facetious pleasantry regarding Mr. Rogers :

" The people of Fair Haven have done well. No man in New York or Massachusetts has had more experience with bad roads than Mr. Rogers, or has met with more success in subduing them. When he first engaged in the petroleum-business on Oil Creek the highways there were rarely navigable for anything on wheels, but were open to navigation by flat-boats most of the year. There was something in the mud of the oil-country at that time which was sure death to the capillary glands. Hairless horses and mules were in the height of fashion. When Mr. Rogers arrived on the strange scene, poling his way up to the hotel on a sawlog, he was at once chosen road-supervisor. In a neat speech, which is still extant, Mr. Rogers thanked the oil-citizens for the confidence reposed in him and then went to work. In the first place, he refined the mud of the highways, taking from it all the merchantable petroleum and converting the residue into stove-polish of an excellent quality. In the next p'ace, he constructed pipe-lines, through which the oil was conveyed, thus keeping it out of the middle of the road, and to-day there is a boulevard along Oil Creek that is hardly surpassed by the Appian Way. Horses are again covered with hair and happiness sits smiling at every hearthstone. The people of Fair Haven have a superintendent of streets to whom they can point with pride."

Dr. J. W. James, of Brady's Bend, who drilled some of the first wells around Oil City and was largely interested in the Armstrong and Bradford fields, in 1858 had a plant near Freeport for extracting coal-oil from shale. At a cost of twelve cents a gallon it produced crude-petroleum, which the company refined partially and sold at a dollar to one-twenty-five. The oil obtained from the rocks by drilling and that distilled from the shale were the same chemically. Dr. James read medicine with Dr. F. J. Alter, who constructed a telegraph Morse journeyed from the east to see before perfecting his own device. Dr. Alter's line extended only from the house about the small yard and back to his study. Full of enthusiasm over its first performance, he cried out to his student, young James : " I believe I could make this thing work a distance of six miles !" Bell's first telephone—a cord stretched between two apple-trees in an orchard at Brantford, Canada—was equally simple and its results have been scarcely less important.

John J. Fisher bought the first thousand barrels of oil in the new exchange at Oil City, on April twenty-third, 1878. Probably the largest purchase was by George Lewis, who took from a syndicate of brokers a block of two-hundred-thousand barrels. The first offer was fifty-thousand, increasing ten-thousand until it quadrupled, with the object of having Lewis cry : " Hold ! Enough !" Lewis wasn't to be bluffed and he merely nodded at each addition to the lot until the other fellow weakened, the crowd watching the pair breathlessly. "Sam" Blakeley, the most eccentric genius in the aggregation, once bid at Parker for a million barrels. Nobody had that quantity to sell and he advanced the bid five cents above the quotations. There was not a response and he offered a million barrels five cents below the ruling price, toying with the market an hour as if it were a foot-ball. He played for big stakes, but none knew who backed him. Coming to Oil City, he reported the market for the

Derrick and cut up lots of shines. One morning he looked glum, oil had tumbled and "Sam" hired an engine to whirl him to Corry. By nightfall he landed in Canada and his oil was sold to square his account in the clearing-house. An hour after his flight William Brough came up from Franklin to take the oil and carry "Sam" over the drop. In the afternoon a sudden rise set in, which would have left Blakeley twenty-thousand dollars profit had he stayed at his post! That was the time "Sam" didn't do "the great kibosh," as he phrased it. For years he has been hanging around New York. He was one of the boys distinguished as high-rollers and extinguished before the shuffle ended.

Telegraph-operators and messenger-boys at the oil-exchanges learned to note the movements of leading speculators and profit thereby. Some of them, with more hope of gain than fear of loss, beginning in a small way by risking a few dollars in margins, coined money and entered the ring on their own account. "Jimmy" Lowe, one of the biggest brokers at Parker and Oil City, slung lightning for the Western-Union when the Oil-City Exchange needed the services of twenty operators and scores of messenger-boys. Among the latter was "Jim" Keene, the Franklin broker. He and John Bleakley often received fifty cents or a dollar for delivering a message to "Johnnie" Steele, who stopped at the Jones House and flew high during his visits to Oil City. Steele and Seth Slocum would dash through the mud on their black chargers, dressed in the loudest style and sporting big diamonds. These halcyon times have passed away and the oil-exchanges have departed. "The glories of our mortal state are shadows."

In January of 1894 the Producers' and Refiners' Oil-Company erected an iron-tank on the hill south-east of Titusville. Lightning destroyed the tank and its contents in May. The second tank was built on the spot in October and on June twelfth, 1895, lightning struck a tree beside it. The burning tree fired the gas and the tank and oil perished. The site is still vacant, the company deciding not to give the electric fluid a chance for a third strike.

George W. N. Yost, who died in New York last year, was once the largest oil-buyer and shipper in the region. He lived at Titusville and removed to Corry, where he built the Climax Mower and Reaper Works, a church, a hand-some residence and blocks of dwellings. Patents of different kinds recouped losses in manufacturing. With Mr. Densmore, of Meadville, he brought out the caligraph. Yost sold to his partner and developed the Yost Typewriter, organized the American Writing-Machine Company and fitted up the shops at Bridge-port, Connecticut, used to manufacture Sharp's rifles during the border-troubles in Kansas. Mr. Yost was a man of striking personality and unflagging energy. He became a strong spiritualist and believed a medium, to whom he submitted completely, put him in communication with his dead relatives and recorded their thoughts on his typewriter.

The men of the oil-region have ever been noted for their commercial honor. It passed into a proverb—"honor of oil." The spirit of the saying, "his word is as good as his bond," has always been lived up to more closely in Oildom than in any other section of the country. The force of business-obligation ran high in the exchanges and among the early dealers in crude. Transactions involving hundreds-of-thousands of dollars occurred every day, without a writ-ten bond or a scrap of paper save a pencil-entry in a memorandum-book. Certificates were borrowed and loaned in this way and the idea of shirking a verbal contract was never thought of. The celerity with which property thus

passed from man to man was one of the striking features of business in the
bustling world of petroleum. And the record is something to be proud of in
these days of embezzlements, defalcations, breaches of trust and commercial
deviltry generally.

The average tank-steamer carries about two-million gallons of oil in bulk
across the Atlantic. In addition to this fleet of steamers, scores of sailing vessels, under charter of the Orient, France, Italy and foreign countries, load
cases and barrels of refined-oil for transport to European ports. American
wooden-ships are chartered sometimes to convey oil to Japan. Thus Russian
competition is met through the instrumentality of pipe-lines to the coast and
transportation by water to points many thousand miles away from the wells that
produced the oil.

The production of crude-petroleum in the United States in 1895, according
to the statistics compiled for the Geological Survey by Joseph D. Weeks, was
fifty-three-million barrels, valued at fifty-eight-million dollars. For 1894 the
figures were fifty-million barrels and thirty-five-million dollars respectively.
All districts except West Virginia and New York shared in the increase. The
total production from the striking of the Drake well in 1859 to the end of 1895
was seven-hundred-and-ten-million barrels. Five-hundred-and-seventeen-mil-
lion barrels of this enormous aggregate represent the yield of the Pennsylvania
and New-York oil-fields. Who says petroleum isn't a big thing?

At Pittsburg you can easily gather a little group of men, such as Charles
Lockhart and Captain Vandergrift, who recall the time when the Tarentum petroleum was termed "a mysterious grease." They had a hand in handling it
when the oil had no commercial name. They watched Samuel M. Kier's efforts
to give it a commercial name and a marketable value. They saw it run to waste
at first, they remember paying a dollar a gallon for it and can tell all about
Drake's visit to Tarentum. They hold their breath when they think of the gold
that changed hands in Venango county after "Uncle Billy" Smith bored the
seventy-foot hole below Titusville, of the wonderful spread of operations and
the dazzling progress of the commodity once despised. They noted the flow
of petroleum toward Europe—how forty casks were sent to France in 1860 as a
curiosity and thirty-nine-hundred in 1863 as a commercial venture. They have
seen this "mysterious grease," that used to flow into the Pennsylvania Canal,
light the world from the Pyramids of Egypt to the salons of Paris, from the
shores of Palestine to the Chinese Wall. They have seen the four salt-and-
oil wells at Tarentum and the solitary oil-well at Titusville multiplied into a
hundred-thousand holes drilled for petroleum and a production almost beyond
calculation. Do the gentlemen composing this little group occupy a position
dramatic in the marvelous events they review? Is petroleum freighted with
interest and a touch of romance at every step of its passage from the well to
the lamp?

THE AMEN CORNER.

Better a kink in the hair than a kink in the character.

Good creeds are all right, but good deeds are the stuff that won't shrink in the washing.

Domestic infidelity does more harm than unbelieving infidelity and hearsay knocks heresy galley-west as a mischief-maker.

> Stick to the right with iron nerve,
> Nor from the path of duty swerve,
> Then your reward you will deserve.

The Baptists of Franklin offered Rev. Dr. Lorimer, the eminent Chicago divine, a residence and eight-thousand dollars a year to become their pastor. How was that for a church in a town of six-thousand population?

" Pray—pray—pray for—" The good minister bent down to catch the whisper of the dying operator, whom he had asked whether he should petition the throne of grace—" pray for five-dollar oil !"

St. Joseph's church, Oil City, is the finest in the oil-region and has the finest altar in the state. Father Carroll, for twenty years in charge of the parish, is a priest whose praises all denominations carol.

> You "want to be an angel?"
> Well, no need to look solemn ;
> If you haven't got what you desire,
> Put an ad. in the want column.

The Presbyterian church at Rouseville, torn down years ago, was built, paid for, furnished handsomely and run nine months before having a settled pastor. Not a lottery, fair, bazaar or grab-bag scheme was resorted to in order to raise the funds.

The Salvation Army once scored a sensational hit in the oil-regions. A lieutenant struck a can of nitro-glycerine with his little tambourine and every house in the settlement entertained more or less Salvation-Army soldier for a month after the blow-up.

> " Like a sawyer's work is life—
> The present makes the flaw,
> And the only field for strife
> Is the inch before the saw."

" What are the wages of sin ?" asked the teacher of Ah Sin, the first Chinese laundryman at Bradford, who was an attentive member of a class in the Sunday-school. Promptly came the answer : "Sebenty-five cents a dozen ; no checkee, no washee !"

The first sound of a church-bell at Pithole was heard on Saturday evening, March 24, 1866, from the Methodist-Episcopal belfry. The first church-bell at Oil City was hung in a derrick by the side of the Methodist church, on the site of a grocery opposite the *Blizzard* office. At first Sunday was not observed. Flowing-wells flowed and owners of pumping-wells pumped as usual. Work went right along seven days in the week, even by people who believed the highest type of church was not an engine-house, with a derrick for its tower, a well for its Bible and a tube spouting oil for its preacher.

> " If you have gentle words and looks, my friends,
> To spare for me—if you have tears to shed
> That I have suffered—keep them not I pray
> Until I hear not, see not, being dead."

Many people regard religion as they do small-pox ; they desire to have it as light as possible and are very careful that it does not mark them. Most people when they perform an act of charity prefer to have it like the measles—on the outside where it can be seen. Oil-region folks are not built that way.

A CLUSTER
OF PIONEER
EDITORS.

COL. LEE M. MORTON.	WALTER R. JOHNS.	L. H. METCALFE.
W. H. LONGWELL.	MAJOR W. W. BLOSS.	C. E. BISHOP.
WARREN C. PLUMER.	J. H. BOWMAN.	HENRY C. BLOSS.
COL. J. T. HENRY.		COL. M. N. ALLEN.

XII.

THE LITERARY GUILD.

Clever Journalists Who Have Catered to the People of the Oil Regions—
Newspapers and the Men Who Made Them—Cultured Writers, Poets and
Authors—Notable Characters Portrayed Briefly—Short Extracts from
Many Sources—A Bright Galaxy of Talented Thinkers—Words and
Phrases that Will Enrich the Language for all Time.

" Reading maketh a full man."—*Bacon*.

"Literature is the immortality of speech."—*Wilmott*.

" This folio of four pages, happy work,
What is it but a map of busy life,
Its fluctuations and its vast concerns?"—*Cowper*.

" News, the manna of a day."—*Green*.

———" And a small drop of ink
Falling like dew upon a thought, produces that
Which makes thousands, perhaps millions, think."—*Byron*.

" Books are men of higher stature, the only men that speak
Aloud for future times to hear."—*Mrs. Browning*.

A. P. WHITAKER.

THIRTY-SIX YEARS have had their entrances and their exits since Col. Drake's little operation on Oil Creek played ducks and drakes with lard-oil lamps and tallow-dips. That seventy-foot hole on the flats below Titusville gave mankind a queer variety of things besides the best light on "this grain of sand and tears we call the earth." With the illuminating blessing enough wickedness and jollity were mixed up to knock out Sodom and Gomorrah in one round. The festive boys who painted the early oil-towns red are getting gray and wrinkled, yet they smile clear down to their boots as they think of Petroleum Centre, Pithole, Babylon, or any other of the rapid places which shed a lurid glare along in the sixties. The smile is not so much on account of flowing wells and six-dollar crude as because of the rollicking scenes which carmined the pioneer period of Petroleum. These were the palmy days of unfathomable mud, swearing teamsters, big barrels, high prices, abundant cash and easy morals, when men left their religion and dress suits "away out in the United States." The air was redolent of oil and smoke and naughtiness, but there was no lack of hearty kindness and the sort of charity that makes the angels want to flap their wings and give "three cheers and a tiger." Even as the city destroyed by fire from heaven boasted one righteous person in the shape of Lot, whose wife was turned into a pillar of salt for being too fresh, so the busy Oil-Dorado had a host of capital fellows, true as steel, bright as a dollar and "quicker'n greas'd lightin'!" Braver, better, nobler, squarer men never doffed a tile to a pretty girl or elevated a heavy boot to the coat-tails of a scoundrel. About the wells, on the streets, in stores and offices could be found gallant souls attracted from the ends of the world by glowing pictures—real oil-paintings—of huge fortunes gained in a twinkling. Ministers, lawyers, doctors, merchants, soldiers, professors, farmers, mechanics and members of every industry were neither few nor far between in the exciting scramble for "the root of all evil."

A host of changes, some pleasing and more unutterably sad, have the swift seasons brought. The scene of active operations has shifted often. The great Bradford region and the rich fields around Pittsburg and Butler have had their innings. Parker, Petrolia, St. Petersburg, Millerstown and Greece City have followed Plumer, Shaffer, Pioneer, Red-Hot and Oleopolis to the limbo of forsaken things. Petroleum Centre is a memory only. Rouseville is reduced to a skeleton. Not a trace of Antwerp, or Pickwick, or Triangle is left. Enterprise resembles Goldsmith's "Deserted Village," or Ossian's "Balaclutha." Tip-Top, Modoc, Troutman, Turkey City, St. Joe, Shamburg, Edenburg and Buena Vista have had their rise and fall. Fagundas has vanished. Pleasantville fails to draw an army of adventurous seekers for oleaginous wealth. Tidioute is an echo of the past and scores of minor towns have disappeared completely. For forms and faces once familiar one looks in vain. Where are the plucky operators who for a half-score years made Oil Creek the briskest, gayest, liveliest spot in America? Thousands are browsing in pastures elsewhere, while other thousands have crossed the bridgeless river which flows into the ocean of eternity.

To keep matters straight and slake the thirst for current literature newspapers were absolutely necessary. Going back to 1859, the eventful year that brought petroleum to the front, Venango county had three weeklies. The oldest of these was the *Spectator*, established at Franklin in 1849, by Albert P. Whitaker. At the goodly age of seventy-eight he wields a vigorous pen. A zealous disciple of Izaak Walton and Thomas Jefferson, he can hook a fish or indite a pungent editorial with equal dexterity. He is an encyclopedia of political lore and racy stories. His *Spectator* is no idle spectator of passing incidents. In 1851 Col. James Bleakley, subsequently a prosperous producer and banker, secured an interest in the plant, selling it in 1853 to R. L. Cochran, who soon became sole proprietor and published the paper seven years. Mr.

R. L. COCHRAN.

Cochran took an active part in politics and agriculture and exerted wide influence. A keen, incisive writer and entertaining talker, with the courage of his convictions and the good of the public at heart, his sterling qualities inspired confidence and respect. Probably no man in North-western Pennsylvania had a stronger personal following. The *Spectator* flourished under his tactful management. It printed the first "oil report," giving a list of wells drilling and rigs up or building in the spring of 1860. Desiring to engage in banking, R. L. Cochran sold the paper to A. P. Whitaker, its founder, and C. C. Cochran. The latter retiring in 1861, Whitaker played a lone hand three years, when the two Cochrans again purchased the establishment. A. P. Whitaker and his son, John H., a first-class printer, bought it back in 1866 and ran the concern four years. Then the elder Whitaker once more dropped out, returning in 1876 and resuming entire control a year later, which closed the shuttlecock changes of ownership that had been in vogue for twenty-five years. Will. S. Whitaker, an accomplished typo and twice the nominee of his party for mayor, has long assisted his father in conducting the staunch exponent of unadulterated Democracy. Col. Bleakley passed away in 1884, leaving a fine estate as a monument of his successful career. He built the Bleakley Block, founded the International Bank, served as City Councilman and was partner in 1842–4 of John W. Shugert in the publication of the *Democratic Arch*, noted for aggressiveness and sarcasm. John H. Whitaker died in Tennessee years ago. R. L. Cochran was killed in June, 1893, on his farm in Sugarcreek Township, by the accidental discharge of a gun. The paper began regular "oil reports" in 1862, prepared by Charles C. Duffield, now of Pittsburg, who would go up the Allegheny to Warren and float down in a skiff, stopping at the wells.

Charles Pitt Ramsdell, a school-teacher from Rockland Township, started the *American Citizen* at Franklin in 1855. Sent to the Legislature in 1858, he sold the healthy chick to William Burgwin and Floyd C. Ramsdell, removed to Delaware and settled in Virginia a few years before his lamented death from wounds inflicted by an enraged bull. J. H. Smith acquired Ramsdell's interest in 1861. The new partners made a strong team in journalistic harness for three years, selling in 1864 to Nelson B. Smiley. He changed the title to *Venango Citizen*. Mr. Burgwin reposes in the Franklin cemetery. Mr. Smith carries on the book-trade, his congenial pursuit for three decades, and is a regular contributor to the religious press. Alexander McDowell entered into partnership with Smiley, buying the entire "lock, stock and barrel" in 1867. His former associate studied law, practiced with great credit, and died at Bradford. Major McDowell, now a banker at Sharon—the number of Venango editors who blossomed into financiers ought to stimulate ambitious quill-drivers—did himself proud in the newspaper lay. His liberality and geniality won hosts of warm friends. He tried his hand at politics and was chosen Congressman-at-Large in 1892, with Galusha A. Grow as running-mate, and Clerk of the House in 1895. A prime joker, he bears the blame of introducing Pittsburg stogies to guileless members of Congress for the fun of seeing the victims cut pigeon-wings doing a sea-sick act. Col. J. W. H. Reisinger purchased the outfit in 1869, guiding the helm skillfully fifteen months. April first—the day had no special significance in this case—1870, E. W. Smiley, the

E. W. SMILEY.

present owner and cousin of Nelson B., succeeded Reisinger. The Colone located at Meadville, where he has labored ably in the journalistic field for a quarter-century. Mr. Smiley steered his craft adroitly, usually "bobbing up serenely" on the winning side. He is a shrewd Republican worker and for twenty years has filled a Senate-clerkship efficiently. What he doesn't know about the inside movements of state and local politics could be jumped through the eye of a needle. His right-bower in running the *Citizen-Press*—the hyphenated name was flung to the breeze in 1884—is his son, J. Howard Smiley, a rising young journalist. The paper toes the mark handsomely, has loads of advertising and does yeoman service for its party. The *Daily Citizen*, the first daily in Oildom, expired on the last day of 1862, after a brief existence of ten issues. A fit epitaph might be Wordsworth's couplet:

> " Since it was so quickly done for,
> Wonder what it was begun for."

Later newspaper ventures at Franklin were refreshingly plentiful. In January, 1876, Hon. S. P. McCalmont launched *The Independent Press* upon the stormy sea of journalism. It was a trenchant, outspoken, call-a-spade-a-spade advocate of the Prohibition cause, striking resolutely at whoever and whatever opposed its temperance platform. Mr. McCalmont wrote the editorials, which bristled with sharp, merciless, unsparing excoriations of the rum-traffic and its aiders and abettors. The paper was worthy of its name and its spirited owner. Neither truckled for favors, cringed for patronage or ever learned to "crook the pregnant hinges of the knee where thrift may follow fawning." Beginning life a poor boy, S. P. McCalmont toiled on a farm, taught school, devoured books, read law and served in the Legislature. For nearly fifty years he has enjoyed a fine practice which brought him well-earned reputation and fortune. Ranking with the foremost lawyers of the state in legal attainments and professional suc-

cess, he does his own thinking, declines to accept his opinions at second-hand and is a first-rate sample of the industrious, energetic, self-reliant American. By way of recreation he works a half-dozen farms, a hundred oil-wells, a big refinery and a coal-mine or two. James R. Patterson, Miss Sue Beatty and Will. S. Whitaker held positions on the *Press*. Mr. Patterson is farming near Franklin and Mr. Whitaker manages the *Spectator*. Miss Beatty, a young lady of rare culture, was admitted to the bar recently.

The Independent Press Association bought the *Press* in 1879. This influential body comprised twelve stockholders, Hon. William R. Crawford, Hon. C. W. Gilfillan, Hon. John

S. P. M'CALMONT.

M. Dickey, Hon. Charles Miller, Hon. Joseph C. Sibley, Hon. S. P. McCalmont, Hon. Charles W. Mackey, James W. Osborne, W. D. Rider, E. W. Echols, B. W. Bredin and Isaac Reineman, whom a facetious neighbor happily termed "the twelve apostles, limited." They enlarged the sheet to a nine-column folio, discarded the bourgeois skirt with long-primer trimmings for a tempting dress of minion and nonpareil and engaged J. J. McLaurin as editor. H. May Irwin, the second editor under the new administration, filled the bill capably until the *Press* and the *Citizen* buried the hatchet and blended into one. Mr. Irwin is not excelled as an architect of graceful, felicitous paragraphs on all sorts of subjects, "from grave to gay, from lively to severe." He possesses in eminent degree the enviable faculty of saying the right thing in the right way, tersely, pointedly and attractively. The *Press* was a model of typographical neatness, newsiness and thorough editing, with a taste for puns and plays on words that added zest

JAMES B. BORLAND.

to its columns. In 1863-4 Mr. Irwin was a compositor and city-editor of the Harrisburg *Patriot and Union,* when Samuel C. Miller was an apprentice, C. M. McDowell set type and John Ferguson was assistant-foreman. He represented the Associated Press and various newspapers in Washington, knew everybody and was on the top wave of popularity.

James B. Borland's *Evening News* appeared in February, 1878, as an amateur daily about six by nine inches. The small seed quickly grew to a lusty plant. James B. Muse became a partner, enlargements were necessary, and to-day the *News* is a seven-column folio, covering the home field

and deservedly popular. Muse retired in 1880, H. May Irwin buying his share and editing the wide-awake paper in capital style. The *News* is independent in politics, very much alive to the welfare of Franklin, brimful of fresh matter and never dull. *Every Evening,* a creditable venture by Frank Truesdell, E. E. Barrackman and A. G. McElhenny, bloomed every evening from July, 1878, to the following March. Truesdell, who went to Titusville, and Barrackman, who went west, sold out to Will. S. Whitaker, McElhenny, a sharp pencil-pusher and horny-handed granger, remaining until the fledgling passed in its checks. H. B. Kantner, a versatile specimen, hatched out the *Morning Star,* Franklin's only morning daily, in 1880. It shone several months and then set

forever and ever. Kantner drifted to Colorado. The *Herald,* the *Penny Press* and *Pencil and Shears* wriggled a brief space and "fell by the wayside." Samuel P. Brigham, an aspiring young lawyer, edited the one-cent *Press* and stirred up a hornet's nest by fiercely assailing the waterworks system and raising Hail Columbia generally. He is at the head of a newspaper in the Silver State. Will. F. Lapsley dished up the second *Every Evening* a short time last summer.

The third weekly Venango boasted in 1859 was the *Allegheny-Valley Echo,* published at Emlenton by Peter O. Conver, a

WILL. S. WHITAKER.

most erractic, picturesque genius. Learning the printing trade in Franklin, the anti-slavery agitation attracted him to Kansas in 1852. He established a paper at Topeka, which intensified the excitement a man of Conver's temperament was not calculated to allay, and it soon climbed the golden stair. Other experiments shared the same fate. Conver roamed around the wild and wooly west several years, returned to Venango county and perpetrated the *Echo* in the fall of 1858. At intervals a week passed without any issue, which the next number would attribute to the sudden departure of the "jour," the non-arrival of white paper, or the absence of the irrepressible Peter on a convivial lark. Sparkling witticisms and "gems of purest ray" frequently adorned the pages of the sheet, although sometimes transgressing the rules of propriety. It was the editor's habit to set up his articles without a manuscript. He would go to the case and put his thoughts into type just as they emanated from his fertile brain. Poetry, humor, satire, invective, comedy, pathos, sentiment and philosophy bunched their hits in a medley of clean-cut originality not even "John Phœnix" could emulate. The printer-editor had a fund of anecdotes and adventures picked up during his wanderings and an off-hand magnetism that insured his popularity. His generosity was limited only by his pocket-book. Altogether he was a bundle of strange contradictions, "whose like we shall not look upon again," big-hearted, impatient of denial, heedless of consequences, indifferent to praise or blame, sincere in his friendships and with not an atom of sham or hypocrisy in his manly fiber. He enlisted in the Fourth Pennsylvania Cavalry when the war broke out, serving gallantly to the close of the struggle at Appomattox.

R. F. Blair, who had taken the *Echo* in 1861, disposed of it in 1863 to J. W. Smullin, by whom the materials were removed to Oil City. Walter L. Porter's *Rising Sun*, W. R. Johns' *Messenger*, Needle & Crowley's *Register*, P. Mc-Dowell's *News*, Col. Sam. Young's *Telegraph*, Hulings & Moriarty's *Times* and Gouchler Brothers' *Critic* in turn flitted across the Emlenton horizon. The *Rising Sun* proved a setting sun, the *Messenger* speeded to the end of its journey very quickly for a messenger, the *Register* did not register enough pay-

ing subscribers, the *News* had too few readers of its news, the *Telegraph* flashed out young, the *Times* was ahead of the times and the *Critic* speedily sank into a critical condition. E. H. Cubbison projected the *Home News* in 1885, enlarged it repeatedly, killed the first section of the name, paid special attention to home news and was rewarded by liberal support. He still holds the fort.

Getting back from the war safe and sound, Conver pitched his tent at Tionesta in 1866 and generated the *Forest Press*. Its peculiar motto —"The first and only paper printed in Forest county and about the only paper of the kind printed anywhere"—indicated the novel stripe of this unique weekly. The crowning feature

COL. J. W. H. REISINGER.

was its department of "Splinters," which included the weird creations of the owner's vivid fancy. The *Press*, after running smoothly a dozen years, did not long survive its eccentric, gifted proprietor, who answered the final roll-call in the spring of 1878, meeting death unflinchingly. He wrote a short will and asked Samuel D. Irwin, his trusted adviser, to prepare his obituary, "sense first, nonsense afterwards." The *Bee*, which Col. Reisinger hived in 1867, sipped honey a season and flew away. Muse's *Vindicator* and Wenk's *Republican* now occupy the field. Mrs. Conver left Tionesta and died in the west. Hosts of old friends who knew and understood Peter O. Conver will be glad to see his characteristic portrait, from a photograph treasured by Judge Proper, and "a nosegay of culled flowers" from his inimitable *Press*:

"That marble slab has arrived at last. Our own beautiful slab, with its polished surface, was manufactured expressly to our order, on which to impose the forms of the Forest *Press*, a fit emblem and unmistakable evidence of the almost unparalleled success of an enterprise started in the very hell of the season and circumstances on a one-horse load of old, good-for-nothing, worn-out, rotten and "bottled" material, taken in payment, etc., and a will to succeed. After we shall have fulfilled our mission through the *Press* and have done with the things of earth, that same slab can be used by the weeping "devils" on which to dance a good-bye to us and our sins, after which they may inscribe with burning charcoal on its polished surface, in letters of transient darkness:

> 'Here
>   lies
>   Pete.
>   The
>   old
>   cuss
>   is
>   dead.'

PETER O CONVER.

"Our mother was a Christian, the best friend we had, and the name of her truant son —your servant—was the last she uttered. We are not a Christian, but when convinced we should be we will be. Never intend to marry or die, if we can help it. In brief, we are a white Indian."

"A promissory note is tuning the fiddle before the performance."

"A man suffering from dyspepsia sees nothing bright in the noonday sun. Another with a rusty liver looks upon a flower-garden as so many weeds. Another with nerves at angles sees nothing lovely in the most beautiful woman. Another with a disordered stomach can utter no word not tinged with acid and fire."

"Smiles are among the cheapest and yet richest luxuries of life. We do not mean the mere retraction of the lips and the exhibition of two rows of masticators—mastiffs, hyenas and the like amiabilities are proficient in that. We do not mean the cold, formal smile of politeness, that plays over the features like moonlight on a glacier—automatons and villains can do that; but we mean the real, genial smile that breaks right out of the heart, like a sunbeam out of a cloud, and lights up the whole face and shines straight into another heart that loves it or needs it."

"Ravishingly rich and gorgeous is our surrounding scenery smiling down upon us in all the dying glory of these autumn days, like the summery landscape in childhood's dreams, impressed on the heart but not described ; like the soul-beam of a good old person passing away. View all the grand and beautiful scenes of earth with the aid of imagination's pencil if you please, and them come to Tionesta in October and behold the masterpiece. It is the finishing touch of beauty from the Master Hand, imparting joy and faith and hope and resignation to the heart of man, which no human pen or pencil may copy and combinations of words have not been discovered to describe ; in fact, we have almost come to the conclusion that he who attempts it is a presuming fool, because there's no language in the dictionary or even invented by the poet to that effect. But if we only live till the sun shines to-morrow, on such another day as this, we'll dig our potatoes, from which patch we can obtain mountain views on every hand alongside of which the Rocky Mountains would appear overgrown and unnatural and Alpine scenery worn-out."

"The first great damper that threw cold water on the Fourth of July was, perhaps, the agitation of the temperance question; then the Sunday-school celebrations gave a mortal blow to its ancient prestige and glory, until now, alas! it has been entirely eclipsed. Bantlings of the third generation are soaring aloft in place of the old gray bird, niggers dancing jubas over the heads of their imperial masters and, great heavens! the very whiskey that we drink at $3 to $7 a gallon in mortal jeopardy. But, seriously speaking, we are in favor of every one following the bent of his or her own inclination in celebrating things. Next week will be our usual occasion for getting full, unless we should accompany a very beautiful young lady hunting, in either of which events the *Press* may also have a celebration of its own and not appear in public on any stage."

"Lieut. Samuel D. Irwin is a rare, original genius, a companion of our boyhood, whose life has been lively and stirring as our own in some respects. He is also a candidate for District Attorney."

"Some people don't care much whether things go endwise or otherwise."

"Next to a feast upon a seventeen-year-old pair of sweet lips, under grapevines, by moonlight, is a foray upon a platter of beans, after fishing for suckers all day."

"One of the greatest bores in the world is he who will persistently gabble about *himself* when you want to talk about *yourself*."

"Pay your debts and shame the devil for an old scoundrel."

"Bright and fair as a Miss in her teens is this beautiful March morning. All nature laughs with gladness. Forest feels glad, the streams sing a glad song in their swim to the sea, Tionesta is glad and the big greyhound Charley Holmes sent Major Hulings wags his sharp tail in token of the gladness and gratitude he cannot otherwise express. He is a gentlemanly, well-bred, $500 purp and got to have his meals regularly."

"Do unto other men as you would have them do unto you and you wouldn't have money enough in two weeks to hire a shirt washed."

"Many a preacher complains of empty pews when they are really not emptier than the pulpit."

"The man who can please everybody hasn't got sense enough to displease anybody."

"To be good and happy kick up your heels and holler Hallelujah!"

"Rev. Brown will preach everybody to hell on the Tubb's Run Flats, Lord willing, next Sunday, between meals."

On the twelfth of January, 1862, Walter R. Johns, who struck the territory four weeks previously, issued the initial number of the Oil-City *Weekly Register*, the first newspaper devoted especially to the petroleum industry, which it upheld tenaciously for five years. The modest outfit, purchased second-hand at Monongahela City, was shipped to Pittsburg by boat, to Kittanning by rail and to its destination by wagons. The editor, publisher, proprietor and compositor —Mr. Johns outdid Pooh-Bah by combining these offices in his own person— accompanied the expedition to aid in extricating the wagons from mud-holes in which they stuck persistently. In 1866 he retired in favor of Henry A. Dow & Co., who fathered the *Daily Register* and soon found the cake dough. Farther on Mr. Johns was identified, editorially or in a proprietary way, with the semi-weekly *Petrolian* and the *Evening Register*, the Parker *Transcript*, the Emlenton *Messenger*, the Lebanon *Republican*, the Clarion *Republican-Gazette* and the Foxburg *Gazette*. Writing with great readiness and heartily in touch with his profession, he took to literary work as a duck takes to water. He and the late Andrew Cone prepared all the petroleum-statistics available in 1862, which, with the gatherings of the years intervening, were published in 1869, under the expressive title of "Petrolia." From Clarion, his home for some years, Mr. Johns returned to Oil City, doing valuable work for the *Derrick* and the *Blizzard*. For seven years he has been employed by the National Transit Company to compile newspaper-clippings and magazine-articles and arrange records of different kinds from every quarter of the oil-regions. The duty is congenial and he fits the place "like der paper mit der wall." Mr. Johns is a son of Louisiana and a hero of two wars. During the Mexican trouble he fought under Zachary Taylor and Winfield Scott, was at the battles of Monterey and Buena Vista and participated in the march from Puebla to the City of Mexico. He served under General Grant in the "late unpleasantness." The death of his estimable wife several years ago was a terrible blow to the Nestor of petroleum journalism, who has gained distinction as printer, editor, author and soldier.

> " Age sits with decent grace upon his visage
> And worthily becomes his silver locks ;
> He bears the marks of many years well spent,
> Of virtue, truth well tried and wise experience."

With the plant of the defunct Emlenton *Echo*, which he had bought from R. F. Blair and boated to Oil City, J. W. Smullin propelled the *Monitor* in 1863. O. H. Jackson, a sort of perambulatory printing-office, and C. P. Ramsdell figured in the ownership at different times. Jackson let go in the fall of 1864 and Jacob Weyand bossed the ranch until it was absorbed by the *Venango Republican*, the first out-and-out political newspaper in the settlement. Smullin farmed in Cranberry township, dispensed justice as "'Squire" and died in 1894. Of Jackson's whereabouts nothing is known. He flaunted the *Sand-Pump* at Oil City, the *Bulletin* at Rouseville, the *Gaslight* at Pleasantville and ephemeral sheets at other points. The outfits of the *Register, Petrolian, Republican* and *Monitor* were consolidated in December, 1867, by Andrew Cone and Dr. F. F. Davis, into the weekly *Times*. The paper was well managed, well edited and well sustained. A syndicate of politicians bought it in 1870, to boom C. W. Gilfillan, of Franklin, for Congress, and George B. Delamater, of Meadville, for State Senator in the Crawford district. A morning daily was tacked on. L. H. Metcalfe, who lost a leg at Gettysburg, had editorial charge. Thomas H. Morrison, of Pleasantville, officiated as manager, W. C. Plumer presided as foreman and A. E. Fay acted as local news-hustler. The daily died with the close of the campaign, a fire that destroyed the establishment hurrying the dissolution. Metcalfe went back to Meadville and was elected county-treasurer. Whole-souled, earnest and trustworthy, he made and retained friends, wrote effectively and "served his day and generation" as a good man should. The grass and the flowers have bloomed above his head for nineteen years. Morrison entered politics, put in a term faithfully as county-treasurer, studied law, practiced at Smethport and was elected judge of the McKean-Potter district.

Hon. Andrew Cone, to whose bounteous purse and willing pen the *Venango Republican* and the Oil City *Times* owed their continuance, was of Puritan descent, born in 1822, near Rochester, N. Y. His father left New England in 1817 and in 1820 married Mary L. Andrews, daughter of Nathanael Andrews of Connecticut, and sister of Mrs. Charles G. Finney, wife of the founder and first president of Oberlin College, Ohio. One of Dr. Finney's daughters is the wife of General Jacob D. Cox, ex-Governor of Ohio, and another the wife of James Monroe, ex-Consul-General of the United States at Rio Janeiro. Bishop Andrews, of the Methodist Episcopal Church, and Hon. Charles Andrews, the distinguished Judge of the Court of Appeals of New York, are cousins of Andrew Cone. Educated at Middleburg Academy, young Andrew married Miss M. L. Hibbard, of Frederick county, Md., and settled on his father's farm. His parents dying, he removed to Michigan, where he lost his young wife and in 1857 married Miss Belinda Morse, of Elmira, N. Y. She died after the birth of two children. In 1862 Mr. Cone came to Oil City and for four years was superintendent of the United Petroleum Farms Association, the powerful corporation that controlled numerous oil-tracts and miles of city-lots. He was also vice-president of the Oil City Savings Bank, organizer of the first Good Templar lodge and among the founders of the Baptist church, superintending the Sabbath-school and giving thousands of dollars towards the erection of the first house of worship. Some of the streets he named from members of the Association, as Bissell and Harriot avenues, in honor of George H. and Harriot Bissell. When he resigned the superintendency—the late Dr. Charles A. Cooper, of happy memory, succeeded him—the company presented Mr. Cone with a corner lot on Bissell avenue, the leading residence street. In 1868 he married Miss Mary Eloisa Thropp, of Valley Forge, a cultured linguist, essayist and magazine-writer. At that time he was publishing the *Times* and collecting the data for his historical "Petrolia," a perfect storehouse of facts and figures pertaining to oil. Appletons printed the book, which involved immense labor and painstaking research. Governor Hartranft appointed Mr. Cone to represent the oil-regions as a State Commissioner to the World's Exposition at Vienna, in 1873. With his wife he traveled over Austria, Italy, Germany, France, Switzerland and Great Britain, Mrs. Cone writing lucid descriptive letters as foreign correspondent of the Philadelphia *Inquirer* and Oil-City *Derrick*. Failing health obliging him to seek a milder climate in 1876, of five consulates offered by President Grant he chose that at Para, Brazil, none the less readily that he had entertained the Brazilian Emperor, Dom Pedro, during his visit to Oil City. He performed his official duties with the same correctness and fidelity that marked his discharge of every obligation. Reappointed by President Hayes to be consul at Pernambuco, he obtained his first leave of absence and returned to New York in September of 1880, hoping to regain his wonted strength. The hope was vain, for on the seventh of November he died, peacefully closing his useful and honorable career as one to whom "well done, good and faithful servant," is spoken through all the centuries.

Mrs. Cone, the eldest of three sisters prominent in literature, still resides in Oil City and contributes occasionally to magazines and newspapers. She warmly seconded her husband's benevolent and literary efforts, accompanied him to Europe and lived five years in Brazil. Her early poems were published in the New York *Knickerbocker*, *Graham's Magazine* and *Godey's Lady's Book*. In 1860 she opened a select school in Philadelphia for young ladies and in 1865 was appointed by the United-States Sanitary Commission, with three other ladies, to distribute supplies to sick and wounded prisoners at Richmond. They were the first Northern ladies to reach the Confederate capital after Lee's surrender. In 1878 she wrote her celebrated poem for the centennial at Valley Forge, her birthplace and the home of her family and ancestors, one of whom, Christian Workizer, was an accomplished German officer under General Wolfe at the siege of Quebec. A brother, Joseph E. Thropp, an impressive reasoner and polished speaker, owns the iron-works at Everett and is married to the eldest daughter of the late Colonel Thomas A. Scott, president of the Pennsylvania railroad, whose master-mind moulded the mightiest railway system the world has ever known. Miss Amelia Thropp, whose "Brazil Papers," poems, stories and "Scenes in Our Village" have been highly commended and extensively copied, lives with her widowed sister, employing her leisure in writing for the press. Mrs. George Porter, the third sister and a resident of Oil City, began to write poetry at ten and at fourteen saw her first prose-sketch—"Winfred Wayne"—in the *Knickerbocker Magazine*. At times she writes for Cleveland, Norristown, Oil City and Philadelphia journals.

Her ballads have been especially admired. The Thropp sisters are esteemed highly for their poetical talents, their Christian character and their sweet unselfishness. "The Wild Flowers of Valley Forge," by Mrs. Cone, will give an idea of their exquisite work :

Blest be the flowers that freely blow
  In this neglected spot,
Anemone with leaves of snow
  And blue Forget-me-not.
God's laurels weave their classic wreath,
  Their pale pink blossoms wave
O'er lowly mounds, where rest beneath
  Our martyrs in their grave.

In white and gold the daisies shine
  All o'er encampment hill;
There wild-rose and the Columbine
  Lift glistening banners still.
Here plumy ferns, an emerald fringe,
  Adorn our stream's bright way ;
And soft grass whence the violet springs,
  With fragrant flowers of May.

Oh, there's a spell around these blooms
  Owned by no rarer flowers ;
They blossomed on our soldiers' tombs
  And they shall bloom on ours.
To us, as to our sires, their tone
  Breathes forth the same glad strain,
"We spring to life when winter's gone,
  And ye shall rise again."

Uncultured 'round our path they grow,
  Smile up before our tread
To cheer, as they did long ago
  Our noble-hearted dead.
Arbutus in the sheltering wood
  Sighs " Here he came to pray,"
And Pansies whisper, " Thus we stood
  When heroes passed away."

Thus every wild-flower's simple leaf
  Breathes in my native vale,
To conscious hearts, some record brief,
  Some true and touching tale.
Wealth's gay parterre in glory stands :—
  I own their foreign claims,
Those gorgeous flowers from other lands,
  Rare plants with wondrous names.

Ye blossomed in our martyr's field
  Beneath the warm spring's sun,
Sprung from the turf where lowly kneeled
  Our matchless Washington.
Ye in our childhood's garden grew,
  Our sainted mother's bowers ;
My grateful heart beats high to you,
  My own wild valley flowers !

The collapse of the *Daily Times* terminated experimental dailies in Oi!

ANDREW CONE.

MRS. CONE.

MISS THROPP.

City. Mr. Gilfillan, F. W. Mitchell, P. R. Gray and other stockholders sold the good-will and smoking ruins to Sheriff H. H. Herpst, who revived the weekly with Dr. Davis at the bellows. It was rather weakly, notwithstanding the doctor's excellent doses of leaded pellets. Advertisers seemed a trifle shy and columns of blank space, by no means nutritious pabulum, were not infrequent. Everybody favored a newer, grander, bolder stride forward. The borough and suburbs had attained the dignity of a city, an oil-exchange had been organized, railroads were coming in and a paper of metropolitan scope was urgently demanded. Usually men adapted to a particular niche turn up and the traditional "long-felt want" is not likely to remain unfilled. Such was the case in this instance.

Coleman E. Bishop and W. H. Longwell landed in Oil City one summer afternoon to "view the landscape o'er," as good Dr. Watts phrased it. They had heard the Macedonian cry and decided to size up the situation. Bishop achieved greatness at Jamestown, N. Y., where he edited the *Journal*, by attacking Commander Cushing, the naval officer who sank the Confederate ram Merrimac, and kicking him down stairs when the indignant marine invaded the sanctum to "horsewhip the editor and pitch him out of the window." Longwell, a brave soldier and sharp man of affairs, had learned the ropes at Pithole and Petroleum Centre. A deal with Mr. Herpst was soon closed, needed material was ordered and a frame building on Seneca street rented. Herpst kept an interest as a silent partner.

The Oil City *Derrick*, ordained to become "the organ of oil," was born on the thirteenth of September, 1871. The name was an inspiration, adopted at the last moment instead of the hackneyed *Times*, which had been agreed upon by the three proprietors. To embody the most conspicuous emblem of the petroleum-business in the head of a newspaper designed to represent the oil-trade suggested itself to the alert editor a few hours before the forms were ready for the press. The venture was a go from the start. People were wakened from their slumber by strong-lunged newsboys shouting, "Derrick, ere's yer Derrick, Derrick!" Their first impulse was to wonder if they had left any derricks out all night, exposed to thieves and marauders, and somebody was bringing them home. The new sheet was scanned eagerly. It had departments of "Spray," "Lying Around Loose" and "Pick-ups," teeming with catchy, piquant, invigorating items. Its advocacy of the producers' cause boomed the paper tremendously. A bitter fight with the Allegheny Valley Railroad increased its circulation and prestige. Bishop's individuality permeated every page and column. He had the sand to continue the railroad war, but a threat to remove the shops from Oil City weakened his partners and they bought him out in 1873. From the "Hub of Oildom" he went to Buffalo to edit the *Express*. Thence he went to Bradford, embarked in oil-operations on Kendall Creek and enlivened the *Chautauqua Herald*, Rev. Theodore Flood's bonanza, one summer. Invited to New York in 1880, he managed the *Merchants' Review* and edited *Judge* until it changed owners in 1885. Leaving the metropolis, he wandered to Dakota and freshened the Rapid City *Republican*. Returning east, he furnished Washington correspondence to various papers. For years he has been disabled by locomotor-ataxia. Mrs. Bishop is a popular teacher of the Delsarte system and has published a book on the subject. Miss Bishop is a talented lecturer. It is not disparaging the galaxy of oil-region journalists to say that C. E. Bishop, the gamest, keenest, raciest member of the fraternity, might be termed a bishop in the congregation of men who have shaped public opinion in the domain of grease. No matter how difficult or delicate the theme, from pre-natal influence to monopoly, from heredity to fishing, from biology to pumpkins, he treated it tersely and charmingly. A thoroughbred from top to toe, his was a Damascus blade and "none but himself can be his parallel."

Captain Longwell—the title was awarded for gallantry in many a hard battle —attended to the business-end with decided success. Buying Herpst's claim, he conducted the whole concern four years and sold out at a steep figure in 1877. He raked in wealth producing and speculating, quitting well-heeled financially. A native of Adams county, he was educated at Gettysburg and learned printing in the office of the Chambersburg *Repository and Whig*, then published by Col. Alexander K. McClure, now the world-famed editor of the Philadelphia *Times*. His mother was a descendant of James Wilson, a signer of the Declaration of Independence. Herpst opened a wall-paper store, removed later to Jamestown and died there in 1884. Square, honest and "straight as a string," he merited the regard of his fellows. Charles H. Morse, the first city-editor, had the snap to corral news at sight and present it toothsomely. Who that knew him in his beardless youth imagined Charley would "get religion" and adorn the pulpit? He entered the ministry and for over twenty years has been pastor of a Baptist church at Mercer. Were he to serve up to his hearers some of the funny experiences he encountered as a reporter, he would discount Talmage's recitals of the slums and Dr. Parkhurst's leap-frog exploits in the Tenderloin! Archie Frazer wrote the market-report, ten or twelve lines at first and a plump column or more ultimately. In November of 1872 it was my luck to engage with the *Derrick* and inaugurate the role of traveling correspondent. Venango and Warren, with Clarion, Armstrong and Butler budding into prominence, covered the oil-fields. Bradford loomed up in the autumn of 1875, extending my mission from the northern line of McKean county to the southern boundary of Butler before the close of the term of five years. These breezy days were crowded with bustle and excitement, adventure and incident. Over the signature of "J. J. M."—possibly remembered by old-timers—fate appointed me to chronicle a multitude of events that played an important part in petroleum-annals. The system of "monthly reports" was arranged methodically, the producing sections were visited regularly and my acquaintance embraced every oil farm and nearly every oil-operator in the rushing, hustling, get-up-and-get world of petroleum.

Orion Clemens, a brother of "Mark Twain," worked on the *Derrick* a few weeks in 1873. The exact opposite of "Mark," his forte was the pathetic. He could write up the death of an insect or a reptile so feelingly that sensitive folks would shed gallons of tears in the wood-shed over the harrowing details. He fairly reveled in the gloomy, somber, tragic element of life. Daily contributions taxed him too severely, as he composed slowly, and his resignation caused no surprise. Frank H. Taylor, a young graduate from the Tidioute *Journal*, succeeded Bishop, vacating the chair to undertake the field-work. Frank can afford to "point with pride" to his career as editor and compiler of statistics. His "Handbook" is an unquestioned authority on petroleum. Once he resigned to float the *Call*, a sprightly Sunday folio, which glistened from the spring of 1877 to October of 1878. "Puts and Calls," the humorous column, had to answer for bursting off tons of vest-buttons. Taylor acquired money and fame as a journalist, was president of select-council, called the turn as a producer and saved a snug competence. During the last Congress he was Hon. J. C. Sibley's secretary, a position demanding remarkable tact and industry. Now he is leasing lands, drilling wells and looking after the oil-properties of Sibley & Co. in Indiana. Oil City is his home and he is as busy as a boy clubbing chestnuts.

Robert W. Criswell, who had forged to the front by his mirth-provoking sketches, followed Taylor as editor in 1877. He fertilized the "Stray-sand," travestied Shakespeare and developed "Grandfather Lickshingle," giving the *Derrick* national celebrity. When the shuffle occurred in 1877 he stepped down and went to the Cincinnati *Enquirer*. Major McClintock, W. J. McCullagh and Frank W. Bowen were on deck about the same time. The Major, now oil

inspector, handled the market-report and press-dispatches. McCullagh held the field-department up to its elevated standard and Bowen ground out first-class local and editorial. Col. Edward Stuck, who had come from York in 1879 to supervise the Bradford *Era*, ran the machine in 1880-2, displaying much ability in the face of manifold hindrances. William Brough and J. M. Bonham of Franklin, gentlemen of high literary attainments, wishing to have a paper of their own, induced Mr. Stuck to leave Bradford, with a view to resurrect the *Sunday Call*. The project was not carried out and he assumed charge of the *Derrick*, with gratifying results. His training was acquired on the York *Democratic Press*, his father's weekly, which Col. Stuck now conducts in connection with the *Daily Age*, established by him after his sojourn in Oil City. He was appointed State Librarian during Governor Pattison's first term and elected Register of Wills of York county in 1889, in recognition of his excellent journalistic services. William H. Siviter, straight from college, was next in order. His polished, scholarly writings were relished by educated people. He paragraphed for the Pittsburg *Chronicle-Telegraph* and for some years has contributed to the comic weeklies. He is responsible for the "High-School Girl," with her Bostonese flavor and highfalutin speech. McCullagh became an operator in the Bradford region, drilled extensively in Ohio, laid by considerable boodle and chose Toledo as his residence. Robert Simpson, who began as "printer's devil" in 1872, remained with the *Derrick* as a writer until the *Blizzard* blew into town, excepting brief respites at Emlenton and Bradford.

P. C. Boyle, whose dash and skill and tireless energy had advanced him steadily, leased the establishment in 1885. He had the vigor and backbone needed to bring the paper back to its pristine strength. By turns a roustabout at Pithole in 1866, a driller, a scout, a reporter, a publisher and an editor, his experience in the oil-country was extensive and invaluable. He published the *Laborer's Voice* at Martinsburg in 1877-8, reported for the *Derrick* and Titusville *Herald* in 1879, for the *Petroleum World* in 1880 and the *Olean Herald* in 1881, conducted the Richburg *Echo* in 1881-2 and scouted all through the developments at Cherry Grove, Macksburg and Thorn Creek in 1882-5. George Dillingham, who had "a nose for news," and J. N. Perrine, gilt-edged and yard-wide in the counting-room, assisted Mr. Boyle in tuning the paper up to high G. The outside fields, daily growing in number and importance, were put in charge of Homer McClintock, the real Homer of oil-reporters. He fattens on timely paragraphs, scents live items in the air and lets no juicy happening escape. The force was augmented as occasion arose, type-setting machines and fast presses were added, the job-office was supplied with the latest and best materials and the *Derrick* is to-day one of the finest, brightest, smartest newspapers that ever edified a community. It is owned by the Derrick Publishing Company, of which Mr. Boyle is president and H. McClintock, J. N. Perrine and Alfred L. Snell are the active members. Mr. Boyle also managed the Toledo *Commercial* and the Bradford *Era*. He is "the Dean of the Fourth Estate" by virtue of eminent services and seniority. Like the lightning, he never needs strike twice in the same spot, because the job is finished at a single lick when he goes "loaded for b'ar."

John B. Smithman, a wealthy operator, to whom Oil City owes its streetrailways and a bridge spanning the Allegheny, in 1880 equipped the *Telegraph*, an evening sunflower, with Philip C. Welch at its head. Isaac N. Pratt, later an advance-agent for Ezra Kendall, had a finger in the pie. The paper was as fetching as a rural maiden in a brand-new calico gown, but two dailies were too rich for the blood of the population and the *Telegraph* wilted at a tender age. Welch tapped a vein of rich humor in the Philadelphia *Call* by originating "Accidentally Overheard," a feature that captured the bakery. It bubbled with actual wit, fragrant as sweet clover and wholesome as morning dew, not revamped and twisted and warmed over. Charles A. Dana, no mean judge of literary merit, recognized the value of the Welch rarebits and secured them for the New York *Sun* at a fixed rate for each, big or little, long or short, large or small. Anon Dana offered him a salary few bank-presidents would refuse and Welch moved to Gotham. The *Sun* that "shines for all" fairly glittered and dazzled. Welch's "Tailor-Made Girl" hit the popular taste and was published in elegant form by the Scribners. Disease preyed upon him, compelling an operation similar to General Grant's. Half the tongue was cut out, affecting his utterance seriously. Weeks and months of patient suffering ended at last in release from earthly pain and sorrow. Mrs. Welch, a noble helpmeet, lives in Brooklyn and is to be credited with the clever, dainty "From Her Point of View," which irradiates the Sunday issues of the New York *Times.* Upon the grave of Philip C. Welch old friends would lay a wreath and drop a sympathetic tear.

> " Alas, Poor Yorick !
> I knew him, Horatio ;
> A fellow of infinite jest,
> Of most excellent fancy."

Frank W. Bowen, a diamond of the first water, H. G. McKnight, the lightning type-slinger, and B. F. Gates, a dandy printer, swarmed from the *Derrick* hive and raised the wind to blow an evening *Blizzard* in 1882. They bought the *Telegraph* stuff and the Richburg *Echo* press, had brains and pluck in abundance and went in to win. The significant motto—" It blows on whom it pleases and for others' snuff ne'er sneezes "—attested the independence of the

free-playing zephyr. Gentle as the summer breezes when dealing with the good, the true and the beautiful, it swept everything before it when a wrong was to be righted, a sleek rascal unmasked or a monopoly toppled over. Bowen's "Little Blizzards" had a laugh in every line. If they stung transgressors by their sharp thrusts, the author didn't lie awake nights trying to load up with mean things. His humor was spontaneous and easy as rolling off a log. Now his friends and admirers—their name is Legion—propose to waft him into the Legislature, a clear case of the office seeking the man. It goes without saying that the *Blizzard* was an instant success. It was no fault of the fond parents that they were built that way and couldn't compel people not to want their exhilarating paper. Place its neat make-up to McKnight's account. Gates flocked by himself to usher in the *Venango Democrat,* which the gods loved so well that it passed through the golden gates in four weeks. Robert Simpson, jocularly styled its "horse editor," was a *Blizzard* trump-card until 1886. He then filled consecutive engagements as exchange-editor, news-editor, night-editor, assistant managing-editor and legislative correspondent of the Pittsburg *Dispatch.* Again he edited the *Derrick* nine months in 1889. Returning to Pittsburg as political-reporter of the *Commercial-Gazette*, he was promoted to legislative-correspondent and lastly to managing-editor, a position of much responsibility. Simpson is among the best all-'round newspaper men in Pennsylvania. E. A. Bradshaw, editor of the Jamestown *Journal*, also held down a stool and manufactured crisp copy in the *Blizzard* sanctum.

The Reno *Times*, an eight-column folio that ranked with the foremost weeklies in the State, was started in 1865 and expired in May of 1866. A department was assigned each kind of news, the matter was classified and set in minion and nonpareil, oil-operations were noted fully and local affairs received due attention. Samuel B. Page, the editor, understood how to glean from exchanges and correspondence. George E. Beardsley, whose parish lay along Oil Creek, about Pithole and the Allegheny River from Franklin to Tidioute, a section thirty miles by seventy, managed the oil-columns admirably. E. W. Mercer kept the books, collected the bills and had general supervision. W. C. Plumer, J. Diffenbach and Edward Fairchilds stuck type and the average edition exceeded ten-thousand copies.

Pithole, the most kaleidoscopic oil-town that ever stranded human lives and bank-accounts, gave birth to the *Daily Record* on the twenty-fifth of September, 1865. It was a five-column folio, crammed with news piping-hot and sold at five cents a copy, or thirty cents a week. Morton, Spare & Co. were the publishers. Col. L. M. Morton—he earned his shoulder-straps in the civil war—edited the *Record*, winning laurels by his wise discernment. He was a manly character, incapable of deceit, a brilliant writer and conversationalist, the soul of honor and courtesy, "a knight without fear and without reproach." He served as postmaster at Milton and spent his closing years as night-editor of the Bradford *Era*, dying at his post, loved and esteemed by thousands of friends. W. H. Longwell, another brave defender of the Union, bought out Spare in May, 1866. Charles C. Wicker and W. C. Plumer were taken into the firm shortly after. In May of 1868, Pithole having crawled into a hole, Longwell changed the base of operations to Petroleum Centre, then at the zenith of its meteoric flight. He sold the paper in 1871 to Wicker, who held on until formidable rivals in Oil City and Titusville forced the *Record* to quit. Generous to a fault and faithful to those who shared his confidence,

CHARLES C. WICKER.

Wicker left the decaying town in 1873, was foreman of the Titusville *Courier*, worked as a compositor at Bradford and died there years ago. He was never satisfied to accept ill-luck without emphatic dissent. He always wore a blue-flannel shirt, a fashion he adopted in the army, and was eccentric in attire.

Charles C. Leonard was "a bright, particular star" in the days of the Pithole *Record*, to which, over the signature of "Crocus," he contributed side-splitting sketches of ludicrous phases of oil-region life. These felicitous word-paintings, with additions and revisions, he published in a volume that had a prodigious sale. He was an Ohioan, born in 1845, and a soldier at sixteen. Arriving at Pithole in 1865, he saw that wonderful place grow from a dozen shanties to a city of fifteen-thousand at a pace distancing Jonah's gourd or Jack-the-Giant-Killer's bean-stalk. In the fall of 1867 he came to theTitusville *Herald*, remaining five years. After short terms with the Cleveland *Leader* and St. Louis *Globe*, he returned to Titusville to write for the *Evening Press*. He went back to St.

Louis and died at Cleveland on the twelfth of March, 1874, wounds received in battle hastening his demise. He was a natural wit, whose keen jokes had the aroma of Attic salt. Mrs. Leonard removed to Detroit, her home at present. One of Charlie's favorite creations was "The Sheet-Iron Cat," written for the Cleveland *Leader.* It passed the rounds of the newspapers and was printed in the *Scientific American.* The sell took immensely, lots of persons sending letters asking the cost of the "cats" and where they could be procured! The article, which will revive many a pleasant memory of "auld lang syne," follows:

CHARLES C. LEONARD.

"A young mechanic in this city, whose friends and acquaintances have heretofore supposed there was "nothing to him," has at last achieved a triumph that will place him at once among the noblest benefactors of mankind. His name will be handed down to posterity with those of the inventors of the "steam man," the patent churn and other contrivances of a labor-saving or comfort-inducing character. His invention, which occurred to him when trying to sleep at night in the sky-parlor of his cheap boarding-house, with the feline demons of midnight clattering over the roof outside, is nothing more than a patent sheet-iron cat with cylindrical attachment, steel-claws and teeth, the whole arrangement being covered with cat-skins, which give it a natural appearance and preserve the clock-work and intricate machinery with which the old thing is made to work. Among the other peculiarities of this ingenious invention are the tail and voice. The former is hollow and supplied by a bellows (concealed within the body) with compressed air at momentary intervals, which causes the appendage to be elevated and distended to three times its natural size, giving to the metallic cat a most warlike and belligerent appearance. By the aid of the same bellows and a tremolo-stop arrangement, the cat is made to emit the most fearful caterwauls and "spitting" that ever awakened a baby, made the head of the family swear in his dreams, or caused a shower of boots, washbowls and other missiles of midnight wrath to cleave the sky.

"Such is the invention. The method of using and the result is as follows: Winding up the patent Thomas-cat, the owner adjusts him upon the house-top or in the back-yard and awaits events. Soon is heard the tocsin of cat-like war in the shape of every known sound that the tribe are capable of producing, only in a key much louder than any live cat could perform in. Every cat within a circle of a half-mile hears the familiar sounds and accepts the challenge, frequently fifty or one hundred appearing simultaneously upon the battle-ground, ready to buckle in. The swelling tail invites combat and they attack old "Ironsides," who no sooner feels the weight of a paw upon his hide than a spring is touched off, his paws revolve in all directions with lightning rapidity and the adversaries within six feet of him are torn to shreds! Fresh battalions come to the scratch only to meet a like fate, and in the morning several bushels of hair, fiddle-strings and toe-nails is all that are seen, while the owner proceeds to wind the iron cat up and set him again.

"But a few pleasant evenings are needed to clean out a common-sized country town of its sleep-disturbers. We understand the inventor will make a proposition next week to the common council to depopulate the city of cats for a moderate sum. We do not intend to endorse any invention or article unless we know that it will perform all that it is claimed to do, and therefore we have not been so explicit in our description as we might have been; but the principle is a good one, and we hope to see every house in the city surmounted with a sheet-iron cat as soon as they are offered for sale, which will be about April the first, the inventor and patentee informs us."

J. H. Bowman and Richard Linn sent forth the *Petroleum Monthly* at Oil City in October, 1870. Their purpose was to treat the oil industry from a scientific stand and present statistics and biographies in magazine style. The *Monthly.* which lasted a year, was ably edited and supplied matter of permanent value. Bowman, a fascinating writer and agreeable companion, went westward and the snows of twenty winters have drifted over his grave. Linn aided in compiling a history of petroleum, spent some years in the east and meandered to Australia.

Pleasantville evolved the *Evening News* in 1869 and the semi-monthly *Commercial-Record* in 1887, both of which sought "the dark realms of everlasting shade," to keep company with J. L. Rohr's Cooperstown *News*, Tom Whitaker's *Gatling Gun*, the Oil City *Critic*, the Franklin *Oil Region*, the Petroleum Centre *Era*, and a score of unwept sacrifices on the altar of Venango journalism. James Tyson, a hardware merchant at Rouseville, in 1872 issued the *Pennsylvanian*, a superior weekly, which subsided with the waning town. Wesley Chambers, the ardent greenbacker and fortunate producer, was a liberal contributor to its well-edited pages. Mr. Tyson migrated to California, living in San Francisco until last year, when he located in Philadelphia. At the age of seventy-eight his faculties are unimpaired and he stands erect. He is an earnest member of the Pennsylvania Historical Society and compiler of a "Life of Washington and the Signers of the Declaration of Independence." This timely

JAMES TYSON.

and interesting work, published in two handsome volumes in 1895, is dedicated to the public schools of the nation. Its instructive narratives are designed to inculcate lessons of patriotism and duty to the country. As an embodiment of facts, collated with extreme care and presented attractively, it serves a laudable purpose and fitly crowns the literary labors of the revered author, who is "only waiting till the shadows are a little longer drawn."

Titusville enjoys the honor of harboring the first petroleum daily that weathered the storm and stayed in the ring. June, 1865, heralded the *Morning Herald* of W. W. and Henry C. Bloss, which possessed the entire field and prospered accordingly. Col. J. H. Cogswell joined the partnership in 1866.

JOHN PONTON.

Major W. W. Bloss, the elder of the two brothers, was a fluent writer and made his mark in journalism. Mastering the details of "the art preservative" at Rochester, N. Y., in 1857 he started a short-lived journal in Kansas, retraced his footsteps to his native heath in 1859, was badly wounded at Antietam, beamed upon Titusville in the spring of 1865 and bought the *Petroleum Reporter*, a moribund weekly. Quitting the *Herald* in six or eight years, in 1873 he unfurled the banner of the *Evening Press*, which did not live to cut its eye-teeth. His next attempt, a tasteful weekly, traveled the road to oblivion. The Major once more headed for Kansas, served in the Legislature and wended his way to Chicago, whence he crossed "to the other side" in the prime of matured manhood.

Henry C. Bloss stuck to the *Herald* "through evil and through good report," steadfastly upholding Titusville and dipping his eagle feather in vitriol when necessary to squelch "a foeman worthy of his steel." He died—the ranks are thinning out sadly—three years ago and his son, upon whom the mantle of his father

21

has descended, is keeping the paper in the van. Col. Cogswell, who dropped out to accept the postmastership, enacting the role of "Nasby" a couple of terms, for eight or ten years has been in the office of the Tidewater Pipe-Line Company. Among the *Herald* force were C. C. Leonard, John Ponton and A. E. Fay. Ponton turned his peculiar talent for invention to electrical pursuits and the giddy telephone. He narrowly missed heading off Prof. Bell in stumbling upon the "hello" machine. Fay forsook the *Herald* for the Oil City *Times*, did a turn on the Titusville *Courier* and hied him to Arizona. He ran a mining-paper, sat in the Legislature, incubated a chicken-nursery that would have dumbfounded Rutherford B. Hayes, farmed a bit and harvested a crop of shekels. "Art" was a nobby youth, popular with "the boys," liked a good cigar and could wing a meaty item on the fly.

The Titusville *Courier*, sprung in 1870 to oppose the *Herald*, was edited by Col. J. T. Henry, an accomplished journalist from Olean, N. Y. In 1871 he bought the *Sunday News*, formerly A. L. Chapman's *Long-Roll*, transferring it in 1872 to W. W. Bloss, who changed it to the *Evening Press*. Col. Henry in 1873 published "Early and Later History of Petroleum," a large volume, replete with information, biographies and portraits. The author speculated profitably in oil, lived at Olean, wrote as the impulse prompted and died at Jamestown in May, 1878. A tear is due the memory of a kingly, chivalrous man, who reflected luster upon his profession and was not fully appreciated until he had reached the haven of eternal rest.

> " A rarer spirit never
> Did steer humanity. * * * The elements
> So mix'd in him that nature might stand up
> And say to all the world, 'This was a man!' "

Warren C. Plumer guided the *Courier* after Col. Henry's retirement. He was no tyro in slinging his quill. Born in Maine in 1835, at fourteen he entered a printing-office, ten years later edited a paper, served three years in the war, set type on the Reno *Times* in 1865 and was editor-journeyman of the Pithole *Record* in the fall of 1866. His "Dedbete" contributions were a striking feature of the *Record*, of which he became joint-owner with Longwell and Wicker in 1867, when Burgess of Pithole, and editor-in-chief upon its removal to Petroleum Centre in 1868. Selling out in 1869, Wicker and Plumer lighted a *Weekly Star* at Titusville that quickly set to rise no more. Plumer was foreman of the Oil City *Times* in 1870-1 and connected with the Tidioute *Journal* in 1872, when offered the editorship of the *Courier*. Elected to the Legislature on the Democratic ticket in 1874, he was defeated for a second term and for Congress as the Greenbackers' candidate in 1878. For a time his political notions were as facile as his Faber and he trained with whatever party chanced to have a vacancy. From 1879 to 1881 he controlled the Meadville *Vindicator*, a soft-money weekly, winding up the latter year on the Richburg *Echo*. In Dakota, his next stamping-ground, he edited Republican papers at Fargo, Bismarck, Aberdeen and Casselton. He stumped several states for Blaine with an eye to an appointment that would have swelled his bank-account to the dimensions of a plumber's. "The Plumed Knight" failed to connect and the plum did not fall into the lap of his eloquent supporter. President Harrison in 1891 appointed him Receiver of the Minot District Land-office, North Dakota, which he resigned last year. As an orator Col. W. C. Plumer—they call him "Colonel" in the Dakotas—trots in the class with Robert G. Ingersoll, Thomas B. Reed and William McKinley and is denominated the "Silver Tongue of the North-west." At the Republican National Conventions in 1884-8 he was unanimously pronounced the finest off-hand speaker in the crowd. He is a finished lecturer and unrivaled story-teller, loves the choicest books, reads the Bible diligently, sticks to his friends and delights to recount his experiences in the Pennsylvania oil-regions.

M. N. Allen, an original stockholder and its last guardian, purchased the *Courier* in 1874. Even his acknowledged skill could not put it on a paying basis and the paper, unsurpassed in quality and appearance, succumbed to the inevitable. Mr. Allen followed Col. Cogswell as postmaster, a proper tribute to his rugged Democracy. Hale and hearty, although "over the summit of life," time has dealt kindly with him and his deft pen has lost none of its vigor. He is editing the *Advance Guard*, the outgrowth of Roger Sherman's departed *American Citizen*, as an intellectual pastime. F. A. Tozer, the champion "fat take," five-feet-four-inches high and four-feet-five-inches around, graduated from the *Courier*, wafted the St. Petersburg *Crude Local* up the flume and was chief-cook of the East-Brady *Times*. His reports were newsy and palatable. He travels for a Pittsburg house and would pay extra fare if passengers were carried by weight.

Graham & Hoag's *Sunday News-Letter* arose from the tomb of the *Evening Press* and the *Sunday News*. J. W. Graham, now of the *Herald*, piloted the trim vessel skilfully. A stock-company of producers, thinking a daily in the family would be "a thing of beauty" and "a joy forever," bought the *News-Letter* and the *Courier* equipment in 1879, to start the *Petroleum World*. James M. Place, a pusher from Pusherville, had solicited the bulk of the subscriptions to the stock and was entrusted with the management. R. W. Criswell edited the paper splendidly. Captain M. H. Butler, who put heaps of ginger into his spicy effusions, and John P. Zane, whose hobby was finance—both have gone the journey that has no return trip—embellished its columns with thoughtful, digestible brain-food. Oil-news, readable locals, dispatches, jaunty selections and bang-up neatness were never lacking. But competition was fierce and the *World* had a hard row to hoe. A committee of stockholders soon took charge. Place, sleepless, indomitable and with the energy of a steam-hammer, opened a big store at Richburg and drove a rattling trade. Setting out to paddle his own canoe as a Corry newsboy at ten, he had run a newsroom at Fagundas, a book-store and the post-office at St. Petersburg, a branch store at Edenburg, large stores at Bradford and Bolivar and won laurels as the greatest newspaper-circulator in the petroleum-diggings. At Harrisburg and Reading he swung Sunday papers and the *Saturday Globe* in New York, his present abode. Samuel Williams, unexcelled as a sprightly writer, and Hon. George E. Mapes, equally competent in the Legislature and the editorial chair, kept the *World* booming until "patience ceased to be a virtue" and the daily ceased to be a sheet. About half the material went to the Oil City *Blizzard* and the rest went to print the *Sunday World* Frank W. Truesdell had determined to originate. The late Hon. A. N. Perrin, ex-Mayor of Titusville, possessing "ample means and ample generosity," backed the project. Truesdell finished his trade as printer in Cleveland and worked at Youngstown and Franklin, settling at Titusville in 1880 to manage the *World* jobbing-room. He was a young man of fine ability and scrupulous integrity. His partnership with Perrin ended in 1887 by his purchase of the entire business. He sold a half-interest in the paper in 1893 and death claimed him in October of 1894. Measured by his thirty-seven years, Frank Willard Trucsdell's life was short; measured by his good deeds, his worthy enterprises, his lofty sentiments and kindly acts, it was longer than that of many who pass the Psalmist's three-score-and-ten. Mrs. Truesdell and her little daughter live in Titusville. F. F. Murray, associated with Walter Izant and W. R. Herbert in the general details, edits the *Sunday World*, which is as frisky as a spring-colt. Born at Buffalo in 1860, Murray was reared in Venango county, whither his father was drawn by the oil-excitement. Correspondence for local papers naturally bore him into the journalistic swim. He whooped it up six years for the *Blizzard*. A regular hummer, he is at home whether flay-

ing monopolists, taking a ruffian's scalp, praising a pretty girl, writing a tearful obituary, dissecting a suspicious job or reeling off a natty poem. "The Old Tramp-Printer," a recent effort, is a fair sample of his quality :

> " Here's a rhyme to the old tramp-printer, who as long as he lives will roam,
> Whose 'card' is his principal treasure and where night overtakes him home ;
> Whose shoes are run over and twisty, whose garments are shiny and thin,
> And who takes a bunk in the basement when the pressman lets him in.

> " It is true there are some of the trampers that only the Angel of Death,
> When he touches them with his sickle, can cure of the 'spirituous breath';
> That some by their fellow-trampers are shunned as unwholesome scamps,
> And that some are just aimless, homeless, restless, typographical tramps.

" But the most of them surely are worthy of something akin to praise,
And have drifted down to the present out of wholesomer, happier days ;
And when, though his looks be as seedy as ever a mortal wore,
Will you find the old tramper minus his marvelous fund of lore?

" What paper hasn't he worked on ? Whose manuscript hasn't he set ?
What story worthy remembrance was he ever known to forget ?
What topics rise for discussion, in science, letters or art,
That the genuine old tramp-printer cannot grapple and play his part ?

" It is true you will sometimes see him when the hue that adorns his nose
Outrivals the crimson flushes which the peony flaunts at the rose ;
It is true that much grime he gathers in the course of each trip he takes,
Inasmuch as he boards all freight-trains between the Gulf and the Lakes.

" Yet his knowledge grows more abundant than many much-titled men's,
Who travel as scholarly tourists and are classed with the upper-tens ;
And few are the contributions these scholarly ones have penned
That the seediest, shabbiest tramper couldn't readily cut and mend.

" He has little in life to bind him to one place more than the rest,
For his hopes in the past lie buried with the ones that he loved the best ;
He has little to hope from Fortune and has little to fear from Fate,
And little his dreams are troubled over the public's love or hate.

" So a rhyme to the old tramp-printer—to the hopes he has cherished and wept,
To the loves and the old home-voices that still in his heart are kept ;
A rhyme to the old tramp-printer, whose garments are shiny and thin,
And who takes a bunk in the basement when the pressman lets him in."

Mr. Mapes gravitated to Philadelphia to write for Colonel McClure's *Times.*
His are the appetizing paragraphs that burnish the editorial page by their subtile
essence. He is a familiar figure at party conventions, which his intimate knowledge of state-politics enables him to gauge accurately. He abhors trickery and
chicanery, deals his hardest blows in exposing corrupt methods, believes taxpayers and voters have rights contractors and bosses are bound to respect and
is a stickler for honest government. Williams also strayed to the Quaker City
as paragrapher for the *Press,* making a phenomenal hit. James G. Blaine com-
plimented Charles Emory Smith upon these tart, peppery nuggets, saying "I
invariably read the *Press* paragraphs before looking at any other paper." This
pleasant tribute added ten dollars a week to Sam's salary, yet he tired of
Philadelphia years ago and glided back to his old home in "the Messer Diocese."
He is now connected with the New-York *Herald.*

R. W. Criswell holds an honorable place among the men who have made
oil-region newspapers known abroad and influential at home. He was born in
Clarion county and educated in Cincinnati. His sketches, signed "Chris,"
introduced him to the public through the medium of the Oil City *Derrick,* the
East Brady *Independent* and the Fairview *Independent,* Colonel Samuel Young's
twin offspring. Retiring from Young's employ at Fairview, he was next heard
of as traveling correspondent of the Cincinnati *Enquirer.* His editorship of the
Derrick in 1877 clinched his fame as a Simon-pure humorist, thirty-six inches
to the yard and one-hundred cents to the dollar. The Shakesperian parodies
and Lickshingle stories, lustrous as the Kohinoor, waltzed the merry round
of the American press and were published in two taking books—"The New
Shakespeare" and "Grandfather Lickshingle." After his departure from the
Petroleum World Criswell renewed his relations with the *Enquirer* as managing-editor. He was John R. McLean's trusty lieutenant and held the great western daily on the topmost rung of the ladder. The New York *Graphic,* the pathfinder of illustrated dailies, needed him and he accepted its flattering offer,

The Cincinnati *Sun* was about to shine on the just and the unjust and he returned to Porkopolis. Colonel John Cockrell coaxed him back to Manhattanville to reconstruct the funny-streak of the overflowing New York *World.* When the Colonel and Joseph Pulitzer disagreed—they "never spoke as they passed by"—he went with Cockrell to the *Commercial Advertiser*, for which he is doing some of the brightest work in the newspaper kingdom. "Mark Anthony's Oration Over Cæsar," from "The Comic Shakespeare," will dispel the gloom and indicate the rare brand of Criswell's vintage:

"Friends, Romans, countrymen! lend me your ears;
I will return them next Saturday. I come
To bury Cæsar because the times are hard
And his folks can't afford to hire an undertaker.

"LEND ME YOUR EARS"

The evil that men do lives after them,
In the shape of progeny, who reap
The benefit of their life insurance.
So let it be with the deceased.
Brutus hath told you that Cæsar was ambitious,
What does Brutus know about it?
It is none of his funeral.
But that it isn't is no fault of the undersigned.
Here under leave of you I come to
Make a speech at Cæsar's funeral.
He was my friend, faithful and just to me;
He loaned me five dollars once when I was in a pinch,
And signed my petition for a post-office.
And Brutus says he was ambitious.
Brutus should chase himself around the block.
Cæsar hath brought many captives home to Rome
Who broke rock on the streets until their ransoms
Did the general coffers fill.
When that the poor hath cried, Cæsar hath wept,
Because it didn't cost anything
And made him solid with the masses.　　[*Cheers.*
Ambition should be made of sterner stuff.
Yet Brutus says he was ambitious.
Brutus is a liar, and I can prove it.

You all did see that on the Lupercal
I thrice presented him a kingly crown,
Which he did thrice refuse, because it did not fit him quite.
Was this ambition? Yet Brutus says he was ambitious.
Brutus is not only the biggest liar in this country,
But he is a politician of the deepest dye.　　[*Applause.*
If you have tears prepare to shed them now.　　[*Laughter.*

You all do know this ulster.
I remember the first time ever Cæsar put it on;
It was on a summer's evening in his tent,
With the thermometer registering ninety degrees in the shade:
But it was an ulster to be proud of,
And it cost him $3 at Marcalus Swartzheimer's,
Corner of Broad and Ferry streets, sign of the red flag.
Old Swartz wanted $40 for it,
But finally came down to $3, because it was Cæsar!

Look! in this place ran Casca's dagger through:
Through this the son of a gun of a Brutus stabbed,
And when he plucked his cursed steel away,
Good gracious how the blood of Cæsar followed it!

[*Cheers, and cries of " Give us something on the Wilson bill !" "Hit him again;" etc.*]

I came not, friends, to steal away your hearts.

I am no thief as Brutus is.
Brutus has a monopoly in all that business,
And if he had his deserts, he would be
In the State prison and don't you forget it.

Kind friends, sweet friends, I do not wish to stir you up
To such a sudden flood of mutiny,
And, as it looks like rain,
The pallbearers will please place the body
    in the wheelbarrow
And we will proceed to bury Cæsar,
Not to praise him."

EDWIN C. DELL.

Edwin C. Bell, a son of the Pine-Tree state, landed at Petroleum Centre in 1886, spent 1869 in the west, returned to Oil Creek in 1870 and for three years punched down oil-wells. In 1874 he started a job-printery at Pioneer, using a press he built from iron-scraps and an oak-rail and learning the trade without an instructor. That fall he transplanted his kit to Titusville and continued in the jobbing line fourteen years, failing health forcing him to recuperate for a spell. Early in 1878 he published the *Leader*, a weekly favoring the "Ohio idea" of paper-currency and the remonetization of silver. Although an artistic and editorial triumph, the *Leader* was a financial failure and petered out in two months. Undismayed by this reverse, Mr. Bell in 1882 flew the flag of the *Republic*, a campaign-oracle of the Greenbackers and supporter of Thomas A. Armstrong for governor. The *Republic*, like the *Argus*, the *Observer* and others of that ilk, didn't attain old age. Bell's first grists—stories and sketches—went into the *Courier* hopper in 1872, suplemented from 1878 to 1882 by bundles of live matter in the Meadville *Vindicator* and the Richburg *Echo*. He edited the *Republican* at Casselton, N. D., in 1882-3, and for the

STEPHEN W. HARLEY.

nine years following his return to Titusville sent a news-letter almost daily to the Oil City *Blizzard*. He has long contributed to the *Sunday World* and in 1888-9 was its assistant-editor. In 1892 he began a history of the Pennsylvania oil-regions, instalments of which the *Derrick* printed, and he hopes to finish the task on a comprehensive scale befitting the subject. His collection of newspaper-files, extracts, jottings, letters, statistics and petroleum-literature cannot be matched anywhere in extent and completeness. No reader ever rang the chestnut-bell on Edwin C. Bell's cheery and instructive tid-bits.

At Tidioute the *Journal*, inaugurated by J. B. Close in 1867, jogged along seven years, competing with White & Co.'s newsy *News*. Part of the time G. A. Needle managed it and W. C. Plumer, Maior McClintock and Frank H. Taylor were attaches. Warren has been blessed with two weeklies, the *Ledger* and the *Mail*, for two generations. Ephraim Cowan founded the *Mail* in 1848 and owned it until his death in 1894. Three dailies vigilantly watch each other and guard the pretty town. Needle, whose sharp lance could prick the fiends of the opposition like a needle, followed the tide to Parker and boosted the *Daily*, which shortly plunged into perpetual night. Its chief contributor was Stephen· W. Harley, who furnished rich budgets of Petrolia odds and ends over the name of " Keno." "Steve" was kindly, obliging, congenial and well-liked. Five summers have come and gone since he was laid beneath the sod. J. Wilson removed the *Oilman's Journal* to Smethport and the weekly *Phœnix* is in undisputed possession of the Parker territory. Clarion county did not escape the frantic rush to stick a newspaper in every mushroom-town. F. H. Barclay inflicted the *Record* on the long-suffering St. Petersburgers, mooring his bark in California when the paper turned up its toes. Tozer's *Crude Local*, which never sported a crude local or editorial, the Fern City *Illuminator*, brighter in name than in substance, the Clarion *Banner*, a species of rag on the bush, the Edenburg *National* and several more slid off the perch with a dull thud, fatal as Humpty Dumpty's tumble.

Frank Herr's *Record* is still making a good record at Petrolia. Colonel Young and the three papers he propagated in Butler county, with a half-dozen elsewhere, have mouldered into dust. He was intensely earnest and industrious, able to maintain his end of a discussion and seldom unwilling to dare opponents knock the chip off his stout shoulder. Rev. W. A. Thorne attempted to reform the race with his Greece City *Review*, hauling the traps to Millerstown upon the depletion of the frontier town. His path was strewn with thorns, mankind resenting his review of everybody and everything. Ex-Postmaster Rattigan braces up the unterrified with his sturdy Chicora *Herald*. St. Joe's bantam crowed mildly and dropped from the roost. The county-seat is fully stocked with political organs, the *Citizen*, the *Eagle* and the *Herald* coaching their respective parties. J. H. Negley & Son are not negligent in their conduct of the *Citizen*. The *Eagle* is the proud bird of Thomas H. Robertson, a trained writer and journalist, now Superintendent of Public Printing in Harrisburg. The *Herald* was for many years the pet of Jacob Zeigler, to whom all Butlerites took off their hats. "Uncle Jake" was the soul of the social circle, a treasury of wit and wisdom, an exhaustless reservoir of pat stories, a mine of practical knowledge and a welcome guest in every corner of Pennsylvania. His soubriquet of "Uncle" fastened upon him in a curious way. At the funeral of a youthful acquaintance the distracted mother, as her boy was consigned to the grave, in a frenzy of grief laid her head upon young Zeigler's breast and exclaimed : "Oh, were you ever a stricken mother?" "No, madam," was the cool reply, "but I expect to be an uncle before sundown to-morrow." Bystanders noted the strange incident and thenceforth the "Uncle" stuck like a fly-blister. His parents are buried in the Harrisburg cemetery, near Joseph Jefferson's father, and whenever he visited the capital he strewed their resting-place with flowers. Who can doubt that the filial son, in whom mingled the strength of a man and the tenderness of a woman, found his loved ones not far away when he entered the pearly gates? Truly "this was the noblest Roman of them all."

Another honored resident of Butler was Samuel P. Irvin, author of "The Oil-Bubble," a pamphlet abounding with delicious satire and bits of personal experience. It was printed in 1868 and produced a sensation. Enjoying very few advantages in his boyhood, Mr. Irvin was emphatically a self-made man. Born in a backwoods-township seventy years ago, his schooling was limited and he toiled "down on the farm." Like Lincoln, Garfield, Simon Cameron and many other country-boys, he rose to distinction by his own exertions. He read assiduously, studied law and stood well at the bar. His literary bent found expression in newspaper articles of very high grade. His charm of manner and speech, his vast store of anecdote and reminiscence, his sincerity and frankness, his warm regard for friends and strong dislike of anything that savored of meanness, combined to render him exceedingly popular. He lived some years at Franklin in the earlier stages of petroleum developments, drilling wells and handling oil-properties on commission. He met death with becoming fortitude, "like one who wraps the drapery of his couch about him, and lies down to pleasant dreams."

The Bradford semi-weekly *New Era*, harbinger of the new era dawning upon McKean county, saw daylight in the spring of 1875. The main object of its founder, Colonel J. K. Haffey, was to invite attention to the possibilities of the locality as a prospective oil-field. Colonel Haffey was a man of varied talents—public-speaker, writer, soldier, surveyor, promoter of oil-enterprises, railroader and expounder of the gospel. Irish by birth, he came to America at fourteen, lived three years in Canada, was licensed to preach and in 1851, at the age of twenty-one, accepted a call to the Baptist church at Bradford, then Littleton.

Marrying Diantha, youngest daughter of Nathan De Golier, in December of 1852, a year later he quit the pulpit, sensibly concluding that the Lord had not called him to starve his family. As surveyor and geologist, he was employed to prospect for coal and iron in McKean and adjacent counties. In 1858-9 he had charge of a gang of men grading the Erie railroad to Buttsville. The first man in Bradford township to enlist in 1861, he raised a force for Colonel Kane's famous "Bucktails," drilled in Harrisburg six weeks, shared in the fighting around Richmond and was honorably discharged because of impaired health, in 1863, with the rank of major. Governor Hartranft appointed him on his staff and the title of colonel resulted. He sold his Bradford home in 1877 and removed to Beverly, N. J., where he published the *Banner*, a temperance paper. His active, helpful career ended on November seventh, 1881, in his fifty-second year. Mrs. Haffey returned to Bradford, to pass the evening of life amid the friends and associations of her girlhood, until summoned to rejoin him who "is not lost, but gone before," whose "good remembrance lies richly in her thoughts."

Ferrin & Weber, of Salamanca, publishers of the *Cattaraugus Republican*, in 1876 bought the *New Era* from Col. Haffey and placed it in charge of Charles F. Persons. He had been in their establishment at Little Valley two years. For nine or ten months he washed rollers, fed presses, carried wood and did the chores allotted to the "printer's devil." His aptitude impressed his employers, who sent him first to Salamanca and then to Bradford, an important post for a youth of twenty-two. Hoping to be an editor some day, he had corresponded for neighboring papers from boyhood on his father's farm, a practice he maintained during his apprenticeship. A few months after reaching Bradford he and the Salamanca firm established the *Daily Era*, with the names of Ferrin, Weber & Persons at the mast-head. Very soon Persons bought out his partners and conducted the paper alone. His ability and energy had full play. The *Era* met the demands of the eager, restless crowds that thronged the streets of Bradford and scoured the hills in quest of territory. Its news was concise and fresh, its oil-reports were not doctored for speculative ends, it had opinions and presented them tersely. Persons sold to W. H. Longwell and W. F. Jordan

early in 1879 and in the fall bought the Olean *Democrat*. The nobby New-York town was feeling the stimulus of oil-operations and he started the *Daily Herald*, enhancing his wallet and well-won reputation. The American Press-Association, which furnishes plate-matter to thousands of newspapers, secured him in 1888 as Local Manager of its New-York office. Two years ago he was promoted to General Eastern-Manager and in 1894 was elected Secretary, Assistant General-Manager and one of the five directors. Mr. Persons occupies a snug home in Brooklyn, with his wife and two little daughters. He is a live representative of the go-ahead, enterprising, sagacious, executive American.

Longwell & Jordan also bought the *Breeze*—it first breathed the oil-laden air of Bradford in 1878 and was edited by David Armstrong, "organizer" of the producers in one of their movements to "get together"—and consolidated it with the *Era*. Col. Edward Stuck, of York, worked the combination successfully some months. Colonel Leander M. Morton was night-editor until his lamented death. Thomas A. Kern attended to the field, preparing the "monthly reports" and posting readers on oil-developments in his bailiwick. Years have flown since poor "Tom," young and enthusiastic, and J. K. Graham, exact and upright, responded to the message that brooks no excuse or postponement. "Musing on companions gone, we doubly feel ourselves alone." Bradshaw, McMullen and others scattered. Jordan, whose first work for papers was done at Petrolia in 1873, went to Harrisburg in 1889. P. C. Boyle secured the *Era* and infused into it much of his own prompt, courageous spirit. David A. Dennison has for years been its efficient editor. His parents removed from Connecticut to a farm south of Titusville when he was a baby. At thirteen David wrote a batch of items, which it tickled him to see in print, without a thought of one day blossoming into a full-fledged "literary feller." Not caring to be a tiller of the soil, he juggled the hammer and lathe in machine-shops to the music of "the Anvil Chorus." A short season on the boards convinced him that he was not commissioned to elevate the stage and wrest the scepter from Edwin Booth, Lawrence Barrett, John McCullough or Alexander Salvini. He whisked to a Bradford shop to strike the iron while it was hot, writing smart descriptions of oil-region scenes for outside papers as a side-issue for several years. A series of his articles on gas-monopoly, in the Elmira *Telegram*, brought reduced rates to consumers and pleasant notoriety to the ironworker, who had proved himself a blacksmith with the sledge and no "blacksmith" with the quill. His name was neither Dennis nor Mud, and the *Daily Oil News*, McMullen & Bradshaw's game-fowl, wanted him forthwith. The salary was not alluring and in the Indian-summer days of 1886 he cast in his lot with the *Era*. Promotion chased him persistently. From reporter he was boosted to city-editor and in 1894 to the editorial management, a flawless selection. He has tussled with all sorts of topics, constructed tales of woe in jingling verse and even tempted fate by firing off a drama, which has not yet run the gamut of publicity. Dennison has been offered good sits in metropolitan offices, but he likes Bradford and clings to the *Era*. He married Miss Katharine Grady in 1883 and three boys gladden the home of the exultant D. A. D. "May his shadow never grow less."

The *Daily News* was converted by Mr. Butler into a Sunday sheet, which Commodore Linderman, *Era* book-keeper and assistant-manager, navigated until it ran ashore in 1894. Butler reeled off the Buffalo *News*, the sharpest, quickest, breeziest afternoon paper in the Bison City. Eben Brewer's *Evening Star* tinted the sky under George Allen's artistic stroke. Allen slid to Buffalo to polish up a railroad periodical and at last accounts was rushing the United Press. H. F. Barber, a man of fine intellect and noble purpose, honed the *Star* for years. Protracted sickness, during which he showed "how sublime it is to suffer and be strong," at last "withered the garlands on his brow." He is dead, but "his speaking dust has more of life than half its breathing moulds." "Judge" Johnson—in 1875 he landed at Bradford, served a term in the Legislature and another as postmaster, operated in oil and died two or three years ago—controlled the *Star* after Barber, whose widow still retains an interest in the paper. Ex-Senator Emery fitted out the *Daily Record*, which seeks to trail the standard of the Standard in the dust and ticket independent producers, refiners and pipe-liners to a petroleum Utopia. "Ed." Jones, the adept who toed the chalk-mark on the Harrisburg *Call*, whirled the emery-wheel so expertly that the *Record* has never approached Davy Jones's locker. It is snappy and full of fight as a shillalah at Donnybrook Fair. Andrew Carr's *Sunday Mail*, freighted with a car of delicate morsels, barked up the wrong tree and went to the bow-wows.

EDWARD C. JONES.

Carr rolled down to Pittsburg to sell buggies, bagging a cargo of ducats. "Tom" L. Wilson —he's as humorous as they make 'em—got out three numbers of *Sunday Morning*, a four-page blanket in size and a ten-course banquet in contents. Col. Ege shut it down for publishing a rank extract from Walt Whitman's "Blades o' Grass." Ege was a banker who hankered to be State Treasurer, banked upon newspaper support, went into bankruptcy, received an appointment in the Philadelphia Mint and traveled westward when Cleveland shuffled the pack for a new deal. Wilson wrote for the oil-region press, handled the Reading branch of a Harrisburg paper, edited the Washington *Review*—Sisterville has a sisterly *Review* now—and rounded up in Buffalo. The *Post*, Bradford's latest Sunday experiment, owes its good looks and good matter to Edward F. McIntyre and George O. Sloan.

One evening in 1877 a young stranger walked into the St. Petersburg post-office, bought a package of stationery at the book-counter and told J. M. Place he was looking for a situation. Place hired him as a clerk. He had come from the homestead farm in Orange county, N. Y., to Cornell University, worked his way and graduated in civil-engineering. Marshall Swartzwelder lectured at St. Petersburg on temperance and Place's clerk sat up all night to report the masterpiece for the *Derrick*. It was his first production in print, a voluntary act on his part, and the article attracted most favorable notice. Its author was at once offered a position on the *Derrick*. He came in contact with oil-statistics and his real genius asserted itself. His painstaking, conscientious reports were accepted as strictly reliable. He would trudge over the hills, wade through miles of mud and ford swollen streams to ascertain the precise status of an important well, rather than approximate it from hearsay. This care and thoroughness gave the highest value to the statistical work of Justus C. McMullen. In 1879 he went to Bradford and worked on the *Breeze*, the *Era* and the *Star*, always with the same devotion that was a ruling maxim of his life. In 1883 he scouted in Warren and Forest counties and became part owner of the *Petroleum Age*. Alfred L. Snell and Major W. C. Armor were associated with him in this admirable monthly, of which he became sole proprietor on the first of December, 1887. A. C. Crum, now on the editorial staff of the Pittsburg *Dispatch*, contributed many a newsy crumb to the *Age*. A newsboy at Pickwick hailed me in front of his stand one cool morning and asked—not in a Pickwickian sense—if it would be worth while to get somebody to send locals to the *Derrick*. "Why not do it yourself?" was my answer. He tried and he succeeded. His work expanded and improved and he adopted journalism permanently. He catered for Oil City and Bradford papers, spun yarns for Pittsburg dailies and was a legislative correspondent several sessions. Snell, a statistical hummer and hard-to-beat purveyor of news, hangs his manuscript on the *Derrick* hook. Armor sponsored a historic book and laid off his armor to second Dr. Egle in the State Library. He has a book-store in Harrisburg and a museum that distances the "Old Curiosity Shop." McMullen established and edited the *Daily Oil-News* in 1886. He died of pleurisy, contracted from exposure in collecting oil-data, on January thirty-first, 1888, cut off at thirty-seven. The *Petroleum Age* did not

stay long behind its unswerving projector. Justus C. McMullen is enshrined in the affections of the people. An unrelenting foe of oppression, he had a warm heart for the poor and pursued his own path of right through thorns or flowers. He married Miss Cora, daughter of Col. L. M. Morton, who lives in Bradford and has one little girl. A brave, grand, exalted spirit passed from earth when J. C. McMullen's light was quenched.

> " On the sands of life
> Sorrow treads heavily and leaves a print
> Time cannot wash away."

Melville J. Kerr, a Franklin boy, son of the senior proprietor of the marble-works, is a popular writer of facetiæ and society small-talk. Possibly "a rose by any other name would smell as sweet," but his cognomen of "Jo Ker" is known to thousands of smiling readers who never heard of Melville. The aspiring youth, believing in the advantages of a big city, journeyed to New York to look for an opportunity that might want a party about his size and style. Unlike Jacob for Rachel, Penelope for Ulysses, the zealots who prayed for Ingersoll's conversion or the Governor of South Carolina for the Governor of North Carolina to "fill 'em up again," he didn't wait long. A soap-mogul liked the ambitious, sprightly young man, introduced him to the swell set and booked him as editor of *The Club*. Kerr's refined humor popped and effervesced with more "bead" than ever. He hobnobbed with millionaires, delighted Ward McAlister and married a lovely girl. Blood will tell as surely as a gossip or a tale-bearer. And that is how the "Jo Ker" is the winning card in one oil-region life.

In the Franklin office of the Galena Oil Works are three successful weavers of rich textures in the literary loom—Dr. Frank H. Johnston, E. H. Sibley and

Samuel H. Gray. Dr. Johnston was born in Canal township, reared on a farm, severely wounded in battling for the Union, studied medicine, practiced at Cochranton and in 1872 located at Petrolia. There "he first essayed to write" for the Oil City *Derrick.* From the very outset his articles were up to concert pitch. Abandoning medicine for letters, he acquired a thorough knowledge of stenography, read the choicest books and wrote in his best vein for the press.

He represented the *Derrick* as its Franklin correspondent with credit to himself and the paper. For sixteen years he has been connected with the Galena Oil Works as secretary of Hon. Charles Miller, a place demanding the superior qualifications with which the doctor is unstintingly endowed.

Edwin Henry Sibley, born at Bath, N. Y., in 1857, is a brother of Hon. Joseph C. Sibley and has resided in Franklin twenty-two years. He was graduated from Cornell University in 1880. For several years he has been treasurer of the Galena Oil-Works and manager of Miller & Sibley's famous Prospect-Hill Stock Farm, positions of responsibility to which his personal address, his training and his business-methods adapt him pre-eminently. Under guise of "Polybius Crusoe Smith, Sage of Cranberry Cross-Roads"—the Smiths are a great family since the by-play of Pocahontas—he contributes to *Puck* and other well-known publications humorous articles and short, quaint, pithy sayings. These display a keen insight into human nature and rare gift of happy, accurate expression. One of his recent effusions—an address welcoming the delegates to an agricultural convention—is a bit of burlesque that deserves to rank with Artemus Ward's brightest efforts or the richest paragraphs in the Biglow Papers. A few buds plucked at random from the flowery mead will serve to illustrate the high-class stamp of Mr. Sibley's work in the field his genius adorns. They are literary nosegays from his terse observations as a philosophic "looker-on in Vienna:"

"The pygmies of Africa are such by nature, but elsewhere they are produced artificially by a diet of petty and envious thoughts."

"'Truth is mighty and will prevail,' but Error generally has the better of it till the seventy-seventh round."

"One of the greatest evils that humanity has to contend with is that so many icebergs have floated down from the North Pole and persist in passing themselves off for men."

"Former loves in making out their title-deeds of the heart to their successors always reserve at least a narrow pathway across a corner."

"Wise men and fools have foolish thoughts; fools tell them, wise men keep them to themselves."

"Parents that haven't time to correct their children when they are small will have time to weep over them when they are grown."

"Affectation (an alias of Deceitfulness) has three picked cronies from whom she is seldom separated. Their names are False Pride, Weakmindedness and Bad Temper."

"If one has too much vitality in his brains he can get rid of it by taking them out and boiling them. If he finds this too much bother, he can accomplish the same result by swallowing a few doses of a decoction of faith-cure, spook-lore and hypnotism."

"The ancient Israelites once worshipped a golden calf, but the modern Americans would worship a golden polecat if they couldn't get the gold in any other form to worship."

"The wife that manages her husband is a genius, the one that bosses him is a tartar, the one that fights with him is a fool, while the one that does none of them is now as much out of fashion as her grandmother's wedding-gown."

"For peace of mind and length of days, put this inscription above the doorway of workshop and home: *Troubles that will not be worth worrying over seven years hence are not worth worrying over now.*"

Samuel H. Gray carries under his hat plenty of the gray-matter that makes bright writers and bright wooers of the Muses. He has been court stenographer of Venango county and holds a confidential position with the firm of Miller & Sibley, applying his spare moments to newspaper writing. His pictures of petroleum-traits and incidents are finished word-paintings, with "light and shade and color properly disposed." Like Silas Wegg, he "drops into poetry" in a friendly way. Such papers as the New York *Truth* strive for his emanations, which savor of Bret Harte and "hold the mirror up to nature" in oleaginous circles. Judge of this " By the Order of the Lord," founded on an actual occurrence in Scrubgrass township:

> "It was back, if I remember, in the year of sixty-five,
> When we formed a part and parcel of that rushin', busy hive
> That extended from Oil City up the crooked crick until
> It reached its other endin' in the town of Titusville;
> When every rock an' hillside was included in a lease,
> An' everyone was huntin' fer the fortune-makin' grease;
> When a poor man pushed and elbowed 'gainst the oily millionaire,
> An' 'the devil take the hindmost' seemed the all-pervadin' prayer.

> "An we hed formed a pardnership, jest Tom an' Jim an' me,
> That was properly recorded as the ' Tough and Hungry Three,'
> An' hed gone an' leased a portion of some hard an' rocky soil
> That we thought looked like the cover of a fountain filled with oil.
> An' we set the drill a'goin' on its long an' greasy quest,
> That meant so much or little to the capital possessed.
> Our money was all in the well, in Providence our trust,
> An' we waited for a fortune, or to liquidate an' bu'st.

> "An' while the drill was chuggin' at its hard an' rocky way
> We three would hold a meetin' at a certain time each day,
> The 'resolves' an' the 'whereases' that the secretary took
> Were properly recorded 'twixt the covers of a book.
> An' we passed a resolution by a vote unanimous
> Thet if Providence would condescend to sorter favor us,
> An' assist the operations on the ' Tough and Hungry' lease,
> We would give to Him a quarter of the total flow of grease.

" Next day the drill broke through into a very oily sand
An' Providence remembered us with strong, unsparin' hand;
The oil came out with steady flow an' loaded up the tanks,
An' the Lord was due rewarded by a solid vote of thanks.
A resolution then came up thet caused the vote to split,
A sort of an amendment, readin' somethin' like, to wit—
' Whereas, a tenth is all the Lord was ever known to crave,
Resolved we give it to Him; but resolved the rest we save.'

"I fit that resolution, an' I fit it tooth an' nail,
Spoke of dangers such proceedin's was most likely to entail;
But two votes were in its favor, an' two votes it only took
Fer to have it due recorded in the resolution-book.
Next day the oil stopped flowin' an' it never flowed no more,
An' the 'Tough and Hungry' combine was a' feelin' blue an' sore,
But they nailed upon the derrick this notice, on a board,
' This well has stopped proceedin's, by the order of the Lord.' "

The Rev. S. J. M. Eaton—his name is ever spoken with reverence—thirty-three years pastor of the Presbyterian church at Franklin, filled a large place in the literary guild.   He loved especially to delve into old books and papers and letters pertaining to the pioneers of Northwestern Pennsylvania.   His faithful

labors in this neglected nook unearthed a troop of traditions and facts which "the world will not willingly let die."   For the "History of Venango County" he furnished a number of leading chapters. His published works include " Petroleum," an epitome of oil-affairs down to 1866, "Lakeside," a tale based upon his father's ministerial experiences in the wilds of Erie county, biographies of eminent divines, sketches of the Erie Presbytery, pamphlets and sermons.   "The Holy City" and " Palestine," embodying his observations in the orient, were issued as text-books by the Chautauqua Circle. Dr. Eaton was my near neighbor for years and hours in his well-stocked library, enriched by his "affluence of discursive talk," are recalled with deep satisfaction.   On the sixteenth of

REV. S. J. M. EATON, D.D.

July, 1889, while walking along the street, he raised his hands suddenly and fell to the pavement, struck down by heart-failure.   "He was not, for God took him" to wear the victor's crown.   Farewell, "until the day dawn and the shadows flee away."

Last June a compact "Life of Napoleon Bonaparte," in harmony with the age of steam and electricity that won't winnow a bushel of chaff for a grain of wheat, which had run through the winter and spring of 1894-5 in *McClure's Magazine*, was published in book-form.   Napoleonic ground had been so plowed and harrowed and raked and scraped and sifted by Hugo, Scott, Abbott, Hazlitt, Bourrienne, Madame Junot and a host of smaller fry that it seemed idle to expect anything new concerning the arbiter of Europe.   Yet the beauty and freshness and acumen of this "Life" surprised and captivated its myriad readers, whose pleasure it increased to learn that the book was the production of a young woman.   The authoress is Miss Ida M., daughter of Franklin S. Tarbell, a wealthy oil-operator.   Her childhood was spent at Rouseville, where her parents lived prior to occupying their present home at Titusville.   The romantic surroundings were calculated to awaken glowing fancies in the acute mind of the little girl. After graduating from Allegheny College, Meadville, she taught in the seminary at Poland, O. Finding the duties of preceptress too burdensome, she returned home to recuperate. The editor of *The Chautauquan* desiring her to aid in preparing Notes on the Readings for the Chautauqua Literary and Scientific Circle, she went to Meadville. She picked up editorial work bit by bit until she found herself associate-editor of the organ of the wonderful Assembly. The magazine required much department-assorting, drudgery apt to become intellectually dissipating and lead ultimately into a rut. Miss Tarbell, therefore, took up a special line of study to vary her routine. Choosing French Revolutionary History, she decided to prepare a series of magazine articles on the women of the dark days of Robespierre, Danton, Marat and Marie Antoinette, holding steadily to the subject until she knew something satisfactory about it. In the spring of 1891 she resigned from the *The Chautauquan* and went abroad to study history, selecting Paris as the place in which to carry out her plans. She remained three years on the continent, giving her time almost exclusively to French historical subjects, writing for American newspapers and contributing special articles to *Scribner's, McClure's* and the *New England Magazine*. This careful preparation fortified her to undertake, upon returning from France in 1894, the "Life of Napoleon" that has brought the gifted daughter of the oil-regions enduring fame. The Scribners will soon publish Miss Tarbell's biographical study of Madame Roland, the heroine of the Revolution. Just now she is residing in New York, editing the articles on Abraham Lincoln in *McClure's*. Her success thus early in her career gives fruitful promise of a resplendent future for the vivacious, winsome biographer of the "Little Corporal."

While many names and terms and phrases peculiar to oil-operations are unintelligible to the tenderfoot as "the confusion of tongues" at Babel, others will be valuable additions to the language. "He has the sand" aptly describes a gritty, invincible character. The fortunate adventurer "strikes oil," the pompous strutter is "a big gasser," foolish anger is "pumping roily" and fruitless enterprise is "boring in dry territory." Misdirected effort is "off the belt," failure "stops the drill," a lucky investment "hits the jugular," a hindrance "sticks the tools" and an abandoned effort "plugs the well." A man or well that keeps at it is "a stayer," one that doesn't pan out is "a duster," one that cuts loose is "a gusher" or "a spouter." Fair promise means "a good show," the owner of pipe-line certificates "has a bundle," fleeced speculators are "shorn lambs"—not limited to Oildom by a large majority—and the ruined operator "shuts 'er down." In a moment of inspiration John P. Zane created "the noble producer," Lewis F. Emery invented "the downtrodden refiner" and Samuel P. Irvin exploited "the Great Invisible Oil-Company." Some of these epigrammatic phrases deserve to go thundering down the ages with Grant's "let us have peace," Cleveland's "pernicious activity" and "a sucker is born every minute."

Nor is the jargon of places and various appliances devoid of interest to the student of letters. Oil City, Petroleum Centre, Oleopolis, Petrolia, Greece City—first spelled G-r-e-a-s-e—Gas City, Derrick City and Oil Springs were named with direct reference to the slippery commodity. From prominent operators came Funkville, Shamburg, Tarr Farm, Rouseville, McClintockville, Fagundas, Prentice, Cochran, Karns City, Angelica, Criswell City, Gillmor, Duke Centre and Dean City. Noted men or early settlers were remembered in Titusville, Shaffer, Plumer, Trunkeyville, Warren, Irvineton, McKean, De Golier, Custer City,

22

Garfield, Franklin, Reno, Foster, Cooperstown, Kennerdell, Milton, Foxburg, Pickwick, Parker, Troutman, Butler, Washington, Mannington and Morgantown. Emlenton commemorates Mrs. Emlon Fox. St. Joe recalls Joseph Oberly, a pioneer-operator in that portion of Butler county. Standoff City kept green a contractor who wished to "stand-off" his men's wages until he finished a well. A deep hole or pit on the bank of the creek, from which air rushed, suggested Pithole. Tip-Top, near Pleasantville, signified its elevated site. Cornplanter, the township in which Oil City is situated, bears the name of the stalwart chief—six feet high and one hundred years old—to whom the land was ceded for friendly services to the government and the white settlers. This grand old warrior died in 1836 and the Legislature erected a monument over his grave, on the Indian reservation near Kinzua. Venango, Tionesta, Conewago, Allegheny, Modoc and Kanawha smack of the copper-hued savage once monarch of the whole plantation. Red-Hot, Hardscrabble, Bullion, Babylon, St. Petersburg, Fairview. Antwerp, Dogtown, Turkey City and Triangle are sufficiently obvious. Sisterville, the centre of activity in West Virginia, is blamed upon twin-islets in the river. Alemagooselum is a medley as uncertain in its origin as the ingredients of boarding-house hash. Diagrams are needed to convey a reasonable notion of "clamps," "seed-bags," "jars," "reamers," "sockets," "centre-bits," "mud-veins," "tea-heads," "conductors," "Samson-posts," "bull-wheels," "band-wheels," "walking-beams," "grasshoppers," "sucker-rods," "temper-screws," "pole-tools," "casing," "tubing," "working-barrels," "standing-valves," "check-valves," "force-pumps," "loading-racks," "well-shooters," "royalty," "puts," "calls," "margins," "carrying-rates," "spot," "regular," "pipage," "storage," and the thousand-and-one things that make up the past and present of the lingo of petroleum.

THE WOMAN'S EDITION.

To raise twenty-five-hundred dollars for an annex to the hospital, the ladies of Oil City, on February twelfth, 1896, issued the "Woman's Edition" of the *Derrick*. It was a splendid literary and financial success, realizing nearly five-thousand dollars. This apt poem graced the editorial page:

Oh! sad was her brow and wild was her mien,
Her expression the blankest that ever was seen;
She was pained, she was hurt at the plain requisition:
"We expect you to write for the Woman's Edition."

Her babies wept sadly, her husband looked blue,
Her house was disordered, each room in a stew;
Do you ask me to tell why this sad exhibition?
She was trying to write for the Woman's Edition.

Oh, what should she write? she had nothing to say;
She pondered and thought all the long weary day;
The question of woman, her life and her mission,
Must all be touched up in the Woman's Edition.

But what could she do—oh, how could she write?
She could bake, she could brew from morning to night;
She had even been known to get up a petition:
But now she must write for "The Woman's Edition."

She felt that she must; her sisters all did it,
Would she fall behind? The saints all forbid it!
If the rest of her life should be spent in contrition,
She felt she must write for the Woman's Edition.

She did it, she wrote it, now read it and ponder;
She treated a subject a little beyond her,
But that was much better than total omission
Of her name from the list on the Woman's Edition.

Now her home is restored, her husband has smiled,
But, alas! that pleased look on his face was beguiled
By her cheerful assent to his simple condition:
That she'll not write again for a Woman's Edition.

WELL FLOWING OIL AFTER TORPEDOING.

XIII.

NITRO-GLYCERINE IN THIS.

EXPLOSIVES AS AIDS TO THE PRODUCTION OF OIL—THE ROBERTS TORPEDO MONOPOLY AND ITS LEADERS—UNPRECEDENTED LITIGATION—MOONLIGHTERS AT WORK—FATALITIES FROM THE DEADLY COMPOUND—PORTRAITS AND SKETCHES OF VICTIMS—MEN BLOWN TO FRAGMENTS—STRANGE ESCAPES—THE LOADED POKKER—STORIES TO ACCEPT OR REJECT AS IMPULSE PROMPTS.

" There is no distinguished Genius altogether exempt from some infusion of Madness."—*Aristotle.*
" Genius must be born and never can be taught."—*Dryden.*
" He who would seek for pearls must dive below."—*Addison.*
" Labor with what zeal we will, something still remains undone."—*Longfellow.*
" Come, bright Improvement, on the car of Time."—*Campbell.*
" Revenge, at first though sweet, bitter ere long, back on itself recoils."—*Milton.*
                " Only these fragments and nothing more !
                Can naught to our arms the lost restore? "—*Ibid.*
" Death itself is less painful when it comes upon us unawares."—*Pascal.*
        " Dead? did you say he was dead? or is it only my brain?
    He went away an hour ago ; will he never come again?"—*Tamar Kermode.*
        " There is no armor against fate."—*Shirley.*
" Dreadful is their doom * * * like yonder blasted bough by thunder riven."—*Beattie.*

NITRO-GLYCERINE LETS GO.

WHEN in 1846 a patient European chemist hit upon a new compound by mixing fuming nitric-acid, sulphuric-acid and glycerine in certain proportions, he didn't know it was loaded. Glycerine is a harmless substance and its very name signifies sweetness. Combining it with the two acids changed the three ingredients materially. The action of the acids caused the glycerine to lose hydrogen and take up nitrogen and oxygen. The product, which the discoverer baptized Nitro-Glycerine, appeared meek and innocent as Mary's little lamb and was readily mistaken for lard-oil. It burned in lamps, consuming quietly and emitting a gentle light. But concussion proved the oily-looking liquid to be a terrible explosive, more powerful than gun-cotton, gunpowder or dynamite. For twenty years it was not applied to any useful purpose in the arts. Strangely enough, it was first put up as a homœopathic remedy for headache, because a few drops rubbed on any portion of the body pained the head acutely. James G. Blaine was given doses of it on his death-bed. An energetic poison, fatalities resulted from imbibing it for whisky, which it resembles in taste. After a time attention was directed unexpectedly to its explosive qualities. A small consignment, sent to this country as a specimen, accidentally exploded in a New-York street. This set the newspapers and the public talking about it and wondering what caused the stuff to go off. Investigation solved the mystery and revealed the latent power of the compound, which had previously figured only as a rare chemical in a half-score foreign laboratories. Miners and contractors gradually learned its value for blasting masses of rock. Five pounds, placed in a stone-jar and suspended against the iron-side of the steamer Scotland, sunk off Sandy Hook, cut a fissure twelve feet long in the vessel. A steamship at Aspinwall was torn to atoms and people stood in mortal terror of the destructive agent. Girls threw away the glycerine prescribed for chapped lips, lest it should burst up and distribute them piecemeal over the next county. Their cotton-padding or charcoal-dentifrice was as dangerous as the glycerine alone, which is an excellent application for the skin. A flame or a spark would not explode Nitro-Glycerine readily, but the chap who struck it a hard rap might as well avoid trouble among his heirs by having had his will written and a cigar-box ordered to hold such fragments as his weeping relatives could pick from the surrounding district. Such was the introduction to mankind of a compound that was to fill a niche in connection with the production of petroleum.

Paraffine is the unrelenting foe of oil-wells. It clogged and choked some of the largest wells on Oil Creek and diminished the yield of others in every quarter of the field. It incrusts the veins of the rock and the pipes, just as lime in the water coats the tubes of a steam-boiler or the inside of a tea-kettle. How to overcome its ill effects was a question as serious as the extermination of the potato-bug or the army-worm. Operators steamed their wells, often with good results, the hot vapor melting the paraffine, and drenched them with benzine to accomplish the same object. A genius patented a liquid that would boil and fizz and discourage all the paraffine it touched, cleaning the tubing and the seams in the sand much as caustic-soda scours the waste-pipe of a sink or closet. These methods were very limited in their scope, the steam condensing, the benzine mixing with the oil and the burning fluid cooling off before penetrating the crevices in the strata any considerable distance. Exploding powder in holes drilled at the bottom of water-wells had increased the quantity of fluid or opened new veins and the idea of trying the experiment in oil-wells suggested itself to various operators. In 1860 Henry H. Dennis, who drilled and stuck the tools in the first well at Tidioute, procured three feet of two-inch copper-pipe, plugged one end, filled it with rifle-powder, inserted a fuse-cord and exploded the charge in presence of six men. The hole was full of water, oil and bits of rock were blown into the air and "the smell of oil was so much stronger that people coming up the hollow noticed it." The same year John F. Harper endeavored to explode five pounds of powder in A. W. Raymond's well, at Franklin. The tin-case holding the powder collapsed under the pressure of the water and the fuse had gone out. William Reed assisted Raymond and W. Ayers Brashear, who had expected James Barry—he put up the first telegraph-line between Pittsburg and Franklin—to fire the charge by electricity. Reed developed the idea and invented the "Reed Torpedo," which he used in a number of wells. A large crowd in 1866 witnessed the torpedoing of John C.

Ford's well, on the Widow Fleming farm, four miles south of Titusville. Five pounds of powder in an earthen bottle, attached to a string of gas-pipe, were exploded at two-hundred-and-fifty feet by dropping a red-hot iron through the pipe. The shock threw the water out of the hole, threw out the pipe with such force as to knock down the walking-beam and samson-post, agitated the water in Oil Creek and "sent out oil." Tubing was put in, the old horse worked the pump until tired out and the result encouraged Ford to buy machinery to keep the well going constantly. This was *the first successful torpedoing of an oil-well!* The Watson well, near by, was similarly treated by Harper, who had brought four bottles of the powder from Franklin and was devoting his time to "blasting wells." For his services at the Ford well he received twenty dollars. Harper, William Skinner and a man named Potter formed a partnership for this purpose. They torpedoed the Adams well, on the Stackpole farm, below the Fleming, putting the powder in a glass-bottle. The territory was dry and no oil followed the explosion. In the fall of 1860 they shot Gideon B. Walker's well at Tidioute. Five torpedoes were exploded in 1860 at Franklin, Tidioute and on Oil Creek. Business was disturbed over the grave political outlook, oil was becoming too plentiful, the price was merely nominal and the torpedo-industry languished.

William F. Kingsbury advertised in 1860 that he would "put blasts in oil-wells to increase their production." He torpedoed a well in 1861 on the island at Tidioute, using a can of powder and a fuse, which ignited perfectly. Mark Wilson and L. G. Merrill lectured on electricity in 1860–61, traveling over the country and exhibiting the principle of "Colt's Submarine Battery," by which "the rock at any distance beneath the surface of the earth may be rent asunder, thereby enabling the oil to flow to the well." Frederick Crocker in 1864 arranged a torpedo to be dropped into a well and fired by a pistol-cartridge inserted in the bottom of the tin-shell. About thirty torpedoes were exploded from 1860 to 1865, all of them in wells filled with water, which served as tamping. Erastus Jones, James K. Jones and David Card exploded them in wells at Liverpool, Ohio. Joseph Chandler handled two or three at Pioneer and George Koch fired one of his own construction in May of 1864. Mr. Beardslee —he struck a vein of water by drilling a hole five feet and exploding a case of powder at the bottom of a well in 1844, near Rochester, N. Y.—came to the oil-region and put in a score of shots in 1865. As long ago as 1808 the yield of water in a well at Fort Regent was doubled by drilling a small hole and firing a quantity of powder. A flowing-well on the lease beside the Crocker stopped when the latter was torpedoed and was rigged for pumping. It pumped "black powder-water," showing that the torpedo had opened an underground connection between the two wells, the effects of the explosion reaching from the Crocker to its neighbor. William Reed made a can strong enough to resist the pressure of the water, let it down the Criswell well on Cherry Run in 1863, failed to discharge it by electricity and exploded it by sliding a hollow weight down a string to strike a percussion-cap.

Notwithstanding these facts, which demonstrated that the yield of oil and water had been increased by exploding powder hundreds of feet under water, in November of 1864 Col. E. A. L. Roberts applied for a patent for "a process of increasing the productiveness of oil-wells by causing an explosion of gunpowder or its equivalent at or near the oil-bearing point, in connection with superincumbent fluid-tamping." He claimed that the action of a shell at Fredericksburg in 1862, which exploded in a mill-race, suggested to him the idea of bombarding oil-wells. However this may be—it has been said he was not at Fredericksburg at the date specified in his papers—the Colonel furnished no drawings and presented no application for Letters Patent for over two years. He constructed six of his torpedoes and arrived with them at Titusville in January of 1865. Captain Mills permitted him to test his process in the Ladies' well, near Titusville, on January twenty-first. Two torpedoes were exploded and the well flowed oil and paraffine. Reed, Harper and three or four others filed applications for patents and commenced proceedings for interference. The suits dragged two years, were decided in favor of Roberts and he secured the patent that was to become a grievous monopoly.

A company was organized in New York to construct torpedoes and carry on the business extensively. Operators were rather sceptical as to the advantages of the Roberts method, fearing the missiles would shatter the rock and destroy the wells. The Woodin well, a dry-hole on the Blood farm, received two injections and pumped eighty barrels a day in December of 1866. During 1867 the demand increased largely and many suits for infringements were entered. Roberts seemed to have the courts on his side and he obtained injunctions against the Reed Torpedo-Company and James Dickey for alleged infringements. Justices Strong and McKennan decided against Dickey in 1871. Producers subscribed fifty-thousand dollars to break down the Roberts patent and confidently expected a favorable issue. Judge Grier, of Philadelphia, mulcted the Reed Company in heavy damages. Nickerson and Hamar, ingenious, clever fellows, fared similarly. Roberts substituted Nitro-Glycerine for gunpowder and established a manufactory of the explosive near Titusville. The torpedo-war became general, determined and uncompromising. The monopoly charged exorbitant prices—two-hundred dollars for a medium shot— and an army of "moonlighters"—nervy men who put in torpedoes at night— sprang into existence. The "moonlighters" effected great improvements and first used the "go-devil drop-weight" in the Butler field in 1876. The Roberts crowd hired a legion of spies to report operators who patronized the nocturnal well-shooters. The country swarmed with these emissaries. You couldn't spit in the street or near a well after dark without danger of hitting one of the crew. Unexampled litigation followed. About *two-thousand* prosecutions were threatened and most of them begun against producers accused of violating the law by engaging "moonlighters." The array of counsel was most imposing. It included Bakewell & Christy, of Pittsburg, and George Harding, of Philadelphia, for the torpedo-company. Kellar & Blake, of New York, and General Benjamin F. Butler were retained by a number of defendants. Most of the individual suits were settled, the annoyance of trying them in Pittsburg, fees of lawyers and enormous costs inducing the operators to make such terms as they could. By this means the coffers of the company were filled to overflowing and the Roberts Brothers rolled up millions of dollars.

The late H. Bucher Swope, the brilliant district-attorney of Pittsburg, was especially active in behalf of Roberts. The bitter feeling engendered by convictions deemed unjust, awards of excessive damages and numerous imprisonments found expression in pointed newspaper paragraphs. Col. Roberts preserved in scrap-books every item regarding his business-methods, himself and his associates. One poetical squib, written by me and printed in the Oil-City *Times*, incensed him to the highest pitch and was quoted by Mr. Swope in an argument before Judge McKennan. The old Judge bristled with fury. Evidently he regretted that it was beyond his power to sentence somebody to the penitentiary for daring hint that law was not always justice. He had not traveled quite so far on the tyrannical road as some later wearers of the ermine, who, "dressed in a little brief authority, play such fantastic tricks before high heaven as make the angels weep" and consign workingmen to limbo for presuming to present the demands of organized labor to employers ! It is not Eugene V. Debs or the mouthing anarchist, but the overbearing corporation-tool on the bench, who is guilty of "contempt of court."

The Roberts patent re-issued in June of 1873, perpetuating the burdensome load upon oil-producers. In November of 1876 suit was brought in the Circuit Court against Peter Schreiber, of Oil City, charged with infringing the Roberts process. Schreiber's torpedo duplicated the unpatented Crocker cartridge and Roberts wanted his scalp. The case was contested keenly four years, coming up for final argument in May of 1879. Henry Baldwin and James C. Boyce, of Oil City, and Hon. J. H. Osmer, of Franklin, were the defendant's attorneys. Mr. Boyce collected a mass of testimony that seemed overwhelming. He spent years working up a masterly defense. By unimpeachable witnesses he proved that explosives had been used in water-wells and oil-wells, substantially in the manner patented by Roberts, years before the holder of the patent had been heard of as a torpedoist. But his masterly efforts were wasted upon Justices Strong and McKennan. They had sustained the monopoly in the previous suits and apparently would not reverse themselves, no matter how convincing the reasons. Mr. Schreiber, wearied by the law's interminable delays and thirty-thousand dollars of expenditure, decided not to suffer the further annoyance of appealing to the United-States Supreme Court. The great body of producers, disgusted with the courts and despairing of fair-play, did not care to provide the funds to carry the case to the highest tribunal and lock it up for years awaiting a hearing. The flood of light thrown upon it by Boyce's researches had the effect of preventing an extension of the patent and reducing the price of torpedoes, thus benefiting the oil-region greatly. Mr. Boyce is now practicing his profession in Pittsburg. He resided at Oil City for years and was noted for his bright wit, his incisive logic, his profound interest in education and his social accomplishments.

Col. Edward A. L. Roberts died at Titusville on Friday morning, March twenty-fifth, 1881, after a short illness. His demise was quite unexpected, as he continued in ordinary health until Tuesday night. Then he was seized with intermittent fever, which rapidly gained ground until it proved fatal. A moment before dissolution he asked Dr. Freeman, who was with him, for a glass of water. Drinking it and staring intently at the doctor, his eyes filled with tears and he said, "I am gone." Pressing back upon the pillow, he expired almost instantly. Col. Roberts was born at Moreau, Saratoga county, New York, in 1829. At seventeen he enlisted as a private, served with commendable bravery in the Mexican war and was honorably discharged after a service of two years. Returning to his native place, he entered an academy and passed several years acquiring a higher education. Subsequently he entered the dental office of his brother at Poughkeepsie, N. Y. Still later he removed to the city and with his brother, W. B. Roberts, engaged in the manufacture of dental material. For his improvements in dental science and articles he was awarded several gold-medals by the American Institute. He patented various inventions that have been of great service and are now in general use. In the oil-region he was best known as the owner of the torpedo-patent bearing his name. He came to Titusville in January of 1865 and the same month exploded two shells in the Ladies' well, increasing its yield largely. From that time to the present the use of torpedoes has continued. The litigation over the patent and infringements attracted widespread attention. The last week of his life Col. Roberts said he had expended a quarter-million dollars in torpedo-litigation. He was responsible for more lawsuits than any other man in the United States. A man of many eccentricities and strong feelings, he was always liberal and enterprising. He left a large fortune and one of the most profitable monopolies in the State. In 1869 he married Mrs. Chase, separated from her in 1877 and lived at the Brunswick Hotel. His widow and two children survived him. Col. Roberts did much to build up Titusville and his funeral was the largest the town has ever witnessed. He sleeps in the pretty cemetery and a peculiar monument, emblematic of the torpedo, marks the burial-plot.

On the palatial Hotel Brunswick, which he built and nurtured as the apple of his eye, Col. Roberts lavished part of his wealth. He decorated and furnished it gorgeously from cellar to roof. The appointments were luxurious throughout. If the landlords he engaged could not meet expenses, the Colonel paid the deficiency ungrudgingly and sawed wood. Finally the house was conducted in business-style and paid handsomely. For years it has been run

by Charles J. Andrews, who was born with a talent for hotel-keeping. "Charlie" is well-known in every nook and corner of Pennsylvania as a "jolly good fellow," keen politician and all-round thoroughbred. He has the rare faculty of winning friends and of engineering bills through the Legislature. He is head of the Liquor League, a tireless worker, a masterly joker and brimming over with pat-stories that do not strike back. He operates in oil and base-ball as a diversion, is a familiar figure in Philadelphia and Harrisburg and popular everywhere.

Dr. Walter B. Roberts, partner of his brother in the torpedo-company, clerked in an Albany bank, taught district-school, studied medicine and rose to eminence in dentistry. Visiting Nicaragua in 1853, he established a firm to ship deer-skins and cattle-hides to the United States and built up a large trade with Central America. Resuming his practice, he and E. A. L. Roberts opened dental-rooms in New York. His brother enlisted and upon returning from the war assigned the Doctor a half-interest in a torpedo for oil-wells he desired to patent. In 1865 Dr. Roberts organized the Roberts Torpedo-Company, was chosen its secretary in 1866 and its president in 1867. He visited

Europe in 1867 and removed to Titusville in 1868, residing there until his death. In 1872 he was elected mayor, but his intense longing for a seat in Congress was never gratified. The oil-producers, whom the vexatious torpedo-suits made hot under the collar, opposed him resolutely. He had succeeded in his profession and his business and his crowning ambition was to go to Washington. The arrow of political disappointment pricked his temper at times, although to the last he supported the Republican party zealously. Dr. Roberts was a man of marked characteristics, tall, stoutly built and vigorous mentally. He did much to advance the interests of his adopted city and was respected for his courage, his earnestness and his benevolence to the poor.

Hon. William H. Andrews managed the campaign of Dr. Roberts, who fancied the adroitness, pluck and push of the coming leader and used his influence to elect him chairman of the Crawford-County Republican Committee. He performed the duties so capably that he served four terms, was secretary of the State Committee in 1887-8 and its chairman in 1890-1. Mr. Andrews was born in Warren county and at an early age entered upon a mercantile career. He established large dry-goods stores at Titusville, Franklin and Meadville, introduced modern ideas and did a tremendous business. He advertised by the page, ran excursion-trains at suitable periods and sold his wares at prices to attract multitudes of customers. Nobody ever heard of dull trade or hard times at any of the Andrews stores. Removing to Cincinnati, he opened the biggest store in the city and forced local merchants to crawl out of the old rut and hustle. But the aroma of petroleum, the motion of the walking-beam, the dash and spirit of oil-region life were lacking in Porkopolis and Andrews returned to Titusville. He engaged in politics with the ardor he had displayed in trade. His skill as an organizer saved the Congressional district from the Greenbackers and won him the chairmanship of the Republican State-Committee. He

WILLIAM H. ANDREWS.

served two terms in the Legislature and was elected to the Senate in 1894. He is chairman of the senatorial committee appointed last session to "Lexow" Philadelphia and Pittsburg. His brother, W. R. Andrews, edits the Meadville *Tribune* and is secretary of the State Committee. Another, Charles J. Andrews, is proprietor of the Hotel Brunswick and an active politician. Senator Andrews rarely wastes his breath on long-winded speeches, wisely preferring to do effective work in committee. No member of the House or Senate is more influential, more ready to oblige his friends, more sought for favors and surer of carrying through a bill. He enjoys the confidence of Senator Quay and his next promotion may be to the United-States Senate as successor of J. Donald Cameron. Mr. Andrews lives at Titusville, has oil-wells on Church Run and a big farm in the suburbs, is prominent in local industries and a representative citizen.

Gradually the quantity of explosive in a torpedo was increased, in order to shatter a wider area of oil-bearing rock. A hundred quarts of Nitro-Glycerine have been used for a single shot. In such instances it is lowered into the well in cans, one resting upon another at the bottom of the hole until the desired amount is in place. A cap is adjusted to the top of the last can, the cord that lowered the Nitro-Glycerine is pulled up, a weight is dropped upon the cap and an explosion equal to the force of a ton of gunpowder ensues. In a few seconds a shower of water, oil, mud and pebbles ascends, saturating the derrick and pelting broken stones in every direction. Frank H. Taylor graphically describes a scene at Thorn Creek :

"On October twenty-seventh, 1884, those who stood at the brick school-house and telegraph-offices in theThorn Creek district and saw the Semple, Boyd & Armstrong No. 2 torpedoed, gazed upon the grandest scene ever witnessed in Oildom. When the shot took effect and the barren rock, as if smitten by the rod of Moses, poured forth its torrent of oil, it was such a magnificent and awful spectacle that no painter's brush or poet's pen could do it justice. Men familiar with the wonderful sights of the oil-country were struck dumb with astonishment, as they beheld the mighty display of Nature's forces. There was no sudden reaction after the torpedo was exploded. A column of water rose eight or ten feet and fell back again, some time elapsed before the force of the explosion emptied the hole and the burnt glycerine, mud and sand rushed up in the derrick in a black stream. The blackness gradually changed to yellow; then, with a mighty roar, the gas burst forth with a deafening noise, like the thunderbolt set free. For a moment the cloud of gas hid the derrick from sight and then, as this cleared away, a solid golden column half-a-foot in diameter shot from the derrick-floor eighty feet through the air, till it broke in fragments on the crown-pulley and fell in a shower of yellow rain for rods around. For over an hour that grand column of oil, rushing swifter than any torrent and straight as a mountain pine, united derrick-floor and top. In a few moments the ground around the derrick was covered inches deep with petroleum. The branches of the oak-trees were like huge yellow plumes and a stream as large as a man's body ran down the hill to the road. It filled the space beneath the small bridge and, continuing down the hill through the woods beyond, spread out upon the flats where the Johnson well is. In two hours these flats were covered with a flood of oil. The hill-side was as if a yellow freshet had passed over it. Heavy clouds of gas, almost obscuring the derrick, hung low in the woods, and still that mighty rush continued. Some of those who witnessed it estimated the well to be flowing five-hundred barrels per hour. Dams were built across the stream, that its production might be estimated ; the dams overflowed and were swept away before they could be completed. People living along Thorn Creek packed up their household-goods and fled to the hill-sides. The pump-station, a mile-and-a-half down the creek, had to extinguish its fires that night on account of gas. All fires around the district were put out. It was literally a flood of oil. It was estimated that the production was ten-thousand barrels the first twenty-four hours. The foreman, endeavoring to get the too's into the well, was overcome by the gas and fell under the bull-wheels. He was rescued immediately and medical aid summond. He remained unconcious two hours, but subsequently recovered fully. Several men volunteered to undertake the job of shutting in the largest well ever struck in the oil-region. The packer for the oil-saver was tied on the bull-wheel shaft, the tools were placed over the hole and run in. But the pressure of the solid stream of oil against it prevented its going lower, even with the suspended weight of the two-thousand-pound tools. One-thousand pounds additional weight were added before the cap was fitted and the well closed. A casing-connection and tubing-lines connected the well with a tank "

Had the owners not torpedoed this well, which they believed to be dry, its value would never have been known. Its conceded failure would have chilled ambitious operators who held adjoining leases and changed the entire history of Thorn Creek.

Torpedoing wells is a hazardous business. A professional well-shooter must have nerves of iron, be temperate in his habits and keenly alive to the fact that a careless movement or a misstep may send him flying into space. James Sanders, a veteran employé of the Roberts Company, fired six-thousand torpedoes without the slightest accident and lived for years after his well-earned retirement. Nitro-Glycerine literally tears its victims into shreds. It is quick as lightning and can't be dodged. The first fatality from its use in the oil-regions befell William Munson, in the summer of 1867, at Reno. He operated on Cherry Run, owning wells near the famous Reed and Wade. He was one of the earliest producers to use torpedoes and manufactured them under the Reed patent. A small building at the bend of the Allegheny below Reno served as his workshop and storehouse. For months the new industry went along quietly, its projector prospering as the result of his enterprise. Entering the building one morning in August, he was seen no more. How it occurred none could tell, but a frightful explosion shivered the building, tore a hole in the ground and annihilated Munson. Houses trembled to their foundations, dishes were thrown from the shelves, windows were shattered and about Oil City the horrible shock drove people frantically into the streets. Not a trace of Munson's premises remained, while fragments of flesh and bone strewn over acres of ground too plainly revealed the dreadful fate of the proprietor. The mangled bits were carefully gathered up, put in a small box and sent to his former home in New York for interment. The tragedy aroused profound sympathy. Mentally, morally and physically William Munson was a fine specimen of manhood, thoroughly upright and trustworthy. He lived at Franklin and belonged to the Methodist church. His widow and two daughters survived the fond husband and father. Mrs. Munson first moved to California, then returned eastward

WILLIAM MUNSON.

and she is now practicing medicine at Toledo, the home of her daughters, the younger of whom married Frank Gleason.

The sensation produced by the first fatality had not entirely subsided when the second victim was added to a list that has since lengthened appallingly. To ensure conparative safety the deadly stuff was kept in magazines located in isolated places. In 1867 the Roberts Company built one of these receptacles two miles from Titusville, in the side of a hill excavated for the purpose. Thither Patrick Brophy, who had charge, went as usual one fine morning in July of 1868. An hour later a terrific explosion burst upon the surrounding country with indescribable violence. Horses and people on the streets of Titusville were thrown down, chimneys tumbled, windows dropped into atoms and for a time the panic was fearful. Then the thought suggested itself that the glycerine-magazine had blown up. At once thousands started for the spot. The site had been converted into a huge chasm, with tons of dirt scattered far and wide. Branches of trees were lopped off as though cut by a knife and hardly a particle could be found of what had so recently been a sentient being, instinct with life and feeling and fondly anticipating a happy career. The unfortunate youth bore an excellent character for sobriety and carefulness. He was a young Irishman, had been a brakeman on the Farmers' Railroad and visited the magazine frequently to make experiments.

On Church Run, two miles back of Titusville, Colonel Davison established a torpedo-manufactory in 1868. A few months passed safely and then the tragedy came. With three workmen—Henry Todd, A. D. Griffin and William Bills —Colonel Davison went to the factory, as was his practice, one morning in September. A torpedo must have burst in course of filling, causing sad destruction. The building was knocked into splinters, burying the occupants beneath the ruins. All around the customary evidences of havoc were presented, although the sheltered position of the factory prevented much damage to Titusville. The mangled bodies of his companions were extricated from the wreck.

while Colonel Davison still breathed. He did not regain consciousness and death closed the chapter during the afternoon. This dismal event produced a deep impression, the extinction of four lives investing it with peculiar interest to the people of Oildom, many of whom knew the victims and sincerely lamented their mournful exit.

Dr. Fowler, the seventh victim, met his doom at Franklin in 1869. He had erected a magazine on the hill above the Allegheny Valley depot, in which large quantities of explosives were stored. With his brother Charles the Doctor started for the storehouse one forenoon. At the river-bridge a friend detained Charles for a few moments in conversation, the Doctor proceeding alone. What happened prior to the shock will not be revealed until all secrets are laid bare, but before Charles reached the magazine a tremendous explosion launched his brother into eternity. A spectator first noticed the boards of the building flying through space, followed in a moment by a report that made the earth quiver. The nearest properties were wrecked and the jar was felt miles away. Careful search for the remains of the poor Doctor resulted in a small lot of broken bones and pieces of flesh, which were buried in the Franklin cemetery. It was supposed that the catastrophe originated from the Doctor's boots coming in contact with some glycerine that may have leaked upon the floor. This is as plausible a reason as can be assigned for a tragedy that brought grief to many loving hearts. The Doctor was a genial, kindly gentleman and his cruel fate was universally deplored.

William A. Thompson, of Franklin, left home on Tuesday morning, August thirteenth, 1870, carrying in his buggy a torpedo to be exploded in a well on the Foster farm. John Quinn rode with him. At the farm he received two old torpedoes, which had been there five or six weeks, having failed to explode,

WILLIAM A. THOMPSON.

to return to the factory. Quinn came up the river by rail. Thompson stopped at Samuel Graham's, Bully Hill, got an apple and lighted a cigar. On leaving he said: "Good-bye, Sam, perhaps you'll never see me again!" Five minutes later an explosion was heard on the Bully-Hill road, a mile from where Dr. Fowler had met his doom. Graham and others hurried to the spot. The body of Thompson, horribly mutilated, was lying fifty feet from the road, the left arm severed above the elbow and missing. The horse and the forewheels of the buggy were found a hundred yards off, the wounded animal struggling that distance before he fell. The body and hindwheels of the vehicle were in splinters. One tire hung on a tree and a boot on another. The main charge of the torpedo had entered the victim's left side above the hip and the face was scarcely disfigured. Mr. Thompson was widely known and esteemed for his social qualities and high character. He was born in Clearfield county, came to Franklin in 1853, married in 1855 and met his shocking fate at the age of thirty-nine. His widow and a daughter live at Franklin.

Thus far the losses of human life were occasioned by the explosion of great quantities of the messengers of death. The next instance demonstrated the amazing strength of Nitro-Glycerine in small parcels, a few drops ending the existence of a vigorous man at Scrubgrass, Venango county, in the summer of 1870. R. W. Redfield, agent of a torpedo-company, hid a can of glycerine in the bushes, expecting to return and use it the following day. While picking berries Mrs. George Fetterman saw the can and handed it to her husband. Thinking it was lard-oil, which Nito-Glycerine in its fluid state resembles closely, Fetterman poured some into a vessel and sent it to his wells. It was used as a lubricant for several days. Noticing a heated journal one morning, Fetterman put a little of the supposed oil on the axle, with the engine in rapid motion. A furious explosion ensued, tearing the engine-house into splinters and partially stunning three men at work in the derrick. Poor Fetterman was found shockingly mangled, with one arm torn off and his head crushed into jelly. The mystery was not solved for hours, when it occurred to a neighbor to test the contents of the oil-can. Putting *one drop* on an anvil, he struck it a heavy blow and was hurled to the earth by the force of the concussion. The can was a common oiler, holding a half-pint, and probably not a dozen drops had touched the journal before the explosion took place. Fetterman was a man of remarkable physical power, weighing two-hundred-and thirty pounds and looking the picture of health and vigor. Yet a quarter-spoonful of nitro-glycerine sufficed to usher him into the hereafter under circumstances particularly distressing.

In the fall a young man lost his life almost as singularly as Fetterman. He attended a well at Shamburg, seven miles south of Titusville. The well was torpedoed on a cold day. To thaw the glycerine a tub was filled with hot water, into which the cans were put. When sufficiently thawed they were taken out, the glycerine was poured into the shell and the torpedoing was done satisfactorily. The tubing was replaced in the well and the young pumper went to turn on the steam to start the engine, carrying a pair of tongs with him. He threw the tongs into the tub of water. In an instant the engine-house was demolished by a fierce explosion. The luckless youth was killed and his body mangled. A small amount of glycerine must have leaked from the cans while they were thawing, as the result of which a soul was hurried into the presence of its Maker with alarming suddenness.

In August of 1871 Charles Clarke started towards Enterprise, a small village in Warren county, ten miles east of Titusville, with a lot of glycerine in a vehicle drawn by one horse. The trip was destined never to be accomplished. By the side of a high hill a piece of very rough road had to be traveled. There the charge exploded. Likely some of the liquid had leaked over the buggy and springs and been too much jolted. The concussion was awful. Pieces of the woodwork and tires were carried hundreds of yards. Half of one wheel lodged near the top of a large tree and for many rods the forest was stripped of its foliage and branches. Part of the face, with the mustache and four teeth adhering, was the largest portion of the driver recovered from the debris. The horse was disemboweled and to numerous trees lots of flesh and clothing were sticking. From the ghastly spectacle the beholders turned away shuddering. The handful of remains was buried reverently at Titusville, crowds of people uniting in the last tribute of respect to "Charlie," whose youth and intelligence had made him a general favorite.

A case similar to Thompson's followed a few weeks after, near Rouseville. Descending a steep hill on his way from torpedoing a well on the Shaw farm, William Pine was sent out of the world unwarned. He had a torpedo-shell and some cans of glycerine in a light wagon drawn by two horses. No doubt, the extreme roughness of the road exploded the dangerous freight. The body of the driver was distributed in minute fragments over two acres and the buggy was destroyed, but the horses escaped with slight injury, probably because the force of the shock passed above them as they were going down the hill. Pine had a premonition of impending disaster. When leaving home he kissed his wife affectionately and told her he intended, should he return safely, to quit the torpedo-business forever next day. He was an industrious, competent young man, deserving of a better fate.

In October of the same year Charles Palmer was blown to pieces at the Roberts magazine, near Titusville, where Brophy died two years before. With Captain West, agent of the company, he was removing cans of glycerine from a wagon to the magazine. He handled the cans so recklessly that West warned him to be more careful. He made thirteen trips from the wagon and entered the magazine for the fourteenth time. Next instant the magazine disappeared in a cloud of dust and smoke, leaving hardly a trace of man or material. West happened to be beside the wagon and escaped unhurt. The horses galloped furiously through Titusville, the cans not taken out bounding around in the wagon. Why they did not explode is a mystery. Had they done so the city would have been leveled and thousands of lives lost. Palmer paid dearly for his carelessness, which was characteristic of the rollicking, light-hearted fellow whose existence terminated so shockingly.

This thrilling adventure decided Captain West, who lived at Oil City, to engage in pursuits more congenial to himself and agreeable to his devoted family. He was finely educated, past the meridian and streaks of gray tinged his dark hair and beard. In November he torpedoed a well for me on Cherry Run. The shell stuck, together we drew it up, the Captain adjusted the cap and it was then lowered and exploded successfully. At parting he shook my hand warmly and remarked: "This is the last torpedo I shall put in for you. My engagement with the company will end next week. Good-bye. Come and see me in Oil City." Three days later he went to shoot a well at Reno, saying to his wife at starting: "This will wind up my work for the company." Such proved to be the fact, although in a manner very different from what the speaker imagined. The shell was lowered into the well, but failed to explode and the Captain concluded to draw it up and examine the priming. Near the surface it exploded, instantly killing West, who was guiding the line attached to the torpedo. He was hurled into the air, striking the walking-beam and falling upon the derrick-floor a bruised and bleeding corpse. He had, indeed, put in his last torpedo. The main force of the explosion was spent in the well, otherwise the body and the derrick would have been blown to atoms. A tear from an old friend, as he recounts the tragic close of an honorable career, is due the memory of a man whose sterling qualities were universally admired.

Early in 1873 two young lives paid the penalty at Scrubgrass. On a bright February morning "Doc" Wright, the torpedo-agent, stopped at the station to send a despatch. The message sent, he invited the telegraph-operator, George Wolfe, to ride with him to the magazine, a mile up the river. The two set out in high spirits, two dogs following the sleigh. Hardly ten minutes elapsed when a dreadful report terrified the settlement. From the magazine on the river-bank a light smoke ascended. Two rods away stood the trembling horse, one eye torn from its socket and his side lacerated. Beside him one dog lay lifeless. Fragments of the cutter and the harness were strewn around promiscuously. Through the bushes a clean lane was cut and a large chestnut-tree uprooted. A deep gap alone remained of the magazine and scarcely a particle of the two men could be found. Dozens of splintered trees across the Allegheny indicated alike the force and general direction of the concussion. A boot containing part of a human foot was picked up fifty rods from the spot. Wright's gold-watch, flattened and twisted, was fished out of the Allegheny, two-hundred yards down the stream, in May. The remains, which two cigar-boxes would have held, were interred close by. A marble shaft marks the grave, which Col. William Phillips, then president of the Allegheny-Valley Railroad, enclosed with a neat iron-railing. It is very near the railway-track and the bank of the river, a short distance above Kennerdell Station. The disaster was supposed to have resulted from Wright's using a hatchet to loosen a can of glycerine from the ice that held it fast. A pet spaniel, which had a habit of rubbing against his legs and trying to jump into his arms, accompanied him from his boarding-house. The animal may have diverted his attention momentarily, causing him to miss the ice and strike the can. The horse lived for years, not much the worse except for the loss of one eye. Wright and Wolfe were lively and jocular and their sad fate was deeply regretted. Many a telegram George Wolfe sent for me when Scrubgrass was at full tide.

One morning in April of 1873 Dennis Run, a half-mile from Tidioute, experienced a fierce explosion, which vibrated buildings, upset dishes and broke windows long distances off. It occurred at a frame structure on the side of a hill, occupied by Andrew Dalrymple as a dwelling and engine-house. He was a "moonlighter," putting in torpedoes at night to avoid detection by the Roberts spotters, and was probably filling a shell at the moment of the explosion. It knocked the tenement into toothpicks and killed Dalrymple, jamming his head and the upper portion of the trunk against an adjacent engine-house, the roof of which was smeared with blood and particles of flesh. One arm lay in the small creek four-hundred feet away, but not a vestige of the lower half of the body could be discovered. A feeble cry from the ruins of the building surprised the first persons to reach the place. Two feet beneath the rubbish a child twenty months old was found unhurt. Farther search revealed Mrs. Dalrymple, badly mangled and unconscious. She lingered two hours. The little orphan, too young to understand the calamity that deprived her of both parents, was adopted by a wealthy resident of Tidioute and grew to be a beautiful girl. Thousands viewed the sad spectacle and followed the double funeral to the cemetery. It has been my fortune to witness many sights of this description, but none comprised more distressing elements than the sudden summons of the doomed husband and wife. Mrs. Dalrymple was the only woman in the oil-region whom Nitro-Glycerine slaughtered.

Is there a sixth sense, an indefinable impression that prompts an action without an apparent reason? At Petrolia one forenoon something impelled me to go to Tidioute, a hundred miles north, and spend the night. Rising from breakfast at the Empire House next morning, a loud report, as though a battery of boilers had burst, hurried me to the street. Ten minutes later found me gazing upon the Dalrymple horror. Was the cause of the impulse that started me from Petrolia explained? An hour sufficed to help rescue the child from the debris, inspect the wreck, glean full particulars and board the train for Irvineton. Writing the account for the Oil-City *Derrick* at my leisure, Postmaster Evans was on hand with a report of the inquest when the evening-train reached Tidioute. The Tidioute *Journal* didn't like the *Derrick* a little bit and the sight of a young man running from its office towards the train, with copies of the paper—not dry from the press—attracted my attention. Mr.

23

Evans said two Titusville reporters had come over during the day. A news-paper-man dearly relishes a "scoop" and it struck me at once that the *Journal* was rushing the first sheets of its edition to the Titusville delegates. Squeez-ing through the jam, A. E. Fay, of the *Courier*, and "Charlie" Morse, of the *Herald*, were pocketing the copies handed them by the *Journal* youth. Fay laughed out loud and said: "Well, boys, I guess the *Derrick's* left this time!" A pat on the shoulder and my hint to "guess again" fairly paralyzed the trio. The conductor shouted "all aboard" and the train moved off. Dropping into the seat in front of Fay, his annoyance could not be concealed. It relieved him to hear me tell of coming through from the north and ask why such a crowd had gathered at Tidioute. He told a fairy-story of a ball-game and his own and Morse's visit to meet a friend! A wish for a glance at the Tidioute paper he parried by answering: "It's yesterday's issue!" Fay was a good fellow and his clumsy falsifying would have shamed Ananias. Keeping him on the rack was rare sport. Clearly he believed me ignorant of the torpedo-accident. The moment to undeceive him arrived. A big roll of manuscript held before his eyes, with a "scare-head" and minute details of the tragedy, prefaced the query: "Do you still think the *Derrick* is badly left?" Many friends have asked me: "In your travels through the oil-region what was the funniest thing you ever saw?" Here is the answer: The dazed look of Fay as he beneld that manuscript, turned red and white, clenched his fists, gritted his teeth and hissed, "Damn you!"

John Osborne, a youth well-known and well-liked, in July of 1874 drove a buckboard loaded with glycerine down Bear-Creek Valley, two miles below Parker. The cargo let go at a rough piece of road in a woody ravine, scatter-ing Osborne, the horse and the vehicle over acres of tree-tops. The concus-sion was felt three miles. Venango, Crawford, Warren and Armstrong coun-ties had furnished nearly a score of sacrifices and Butler was to supply the next. Alonzo Taylor, young and unmarried, went in the summer of 1875 to torpedo a well at Troutman. The drop-weight failed to explode the percussion-cap and Taylor drew up the shell, a process that had cost Captain West his life and was always risky. He got it out safely and bore the torpedo to a hill to exam-ine the priming. An instant later a frightful explosion stunned the neighbor-hood. Taylor was not mangled beyond recognition, as the charge was giant-powder instead of Nitro-Glycerine. Nor was the damage to surrounding ob-jects very great, owing to the tendency of the powder to expend its strength downward. This was the only torpedo-fatality of the year, the number of cas-ualties having induced greater caution in handling explosives.

One of the first persons to reach the spot and gather the remains of Wil-liam Pine was his friend James Barnum, who died in the same manner at St. Petersburg eighteen months later. Barnum was the Roberts agent in Clarion county. On February twenty-third, 1876, he drove to Edenburg for three-hun-dred pounds of glycerine, to store in the magazine a mile from St. Petersburg. A fearful concussion, which the writer can never forget, broke hundreds of windows and rocked houses to their foundations at six o'clock that evening. To the magazine, on a slope sheltered by trees, people hastened. A huge iron-safe, imbedded in a cave dug into the hill, was the repository of the ex-plosives. Barnum had tied his team to a small tree and must have been taking the cans from the wagon to the safe. A yawning cavity indicated the site of the magazine. Both horses lay dead and disemboweled. The biggest piece of the luckless agent would not weigh two pounds. One of his ears was found next morning a half-mile away. The few remnants were collected in a box and buried at Franklin. A wife and several children mourned poor "Jim," who was a lively, active young man and had often been warned not to be so careless with the deadly stuff. Mrs. Barnum heard the explosion, uttered a piercing shriek and ran wildly from her house towards the magazine, sure her husband had been killed.

W. H. Harper, who received a patent for improvements in torpedoes, went to his doom at Keating's Furnace, two miles from St. Petersburg, in July of 1876. Drawing an unexploded shell from a well, precisely as West and Taylor had done, he stooped down to examine the priming. The contents exploded and drove pieces of the tin-shell deep into his flesh and through his body. How he survived nine days was a wonder to all who saw the dreadful wounds of the unlucky inventor.

McKean county supplied the next instance. Repeated attempts were made to rob a large magazine on the Curtis farm, two miles south of Bradford. Incredible as it may seem, the key-hole of the ponderous iron-safe in the hillside was several times stuffed with Nitro-Glycerine and a long fuse and a slow match applied to burst the door. None of these foolhardy attempts succeeding, on the night of September fifteenth, 1877, A. V. Pulser, J. B. Burkholder, Andrew P. Higgins and Charles S. Page, two of them "moonlighters," it is supposed tried pounding the lock with a hammer. At any rate, they exploded the magazine and were blown to fragments, with all the gruesome accompaniments incident to such catastrophes. That men would imperil their lives to loot a safe of Nitro-Glycerine in the dark beats the old story of the thief who essayed to steal a red-hot stove. In this case retribution was swift and terrible, but a magazine at St. Petersburg was broken open and plundered successfully.

Seventeen days later J. T. Smith, of Titusville, who had charge of a magazine on Bolivar Run, four miles from Bradford, lost his life experimenting with glycerine. Col. E. A. L. Roberts and his nephew, Owen Roberts, stood fifty yards from the magazine as Smith was thrown into the air and frightfully mangled. They escaped with slight bruises, a lively shaking up and a hair-raising fright.

The summer of 1878 was a busy season in the northern field. Foster-Brook Valley was at the hey-day of activity, with hundreds of wells drilling and well-shooters very much in evidence. Among the most expert men in the employ of the Roberts Company was J. Bartlett, of Bradford. He went to Red Rock, an ephemeral oil-town six miles north-east of Bradford, to torpedo a well in rear of the McClure House, the principal hostelry. Although Bartlett's recklessness was the source of uneasiness, he had never met with an accident and was considered extremely fortunate. It was a rule to explode the cans that had held the glycerine before pouring it into the shell. Bartlett torpedoed the well, piled wood around the empty cans and set it on fire. He and a party of friends waited at the hotel for the cans to explode. The fire had burned low and Bartlett proceeded to investigate. He lifted a can and turned it over, to see if it contained any glycerine. The act was followed by an explosion that shook every house in the town and shattered numberless windows. Bartlett's companions were knocked senseless and the shooter was blown one-hundred feet. When picked up by several men, who hurried to the scene, he presented a horrible sight. His clothing was torn to ribbons and his body riddled by pieces of tin. The right arm was off close to the shoulder and the right leg was a pulp. He was removed to a boarding-house and died in great agony three hours after.

Stories of hapless "moonlighters" scattered to the four winds of heaven were recounted frequently. Their business, done largely under cover of darkness, was exceptionally dangerous. The "moonlighter" did not haul his load in a wagon openly by daylight. He would place two ten-quart cans of glycerine in a meal-sack, sling the bag over his shoulder and walk to the scene of his intended operations, generally at night. One evening in the spring of 1879 a "moonlighter" named Reed appeared at Red Rock, somewhat intoxicated and bearing two cans of glycerine in a bag. He handled the bag in a style that struck terror to the hearts of all onlookers, many of whom remembered poor Bartlett. It was unsafe to wrest it from him by force and the Red-Rockers heaved a sigh of relief when he started to climb the hill leading to Summit City. Scores watched him, expecting an accident. At a rough spot Reed stumbled and the cans fell to the ground. A terrific explosion shook the surrounding country. A deep hole, ten feet in diameter, was blown in the earth and houses in the vicinity were badly shaken. The explosion occurred directly under a tree. When an attempt was made to gather up Reed's remains the greater portion of the body was in the tree, scraps of flesh of various sizes hanging from its branches. The concussion passed above Red Rock, hence the damage to property was small. Reed was dispersed over an acre of brush, a fearful illustration of the incompatibility of whisky and Nitro-Glycerine.

W. O. Gotham, John Fowler and Harry French went to their usual work at Gotham's Nitro-Glycerine factory, near Petrolia, on the morning of October twenty-seventh, 1878. An explosion during the forenoon tore Fowler to shreds, mutilated French shockingly and landed Gotham's dead body in the stream with hardly a sign of injury. Petrolia never witnessed a sadder funeral-procession than the long one that followed the unfortunate three to the tomb. Gotham had a family and was widely known ; the others were strangers, far from home and loved ones.

On February twentieth, 1880, James Feeney and Leonard Tackett started in a sleigh with six cans under the seat to torpedo a well at Tram Hollow, eight miles east by north of Bradford. The sleigh slipped into a rut on a rough side-hill and capsized. The glycerine exploded, throwing Tackett high in the air and mangling him considerably. Feeney lay flat in the rut, the violence of the shock passing over him and covering him with snow and fence-rails. His face was scorched and his hearing destroyed, but he managed to crawl out, the first man who ever emerged alive from the jaws of a Nitro-Glycerine eruption. He is still a resident of Bradford. A dwelling close to the scene was wrecked, the falling timbers seriously injuring two of the inmates.

At two o'clock on the morning of December twenty-third, 1880, a powerful concussion startled the people of Bradford from their slumbers, caused by a glycerine-explosion just below the city-limits. Alvin Magee was standing over the deadly compound, which had been put in a tub of hot water to thaw. Usually the subtle stuff is stored in a cold place, to congeal or freeze until needed. Magee and the derrick were blown into space, only a few bits of flesh and bone and splintered wood remaining. His two companions were in the engine-house and got off with severe bruises and permanent deafness. Two men named Cushing and Leasure were killed the same way in January, at a well near Limestone. Cushing came to see the torpedo put into the well and was standing near the engine-house, into which Leasure had just gone, when the accident occurred. The glycerine was in hot water to thaw and a jet of steam turned on, with the effect of sending it off prematurely. Cushing's body did not show a mark, his death probably resulting from concussion, while Leasure was torn to fragments.

E. M. Pearsall, of Oil City, died on July fourteenth, 1880, from the effect of burns a few hours before. In company with two other men he went to torpedo one of his wells on the Clapp farm. The tubing had been drawn out and a large amount of benzine poured into the hole. The torpedo was exploded, when the gas and benzine took fire and enveloped the men and rig in flames. The clothes of Pearsall, who was nearest to the derrick, caught fire and burned from his body. His limbs, face and breast were a fearful sight. His intense suffering he bore like a hero, made a will and calmly awaited death, which came to his relief at nine o'clock in the evening. Pearsall was dark-haired, dark-eyed, slender, wiry and fearless.

J. Plumer Mitchell—we called him "Plum"—worked for me on the *Independent Press* at Franklin in 1879-80. Everybody liked the bright, genial, capable young man who set type, read proof, wrote locals, solicited advertisements and won golden opinions. He married and was the proud father of two winsome children. Meeting me on the street one day shortly after quitting the *Press*, we chatted briefly.

"I am through with sticking type," he said.

"What are you driving at now?"

"Torpedoing wells. I started on Monday."

"Well, be sure you get good pay, for it's risky business, and don't furnish a thrilling paragraph for the obituary-column."

"I shall do my best to steer clear of that. Good-bye."

That was our last meeting. He met the fate that overtook West, Taylor and Harper, shooting a well at Galloway. The shattered frame rests in the cemetery and the widow and fatherless daughters of the lamented dead reside at Franklin. Poor "Plum!"

J. PLUMER MITCHELL.

T. A. McClain, an employé of the Roberts Company, was hauling two-hundred quarts of glycerine in a sleigh from Davis Switch to Kinzua Junction, on February fourteenth, 1881. The horses frightened and ran off. The sleigh is supposed to have struck a stump and the cargo exploded. Hardly a trace of McClain could be found and a bit of the steel-shoeing was the only part of the sleigh recovered. Obliteration more complete it would be difficult to imagine.

The most destructive sacrifice of life followed on September seventh. Charles Rust, a Bradford shooter, drove to Sawyer City to torpedo a well on the Jane Schoonover farm. It is alleged that Rust had domestic trouble, wearied of life and told his wife when leaving that morning he would never return. A small crowd assembled to witness the operation. William Bunton, owner of several adjacent wells, Charles Crouse, known as "Big Charlie the Moonlighter," James Thrasher, tool-dresser, and Rust were on the derrick-floor. Rust filled the first shell, fixed the firing-head and struck the cap two sharp blows with his left hand. There was a blinding flash, then a deafening report. Dust, smoke and missiles filled the air. The derrick was demolished and pieces of board flew hundreds of yards with the force of cannon-balls. One hit Crouse in the center of the forehead and passed through the skull. His face was terribly lacerated and the clothing stripped off his body. Bunton and Thrasher were not mangled beyond recognition, while Rust was thrown a hundred yards. His legs were missing, the face was battered out of the semblance of humanity and not a vestige of clothing was left on the mutilated trunk. Frederick Slatterly, a lad on his way to school, was hit by a piece of the derrick, which ripped his abdomen and caused death in three hours. Three boys walking behind young Slatterly were thrown down and hurt slightly. Mr. Bunton gasped when picked up and lived five minutes. He was an estimable citizen, an elder in the Presbyterian church, intelligent and broad-minded. Thrasher and Crouse were industrious workmen. Edward Wilson, a gauger, standing ten rods away, was perforated by slivers and pieces of tin, his injuries confining him to bed several months. Thomas Buton and John Sisley were at the side of the derrick, within six feet of Rust, yet escaped with trifling injuries. The tragedy produced a sensation, all the more fearful from the belief of some who witnessed it that Rust intended to commit suicide and in compassing his own death killed four innocent victims.

The Roberts magazine on the Hatfield farm, two miles south of Bradford, blew up on the night of October thirteenth, 1881. Nobody doubted it was the work of "moonlighters" attempting to steal the glycerine. Traces of blood and minute portions of flesh on the stones and ties indicated that two persons at least were engaged in the job. Who they were none ever learned.

John McCleary, a Roberts shooter, had a remarkable escape on December twenty-seventh, 1881. While filling the shell at a well near Haymaker, in the lower oil-field, the well flowed and McCleary left the derrick. The column of oil threw down the shell and the glycerine exploded promptly, wrecking the derrick and tossing the fleeing man violently to the ground. He rose to his feet as four cans on the derrick-floor cut loose. McCleary was borne fifty feet through the air, jagged splinters of tin and wood pierced his back and sides and he fell stunned and bleeding. He was not injured fatally. Like Feeney, Buton and Sisley, he survives to tell of his close call. Less fortunate was Henry W. McHenry, who had torpedoed hundreds of wells and was blown to atoms near Simpson Station, in the southern end of the Bradford region, on February fifth, 1883. His fate resembled West's, Taylor's, Harper's and Mitchell's.

In the summer of 1884 Lark Easton went to torpedo a well at Coleville, seven miles southeast of Bradford. He tied his team in the woods, carried some cans of glycerine to the well and left four in the wagon. A storm blew down a tree, which fell on the wagon and exploded the glycerine, demolishing the vehicle and killing one horse. It was a lucky escape, if not much of a lark, for Easton.

A peculiar case was that of "Doc" Haggerty, a teamster employed to haul Nitro-Glycerine to the magazine near Pleasantville. In December of 1888 he took fourteen-hundred pounds on his wagon and was seen at the magazine twenty minutes before a furious explosion occurred. Pieces of the horses and wagon were found, but not an atom of Haggerty. He had disappeared as completely as Elijah in the chariot of fire. An insurance-company, in which he held a five-thousand-dollar policy, resisted payment on the ground that, as no remains of the alleged dead man could be produced, he might be alive! Some pretexts for declining to pay a policy are pretty mean, but this certainly capped the climax. Experts believed the heat generated by the explosion was sufficient to cremate the body instantaneously, bones, clothes, boots and all.

James Woods and William Medeller, two experienced shooters, were ushered into eternity on December tenth, 1889, by the explosion of the Humes Torpedo-Company's magazine at Bean Hollow, two miles south of Butler. They had gone for glycerine and that was the end of their mortal pilgrimage. Six years later, on December fourth, 1895, at the same place and in the same way, George Bester and Lewis Black lost their lives. Bester was blown to atoms and only a few threads of his clothing could be picked up. The lower part of Black's face, the trunk and right arm remained together, while other portions of the body were strewn around. The left arm was in a tree three-hundred yards distant. Huge holes marked the site of the two Humes magazines, a hundred feet apart. The mangled horse lay between them, every bone in his carcass broken and the harness cut off clean. The buggy was in fragments, with one tire wrapped five times about a small tree. Not a board stayed on the boiler-house and the boiler was moved twenty feet and dismantled. The factory, two-hundred feet from the magazines, was utterly wrecked. The young men left Butler early in the morning, Black going for company. The supposition is that Bester was removing some of the cans from the shelf, intending to take them out, and that he dropped one of them. About seven-hundred pounds of glycerine were stored in one of the magazines and a less amount in the other. George Bester was twenty-eight and had a wife and two small children. He was industrious, steady and one of the best shooters in the business. Black was twenty and lived with his parents. The concussion jarred every house in Butler, broke windows and loosened plaster in the McKean school-building, causing a panic among the children.

W. N. Downing's death, on January second, 1891, at the Victor Oil-Company's well, in the Archer's Forks oil-field, near Wheeling, West Virginia, was very singular. The glycerine used to torpedo the well the previous day had been thawed in a barrel of warm water. Next day two of the owners drove out to see the well and talk with Downing, who was foreman of the company. On their way back to Wheeling they heard an explosion, conjectured the boiler had burst and returned to the lease. Mr. Downing's body lay near where the barrel of water stood. The barrel had vanished and a large hole occupied its place. The victim's head was cut off on a line with the eyes. The only explanation of the accident was that the glycerine had leaked into the barrel and a sudden jar had caused the stuff to explode. Beside the well, in the fence-corner, were twelve cans of glycerine not exploded. Downing lived at Siverlyville, above Oil City, whither his remains were brought for interment.

Letting a torpedo down a well at Bradford in September of 1877, a flow of oil jerked it out, hurled the shell against the tools, which were hanging in the derrick, and set off the nitro on the double quick. The shooter jumped and ran at the first symptoms of trouble, the derrick was sliced in the middle and set on fire. The rig burned and strenuous exertions alone saved neighboring wells. The fire was a novelty in the career of the explosive.

Occasionally Nitro-Glycerine goes off by spontaneous combustion without apparent provocation. On December fifth, 1881, two of the employés noticed a thin smoke rising from the top row of cans in the Roberts magazine at Kinzua Junction. They retreated, came back and removed eighty cans, observed the smoke increasing in density and volume and decided to watch further proceedings from a safe distance. Twelve-hundred quarts exploded with such vigor that the earth jarred for miles and a big hole was ploughed in the rock. In November of 1885 the Rock Glycerine-Company's factory on Minard Run, four miles south of Bradford, was wrecked for the fourth time. O. Wood and A. Brown were running the mixture into "the drowning tank," to divest it of the acid. The process generates much heat and acid escaping from a leak in the tank fired the wood-work. Wood and Brown and a carpenter in the building, knowing their deliverance depended upon their speed, took French leave. Samuel Barber, a teamster, was unloading a drum of acid in front of the building and joined the fugitives in their flight. The glycerine obligingly waited until the four men reached a safe spot and then reduced the factory to kindling-wood. Barber's horses and wagon were not hurt mortally, the animals bleeding a little from the nose. Next evening Tucker's factory at Corwin Centre, six miles north-east of Bradford, followed suit. Griffin Rathburn, who was making a run of the fluids, fled for his life as the mass emitted a flame. He saved himself, but the factory and a thousand pounds of the explosive went on an aërial excursion.

Men in Ohio, West Virginia and Indiana have added to the dismal roll of those who, leaving home happy and buoyant in the morning, ere the sun set were dispersed over acres of territory. Yet all experiences with the dread compound have not been serious, for at intervals a comic incident brightens the page. Robert L. Wilson, a blacksmith on Cherry Run in 1869-70, was a first-class tool-manufacturer. Joining the Butler tide, he opened a shop at Modoc. A fellow of giant-build entered one day, bragged of his muscle as well as his stuttering tongue would permit and wanted work. Something about the fellow displeased Wilson, who was of medium size and thin as Job's turkey, and he decided to have a little fun at the stranger's expense. He asked the burly visitor whether he could strike the anvil a heavier blow than any other man in the shop. The chap responded yes and Wilson agreed to hire him if he proved his claim good. Wilson poured two or three drops of what looked like lard-oil on the anvil and the big 'un braced himself to bring down the sledge-hammer with the force of a pile-driver. He struck the exact spot. The sledge soared through the roof and the giant was pitched against the side of the building hard enough to knock off a half-dozen boards. When he extracted himself from the mess and regained breath he blurted out : "I t-t-told you I co cou-could hi-hi-hit a he-he-hell of a b-bl-blow !" "Right," said Wilson, "you can beat any of us ; be on hand to-morrow morning to begin work." The man worked faithfully and did not discover for months that the stuff on the anvil was Nitro-Glycerine.

The farm-house of Albert Jones, three miles from Auburn, Illinois, was demolished on a Sunday afternoon in November of 1885. Jones had procured some Nitro-Glycerine to remove stumps and set the can on the floor of the dining-room. After dinner the family visited a neighbor, locking up the house. About three o'clock a thundering detonation alarmed the Auburnites, who couldn't understand the cause of the rumpus. A messenger from the country enlightened them. The Jones domicile had been wrecked mysteriously and the family must have perished. Excited people soon arrived and the Joneses put in an appearance. The house and furniture were scattered in tiny tidbits over an area of five-hundred yards. Half the original height of the four walls was standing, with a saw-tooth and splintered fringe all around the irregular top of the oblong. Two beds were found several hundred yards apart, in the road in front of the house. A sewing-machine was buried head-first in the flower-garden. Wearing-apparel and household-articles were strewn about the place.

While Mr. Jones and a circle of friends were viewing the wreck and wondering how the Nitro-Glycerine exploded a faint cry was heard. A search resulted in finding the family-cat in the branches of a tree fifty feet from the dwelling. It was surmised the cat caused the disaster by pushing from the table some article sufficiently heavy to explode the glycerine on the floor. The New York *Sun's* famous grimalkin should have retired to a back-fence and begun his final caterwauling over the superior performance of the Illinois feline. Jones and his friends unanimously endorsed the verdict : " It was the cat."

The first statement coupling a hog and Nitro-Glycerine in one package was written by me in December of 1869, at Rouseville, and printed in the Oil-City *Times*. The item went the rounds of the press in America and Europe, many papers giving due credit and many localizing the narrative to palm it off as original. One of the latter was " Brick " Pomeroy's La Crosse *Democrat*, which laid the scene in that neck of woods. The tale has often been resurrected and it was reported in a New-Orleans paper last month. The original version of "The Loaded Porker " read thus :

" Rouseville furnishes the latest unpatented novelty in connection with Nitro-Glycerine. A torpedo-man had taken a small parcel of the dangerous compound from the magazine and on his return dropped into an engine-house a few minutes, leaving the vessel beside the door. A rampant hog, in search of a rare Christmas dinner, discovered the tempting package and unceremoniously devoured the entire contents, just finishing the last atom as the torpedoist emerged from the building ! Now everybody gives the greedy animal the widest latitude. It has full possession of the whole sidewalk whenever disposed to promenade. All the dogs in town have been placed in solitary confinement, for fear they might chase the loaded porker against a post. No one is sufficiently reckless to kick the critter, lest it should unexpectedly explode and send the town and its total belongings to everlasting smash ! The matter is really becoming serious and how to dispose safely of a gormandizing swine that has imbibed two quarts of infernal glycerine is the grand conundrum of the hour. When he is killed and ground up into sausage and head-cheese a new terror will be added to the long list that boarding-houses possess already."

Charles Foster, of the High-Explosive Company, had an adventure in March of 1896 that he would not repeat for a hatful of diamonds. He loaded five-hundred quarts of glycerine at the magazine near Kane City. On Rynd Hill the horses slipped and one fell. The driver jumped from his seat to hold the animal's head that it might not struggle. He cut the other horse free from the harness, as the road skirted a precipice and the frightened beast's rearing and plunging would almost certainly dump the wagon and outfit over the steep bank. Nobody was in sight, the driver had no chance to block the wheels and the wagon started down the hill backward. The vehicle, with its load of condensed destruction, kept the road a few yards and pitched over the hill, turning somersaults in its descent. It brought up standing on the tongue in a heap of stones. The covers were torn off the wagon and the cans of the explosive were widely scattered. Seven in one bunch were picked up ten yards below the road. A three-cornered hole had been jammed in the bottom of one of the eight-quart cans and the contents were escaping. Darkness came on before the glycerine could be removed to a place of safety. Foster secured a rig and drove home, after arranging to have the stuff taken to the factory next morning. How the explosive, although congealed, stood the shock of going over the hill and scattering about without soaring skyward is one of the unfathomed mysteries of the Nitro-Glycerine business.

A Polish resident of South Oil City carried home what he took to be an empty tomato-can. His wife chanced to upset it from a shelf in the kitchen. A few drops of glycerine must have adhered to the tin. The can burst with fearful violence, blowing out one side of the kitchen, destroying the woman's eyes and nearly blinding her little daughter. A woman at Rouseville poured glycerine, mistaking it for lard-oil, into a frying-pan on the stove, just as her husband came into the kitchen. He snatched up the pan and landed it in a snow-bank so quickly the stuff didn't burst the combination. The wife started to scold him, but fainted when he explained the situation.

The wonderful explosion at Hell-Gate in 1876, when General Newton fired two-hundred tons of dynamite and cleared a channel into New-York harbor for the largest steamships, brought to the front the men who always tell of something that beats the record. A group sat discussing Newton's achievement at the Collins House, Oil City, as a Southerner with a military title entered. Catching the drift of the argument he said :

"Talk about sending rocks and water up in the air! I knew a case that knocked the socks clear off this little ripple at New York !"

"Tell us all about it, Colonel," the party chorused.

"You see I used to live down in Tennessee. One day I met a farmer driving a mule that looked as innocent as a cherub. The farmer had a whip with a brad in the end of it. Just as I came up he gave the mule a prod. Next moment he was gone. It almost took my breath away to see a chap snuffed out so quick. The mule merely ducked his head and struck out behind. A crash, a cloud of splinters and the mule and I were alone, with not a trace of farmer or wagon in sight. Next day the papers had accounts of a shower of flesh over in Kentucky and I was the only person who could explain the phenomenon. No, gentlemen, the dynamite and Nitro-Glycerine at Hell-Gate couldn't hold a candle to that Tennessee mule !"

The silence that followed this tale was as dense as a London fog and might have been cut with a cheese-knife. It was finally broken by a *Derrick* writer, who was a newspaper man and not easily taken down, extending an invitation to the crowd to drink to the health of Eli Perkins's and Joe Mulhatten's greatest rival.

William A. Meyers, whom every man and woman at Bradford knew and admired, handled tons of explosives and shot hundreds of wells. He had escapes that would stand a porcupine's quills on end. To head off a lot of fellows who asked him for the thousandth time concerning one notable adventure, he concocted a new version of the affair. "It was a close call," he said, "and no mistake. In the magazine I got some glycerine on my boots. Soon after coming out I stamped my heel on a stone and the first thing I knew I was sailing heavenward. When I alighted I struck squarely on my other heel and began a second ascension. Somehow I came down without much injury, except a bruised feeling that wore off in a week or two. You see the glycerine stuck to my boot-heels and when it hit a hard substance it went off quicker than Old Nick could singe a kiln-dried sinner. What'll you take, boys?"

So the darkest chapter in petroleum history, a flood of litigation, a mass of deception, a black wave of treachery and a red streak of human blood, must be charged to the account of Nitro-Glycerine.

HITS AND MISSES.

A Bradford minister, when the Academy of Music burned down, shot wide of the mark in attributing the fire to "the act of God." Sensible Christians resented the imputation that God would destroy a dozen houses and stores to wipe out a variety-theater, or that He had anything to do with building up a trade in arson and figuring as an incendiary.

> He struck a match and the gas exploded;
> An angel now, he knows it was loaded.

"Mariar, what book was you readin' so late last night?" asked a stiff Presbyterian father at Franklin. "It was a novel by Dumas the elder." "'Elder!' I don't believe it. What church was he elder on, Ish'd like to know, and writ novels? Go and read Dr. Eaton's Presbytery uv Erie."

> Said a Warren young maiden: "Alas, Will,
> You come every night,
> And talk such a sight,
> And burn so much light,
> My papa declares you're a Gas Bill!"

Hymn-singing is not always appropriate, or a St. Petersburg leader would not have started "When I Can Read My Title Clear" to the minstrel melody of "Wait for the Wagon and We'll All Take a Ride!" At an immersion in the river below Tidioute, as each convert, male or female, emerged dripping from the water, the people interjected the revivalist chorus:

> "They look like men in uniform,
> They look like men of war!"

Mr. Gray, of Boston, once discovered a "non-explosive illuminating gasoline." To show how safe the new compound was, he invited a number of friends to his rooms, whither he had taken a barrel of the fluid, which he proceeded to stir with a red-hot poker. As they all went through the roof he endeavored to explain to his nearest companion that the particular fluid in the barrel had too much benzine in it, but the gentleman said he had engagements higher up and could not wait for the explanation. Mr. Gray continued his ascent until he met Mr. Jones, who informed him that there was no necessity to go higher, as everybody was coming down; so Mr. Gray started back to be with the party. Mr. Gray's widow offered the secret for the manufacture of the non-explosive fluid at a reduced rate, to raise money to buy a silver-handled coffin with a gilt plate for her departed husband.

> The speech of a youth who goes courting a lass,
> Unless he's a dunce at the foot of the class,
> Is sure to be se won'd with natural gas.

Grant Thomas, train-dispatcher at Oil City of the Allegheny Valley Railroad, is one of the jolliest jokers alive. When a conductor years ago a young lady of his acquaintance said to him: "I think that Smith girl is just too hateful; she's called her nasty pug after me!" "Oh," replied the genial ticket-puncher, in a tone meant to pour oil on the troubled waters, "that's nothing; half the cats in Oil City are called after me!" The girl saw the point, laughed heartily and the angel of peace hovered over the scene.

> "What's in a name?" so Shakespeare wrote.
> Well, a good deal when fellows vote,
> Want a check cashed, or sign a note;
> And when an oilman sinks a well,
> Drv as the jokes of Digby Bell,
> Dennis or Mud fits like a shell.

STANDARD BUILDING, 26 BROADWAY, NEW YORK.

XIV.

THE STANDARD OIL-COMPANY.

Growth of a Great Corporation—Misunderstood and Misrepresented—Improvements in Treating and Transporting Petroleum—Why Many Refineries Collapsed—Real Meaning of the Trust—What a Combination of Brains and Capital Has Accomplished—Men Who Built Up a Vast Enterprise that has no Equal in the World.

"Genius is the faculty of growth."—*Coleridge.*
"In union there is strength."—*Popular Adage.*
"Success affords the means of securing additional success."—*Stanislaus.*
"We must not hope to be mowers
Until we have first been sowers."—*Alice Cary*
"Fortune, success, position, are never gained but by determinedly, bravely striking, growing, living to a thing."—*Townsend.*
"The goal of yesterday will be the starting-point of to-morrow."—*Ibid.*
"That thou art blam'd shall not be thy defect."—*Shakespeare.*
"Amongst the sons of men how few are known
Who dare be just to merit not their own."—*Churchill.*
"The keen spirit seizes the prompt occasion."—*Hannah More.*

JOHN D. ROCKEFELLER.

OMPARED with a petroleum-sketch which did not touch upon the Standard Oil-Company, in different respects the greatest corporation the world has ever known, Hamlet with "the melancholy Dane" left out would be a masterpiece of completeness. Perhaps no business-organization in this or any other country has been more misrepresented and misunderstood. To many well-meaning persons, who would not willfully harbor an unjust thought, it has suggested all that is vicious, grasping and oppressive in commercial affairs. They picture it as a cruel monster, wearing horns and cloven-hoofs and a forked-tail, grown rich and fat devouring the weak and the innocent. Its motives have been impugned, its methods condemned and its actions traduced. If a man in Oildom drilled a dry-hole, backed the wrong horse, lost at poker, dropped money speculating, stubbed his toe, ran an unprofitable refinery, missed a train or couldn't maintain champagne style on a lager-beer income, it was the fashion for him to pose as the victim of a gang of conspirators and curse the Standard vigorously and vociferously !

The reasons for this are various. The Standard was made the scapegoat of the evil deeds alleged to have been contemplated by the unsavory South-Improvement Company. That odious combine, which included a number of railroad-officials, oil-operators and refiners, disbanded without producing, refining, buying, selling or transporting a gallon of petroleum. "Politics makes strange bedfellows" and so does business. Among subscribers for South-Improvement stock were certain holders of Standard stock and also their bitterest opponents ; among those most active in giving the job its death-blow were prominent members of the Standard Oil-Company. The projected spoliation died "unwept, unhonored and unsung," but it was not a Standard scheme.

Envy is frequently the penalty of success. Whoever fails in any pursuit likes to blame somebody else for his misfortune. This trick is as old as the race. Adam started it in Eden, Eve tried to ring in the serpent and their posterity take good care not to let the game get rusty from disuse ! Its aggregation of capital renders the Standard, in the opinion of those who have "fallen outside the breastworks," directly responsible for their inability to keep up with the procession. Sympathizers with them deem this "confirmation strong as proof of Holy Writ" that the Standard is an unconscionable monopoly, fostered by crushing out competition. Such reasoning forgets that enterprise, energy, experience and capital are usually trump-cards. It forgets that "the race is to the swift," the battle is to the mighty and that "Heaven is on the side with the heaviest artillery." Carried to its logical conclusion, it means that improved methods, labor-saving appliances and new processes count for nothing. It means that the snail can travel with the antelope, that the locomotive must wait for the stage-coach, that the fittest shall not survive. In short, it is the double-distilled essence of absurdity.

Any advance in methods of business necessarily injures the poorest competitor. Is this a reason why advances should be held back? If so, the public could derive no benefit from competition. The fact that a man with meagre resources labors under a serious disadvantage is not an excuse for preventing stronger parties from entering the field. The grand mistake is in confounding combination with monopoly. By combination small capital can compete successfully with large capital. Every partnership or corporation is a combination, without which undertakings beyond individual reach would never be accomplished. Trunk railroads would not be built, unity of action would be destroyed, mankind would segregate as savages and the trade of the world would stagnate. Combinations should be regulated, not abolished. Rightful competition is not a fierce strife between persons to undersell each other, that the one enduring the longest may afterwards sell higher, but that which furnishes the public with the best products at the least cost. This is not done by selling below cost, but by diminishing in every way possible the cost of producing, manufacturing and transporting. The competition which does this, be it by an individual, a firm, a corporation, a trust or a combination, is a public benefactor. This kind of competition uses the best tools, discards the sickle for the cradle and the cradle for the reaper, abandons the flail for the threshing-machine and adopts the newest ideas wherever and whenever expenses can be lessened. To this end unrestricted combination and unrestricted competition must go hand-in-hand. A small profit on a large volume of business is better for the consumer than a large profit on a small business. The man who sells a million dollars' worth of goods a year, at a profit of five per cent., will become rich, while he who sells only ten-thousand dollars' worth can get a bare living. If the builder of a business of one hundred-thousand dollars deserve praise, why should the builder of a business of millions be censured? Business that grows greater than people's limited notions should not for that cause be fettered or suppressed. When business ceases to be local and has the world for its market, capital must be supplied to meet the increasing demand and combination is as essential as fresh air. Thus large establishments take the place of small ones and men acting in concert achieve what they would never attempt separately. The more perfect the power of association the greater the power of production and the larger the proportion of the product which falls to the laborer's share. The magnitude of combinations must correspond with the magnitude of the business to be done, in order to secure the highest skill, to employ the latest devices, to pay the best wages, to invent new appliances, to improve facilities and to give the public a cheaper and finer product. This is as natural and legitimate as for water to run down hill or the fleet greyhound to distance the slow tortoise.

How has the Standard affected the consumer of petroleum-products? What has it done for the people who use illuminating oils? Has it advanced the price and impaired the quality? The early distillations of petroleum were unsatisfactory and often dangerous. The first refineries were exceedingly primitive and their processes simple. Much of the crude was wasted in refining, a business not financially successful as a rule until 1872, notwithstanding the high prices obtained. Methods of manufacture and transportation were expensive and inadequate. The product was of poor quality, emitting smoke and unpleasant odor and liable to explode on the slightest provocation. In 1870 a few persons, who had previously been partners in a refinery at Cleveland, organized the Standard Oil-Company of Ohio, with a capital of one-million dollars, increased subsequently to three-and-a-half millions. For years the history of refining had been mainly one of disaster and bankruptcy. A Standard Oil-Company had been organized at Pittsburg by other persons and was doing a large trade. The Cleveland Standard Refinery, the Pittsburg Standard Refinery, the Atlantic Refining Company of Philadelphia and Charles Pratt & Co. of New York were extensive concerns. Because of the hazardous nature and peculiar conditions of the refining industry, the need of improved methods and the manifold advantages of combination, they entered into an alliance for their mutual benefit. Refineries in the oil-regions had combined before, hence the association of these interests was not a novelty. The cost of transporation and packages had been important factors in crippling the industry. Crude was barreled at the wells and hauled in wagons to the railroads prior to the system of transporting it by pipes laid under ground. Railroad-rates were excessive and irregular. Refiners who combined and could throw a large volume of business to any particular road secured favorable rates. The rebate-system was universal, not confined to oil alone, and possibly this fact had much to do with the combination of refiners afterwards known as the Standard Oil-Company.

Very naturally the Standard endeavored to secure the lowest transportation-rates. Quite as naturally railroad-managers, in their eagerness to secure the traffic, vied with each other in offering inducements to large shippers of petroleum. The Standard furnished, loaded and unloaded its own tank-cars, thereby eliminating barrels and materially cheapening the freight-service. This reduction of expense reduced the price of refined in the east to a figure which greatly increased the demand and gave oil-operations a healthy stimulus. Still more important was the introduction of improvements in refining, which yielded a larger percentage of illuminating-oil and converted the residue into merchantable products. Chemical and mechanical experts, employed by the combined companies to conduct experiments in this direction, aided in devising processes which revolutionized refining. The highest quality of burning-oil was obtained and nearly every particle of crude was utilized. Substances of commercial value took the place of the waste that formerly emptied into the streams, polluting the waters and the atmosphere. In this way the cost was so lessened that kerosene became the light of the nations. Consumers, whose dime now will buy as much as a dollar would before the "octopus" was heard of, are correspondingly happy.

Since consumers have fared so well, how about refiners outside the Standard? That smaller concerns were unable to compete with the Standard under such circumstances was no reason why the public should be deprived of the advantages resulting from concentration of capital and effort. Many of these, realizing that small capital is restricted to poor methods and dear production, either sold to the Standard or entered the combination. In not a few cases wide-awake refiners took stock for part of the price of their properties and engaged with the company, adding their talents and experience to the common fund for the benefit of all concerned. Others, not strong enough to have their cars and provide all the latest improvements, made such changes as they could afford to meet the requirements of the local trade, letting the larger ones attend to distant markets. Some continued right along and they are still on deck as independent refiners, always a respectable factor in the trade and never more active than to-day. Those who would neither improve, nor sell, nor combine, sitting down placidly and believing they would be bought out later on their own terms, were soon left far behind, as they deserved to be. Let it be said positively that the Standard, in negotiating for the purchase or combination of refineries, treated the owners liberally and sought to keep the best men in the business. A number who put up works to sell at exhorbitant prices, failing in their design, howled about "monopoly" and "freezing out" and tried to pass as martyrs. It is true hundreds of inferior refineries have been dismantled, not because they were frozen out by a crushing monopoly, but because they lacked requisite facilities. The refineries in vogue when the Standard was organized could not stay in business a week, if resurrected and revived. A team of pack-mules might as well try to compete with the New York Central Railroad as these early refineries to meet the requirements of the petroleum-trade at its present stage of perfection. They were "frozen out" just as stage-coaches were "frozen out" by the iron-horse or the sailing-vessel of our grandfathers' time by the ocean-liner that crosses the Atlantic in six days. Every labor-saving invention and improvement in machinery throws worthy persons out of employment, but inventions and improvements do not stop for any such cause. Business is a question of profit and convenience, not a matter of sentiment. The manufacturer who, by an improved process, can save a fraction of a cent on the yard or pound or gallon of his output has an enormous advantage. Must he be deprived of it because other manufacturers cannot produce their wares as cheaply? Refining petroleum is no exception to the ordinary rule and a transformation in its methods and results was as inevitable as human progress and the changes of the seasons.

Over-production is justly chargeable with the low price of crude that wafted many producers into bankruptcy. Regardless of the inexorable laws of supply and demand, operators drilled in Bradford and Butler until forty-million barrels were above ground and the price fell to forty cents. Time and again the wisest producers sought to stem the tide by stopping the drill, which started with renewed energy after each brief respite. With the stocks bearing the market the dropping of crude to a price that meant ruin to owners of small wells was as certain as death and taxes. Gold-dollars would be as cheap as pebbles if they were as plentiful. Forty-million barrels of diamonds stored in South Africa would bring the glistening gems to the level of glass-beads. The Standard, through the National-Transit Company, erected thousands of tanks to husband the enormous surplus, which the world could not consume and would not have on any terms. Hosts of operators were kept out of the sheriff's grasp by this provision for their relief, using their certificates as collateral during the period of extreme depression. The richest districts were drained at length, consumption increased and production declined, stocks were reduced and prices advanced. Then a number of oil-operators, foremost among whom were some of the men whom the Standard had carried over the grave crisis, thought the National-Transit was making too much money storing crude and tried to secure legislation that was hardly a shade removed from confiscation. The legislature refused to pass the bills, the company voluntarily reduced its charges and the agitation subsided. Thousands of producers sold or entered large companies, into whose hands a good share of the development has fallen, mainly because of the great expense of operating in deep territory and the wisdom of dividing the risk attendant upon seeking new fields. Operators who had to retire were "frozen out" by excessive drilling, nothing more and nothing less!

The highest efficiency in all fields of economical endeavor is obtained by the greatest degree of organization and specialization of effort. To attack large concerns as monopolies, simply because they represent millions of dollars under a single management, is as stupid and unjust as the narrow antagonism of ill-balanced capitalists to organized labor. If organized capital means better methods, greater facilities and improved processes, organized labor means better wages, greater recognition and improved industrial conditions. Hence both deserve to be encouraged and both should work in harmony. The Standard Oil-Company established agencies in different states for the sale of its products. As the business grew it organized corporations under the laws of these states, to carry on the industry under corporate agencies. Manufactories were located at the seaboard for the export-trade. It was easier and cheaper to pipe crude to the coast than to refine it at the sources of supply and ship the varied products. Thus the refining of export-oil was done at the seaboard, just as iron is manufactured at Pittsburg instead of at the ore-beds on Lake Superior. The company aimed to open markets for petroleum by reducing the cost of its transportation and manufacture and bettering its quality. It manufactured its own barrels, cans, paints, acids, glue and other materials, effecting a vast saving. On January second, 1882, the forty persons then associated in the Standard owned the entire capital of fifteen corporations and a part of the stock of a number of others. Nine of these forty controlled a majority of the stocks so held, and it was agreed on that date that all the stocks of the corporations should be placed in the hands of these nine as trustees. The trustees issued certificates showing the extent of each block of stock so surrendered, and agreed to conduct the business of the several corporations for the best interests of all concerned. This was the inception of the Standard Oil-Trust,

24

the most abused and least understood business-organization in the history of the race.

The Standard Trust, which demagogues lay awake nights coining language to denounce, did not unite competing corporations. The corporations were contributory agencies to the same business, the stock owned by the individuals who had built up and carried on the business and held the voting power. These individuals had combined not to repress business, but to extend it legitimately, by allying various branches and various corporations. The organization of the Trust was designed to facilitate the business of these corporations by uniting them under the managment of one Board of Trustees. This object was business-like and laudable. It had no taint of a scheme to "corner" a necessity of life and elevate the price at the expense of the masses. On the contrary, it was calculated to enlarge the demand and supply it at the minimum of profit. For ten years the Standard Trust continued in existence, dissolving finally in 1892. During this term its stockholders increased from forty to two thousand. Many of the most skillful refiners and experienced producers joined the combination and were retained to manage their properties. Each corporation was managed as though independent of every other in the Trust, except that the rivalry to show the best record stimulated them to constant improvement. Whatever economy one devised was adopted by all. The business was most systematic and admirably managed in every detail, running as harmoniously as the different parts of a watch. Clerks, agents and employés who could save a few hundred dollars purchased Trust Certificates and thus became interested in the business and gains. If it is desirable to multiply the number who enjoy the profits of production, how can it be done better than through ownership of stock in industrial associations? The problem of co-operation and profit-sharing can be solved in this way. The Standard Trust was a real object-lesson in economics, which illustrated in the fullest measure the benefits of an association in business that affected consumers and producers of a great staple alike favorably.

Misrepresentation is as hard to eradicate as the Canada thistle or the English sparrow. Once fairly set going, it travels rapidly. "A lie will travel seven leagues while Truth is pulling on its boots." The Standard is the target at which invidious terms and bitter invective have been hurled remorselessly, often through downright ignorance. Although reputable editors might be misled, in the hurry and strain of daily journalism, to give currency to deliberate falsehoods against corporations or capitalists, reasonable fairness might be expected from the author of a pretentious book. Henry D. Lloyd, of Chicago, last year published "Wealth Against Commonwealth," an elaborate work, which is devoted mainly to an assault upon the Standard Oil-Company. The book, notable for its distortion of facts and suppression of all points in favor of the corporation it assails, caters to the worst elements of socialism. The author views everything through anti-combination glasses and, like the child with the bogie-man, sees the monopoly-spook in every successful aggregation of capital. He confounds the South-Improvement Company with the Standard and charges to the latter all the offenses supposed to lie at the door of the organization that died at its birth. One thrilling story is cited to show that the Standard robbed a poor widow. The narrative is well calculated to arouse public resentment and encourage a lynching-bee. It has been repeated times without number. Within the past month two Harrisburg ministers have referred to it as a startling evidence of the unscrupulous tyranny of the Standard millionaires. To make the case imposing Mr. Lloyd informs mankind that the husband of this widow had been "a prominent member of the Presbyterian church, president of a Young Men's Christian Association and active in all religious and benevolent enterprises." After his death she continued the business until she was finally coerced into selling it to the Trust at a ruinously low price—a mere fraction of its actual value. Mr. Lloyd states her hopeless despair as follows:

"Indignant with these thoughts and the massacred troop of hopes and ambitions that her brave heart had given birth to, she threw the letter—a letter she had received from the Standard regarding the sale of her property—into the fire, where it curled up into flames like those from which a Dives once begged for a drop of water. She never reappeared in the world of business, where she had found no chivalry to help a woman save her home, her husband's life-work and her children."

Is this harrowing statement true? The widow continued the business four years after her husband's death. Competition increased, prices tumbled, the margin of profit was constantly narrowing, new appliances simplified refining-processes and the widow's plant was no longer adapted to the business. She sold for sixty-thousand dollars, the Standard paying twice the sum for which a refinery better suited to the purpose could be constructed. Foolish friends afterwards told her she had sold too low and the widow wrote a severe letter to the president of the Standard. The company had bought the property to oblige her and at once offered it back. She declined to take it, or sixty-thousand dollars in Standard stock, evidently realizing that the refinery had lost its profit-earning capacity and that even the new management might not be able to make it pay. This will serve to illustrate the unfairness of "Wealth Against Commonwealth," which has been widely quoted because of its presumed reliability and the high standing of the publishers. Yet this story of imaginary wrong has been worked into speeches, sermons and editorials of the fiercest type! In its treatment of the widow the Standard was truly magnanimous. Few business-men would consent to undo a transaction and have their labor for naught, simply because the other party had become dissatisfied. Possibly Mr. Lloyd would not be as generous if there was any profit in the transaction. If the Standard cut prices to ruin the widow and other competitors, would not oil have gone up again when they were disposed of? No such upward movement occurred. The widow disappeared. Many small refineries disappeared. Monopoly railroad-contracts, if such ever existed, have disappeared, but the price of refined-oil has been falling steadily for twenty years, declining from an average of nineteen cents a gallon in 1876 to five cents in 1895. The potent fact in this connection is that the Standard has continued to make profits with the declining price of oil. This conclusively demonstrates that the decline was due to economic improvements in the productive methods and not to a malicious cut to ruin a widow or anybody else, as Mr. Lloyd assumes. Otherwise a profit accompanying the fall in price would have been impossible and the Standard would have been sold out by the sheriff long years ago.

All the dealers in slander from Lloyd down to the chronic kicker who has attempted to make money by annoying the Standard have played the Rice case as a trump-card. According to their version, Mr. Rice was an angelic Vermonter, whose success inspired the Standard with devilish enmity and it determined to compass his ruin. Rice had operated at Pithole and at Macksburg and owned a small refinery at Marietta. It was alleged that the Cleveland & Marietta Railroad discriminated against him, doubling his freight-charge and giving the Standard a drawback on all the oil that went over the road. This was an iniquitous arrangement, entered into by the receiver of the road and cancelled by the Standard whenever a report of what was done reached New York. Mr. Rice had paid two-hundred-and-fifty dollars wrongfully, the money was at once refunded and Mr. Rice did not harass the company into buying his twenty-thousand dollar refinery for a half-million. This will serve as an example of the dishonest mistatements that had wrought lots of good people up to white heat. The sins of the trusts may be very scarlet and very numerous, but economic literature should not pollute the sources of information and the foundations of public opinion.

When the history of this wonderful century is written it will tell how an American boy, born in New York sixty years ago, clerked in a country-store, kept a set of books, started a small oil-refinery at Cleveland and at forty was the head of the greatest business in the world. This is, in outline, the story of John D. Rockefeller's successful career. Yesterday, as it were, a youth with nothing but integrity, industry and ambition for capital—a pretty good outfit, too—to-day he is one of the half-dozen richest men in Europe or America. Better than all else, integrity that is part and parcel of his moral nature, industry that finds life too fruitful to waste it idly and ambition to excel in good deeds as well as in business are his rich possession still. Gathering the largest fortune ever accumulated in twenty-five years has not blunted his fine sensibilities, dwarfed his intellectual growth, stifled his religious convictions or absorbed his whole being. Increasing wealth brought with it a deep sense of increasing responsibility and he is honored not so much for his millions as for the use he makes of them. Even in an age unrivalled for money-getting and money-giving, Mr. Rockefeller's keen foresight, executive ability and wise liberality have been notably conspicuous. His faith in the future of petroleum and his desire to benefit humanity he has shown by his works. Believing in the power of united effort to develop an infant-industry, his genius devised the system of practical co-operation that developed into the Standard Oil-Trust, against which prejudice and ignorance have directed their fiercest fire. Believing in education, his magnificent endowment of Chicago University—eight to ten-million dollars—ranks him with the foremost contributors to the foundation of a seat of learning since schools and colleges began. Believing in fresh air for the masses, he donated Cleveland a public park and a million to equip it superbly. Believing in spiritual progress, he builds churches, helps weak congregations and aids in spreading the gospel everywhere. Believing in the claims of the poor, his charities amount to hundreds-of-thousands of dollars yearly, not to encourage pauperism and dependence, but to relieve genuine distress, diminish human suffering and put struggling men and women in the way to improve their condition. He has differed from nearly all other eminent public benefactors by giving freely, quietly and modestly during his active life, without seeking the popular applause his munificence could easily obtain.

Mr. Rockefeller is a strict Baptist, a regular attendant at church and prayer-meeting, a teacher in the Sunday-school and a staunch advocate of aggressive Christianity. His advancement to commanding wealth has not changed his ideas of duty and personal obligation. He realizes that the man who lives for himself alone is always little, no matter how big his bank-account. He and his family walk to service or ride in a street-car, with none of the trappings befitting the worship of Mammon rather than the glory of God. Earnest, positive and vigorous in his religion as in his business, he takes no stock in the dealer who has not stamina or the profession of faith that is too destitute of backbone to have a denominational preference. The president of the Standard Oil-Company impresses all who meet him with the idea of a forceful, decisive character. He looks people in the face, his eyes sparkle in conversation and he relishes a bright story or a clever narration. You feel that he can read you at a glance and that deception and evasion in his presence would be utterly futile. The flatterer and sycophant would make as little headway with him as the bunco-steerer or the green-goods vendor. His estimate of men is rarely at fault and to this quality some measure of the Standard's success must be attributed. As if by instinct, its chief officer picked out men adapted to special lines of work—men who would not be misfits—and secured them for his company. The capacity and fidelity of the Standard corps are proverbial. Whenever Mr. Rockefeller wishes to enjoy a breathing-spell at his country-seat up the Hudson or on his Ohio farm, he leaves the business with perfect confidence, because his lieutenants are competent and trustworthy and the machine will run along smoothly under their watchful care. He has not accumulated his money by wrecking property, but by building up, by persistent improvement and by rigidly adhering to the policy of furnishing the best articles at the lowest price. Fair-minded people are beginning to understand something of the service rendered the public by the man who stands at the head of the petroleum-industry and more than any other is the founder of its commerce. He has invested in factories, railroads and mines, giving thousands employment, developing the resources of the country and adding to the wealth of the nation. He is human, therefore he sometimes errs ; he is fallible, therefore he makes mistakes, but the world is learning that John D. Rockefeller has no superior in business and that the Standard Oil-Company is not an organized conspiracy to plunder producers or consumers of petroleum. It is time to dismiss the idea that ability to build up and maintain a large business is discreditable, that marvellous success is blameworthy and that business-achievements imply dishonesty.

William Rockefeller, who resembles his brother in business skill, is a leader in Standard affairs and has his office in the Broadway building. He was a member of the first Board of Trustees and bore a prominent part in organizing and developing the Oil-Trust. He is largely interested in railroads, belongs to the best clubs, likes good horses and contributes liberally to worthy objects. The Standard folks don't lock up their money, loan it on mortgages at extravagant rates, spend it in Europe or try to get a gold squeeze on the government. They employ it in manufactures, in railways, in commerce and in enterprises that promote the general welfare.

From the days of the little refinery in Cleveland, the germ of the Standard, Henry M. Flagler and John D. Rockefeller have been closely associated in oil. Samuel Andrews, a practical refiner and for some time their partner, retired from the firm with a million dollars as his share of the business. The organization of the Standard Oil-Company of Cleveland was the first step towards the greater Standard Oil-Company of which all the world knows something. Its growth surprised even the projectors of the combination, who "builded better than they knew." Mr. Flagler devotes his time largely to beneficent uses of his great wealth. He recognizes the duty of the possessor of property to keep it from waste, to render it productive and to increase it by proper methods. A vast tract of Florida swamp, yielding only malaria and shakes, he has converted into a region suited to human-beings, producing cotton, sugar and tropical fruits and affording comfortable subsistence to thousands of provident settlers. He has transformed St. Augustine from a faded antiquity into a modern town, with the magnificent Ponce de Leon Hotel, paved streets, elegant churches, public halls, and all conveniences, provided by this generous benefactor at a cost of many millions. He has constructed new railroads, improved lines built previously, opened interior counties to thrifty emigrants and performed a work of incalculable advantage to the New South. He and his family attend the West Presbyterian Church, of which the Rev. John R. Paxton, formerly of Harrisburg, was pastor until 1894. Mr. Flagler is of average height, slight build and erect figure. His hair is white, but time has not dealt harshly with the liberal citizen whose career presents so much to praise and emulate.

John D. Archbold, vice-president of the Standard Oil-Company and its youngest trustee during the entire existence of the Oil-Trust, has been actively connected with petroleum from his youth. No man is better known and better

liked personally in the oil-regions. From his father, a zealous Methodist minister, and his good mother, one of the noble women to whom this country owes an infinite debt of gratitude, he inherited the qualities of head and heart that achieved success and gained multitudes of friends. A mere lad when the reports of golden opportunities attracted him from Ohio to the land of petroleum, he first engaged as a shipping-clerk for a Titusville refinery. His promptness, accuracy, and pleasant address won him favor and promotion. He soon learned the whole art of refining and his active mind discovered remedies for a number of defects. Adnah Neyhart induced him to take charge of his warehouse in New York City for the sale of

JOHN D. ARCHBOLD.

refined-oil. His energy and rare tact increased the trade of the establishment steadily. Mr. Rockefeller met the bright young man and offered him a responsible position with the Standard. He was made president of the Acme Refining Company, then among the largest in the United States. He improved the quality of its products and was entrusted with the negotiations that brought many refiners into the combination. He had resided at Titusville, where he married the daughter of Major Mills, and was the principal representative of the Standard in the producing section. When the Trust was organized he removed to New York and supervised especially the refining-interest of the united corporations. His splendid executive talent, keen perception, tireless energy and honorable manliness were simply invaluable. Mr. Archbold is popular in society, has an ideal home, represents the Standard in the directory of different companies and merits the high esteem ungrudgingly bestowed by his associates in business and his acquaintances everywhere.

The personal traits and business-successes of Charles Pratt, an original member of the Standard Trust, were typical of American civilization. The son of poor parents in Massachusetts, where he was born in 1830, necessity compelled him to leave home at the early age of ten and seek work on a farm. He toiled three years for his board and a short term at school each winter. For his board and clothes he next worked in a Boston grocery. His first dollar in money, of which he always spoke with pride as having been made at the workbench, he earned while learning the machinist-trade at Newton, in his native state. With the savings of his first year in the machine-shop he entered an academy, studying diligently twelve months and subsisting on a dollar a week.

Then he entered a Boston paints-and-oil store, devoting his leisure hours to study and self-improvement. Coming to New York in 1851, he clerked in Appleton's publishing-house and later in a paint-store. In 1854 he joined C. T. Reynolds and F. W. Devoe in a paints-and-oil establishment. Petroleum refining became important and the partners separated in 1867, Reynolds controlling the paints-department and Charles Pratt & Co. conducting the oil-branch of the business. The success of the latter firm as oil-refiners was extraordinary. Astral-oil was in demand everywhere. The works at Brooklyn, continuous and surprising as was their expansion, found it difficult to keep pace with the consumption. The firm entered into

CHARLES PRATT.

the association with the Cleveland, Pittsburg and Philadelphia companies that culminated in the Standard Oil-Trust, Mr. Pratt holding the relation of president of the Charles-Pratt Manufacturing Company. He lived in Brooklyn and died suddenly at sixty-three, an attack of heart-disease that prostrated him in his New-York office proving fatal in three hours. For thirty years he devoted much of his time to the philanthropies with which his name will be perpetually identified. He built and equipped Pratt Institute, a school of manual arts, at a cost of two-million dollars. He spent a half-million to erect the Astral Apartment Buildings, the revenue of which is secured to the Institute as part of its endowment. He devoted a half-million to the Adelphia Academy and a quarter-million towards the new edifice of Emanuel Baptist Church, of which he was a devout, generous member. His home-life was marked by gentleness and affection and he left his family an estate of fifteen to twenty-millions. Charles Pratt was a man of few words, alert, positive and unassuming, sometimes blunt in business, but always courteous, trustworthy and deservedly esteemed for liberality and energy.

Jabez A. Bostwick, a member of the Standard Trust from its inception, was born in New York State, spent his babyhood in Ohio, whither the family moved when he was ten years old, and died at sixty-two. His business-education began as clerk in a bank at Covington, Ky. There he first came into public notice as a cotton-broker, removing to New York in 1864 to conduct the same business on a larger scale. He secured interests in territory and oil-wells at Franklin in 1860, organized the firm of J. A. Bostwick & Co. and engaged extensively in refining. The firm prospered, bought immense quantities of crude and increased its refining capacity extensively. Mr. Bostwick was active in forming the Standard Oil-Trust and was its first treasurer. He severed his connection with his oil-partner, W. H. Tilford, who also entered the Standard Oil-Company. Seven years before his death he retired from the oil-business to accept the presidency of the New York & New England Railroad. He held the position six years and was succeeded by Austin Corbin. Injuries during a fire at his country-seat in Mamaroneck caused his death. The fire started in Frederick A. Constable's stables, in rear of Mr. Bostwick's. Unknown to his coachman, who was pushing behind it, Mr. Bostwick seized the whiffletrees of a carriage. Suddenly the vehicle swerved and the owner was violently jammed against the side of the stable. The coachman saw his peril and pulled the carriage back. Mr. Bostwick reeled forward, his face white with pain and sank moaning upon a buckboard. "Don't leave me, Mr. Williams," he whispered to his son's tutor, "I fear I am badly hurt." The sufferer was carried to the house, became unconscious and died in ten minutes, surrounded by members of his household and his neighbors. In 1866 Mr. Bostwick married a daughter of Ford Smith, a retired Cincinnati merchant, who removed to New York during the war. They had a son and two daughters. The daughters married and were in Europe when their father met his tragic fate. The widow and children inherited an estate of twelve millions. Mr. Bostwick was liberal with his wealth, giving largely without ostentation. Forrest College, in North Carolina, and the Fifth Avenue Baptist Church of New York were special recipients of his bounty, while his private benefactions amounted to many thousands yearly. He was strict almost to sternness in his dealings, preferring justice to sentiment in business.

These were the six trustees of the Standard Oil-Trust as first constituted of whom the world has heard and read most. Many of the two-thousand stockholders of the Standard Oil-Company are widely known. Benjamin Brewster, president of the National-Transit Company, retired with an ample fortune. His successor, H. H. Rogers, the present head of the pipe-line system, is noted alike for business-sagacity and sensible benefactions. The great structure at No. 26 Broadway, the largest office-building in New York occupied by one concern, is the Standard headquarters. Each floor has one or more departments, managed by competent men and all under supervision of the company's chief officials. From the basement, with its massive vaults and steam-heating plant, to the roof every inch is utilized by hundreds of book-keepers, accountants, stenographers, telegraphers, clerks and heads of divisions. Everything moves with the utmost precision and smoothness. President Rockefeller has his private offices on the eighth floor, next the spacious room in which the Executive Committee meets every day at noon for consultation. Mr. Flagler, Mr. Archbold and Mr. Rogers are located conveniently. The substantial character of the building and the business-like aspect of the departments impress visitors most favorably. There is an utter absence of gingerbread and cheap ornamentation, of confusion and perplexing hurry. The very air, the clicking of the telegraph-instruments, the noiseless motion of the elevators and the prompt dispatch of business indicate solidity, intelligence and perfect system. From that building the movements of a force of employés, numbering twice the United States army and scattered over both hemispheres, are directed. The sails of the Standard fleet whiten every sea, its products are marketed wherever men have learned the value of artificial light and its name is a universal synonym for the highest development of commercial enterprise in any age or country.

Business-men recall with a shudder the frightful stringency in 1893. All over the land industries drooped and withered and died. Raw material, even wool itself, had no market. Commerce languished, wages dwindled, railroads collapsed, factories suspended, and myriads of workmen lost their jobs. Merchants cut down expenses to the lowest notch, loans were called in at a terrible sacrifice, debts were compromised at ten to fifty cents on the dollar, the present was dark and the future gloomy. The balance of trade was heavily against the United States. Government securities tumbled and a steady drain of gold to Europe set in. The efforts of Congress, the Treasury Department and syndicates of bankers to stem the tide of disaster were on a par with Mrs. Partington's attempt to sweep back the ocean with a sixpenny-broom. Amid the general demoralization, when the nation seemed hastening to positive ruin, one splendid enterprise alone extended its business, multiplied its resources and was largely instrumental in restoring public confidence.

The Standard Oil-Company, unrivalled in its equipment of brains and skill and capital, not merely breasted the storm successfully, but did more than all other agencies combined to avert widespread bankruptcy. Through the sagacity and foresight of this great corporation crude oil advanced fifty per cent., thereby doubling and trebling the prosperity of the producing sections, without a corresponding rise in refined. By this wise policy, which only men of nerve and genius could have carried out, home consumers were not taxed to benefit the oil-regions and the exports of petroleum-products swelled enormously. As the result, while the American demand increased constantly, millions upon millions of dollars flowed in from abroad, materially diminishing the European drainage of the yellow metal from this side of the Atlantic. The salutary, far-reaching effects of such management, by reviving faith and stimulating the flagging energies of the country, exerted an influence upon the common welfare words and figures cannot estimate. Petroleum preserved the thread of golden traffic with foreign nations.

Hon. Samuel C. T. Dodd, one of the ablest lawyers Pennsylvania has produced, is general solicitor of the Standard and resides in New York. His father, the venerable Levi Dodd, established the first Sunday-school and was president of the second company that bored for oil at Franklin, the birthplace of his son in 1830. Young Samuel learned printing, graduated from Jefferson College in 1857, studied law with James K. Kerr and was admitted to the Venango Bar in August of 1859. His brilliant talents, conscientious application and legal acquirements quickly won him a leading place among the successful jurists of the state. During a practice of nearly twenty-two years in the courts of the district and commonwealth he stood in the front rank of his profession. He served with credit in the Constitutional Convention of 1873, framing some of its most important provisions. He traveled abroad and wrote descriptions of foreign lands so charming they might have come from Washington Irving and N. P. Willis. His selection by the Standard Oil-Trust in 1881 as its general solicitor was a marked recognition of his superior abilities. The position, one of the most prominent and responsible to which a lawyer can attain, demanded exceptional qualifications. How capably it has been filled the records of all legal matters concerning the Standard abundantly demonstrate. Mr. Dodd's profound knowledge of corporation-law, eminent sense of justice, forensic skill, rare tact and clear brain have steered the great company safely and honorably through many suits involving grave

SAMUEL C. T. DODD.

questions of right and millions of money. The papers he prepared organizing the Standard Trust have been the models for all such documents since they left his desk. Terse logic, sound reasoning, pointed analysis and apposite expression distinguish his legal opinions and arguments, combining the vigor of a Damascus blade with the beauty of an epic. He is a delightful conversationalist, sincere friend and prudent counsellor, kindly, affable and thoroughly upright. His home, brightened by a loving wife and devoted family, is singularly happy. Amid the cares and anxieties incident to professional life he has cultivated his fine literary-taste, writing magazine-articles and wooing the muses at intervals of leisure only too far apart. He has the honor of writing the first poem on petroleum that ever appeared in print. It was a rich parody on Byron's "Isles of Greece" and was published in the spring of 1860, as follows:

The land of Grease! the land of Grease!
　Where burning Oil is loved and sung;
Where flourish arts of sale and lease,
　Where Rouseville rose and Tarville sprung;
Eternal summer gilds them not,
But oil-wells render dear each spot.

The ceaseless tap, tap of the tools,
　The engine's puff, the pump's dull squeak,
The horsemen splashing through the pools
　Of greasy mud along the Creek,
Are sounds which cannot be suppress'd
In these dear Ile-lands of the Bless'd.

Deep in the vale of Cherry Run
　The Humboldt Works I went to see,
And sitting there an oil-cask on
　I found that Grease was not yet free;
For busily a dirty carl
Was branding " bonded " on each barrel.

I sat upon the rocky brow
　Which o'erlooks Franklin—far-famed town;
A hundred derricks stood below
　And many a well of great renown;
I counted them at break of day,
And when the sun set where were they?

They were still there. But where art thou,
　My dry-hole? On the river-shore
The engine stands all idle now,
　The heavy auger beats no more;
And must a well of so great cost
Be given up and wholly lost?

'Tis awful when you bore a well
　Down in the earth six-hundred feet,
To find that not a single smell
　Comes up your anxious nose to greet;
For what is left the bored one here?
For Grease a wish; for Grease a tear!

Must I but wish for wells more bless'd?
　Must I but weep? No, I must toil!
Earth, render back from out thy breast
　A remnant of thy odorous oil!
If not three-hundred, grant but three
Precious barrels a day to me.

What! silent still? and silent all?
　Ah no! the rushing of the gas
Sounds like a distant torrent's fall
　And answers, bore ahead, you ass,
A few feet more; you miss the stuff
Because you don't go deep enough!

In vain! in vain! Pull up the tools!
　Fill high the cup with lager-beer!
Leave oil-wells to the crazy fools
　Who from the East are flocking here.
See at the first sight of the can
How hurries each red-shirted man!

Fill high the cup with lager-beer!
　The maidens in their promenade
Towards my lease their footsteps steer
　To see if yet my fortune's made;
But sneers their pretty faces spoil
To find I have not yet struck oil.

Place me in Oil Creek's rocky dell,
　Though mud be deep and prices high;
There let me bore another well
　And find petroleum or die.
No more I'll work this dry-hole here;
Dash down that cup of lager-beer.

A pretty little story is told of Miss Edith Rockefeller while at boarding-school, illustrative of the manner in which she was trained by her father to consider herself as no more than moderately wealthy. Miss Edith, with a party of girls from her class, presented herself at a furniture-dealer's to choose a gift for a favorite teacher. The price of the pretty writing-desk was more than the sum of money in their possession. The girls suggested that, if the desk were sent, they would forward the balance as soon as possible. The furniture-dealer very politely, but also very decidedly, informed the girls that he could not do as they asked. "But," he said, "if you can think of any New-York business-man with whom any of your fathers is acquainted and who will vouch for you, the matter may be arranged."

"Why," said the daughter of the petroleum-magnate, "I think my papa has an office away down on Broadway; possibly we can get the money there."

"Who is your father?" queried the dealer.

"His name is Rockefeller," replied the girl; "John D. Rockefeller; he is in the oil-business."

The merchant gasped and looked at the girl in amazement. "John D. Rockefeller your father? Is John D. Rockefeller good for twenty-five dollars?" he repeated. Then he recovered his presence of mind sufficiently to order the desk packed up and sent immediately, while Miss Edith, very much aston-ished at his unwonted excitement, thanked him with pretty and simple grace.

Although the Standard pays the highest wages in the world and has never had a serious strike in its grand army of forty-thousand men, not one cent of a reduction was ordered during the panic. No works stopped and no employés were turned adrift to beg or starve. On the contrary, improvements and addi-tions were made continually, the force of workmen was augmented, cash was paid for everything bought, no claims remained unsettled and nobody had to wait an hour for money justly due. These are facts for the toiling masses, whom prejudice against big corporations sometimes misleads, to understand and consider.

Russian competition, the extent and danger of which most people do not begin to appreciate, was met and overcome by sheer tenacity and superior gen-eralship. The advantages of capable, courageous, intelligent concentration of the varied branches of a great industry were never manifested more strongly. Deprived of the invincible bulwark the Standard offered, the oil-producers of Pennsylvania, New York, West Virginia, Ohio and Indiana would have been utterly helpless. The Muscovite bear would have gobbled the trade of Europe and Asia, driving American oil from the foreign markets. Local consumption would not have exhausted two-thirds of the production, stocks of crude would have piled up and the price would have fallen proportionately. Instead of rank-ing with the busiest, happiest and most prosperous quarters of the universe, as they are to-day, the oil-regions of five states would have been irretrievably ruined, dragging down thousands of the brightest, manliest, cleverest fellows on God's footstool! Instead of bringing a vast amount of gold from England, France and Germany for petroleum produced on American soil, refined by American workmen paid American wages and exported by an American com-pany in American vessels, the trade would have been killed, the cash would have stayed across the waters and the country at large would have suffered incalculably! These are things to think of when some cheap agitator, with a private axe to grind, a mean spite to gratify or a selfish object to attain, raises a howl about monopoly and insists that the entire creation should "damn the Standard!"

XV.

JUST ODDS AND ENDS.

How Natural Gas Played Its Part—Fire and Water Much in Evidence—Changes in Methods and Appliances—Deserted Towns—Peculiar Coincidences and Fatalities—Railroad Episodes—Reminiscences of Bygone Scenes—Practical. Jokers—Sad Tragedies—Lights and Shadows Intermingle and the Curtain Falls Forever.

"Variety's the very spice of life."—*Cowper.*
"Fuss and feather, wind and weather, varied items strung together."—*Oil City Derrick.*
"Laugh when we must, be candid when we can."—*Pope.*
    "'A picker-up of unconsidered trifles'
    From many sources facts and fancies rifles."—*Ibid.*
"Every house should have a rag-bag and a general storeroom."—*Miss Parloa.*
"A little nonsense now and then is relished by the wisest men."—*Holmes.*
    "Let days pass on, nor count how many swell
    The episode of life's back chronicle."—*Lytton.*
"Fond memory brings the light of other days around me."—*Ibid.*
"Close up his eyes and draw the curtain close."—*Shakespeare.*

ATURAL-GAS, the cleanest, slickest, handiest fuel that ever warmed a heart or a tenement, is the right bower of crude-petroleum. It is the one and only fuel that mines, transports and feeds itself, without digging every spoonful, screening lumps, carting, freighting and shoveling into the stove or furnace. Getting it does not imperil the limbs and lives of poor miners—the most overworked and underpaid class in Pennsylvania—in the damp and darkness of death-traps hundreds of feet beneath the surface of the ground. You drill a hole to the vital spot, lay a pipe from the well to the home or factory, turn a stop-cock to let out the vapor, touch off a match and there it is—the brightest, cleanest, steadiest, hottest fire on earth. Not a speck of dust, not an atom of smoke, not a particle of cinder, not a taint of sulphur, not a bit of ashes vexes your soul or tries your temper. There is no carrying of coal, no dumping of choked grates, no waiting for kindling to catch or green wood to burn, no scolding about sulky fires, no postponement of heat because the wind blows in the wrong direction. Blue Monday is robbed of all its terrors, the labor of housekeeping is lightened and husbands no longer object to starting the fire on cold mornings. A nice blaze may be let burn all night in winter and kept on tap in summer only when needed. It is lighted or extinguished as readily as the gas-jet in the parlor. It melts iron, fuses glass, illumines mills and streets, broils steaks to perfection and does away with many a fruitful source of family-broils. It saves wear and tear of muscle and disposition, lessens the production of domestic quarrels, adds to the pleasure and satisfaction of living and carries the spring-time of existence into the autumn of old age. Set in a dainty metal frame, with background of asbestos and mantel above, its glow is cheerful as the hickory-fire in the hearth. It gives us the ingle-nook modernized and improved, the chimney-corner brought down to date. It glides through eighty-thousand miles of pipes in Pennsylvania, Ohio, West Virginia, Indiana and New York and employs a hundred-million dollars to supply it to people within reach of the bounteous reservoirs the kindly earth has treasured all through the centuries. If it be not a blessing to humanity, the fault lies with the folks and not with the stuff. The man who spouts gas is a nuisance, but the well that spouts gas is something to prize, to utilize and be thankful for. Visitors to the oil-region or towns near enough to enjoy the luxury, beholding the beauty and adaptability of natural-gas, may be pardoned for breaking the tenth commandment and coveting the fuel that is Nature's legal-tender for the comfort and convenience of mankind.

The pretty town of Fredonia, in New York state, three miles from Lake Erie and forty-five south-west of Buffalo, enjoys the distinction of first using natural-gas for illuminating purposes. It is a beautiful place, famous for fine roads, fine scenery and fine vineyards. Canodonay Creek, a small but rapid stream, passes through it to the lake. Opinions vary as to the exact date when the gas was utilized, some authorities making it 1821, others 1824 and a few 1829. The best information fixes it at 1824, when workmen, in tearing down an old mill, observed bubbles on the water that proved to be inflammable. The hint was not lost. A company bored a hole *one-inch-and-a-half* in diameter into the limestone-rock. The gas left its regular channel, climbed the hole, lighted a new mill and was piped to a hundred houses in the village at a cost of one-fifty a year for each. The flame was large and strong and for years Fredonia was the only town in America lighted by "nature-gas." A gasometer was constructed, which collected *eighty-eight* cubic feet in twelve hours. The inhabitants didn't keep late hours. A mile nearer Lake Erie many gas-bubbles gamboled on the stream. Efforts to convey the gas to the light-house at Dunkirk failed, as it was only half the weight of air and would not descend the difference in elevation.

A light-house at Erie was lighted by natural-gas in 1831, "the Burning Spring," a sheet of water through which the vapor bubbled, furnishing the supply. A tower erected over the spring held the gas that accumulated during the day and wooden-pipes conveyed it at night to the light-house.

Dr. Charles Oesterlin, a young German physician, sixty years ago unpacked his pill-boxes and hung out his little sign at Findlay, in Northwestern Ohio. He was an expert geologist and mineralogist, but the flat Black Swamp afforded poor opportunities to study the rocks underlying the limestone. The young physician detected the odor of sulphuretted hydrogen in the town and along the banks of the Blanchard River. It puzzled him to guess the source of the odor. He spoke to the farmers, who smelled the stuff, knew nothing and cared less about its origin or properties. The Doctor searched for a sulphur-spring. In October of 1836 the solution came. A farmer was digging a well three miles from town. A spring was tapped and the water "boiled," as the diggers expressed it. Debating what to do, they were called to supper, returned after dark and lighted a torch to examine the well. Holding the torch over the well an explosion startled them and a flame ascended that lasted for days. Nobody was seriously hurt, but all thought the devil had a finger in the pie. Dr. Oesterlin connected the incident with the odor and it confirmed his theory of a gas that would burn and might serve as fuel. At a stone-quarry he made a cone of mud over a fissure, covered it with a bucket and applied a light. When the Doctor picked himself up in an adjoining corn-field the bucket was still sailing north towards Toledo. Daniel Foster, another Findlay farmer, dug a well in 1838. Gas issued from the hole before water was seen. Foster had a practical mind. He inverted a copper-kettle over the hole, rigged a wooden pump-stock beneath the kettle, plastered around it with clay, joined more pump-stocks together, stuck an old gun-barrel in the end of the last one, lighted the gas in his kitchen and by means of the flame boiled water, roasted coffee and illumined the apartment. Then Dr. Oesterlin declared Findlay was right over a vast caldron of gas. People laughed at him, adhered to tallow-dips and positively refused to swallow such a dose. Petroleum-developments in Pennsylvania fortified his faith and he sought to interest the public in a company to "bore a hole twenty inches across." Sinners in Noah's day were less impervious. Business-men scoffed and declined to subscribe for stock. He tried again in 1864 and 1867 with the same result. A company was organized to manufacture coal-gas. He talked of the absurdity of *making* gas at Findlay as equal to setting up a manufactory of air or water. It was no use. At last the triumph of natural-gas in Pennsylvania was manifested too strongly for the obtuse Findlayites to ignore it. In 1884 the Doctor managed to enlist four-thousand dollars of capital and start a well in a grove a mile east of town, where the odor was pungent and gas flowing through a tile-pipe he planted in the ground burned for weeks. He watched the progress of the work with feverish anxiety. The hopes of fifty long years were to be grandly realized or dashed forever. Sleepless nights succeeded restless days as the veteran's heart-beats kept time with the rhythmic churning of the drill. At five, six and seven-hundred feet morsels of gas quickened the expectations of success. At eleven-hundred feet, in the Trenton limestone, on November tenth, 1884, gas burst forth with terrific force. The well was drilled sixteen-hundred feet and encountered salt-water. It was plugged below the gas-vein, the gas was lighted, an immense flame shot up and for months a quarter-million feet a day burned in the open air. Findlay grew from five-thousand to fifteen-thousand population and manufacturing flourished. Dr. Oesterlin, slight of frame, infirm with age, his thin locks and beard white as snow, had waited fifty years for his vindication. It came when he had reached four-score, full, complete and overwhelming. He bore his honors meekly, lived to round out eighty-two and nowhere is it recorded that he even once yielded to the temptation of remarking : "I told you so !"

Gas was used as fuel at pumping-wells on Oil Creek in 1862. It was first collected in "gas-barrels," one pipe leading from the well to the receptacle and another from the barrel to the boiler. Many fires originated from the flame, when the pressure of gas was small, running back to the barrel and exploding it. A pumper at Rouseville, seated on a gas-barrel at such a moment, went skyward and may be ascending yet, as he never returned for his week's wages. D. G. Stillwell, better known as "Buffalo Joe," drilled a gasser in 1867 at Oil City, on the site of the Greenfield Lumber-Company's office. He piped the gas to several houses, but the danger from constant changes of pressure led to its abandonment. This is the first authentic record of the use of "the essence of Sheol" for cooking food and heating dwellings. In 1883 the Oil-City Fuel-Supply Company laid a six-inch gas-line to wells at McPherson's Corners, Pine-grove township, eight miles distant. The gas was produced from the second and third sands, at a depth of nine to ten-hundred feet and a pressure not exceeding two-hundred pounds to the square inch. In 1885 the late Samuel Speechly started a well on his farm near McPherson's, intending to drill three-thousand feet in search of the Bradford sand. Oil-bearing strata dip twenty feet to the mile southward and Speechly believed the northern rocks existed far beneath the ordinary third-sand in Venango county. On April thirteenth, at nine-teen-hundred feet, the drill penetrated what has since been called the "Speech-ly sand," the most extraordinary and valuable fuel-sand as yet discovered. In this sand at three feet pressure of gas became entirely too great to keep jerk-ing the tools. The gas company leased the well and turned it into the line with-out being able to gauge it on account of the high volume.

Speechly commenced a second well and the company, having previously laid a new ten-inch line to Oil City, constructed branches to Franklin and Titusville. The second well proved to be the largest to the present time, excepting the Big Moses in West Virginia. For a time it could not be controlled. The roar of the escaping gas could be heard for miles. Eventually it was tubed and the pressure was six-hundred pounds. Many wells in other fields have had greater pressure, but the large volume of the Speechly well made it a wonder. One day all the other wells connected with the main-line were discontinued from the line temporarily and the Jumbo turned in. The flow was sufficient to supply Oil City, Titusville and Franklin with all the gas required. Hundreds of wells have been drilled to the Speechly sand and the field now reaches from the southern part of Rockland township, Venango county, to Tionesta township, Forest county. It is about thirty miles long, with an average width of three miles, while the sand ranges in thickness from fifty to one-hundred feet. The pressure gradually diminishes. It requires constant drilling to keep up the supply, the Oil-City Company alone having about four-hundred wells.

Samuel Speechly died on Sunday night, January ninth, 1893, aged sixty-one, at his home in the gas-district bearing his name. His life was notably eventful, adventurous and fortunate. Born in England in 1832, at fourteen he began to learn locomotive-building and marine-engineering at Newcastle-on-Tyne. At twenty Robert Stephenson & Co. sent him to China to join a steamer engaged in the opium-trade. In 1855 he entered the service of the Chinese government to suppress piracy on the coast, and in 1857 started at Hong Kong the first engineering-business in the vast empire ruled by the pig-tailed Brother of the Sun. He visited America in 1872 and lived in Philadelphia. Wanting plenty of room, he went to Northwestern Pennsylvania, resided a year in Cranberry township, concluded to stay and settled on what subsequently became the famous Speechly farm. The well he drilled in 1885 had neither oil nor gas in the usual formations. Veteran operators advised him to abandon it, but Speechly entertained a notion of his own and the world knows the sequel. He was married in China in 1864 to Miss Margaret Galbraith, who survives him, with two daughters, Emily, born in China, and Adelaide, born in America. His widow and children occupy the old home on the farm.

Bishop Potter, stopping at Narrowsburg in 1854, noticed jets of gas exuding from the bank of the Delaware river at Dingman's Ferry, forty miles above Easton, and published an article on the subject. A company in 1860 bored three wells, but the result was not encouraging, as politicians are the most gaseous bodies Northampton county has produced for thirty years. A gas-well at Erie attracted considerable attention in 1860 and was followed by a number more, which from a shallow depth yielded fuel to run several factories. East Liverpool, Ohio, put the product to practical use early in the seventies as a substitute for coal. The first well, drilled in 1860, caught fire and destroyed the rig. Geologists say natural-gas is the disembodied spirits of plants that grew in the sunshine of ages long before the foundations of the buried coal-measures were laid, so long ago shut up and forsaken by the light-hearted sun that it is a wonder they hadn't forgotten their former affinity. But they hadn't. They rushed out to the devouring kiss of their old flame at the first tap of the drill on their prison-house, like a foolish girl at the return of a fickle lover. They found Old Sol flirting with their younger sister, playing sweet to a lot of new vegetation. Before they had time to form a sewing-circle and resolve that all the male sex are horrid, they took fire with indignation at his fickleness and the tool-dresser's forge and burst with a tremendous explosion. The fire was quenched and gas poured out of the pioneer-well fifteen years. Street-lamps were left burning all day, which was cheaper than to bother putting them out, and East Liverpool prospered as a hive of the pottery-industry. The celebrated well at East Sandy, Venango county, which gave birth to Gas City in 1869, burned a year with a roar audible three miles. Becoming partially exhausted, the fire was put out and the product was used for fuel at numerous wells. The famous Newton well, on the A. H. Nelson farm, was struck in May of 1872 and piped in August to Titusville, five miles south west. Its half-million cubic-feet per day supplied three-hundred firms and families with light and fuel. Henry Hinckley and A. R. Williams organized the company, one of the very first in Pennsylvania to utilize natural-gas on an extensive scale. The same year gas from the Lambing well was piped to Fairview and Petrolia. The Waugh well at Millerstown and the Berlin at Thompson's Corners, Butler county, were the next big gassers. The great Delamater No. 2, near St. Joe, finished in 1874, for months was the biggest gas-well in the world. Its output was conveyed to the rolling-mills at Sharpsburg. The first gas-well in Butler county is credited to John Criswell, of Newcastle, who drilled for salt-water in 1840 near Centreville, struck a vein of the vapor at seven-hundred feet and fired it to heat his evaporating-pans.

25

At Leechburg and Apollo natural-gas has been used in puddling-furnaces since 1872. It will supply the huge mills at Vandergrift, the model town that is to be the county-seat of Vandergrift county, which the next Legislature will set off from Armstrong, Westmoreland and contiguous districts. It was the fuel of the cutlery-works at Beaver Falls from 1876 until the wells ceased producing in 1884. In 1875 Spang & Chalfant piped it from Butler to their mills in the suburbs of Pittsburg. Though Pittsburgers knew of its value in the oil-region for twenty years, they regarded it as a freak and not calculated to affect their interests favorably. Iron manufactured by its means was of superior quality, owing to the absence of sulphur and the intensity of the heat. In 1877 the Haymaker well opened the Murraysville gas-field, but that immense storehouse of potential energy lay dormant until Pew & Emerson piped the product to Pittsburg. In June of 1884 George Westinghouse, inventor of the air-brake and of various electric-appliances, struck a gas-well near his residence in Pittsburg. From that date the development was enormous. Wells producing from two to twenty-million cubic-feet a day were in order. The Philadelphia Company— Westinghouse was its president—alone tied up forty-thousand acres of gas-territory, drilled hundreds of wells and laid thousands of miles of pipes. Hon. James M. Guffey headed big corporations that supplied Wheeling, a portion of Pittsburg and dozens of smaller towns. The coal displacement in Pittsburg equaled thirty-thousand tons daily. Twenty and twenty-four-inch mains intersected the city. Iron, brass, steel and metal-working establishments consumed it. Glass-factories turned out by its aid plate-glass such as mankind had never seen before. The flaming breath of the new demon transformed the appearance and revolutionized the iron-manufacture of the Birmingham of America. The Smoky City was a misnomer. Soot and dirt and smoke and cinders disappeared. People washed their faces, men wore "biled shirts" and girls dressed in white. The touch of a fairy-wand could not have made a more resplendent change. Think of green grass, emerald hues, clear sunlight and clean walls in Pittsburg! At first timid folks feared to introduce it, because the pressure could not be regulated. All this has been remedied. The roaring, hissing monster that almost bursts the gauge at the well is tamed and subjugated to the meekness of a dove by valves and gasometers, which can reduce the pressure to a single ounce. Queer, isn't it, that Pittsburg should be metamorphosed by natural-gas—the fires of hell as it were—into a city of delightful homes, an industrial paradise?

Gas-wells of high pressure were found in Ohio by thousands, as though striving to vie with the oil-wells which, beginning at Mecca in 1860 and ending at Lima, stocked up twenty-million barrels of crude. Over three-hundred companies were chartered in a year to supply every town from Cincinnati to Ashtabula. Natural-gas raged and blistered and for a term was the genuine "Ohio idea." For thirty years wells at New Cumberland, West Virginia, have furnished fuel to burn brick. The same state has the biggest gassers in existence and lines to important cities are projected. If "the mountain won't come to Mohammed, Mohammed must go to the mountain." Indiana has gas and oil in four counties, with Gas City as headquarters and lots of fuel for houses and factories in Indianapolis and the chief cities. The Hoosiers have carried out the principle of Edward Eggleston's Mrs. Means: "When you're a-gittin' git plenty, I say." Illinois had a morsel of oil and gas in wells at Litchfield. Kentucky and Tennessee are blessed with "a genteel competence" and Kansas has not escaped. Michigan has gas-wells at Port Huron and St. Paul once boasted a company capitalized at a half-million. Buffalo inhaled its first whiff of natural-gas, piped from wells in McKean county, on December first, 1886. Youngstown was initiated next day, from wells in Venango. A Mormon company bored wells at Salt Lake, but polygamy was not supplanted by any odor more unsavory. In Canada gas is abundant and Robert Ferguson, now a well-to-do farmer near Port Sarnia, first turned it into an engine cylinder as a joke on the engineer at the pump station in Enniskillen township. Steam was low, the engineer was absent, Ferguson cut the pipe leading from the boiler, connected it with one from a gas-well near-by, opened the throttle and, to his astonishment, found the pressure greater than steam. Natural-gas, a gift worthy of the immortal gods, worthy of the admiration of Vulcan, worthy of the praise of poets and historians, the agent of progress and saver of labor, is not a trifle to be brushed off like a fly or dismissed with a contemptuous sneer.

Pittsburg iron-works and rolling-mills received natural-gas at about two-thirds the cost of coal. The coal needed to produce a ton of metal cost three dollars, the gas that did the same service cost one-ninety. Besides this important saving, the expense of handling the fuel, hauling away cinders and waiting for furnaces to heat or cool was avoided. Gas-heat was uniform, stronger, more satisfactory, could be regulated to any temperature, turned on at full head or shut off instantly. Thus Pittsburg possessed advantages that boomed its manufactories immensely and obliged many competitors less favored to retire. In this way the anomaly of freezing out men by the use of greater, cheaper heat was presented.

On March seventeenth, 1886, at Pittsburg, Milton Fisher, of Columbus, was the first person to be incinerated in a natural-gas crematory. In fifty minutes the body was reduced to a handful of white powder. The friends of the deceased pronounced the operation a success, but Fisher was not in shape to express his opinion.

A singular accident occurred near Hickory, Washington county, on the night of December fourth, 1886. Alfred Crocker, an employé of the Chartiers Gas-Company, had been at the tanks on the McKnight farm and was going toward the well. The connecting-pipe between the well and tank burst with terrible force, striking Crocker on the left leg, blowing the foot and ankle completely off and injuring him about the body. The explosion hurled the large gas-tank a hundred feet. The young man died next morning.

The steam tow-boat Iron City once grounded near the head of Herr's Island, above Pittsburg. The stern swung around and caught on a pipe conveying natural gas across the Allegheny river. In trying to back the vessel off the pipe broke, the escaping gas filled the hold and caught fire from the furnace. An explosion split the boat from stem to stern, blew off the deck and blew the crew into the river. The boat burned to the water's edge.

Near Halsey, in the Kane field, James Bowser was standing on a gas-tank, while a workman was endeavoring to dislodge an obstruction in the pipe leading from the well. The removal of the obstruction caused the pent-up gas to rush into the tank with such force that the receptacle exploded, hurling Bowser high in the air. He alighted directly in front of the heavy volume of gas escaping through the broken pipe. Before he could be rescued he was denuded of all clothing, except one boot. His clothing was torn off by the force of the gas and his injuries were serious.

Workmen laying pipe to connect with the main at Grapeville were badly flustered one frosty morning. By mistake the gas was turned on, rushing from the open end with great force. It ploughed up the earth and pebbles and ignited, the flinty stones producing a spark that set the whole thing in a blaze. Gas-wells yield liberally at Grapeville, supplying the glass-works at Jeannette and houses at Johnstown, the farthest point east to which the vapor-fuel has been piped.

J. S. Booker, an Ohio man, claimed to spot gas. His particular virtue lay in the muscles at the back of the neck, which rise up and irritate him in the presence of natural-gas. This is ahead of rheumatism as a rain-indicator. Booker's own story is that an attack of asthma left him in a sensitive state, so that when he passes over a vein of gas the electricity runs through his legs, up his spine and knots the muscles of the neck. The story deserves credit for its rare simplicity. With the whole realm of fiction at his command, Booker chose only a few simple details and was content to pass current as a sort of human witch-hazel.

At Economy, where a hundred stand-pipes for natural-gas illuminate the streets, bugs and fruit-vermin were slaughtered wholesale. In the mornings there would be a fine carpet of bugs around every post. Chickens and turkeys would have a feast and a foot-race from the roosts to see which would get to the already-cooked breakfast first. The trees came out in bloom earlier and healthier than formerly, because the vermin were destroyed and the frosts kept from settling by the gas-lights, which burn constantly. As a promoter of vegetation natural-gas beats General Pleasanton's blue-glass out of sight.

Samuel Randall, the Democratic statesman, visited the gas-wells at Murraysville with Hon. J. M. Guffey. From a safe distance the visitor threw a Roman candle at a huge column of vapor, which blazed quicker than a church-scandal, to Mr. Randall's great delight. President and Mrs. Cleveland were afforded a similar treat by Mr. Guffey. The chivalrous host chartered a train and had a big well fired for the distinguished visitors. The lady of the White House was in ecstacies and the President evidently thought the novel exhibition knocked duck-shooting silly. Could a mind-reader have X-rayed his thinking-department it would likely have assumed this form : "Mr. Guffey, *you* have a tremendous body of gas here, but *I* have Congress on my hands !"

Eli Perkins lectured at St. Petersburg one night and next day rode with me through part of the district. He wanted points regarding natural-gas and smilingly jotted down a lot of Munchausenisms current in the oil-region. A week later he sent me a marked copy of the New-York *Sun*, with columns of delicious romance concerning gas-wells. Eli was no slouch at drawing the long-bow, but he fairly surpassed himself, Jules Verne and Rider Haggard on this occasion. His vivid stories of tools hurled by gas a thousand feet, of derricks lifted up bodily, of men tossed to the clouds and picturesque adventures generally were marvels of smooth, easy, fascinating exaggeration. Perhaps "if you see it in the *Sun* it's so," but not when Eli Perkins is the chronicler and natural-gas the subject.

Fire and water have scourged the oil-region sorely. A flood in March of 1865 submerged Oil City, floated off hundreds of oil-tanks and small buildings and did damage estimated at four-millions of dollars. Fire in May of 1866 wiped out half the town, the loss footing up a million dollars. The most appalling disaster occurred on Sunday, June fifth, 1892. Heavy rains raised Oil Creek to such a height that mill-dams at Spartansburg and Riceville gave way, precipitating a vast mass of water upon Titusville during Saturday night. With a roar like thunder it struck the town. Sleepers were awakened by the resistless tide and drowned. Refineries and tanks of oil caught fire and covered acres of the watery waste with flames. Helpless men, women and children tottered and tumbled and disappeared, the death-roll exceeding fifty. The two elements seemed to strive which could work the greater destruction. Above Oil City a huge tank of benzine was undermined and upset on Friday morning. The combustible stuff floated on the creek, which had risen four feet over the floors of houses on the flats. The boiler-fire at a well near the Lake-Shore tunnel ignited the cloud of benzine. An explosion followed such as mortal eyes and ears have seldom seen and heard. The report shook the city to its foundations. A solid sheet of flame rose hundreds of feet and enveloped the flats in its fatal embrace. Houses charred and blazed at its deadly touch and fifty persons perished horribly. The sickening scene reminded me of the Johnstown carnage in

SCENE AT OIL CITY AFTER THE DISASTER ON JUNE 5, 1892.

1889, with its miles of flooded ruins and dreadful blaze at the railroad-bridge. Whole families were blotted out. Edwin Mills, his wife and their five children died together. Heroic rescues and marvelous escapes were frequent. John Halladay Gordon saved forty people in his boat, rowing it amid the angry flames and swirling waters at imminent risk. The recital of brave deeds and thrilling experiences would fill a volume. That memorable Sunday was the saddest day Oil City and Titusville ever witnessed. The awful grandeur of the spectacle at both places has had no parallel.

Sweeping into the yards of a refinery at the upper end of Titusville, the water tore open a tank containing five thousand gallons of gasoline. Farther down an oil-tank and a gasoline-tank were rent in twain. Water covered the streets and shut people in their houses. The gas-works and the electric-plant were submerged and the city was in darkness. At midnight a curious mist lay thick and dense and white for a few feet above the water. It was the gasoline-vapor, a cartridge a half-mile long, a quarter-mile wide and two yards thick, with a coating of oil beneath, waiting to be fired.   One arm of the mist reached into the open furnaces of the Crescent Works and touched the live coals on the grate. There was a flash as if the heavens had been split asunder.   Then the explosion came and death and havoc reigned.   And the horror was repeated at Oil City, until people wondered if the Day of Judgment could be more terrifying.   The infinite pity and sadness of it all !

The fire that desolated St. Petersburg started in Fred Hepp's beer-saloon. Hepp had a sign representing a man attempting to lift a schooner of lager as big as himself and remarking, "Oxcuse me ov you bleese."   The fire "oxcused" him from further exertion.

The burning of the Acme Refinery at Titusville, on June eleventh, 1880, entailed a loss of six-hundred-thousand dollars.   It caught from a tank light-

RUINS OF ACME REFINERY, TITUSVILLE, AFTER FIRE ON JUNE 11, 1880.

ning had struck.   By great efforts the railroad-bridge and the Octave Refinery were saved.   The fire raged three days and nights and the departments from Warren, Corry and Oil City were called to render assistance.   Hardly a town in the oil-regions has been unharmed by fire or flood, while many have been ravaged by both.

Charles Highberger, who had lost a leg, was elected a justice of the peace at Pithole in 1866.   Attorney Ruth, who came from Westmoreland county, was urging the conviction of a miserable whelp when he noticed Highberger had fallen asleep, as was his custom during long arguments.   Mr. Ruth aroused him and remarked :   " I wish your honor would pay attention to the points which I am about to make, as they have an important bearing on the case." Highberger opened his eyes, glared around the room and rose on his crutches in great wrath, exclaiming : "There has been too much blamed chin-whacking in this case ; you have been talking two hours and I haven't seen a cent of costs. The prisoner may consider himself discharged. The court will adjourn to Andy Christy's drug-store." That was the way justice was dispensed with in those good old days at Pithole.

"You're not fit to sit with decent people ; come up here and sit along with me !" thundered a Butler teacher who sat at his desk hearing a recitation, as he discovered at a glance the worst boy in school annoying his seatmate.

John Wallace, an early operator at Rouseville and merchant at Rynd, died in 1880. Born in Great Britain, he served in the English army, participated in the Crimean war and was one of the "Gallant Six Hundred" in the desperate charge at Balaklava immortalized by Tennyson. His shrewdness and enterprise were rewarded with a snug competence.

J. W. Sherman, who owned the Sherman well on Oil Creek, died at Cleveland lately. The ceaseless march to the tomb is rapidly thinning the ranks of the men who bore the burden of pioneer-operations.

Van Buren, Indiana, is experiencing an oil-boom that brings comfort and joy to the unsophisticated inhabitants. The biggest well in the Indiana field was struck recently near the town. Oil spurted fifty feet above the derrick. A local paper says : "The strike has given the town a tremendous boom. Several real-estate offices have opened and the town-council has raised the license for faro-banks from five dollars a year to twelve dollars." At this rate Van Buren ought soon to be in the van.

John Jeffersey, an Indian pilot, died at Tionesta in 1894. One dark night he plunged into the Allegheny, near Brady's Bend, to grasp a skiff loaded with cans of glycerine that brushed past his raft. Jacob Barry and Richard Spooner jumped from the skiff as it touched the raft, believing an explosion inevitable, and sank beneath the waters. As "Indian John" caught the boat he yelled : "Me got it him ! Me run it him and tie !" He guided the craft through the pitchy darkness and anchored it safely. Had it drifted down the river a sad accident might have been the sequel. Happily Americanite, quite as powerful and much safer, is displacing Nitro-Glycerine.

Andrew Dalrymple, who perished at Tidioute, was at his brother's well ten minutes before the fatal explosion and said to the pumper : "I have five-hundred dollars in my trousers and next week I'm going west to settle on a farm." Man and wife and money were blotted out ruthlessly and the trip west was a trip into eternity instead.

Frequently loads of explosives are hauled through the streets of towns in the oil-region, despite stringent ordinances and lynx-eyed policemen. Once a well-known handler of glycerine was arrested and taken before the mayor of Oil City. He denied violating the law by carrying the stuff in his buggy. An officer bore a can at arm's length and laid it tenderly on the floor. "Now, you won't deny it?" interrogated the mayor. "No," replied the prisoner, "there seems to be a lot of it." Then he hit the can a vicious kick, sending it against the wall with a thud. The spectators fled and the mayor tried to climb through the back-window. The can didn't explode, the agent put it to his lips, took a hearty quaff and remarked : "Mr. Mayor, try a nip ; you'll find this whisky goes right to the ticklish spot !"

Womanly intuition is a hummer that discounts science, philosophy and rep-tape. Mrs. Katherine E. Reed died at Sistersville in June of 1896. Her foresight secured fortunes for herself and many others in Tyler county. Left a widow five years ago, with eight children and a farm that would starve goats to death, she leased the land for oil-purposes. The test-well proving dry, Mrs. Reed implored the men to try again at a spot she had proposed for the first venture. The drillers were hard up, but consented to make a second trial when the good woman agreed to board them for nothing in case no oil was found. The well was the biggest gusher in the bundle. To-day it is producing largely and is known all over West Virginia as "The Big Kate." Mrs. Reed cleared two-hundred-thousand dollars from the sterile tract, which would sell for as much more yet, and her children and neighbors are independent for life.

A young lady at Sawyer City accepted a challenge to climb a derrick on the Hallenback farm, stand on top and wave her handkerchief. She was to receive a silk-dress and a ten-dollar greenback. The feat was performed in good shape. It is probably the only instance on record where a woman had the courage to climb an eighty-foot derrick, stand on top and wave her handkerchief to those below. It was done and the enterprising girl gathered in the wager.

Mrs. Sands, formerly a resident of Oil City, built the Sands Block and owned wells on Sage Run. McGrew Brothers, of Pittsburg, struck a spouter in 1869 that boomed Sage Run a few months. A lady at Pleasantville, who had coined money by shrewd speculations in oil-territory, purchased two-hundred acres near the McGrew strike, while the well was drilling and nobody thought it worth noticing. The lady was Mrs. Sands, who enacted the role of "a poor lone widow," anxious to secure a patch of ground to raise cabbage and garden-truck, to get the property. She worked so skillfully upon the sensibilities of the Philadelphians owning the land that they sold it for a trifle "to help a needy woman!" Her first well, finished the night before the "thirty-day shut down," flowed five-hundred barrels each twenty-four hours. The "poor lone widow" valued the tract at a half-million dollars and at one time was rated at six-hundred-thousand, all "earned by her own self." Yet weak-minded men and strong-minded women talk of the suppressed sex!

A Franklin lady asked her husband one morning to buy five-thousand barrels of oil on her account, saying she had an impression the price would advance very soon. To please her he promised to comply. At dinner she inquired about it and was told the order had been filled by an Oil-City broker. In the afternoon the price advanced rapidly. Next morning the lady asked hubby to have the lot sold and bring her the profits. The miserable husband was in for it. He dared not confess his deception and the only alternative was to pay the difference and keep mum. His sickly smile, as he drew fifteen-hundred dollars out of the bank to hand his spouse, would have cracked a mirror an inch thick. Solomon got a good deal of experience from his wives and that Franklin husband began to think "a woman might know something about business after all."

Mrs. David Hanna, of Oil City, is not one of the women whose idea of a good time is to go to a funeral and cry. She tried a bit of speculation in certificates and the market went against her. She tried again and again, but the losses exceeded the profits by a large majority. The phenomenal spurt in April of 1895 was her opportunity. She held down a seat in the Oil-Exchange gallery three days, sold at almost the top notch and cleared twelve-thousand dollars. People applauded and declared the plucky little woman "had a great head."

A little girl at Titusville, when she had prayed to have herself and all of her relations cared for during the night, added : "And, dear God, do try and take good care of yourself, for if anything should happen to you we would all go to pieces."

A Franklin mother was putting her three children to bed. They knelt down to say their prayers. The elder of the two girls struck a snag in "Now I lay me down to sleep." Three lines went through all right, but she stuck on the fourth and kept repeating, "If I should die before I wake, If I should die, If I"——. At last she turned to her brother, the eldest of the trio, and inquired : "What comes after "If I should die?" Quick as a flash and direct as a rifle-ball came the unexpected answer : "Why, a funeral, you darned fool !" This broke up the prayer-meeting in short metre.

Hon. Thomas W. Phillips, the wealthy oil-producer, who declines to serve a third term in Congress, labored zealously to secure legislation that would settle differences between employers and employés by arbitration. He offered to pay a quarter-million dollars to meet the expense of a thorough Congressional inquiry into the condition of labor, with a view to the presentation of an authoritative report and the adoption of measures calculated to prevent strikes and promote friendly relations. When the suspension of drilling in the oil-region deprived thousands of work for some months, Mr. Phillips was especially active in effecting arrangements by which they received the profits upon two-million barrels of crude set apart for their benefit. The Standard Oil-Company, always considerate to labor, heartily furthered the plan, which the rise in oil rendered a signal success. This was the first time in the history of any business that liberal provision was made for workmen thrown out of employment by the stoppage of operations. What a contrast to the grinding and squeezing and shooting of miners and coke-workers by "coal-barons" and "iron-kings !" When you come to size them up the oil-men don't have to shrink into a hole to avoid close scrutiny. They pay their bills, are just to honest toil, generous to the poor and manly from top to toe. They may not relish rheumatism, but this doesn't compel them to hate the poor fellow it afflicts. As Uncle Toby observed : "God bless us every one !"

Thirty miles from Cripple Creek, Colorado, the new town of Guffey is the focus of a silver-development that completely overshadows its golden neighbor. The place is fitly named for Hon. James M. Guffey, the successful Pennsylvania oil-producer and political leader. Purchasing an option on an undeveloped mine near the famous Trade Dollar, he put in machinery and spent a large sum to test the claim. The result is one of the richest silver-mines on earth—millions have been offered for it—floods of congratulations for the clever owners and a deluge of good wishes for the dazzling town that is to-day the most interesting spot in the Centennial State.

You may meet them at Oshkosh or Kalamazoo, in New York or Washington, around Chicago or San Francisco, about New Orleans or Mexico, but not a few men conspicuously successful in finance, manufactures, literature or politics have been mixed up with oil some time in their career. Commodore Vanderbilt, Jay Gould, James Fisk, Thomas A. Scott, John A. Garrett and A. J. Cassatt profited largely from their oil-interests. Mr. Cassatt, superintending the Warren & Franklin Railroad, acquired the knowledge of oil-affairs he turned to account in shaping the transportation-policy of the Pennsylvania Railroad. Besides the colossal gains of the Standard Oil-Company, petroleum won for such men as Captain J. J. Vandergrift, J. T. Jones, J. M. Guffey, John

McKeown, John Galey, J. J. Carter, Charles Miller, Frederic Prentice, S. P. McCalmont, William Hasson, George V. Forman, Thomas W. Phillips, John Satterfield, H. L. Taylor, John Pitcairn, Theodore Barnsdall, E. O. Emerson, Dr. Roberts, George K. Anderson, Jonathan Watson, Hunter & Cummings, Greenlee & Forst, the Grandins, the Mitchells, the Fishers, the McKinneys, the Plumers, the Lambertons and a host of others from one to ten-millions apiece. Certainly coal, cotton or iron, or all three combined, can show no such list. Oil augmented the fortunes of Stephen Weld, Oliver Ames and F. Gordon Dexter, the largest in New England. It put big money into the pockets of Andrew Carnegie, William H. Kemble and Dr. Hostetter. To it the great tube-works, employing thousands of men, and multitudes of manufacturing-plants owe their existence and prosperity. Some of the brightest newspaper-writers in New York, Philadelphia and Chicago learned force and directness amid the exciting scenes of Oildom. Several are authors of repute and contributors to magazines. Grover Cleveland, while mayor of Buffalo, imbibed business-wisdom and notions of sturdy independence from his acquaintance with Bradford oil-operators. Governor Curtin was a large stockholder in oil-companies on Cherry Run and Governor Beaver may claim kin with the fraternity as the owner of oil-wells in Forest county. No member of Congress for a generation made a better record than J. H. Osmer, Dr. Egbert, J. C. Sibley, C. W. Stone and Thomas W. Phillips. Galusha A. Grow was president of the Reno Oil-Company. Mr. Sibley was tendered the second place on the Democratic ticket at Chicago and could have been nominated for president, instead of William J. Bryan, but for the stupid hostility of a Pennsylvania boss. More capable, influential members than W. S. McMullan, Lewis Emery, J. W. Lee, W. R. Crawford, William H. Andrews, Captain Hasson, Willis J. Hulings, Henry F. James and John L. Mattox never sat in the State Senate or the Legislature. And so it goes in every part of the country, in every profession, in every branch of industry and in every business requiring vigor and enterprise.

B. D. J. McKeown is probably the only millionaire ball-player in the United States. He belongs to the Washington team, which is a member of the Pennsylvania State-League, and has played first base with the nine the entire season. He is a son of the late John McKeown and a keen man of affairs. A clean fielder, heavy batter and swift base-runner, he has a fine reputation as a ball-player. As a drawing attraction he is valuable, many who take little interest in the game going when his club plays, simply to see a young fellow with two or three-million dollars at his command in the diamond.

Jokers who cultivate a musty taste and dense ignorance poke fun at Philadelphia as a city of the snail-pace variety. They hint that it moves like an ice-wagon and that a man, tumbling from the roof of a thirteen-story building, descended too slowly upon the granite-pavement to be jarred by the fall. They forget that Philadelphia is the greatest manufacturing city in America, that it boasts the finest municipal-building in the world, that it has unrivaled parks, that it contains more owners of their homes than New York and Chicago combined, that it had cable-cars and asphalt-streets before such things were dreamed of in Gotham or Washington, that it possesses some of the greatest merchants on earth and sports trick-politicians who could give Tammany no end of fresh points on the game. One of the busiest spots in the Quaker City, north-west corner of Tenth and Market streets, is Martindale & Co.'s. Thomas Martindale, who leads the retail-grocery trade, brought with him to Philadelphia twenty years ago the vim and energy that gained him fame and fortune in the oil-region. He clerked for years in a Boston dry-goods store. Tapes and ribbons and nipping off samples for shoppers were not adapted to the ambitious young man, who had a soul for something larger and better. He quit Massachusetts for Pennsylvania, first looking about Pittsburg and landing at Oil City in 1869. Business was lively and he liked the style of the place. He took the first job that offered—grubbing out a road to his wells for John S. Rich. He used eyes and brain and soon knew how to "run engine" and manage a well. Three dollars a day as a pumper was the first promotion. He boarded at a house on Charley Run whose proprietor sold vegetables and green groceries. Martindale bought a half-interest and moved his trunk into the dusty shop. He had struck his gait and the result of his advent was soon apparent. His partner sold him the whole concern. He brightened the premises and painted the front red, white and blue. The "Checkered Store," a frame in the Third Ward, became noted for excellent wares and

THOMAS MARTINDALE.

moderate prices. A delivery-system was introduced which quickly grew from a few packages in a hired-dray to double-teams and handsome wagons. The "Blue Store," bigger and finer, was rented to secure necessary room. Every day the *Derrick* printed a new "ad." People read it eagerly and waited for its fresh announcement impatiently. A big brick-store at the river-bridge was the next step. Trade expanded and customers came from the whole region to "the Mammoth." A good offer for the establishment was accepted and the store is still conducted by Steffee & Co. Martindale removed to Philadelphia and infused new life into the grocery-trade. He opened the first California store. It was a revelation to the citizens to be able to get wines and fruits straight from the Pacific coast. They realized that a live merchant had come to town and patronized him liberally. The business spread out and partners were taken in. The firm bought and sold for cash only. Long credits and bad debts had no place in its system. The senior member took a leading part in the Trades League, in the Grocers' Association and in every movement to improve the business and benefit the community. He was active in advancing the interests of the trade and the public. He wrote and spoke against civic abuses and the exactions of would-be monopolists. He imbued those about him with something of his own earnest energy and was recognized everywhere as the brightest, hardest-working, most progressive and public-spirited grocer in William Penn's great city. If there be a place where the grocery-trade is on a higher, better plane than in Philadelphia it is not in Pennsylvania, not in the United States and not on this planet. And this satisfactory result is largely owing to the ability and enthusiasm of the wide-awake merchant who caught the inspiration of five-dollar oil. His talents and his services are appreciated by good citizens, thousands of whom intend to see that the next mayor of Philadelphia spells his name Thomas Martindale.

William H. Vanderbilt and a party of friends visited Buffalo in 1880 to witness the trial-race of a famous horse. A special brought the distinguished visitors, who were "the observed of all observers" as they appeared in the

judge's stand. Mr. Vanderbilt was in excellent humor, shook hands pleasantly with everybody introduced and enjoyed the sport amazingly. To one he
remarked just before the grand event of the day :

"So you are from the oil-region?"

"Yes."

"Let me tell you a little story of my experience."

"I should like to hear it."

"I never visited your part of the country but once. I went to Titusville to
confer with some gentlemen and stayed there all night. The country and the
people interested me greatly, but how much do you suppose that visit cost me?"

"I couldn't begin to guess."

"Just two-and-a-half-million dollars !"

"How was that?"

"I agreed to aid in building a railroad projected from Titusville to Dunkirk and lost this amount before I got out of the enterprise."

At this instant the horse approached the starting-point and the conversation
was interrupted, never to be renewed. The noble animal beat his previous record, amid the breathless attention of the occupants of the stand. Each quarter-
second was noted carefully by the holders of a half-dozen stop-watches. When
it became evident that the horse would do what his owner predicted, Mr. Vanderbilt turned to William Rockefeller and said : "Take him. He's cheap at
forty-thousand !" Ten minutes later the New-Yorkers left for Cleveland.

Nine years ago Adolph Schreiner died in a Vienna hospital, destitute and
alone. Yet he was the only son of a man known in Galicia as "the Petroleum
King" and founder of the great industry of oil-refining. The father shared the
lot of many inventors and benefactors, increasing the world's wealth untold
millions and poverty-stricken himself in his last days. Schreiner owned a piece
of ground near Baryslaw from which he took a black, tarry muck the peasants
used to heal wounds and grease cart-axles. He kneaded a ball from the slime,
stuck a wick into it and a red flame burned until the substance exhausted.
This was *the first petroleum-lamp!* Later Schreiner heard of distillation, filled
a kettle with the black earth and placed it on the fire. The ooze boiled over
and exploded, shivering the kettle and covering the zealous experimenter with
deep scars. He improved his apparatus, produced the petroleum of commerce
and sold bottles of the fluid to druggists in 1853. He drilled the first Galician
oil-well in 1856 and built a real refinery, which fire destroyed in 1866. He rebuilt the works on a larger scale and fire blotted them out, ruining the owner.
Gray hairs and feebleness had come, he ceased the struggle, drank to excess
and died in misery. His son, from whom much was expected, failed as a merchant and peddled matches in Vienna from house to house, just as the aged
brother of Signor Blitz, the world-famed conjuror, is doing in Harrisburg to-
day. Dying at last in a public hospital, kindred nor friends followed the poor
outcast to a pauper's grave. "Vanity of vanities, all is vanity and vexation of
spirit."

> Life's page holds each man's autograph—
> Each has his time to cry or laugh,
> Each reaps his share of grain or chaff,
> But all at last the dregs must quaff—
> The tombstone holds their epitaph.

The irrepressible "Sam" Blakely originated the term "shuffle," which he
often practiced in his dealings in the oil-exchanges, and the phrase, "Boys,
don't take off your shirts !" This expression spread far and wide and was

actually repeated by Osman Pasha—if the cablegrams told the truth—at the battle of Plevna, when his troops wavered an instant in the face of a dreadful rain of bullets. "Sam" also inaugurated the custom of drinking Rhine-wine. Once he constituted himself a committee of one to celebrate the Fourth of July at Parker. He printed a great lot of posters, which announced a celebration on a gorgeous scale—horse-races, climbing the greased pole, boat-races, orations, fireworks and other attractions. These were posted about the city and on barns and fences within a radius of ten miles. A friend asked him how his celebration was likely to come off. "Oh," he said, "we're going to get all the hayseeds in here and then we'll give them the great kibosh." On the glorious day "Sam" mounted a box in front of the Columbia hose-house and delivered an oration before four-thousand people, who pronounced it the funniest thing they ever heard and accepted the situation good-naturedly. Some impromptu games were got up and the day passed off pleasantly.

Ruel A. Watson, an active broker, as he lay gasping for breath, raised his head, asked an attendant "What's the market?" sank back on his pillow and expired. "The ruling passion is strong in death."

When men went crazy at Pithole and outsiders thought the oil-country was "flowing with milk and honey" and greenbacks, a party of wags thought to put up a little joke at the expense of a new-comer from Boston. They arranged with the landlord for some coupon-bonds to use in the dining-room of the hotel and to seat the youth at their table. The New-Englander was seated in due course. The guests talked of oil-lands, fabulous strikes and big fortunes as ordinary affairs. Each chucked under his chin a five-twenty government-bond as a napkin. One lay in front of the Bostonian's plate, folded and creased like a genuine linen-wiper. Calmly taking the "paper" from its receptacle, the chap from The Hub wiped his brow and adjusted the valuable napkin over his shirt-bosom. A moment later he beckoned to a servant and said: "See here, waiter, this napkin is too small; bring me a dish of soup and a 'ten-forty.'" The jokers could not stand this. A laugh went around the festive board that could have been heard at the Twin Wells and the matter was explained to the bean-eater. He was put on the trail of "a soft snap" and went home in a month with ten-thousand dollars. "Bring me a ten-forty" circulated for a twelve-month in cigar-shops and bar-rooms.

A jovial Parker merchant purchased a new hat, which he invited a half-dozen jolly brokers to visit a saloon to help him fit properly. The hat was handed around for all to admire. One of the meanest jokers that never held a public-office, while the boys were looking at the ceiling through tilted beer-glasses, slipped a thin slice of Limburger cheese under the sweat-band of the tile. The merchant returned to his store, laid his hat on the desk in the office and began to answer letters. He thought he detected a smell. His partner asked if he felt quite well and a clerk hinted somebody's feet needed washing. The hat-owner said he would go home and rest. On the street people held their noses as he drew near and a friend remarked that the air was full of miasma. At the door his wife inquired what was wrong. He told her he feared mortification had set in and she agreed with him. She remarked that if any disease that smelled like that had got hold of him he would be a burden to himself if he lived very long. She got his clothes off, soaked his feet in mustard-water and he slept. The children would come in and get a smell of the hat, look at each other with reproachful glances and go out and play. The man dreamed a small-pox flag was hung in front of his house and that he was riding in a butcher-wagon to the pest-house. The wife sent for a doctor. The doctor picked up the patient's hat, tried it on and got a whiff. He said the hat was picked before it was ripe. Then the doctor and wife held a post-mortem examination of the hat and found the Limburger. Few and short were the prayers they said. They awoke the patient and the doctor asked him if his worldly affairs were in a satisfactory condition. He gasped and said they were. The doctor asked him if he had made his will. He said that he had not, but that he wanted a lawyer at once. The doctor asked him if he felt as though he was ready to shuffle off. The man said he had always tried to lead a different life and to do as he would be done by, but that he might have made a mis-deal in some way and he would like to have a minister sent for to take an account of stock. Then the doctor brought to the bedside the hat, opened up the sweat-leather and showed the dying man what it was that smelled so. The patient pinched himself to see if he was alive, jumped out of bed, called for his revolver and the doctor couldn't keep up with him on his way to the saloon to bribe the bar-tender to tell what son of a pelican had planted the odoriferous cheese in his hat-lining. The story went the giddy rounds and was trimmed to suit various localities, but Col. Robert B. Allen, "Jimmy" Lowe and "Sam" Blakely were the only men who saw the first perpetrator of the Limburger act perform the deed.

In 1863-4 J. B. Allen, of Michigan, a first-class scholar and chemist, had charge of the prescription-department in Dr. R. Colbert and Dr. Egbert's drug-store. He could read Greek as readily as English, declaim in Latin by the hour, quote from any of the classics and speak three or four modern languages. To raise money to pay off a mortgage on his father's farm he walked across the Allegheny on a wire thirty feet above the water. He carried a large flag, attached to a frame mounted on a pulley-wheel, which he shoved with one hand, holding a balance-pole in the other. It was a feat Blondin could not excel. Allen was decidedly eccentric and the hero of unnumbered stories. Once a mud-bespattered horseman rushed into the store with a prescription that called for a deadly poison. The horseman was informed it was not safe to fill it, but he insisted upon having it, saying it bore a prominent doctor's signature and there could be no mistake. Allen filled it and wrote on the label: "Caution—If any damphool takes this prescription it will kill him as dead as the devil!"

Lillian Edgarton, the plump and talented platform-speaker, was billed to appear at Franklin. She traveled from Pittsburg by rail. A Parker broker was a passenger on the train and wired to the oil-exchange that Josie Mansfield was on board. The news flew and five-hundred men stood on the platform when the train arrived. The broker jumped off and said the lady had a seat near the center of the coach he had just left. The boys climbed on the car-platform, opened the door and marched in single file along the aisle to get a look at "Josie." The conductor tore his hair in anguish that the train would not carry such a crowd as struggled to get on, but he was dumbfounded when the long procession began to get off. The sell was not discovered until next morning, by which time the author of the joke had started on his summer-vacation and could not be reached by the vigilance-committee.

Hon. Reuben Carroll, a pioneer operator, was born in Mercer county in 1823, went to Ohio to complete his education, settled in the Buckeye State and was a member of the Legislature when developments began on Oil Creek. Solicited by friends to join them in an investment that proved fortunate, he removed to Titusville and cast his lot with the producers. He operated extensively in the northern fields, residing at Richburg during the Allegany excitement. He took an active interest in public affairs and contributed stirring articles on politics, finance and good government to leading journals. He opposed Wall-street domination and vigorously upheld the rights of the masses. Upon the decline of Richburg he located at Lily Dale, New York, where his active mind finds congenial exercise maintaining the cause of the people against encroachments of the money-power. Mr. Carroll, as a representative producer, was asked to become a member of the South Improvement Company in 1872. The offer aroused his inflexible sense of justice and was indignantly spurned. With schemes of spoliation a man of his character and temperament could have no sympathy. He knew the sturdy quality and large-heartedness of the Oil-Creek operators and did not propose to assist in their destruction. He resolutely resisted the torpedo monopoly and the bogus claims of the fraudulent

REUBEN CARROLL..

inventors who sought to levy tribute from the oil-interests. At seventy-three Mr. Carroll is vigorous and well-preserved, ready to combat error and champion truth with tongue and pen. An intelligent student of the past and of current events, a close observer of the signs of the times and a keen reasoner, Reuben Carroll is a fine example of the men who are mainly responsible for the birth and growth of the petroleum-development.

There is much uncertainty as to the youngest soldier in the civil-war, the oldest Mason, the man who first nominated McKinley for President and who struck Billie Patterson, but none as to the youngest dealer in oil-well supplies in the oil region. This distinction belongs to Ralph W. Carroll, a native of Youngstown, Ohio, and son of Hon. Reuben Carroll. Born in 1860, at eighteen he was at the head of a large business at Rock City, in the Four-Mile District, five miles south-west of Olean. His three brothers were associated with him under the firm name of Carroll Brothers. The firm was the first to open a supply-store at Richburg, with a branch at Allentown, four miles east, and an establishment later at Cherry Grove. In 1883 Ralph W. succeeded the firm, his brothers retiring, and located at Bradford. He carried on a large trade as oil-region agent of the American Tube and Iron Company, the Ball Engine Company and other manufacturers of all goods for oil-wells and gas-wells. At Bradford he issued the

RALPH W. CARROLL.

first net price-list of oil-well supplies ever published. The innovation pleased the producers, whose restless energy did not relish sitting down to figure out discounts. Mr. Carroll was the first oil-country representative to open a supply-store at Lima. In 1886, the scene of activity having shifted to Butler, Allegheny and Washington counties and West Virginia, he opened offices and ware-houses at Pittsburg. The pipe-business had attained such proportions as to de-mand his entire efforts until 1894, when he removed to New York to engage in placing special investments. The young merchant was secretary of the Pro-ducers' Protective Association, organized at Richburg in 1881, and a member of the executive committee that conducted the fight against the Roberts Tor-pedo Company. Hon. David Kirk, Asher W. Milner, J. E. Dusenbury and "Farmer" Dean were his four associates on this important committee, which gained the victories that resulted in a final compromise and great reduction in the price of torpedoes. Roscoe Conkling, for the Roberts side, and General Butler, for the Producers' Association, measured swords in this legal warfare. Retaining the excellent traits that made him popular and influential at Rich-burg, Bradford and Pittsburg, Mr. Carroll has a warm welcome for his oil-region friends, a class of men the like of whom for geniality, sociability, liberality and enterprise the world can never duplicate.

Michael Geary, whose death last year was a severe blow to Oil City, for-cibly illustrated what energy and industry may accomplish. He was a first-class boiler-maker and machinist, self-reliant, stout-hearted and strong men-tally and physically. In 1876 he started the Oil-City Boiler-Works in a small building, Daniel O'Day and B. W. Vandergrift furnishing the money and tak-ing an interest in the business. O'Day and Geary became sole owners in 1882. The plant was enlarged, the tube-mills were added, acres of buildings dotted the flats and a thousand men were employed. Engines, tanks, stills, tubing, casing and boilers of every description were manufactured. The machinery comprised the latest and fullest equipment. The business grew amazingly. Joseph Seep was admitted to partnership and branch-offices were established in New York, Chicago, Pittsburg and at various points in the oil-producing states. The firm led the world as tank-builders, actually constructing one-third the total iron-tankage in the United States. Mr. Geary bought and re-modeled the Arlington Hotel, fostered local enterprises and was a most pro-gressive citizen. He died in the vigor of manhood. The splendid industries he reared and the high place he held in public esteem are his enduring monument.

"The Fredonia Gas-Light and Water-Works Company," which obtained a special charter in 1856, was undoubtedly the first natural-gas company in the world. Its object was, "by boring down through the slate-rock and sinking wells to a sufficient depth to penetrate the manufactories of nature, and thus collect from her laboratories the natural-gas and purify it, to furnish the citizens with good cheap light." The tiny stream of gas first utilized at the mill yielded its mite forty years. When Lafayette remained a night at Fredonia in 1824, on his triumphal visit to the United States, "the village-inn was lighted with gas that came from the ground." The illustrious Frenchman saw nothing in his travels that interested and delighted him more than this novel illumination.

Col. J. A. Barrett, for many years a citizen of Illinois and law-partner of Abraham Lincoln, in 1886 removed to a tract of five-thousand acres on Tug Fork, near the quiet hamlet of Warfield. Gas issued from the soil and tradi-tion says George Washington fired the subtle vapor at Burning Spring while surveying in West Virginia before the Revolution. Captain A. Allen, who pioneered the oil-business on Little Kanawha, leased the tract from Col. Bar-rett and struck a vast reservoir of gas at two-thousand feet.

A Bradford youth, whose doting mother presented him with eighteen-thousand dollars on his twenty-first birthday, to begin business for himself, went to New York and returned in six weeks without a cent. He didn't gamble in stocks, but he saw New York by gaslight and made everything hot for the Gotham bloods.

The Anchor Oil-Company's No. 1, the first well finished near "646," in Warren county, flowed two-thousand barrels a day on the ground until tanks could be provided. It burned when flowing a thousand barrels and for ten days could not be extinguished. One man wanted to steam it to death, another to drown it, another to squeeze its life out, another to smother it with straw, another to dig a hole and cut off the flow, another to roll a big log over it, another to blow out its brains with dynamite, another to blind it with carbolic acid, another to throw up earth-works and so on until the pestered owners wished five-hundred cranks were in the asylum at North Warren. Pipes were finally attached in such a way as to draw off the oil and the flame died out.

Col. Drake used the first driving-pipe to reach the bed-rock.

The first tubing used in oil-wells was manufactured at Pittsburg and was the same as used in salt-wells at Tarentum. Some of it was made with brass screw-joints, eight threads to the inch, and soldered on the pipe.

William A. Smith, who drilled the Drake well, made the first rimmer. While enlarging a well with a bit the point broke off, after which greater progress was made. This accident suggested the rimmer.

Drilling with a cable attached to the tools was invented by the Chinese many years since, but was introduced into the oil-country by the Tarentum drillers. Pole-tools for drilling were little used in the oil-regions, except to fish out lost tools.

Early well-owners found the tools and fuel, paid all expenses but labor and paid three-dollars-and-fifty-cents per foot to the contractor, yet so many contractors failed that a lien-law was passed. George Koch, in November of 1873, took out a patent on fluted drills, which did away with the rimmer, reduced the time of drilling a well from sixty days to twenty and reduced the price from three dollars per foot to fifty cents

Sam Taft was the first to use a line to control the engine from the derrick, at a well near McClintockville, in 1867. Henry Webber was the first to regulate the motion of the engine from the derrick. He drilled a well near Smoky City, on the Porter farm, in 1863, with a rod from the derrick to the throttle-valve. He also dressed the tools, with the forge in the derrick, perhaps the first time this was done. He drilled this well six-hundred feet with no help. Near this well was the first plank-derrick in the oil-country.

The first derricks were of poles, twelve feet base and twenty-eight to thirty feet high. The ladder was made by putting pins through a corner of a leg of the derrick. The Samson-post was mortised in the ground. The band-wheel was hung in a frame like a grindstone. A single bull-wheel, made out of about a thousand feet of lumber, placed on the side of the derrick next to the band-wheel, with a rope or old rubber-belt for a brake, was used. When the tools were let down the former would burn and smoke, the latter would smell like ancient codfish.

"Ivry gintleman will soon go horseback on his own taykittle" was the inspired exclamation of an Irish baronet upon beholding the initial trip of the first locomotive. Vast improvements in the application of power have been effected since Stephenson's grand triumph, nowhere more satisfactorily than in the oil-regions. Producers who remember the primitive methods in vogue along Oil Creek can best appreciate the wonderful progress made during three decades. The tedious process of drilling wet-holes with light tools has gone where the woodbine twineth. Casing has retired the seed-bag permanently and from the polish-rod to the working-barrel not the smallest detail remains unimproved. Having a portable engine and boiler at each well has given place to the cheaper plan of coupling a host of wells together, two men thus doing the work that once required twenty or thirty. Pipe-lines have superseded greasy barrels and swearing teamsters and even tank-cars are following the flat-boats of pioneer times to oblivion. In short, labor-saving systems have revolutionized the business so completely that the fathers of the early styles would utterly fail to recognize their offspring in the petroleum-development as conducted now-a-days.

Sad accidents happened before drillers learned how to manage a flowing oil-well with casing in it. At Frank Fertig's well, Antwerp, a man was burned to death. The burning of the Shoup & Vensel well at Turkey City cost three lives and led to an indignation-meeting at St. Petersburg to protest against casing. Danger from its use was soon removed by Victor Gretter's invention of the oil-saver. Gretter, a small, dark-haired, dark-eyed man, lived at St. Petersburg. He was an inventive genius and a joker of the first water. His oil-saver doubtless saved many lives, by preventing gas and oil from escaping when a vein was tapped and coming in contact with the tool dresser's fire in the derrick.

The first three-and-a-quarter casing was used on the Tarr farm, in March of 1865. Every well on the farm was flooded with water, not a barrel of oil was produced and small casing was introduced. The production rose to a thousand barrels per day, which insured the success of the method. Cleaning out could be done very rapidly, which gave rise to the idea of drilling through casing. This was first done on Benninghoff Run, in the summer of 1868. One of the greatest inventions connected with the oil business, casing was not patented.

Before casing was introduced it was often difficult to tell if oil was found. Oilmen would examine the sand, look for "soot" on the sand-pumpings and place a lighted match to the sand-pump immediately after it was drawn from the well, as a test for gas. If the driller was sure the drill dropped two or three feet, with "soot" on the sand-pumpings, the show was considered worth testing. A seed-bag was put on the tubing and the well was allowed to stand a day or two to let the seed swell. To exhaust the water sometimes required weeks, but when all hope of a producer was lost and the last shovel of coal was in the boiler the oil might come. There seemed to be a virtue in that last shovel of coal. The shoemaker who could make a good seed-bag was a big man. The man who tied on the seed-bag for a well that proved a good producer was in demand. If, after oil showed itself, flax-seed was seen coming from the pipe the well-owner's heart could be found in his boots. The bag was burst, the water let in and the operator's hopes let out.

Mud-veins in the third sand on Oil Creek and at Pithole would often stick the tools effectually. On Bull Run three wells in one derrick were abandoned with tools stuck in the third sand. The theory was that the mud-vein was a stratum of slate in the sand, which became softened and ran into the well when water came in contact with it. Casing has robbed it of its terrors.

All kinds of engines, from one to fifty horse-power, were used on Oil Creek in the sixties. The old "Fabers," with direct attachment, will recall many a broad grin. The boys called them "Long Johns." The Wallace-engine had hemp-packing on the piston, and the inside of the cylinder, rough as a rasp, soon used it up and leaked steam like a sieve. The Washington-engine was the first to come into general use. C. M. Farrar, of Farrar & Trefts, whose boilers and engines have stood every test demanded by improvements in drilling, made the drawing for the first locomotive-pattern boiler on a drilling well—a wonderful stride in advance of the old-time boiler. Trefts made the castings for the engine that pumped the Drake well and was the first man, in company with J. Willard, to use ropes on Oil Creek in drilling. This was on the Foster farm, near the world-famed Empire well, in 1860. Willard made the second set of jars on the creek. Senator W. S. McMullan was a stalwart blacksmith, who made drilling-tools noted for their enduring quality.

The Beardsleys, Fishers, Dollophs and Fosters were the first inhabitants in the wilds of Northern McKean. Henry Bradford Dolloph, whose house above Sawyer City was shattered by a glycerine explosion, was the first white child who saw daylight and made infantile music in the Tuna Valley. One of the first two houses where Bradford stands was occupied by the Hart family, parents and twelve children. When the De Golias settled up the East Branch a road had to be cut through the forest from Alton. Hon. Lewis Emery's No. 1, on the Tibbets farm, the first good well up the Branch, produced oil that paid two or three times the cost of the entire property.

By the side of the romance, the pathos, the tragedy and the startling incidents of the oil-regions thirty years ago the gold-excitements of California and Australia and the diamond-fever of South Africa are tame and vapid. Prior to the oil-development settlers in the back townships lived very sparingly. Children grew up simple-minded and untutored. The sale of a pig or a calf or a turkey was an event looked forward to for months. Petroleum made not a few of these rustics wealthy. Families that had never seen ten dollars suddenly owned hundreds-of-thousands. Lawless, reckless, wicked communities sprang up. The close of the war flooded the region with paper-currency and bold adventurers. Leadville or Cheyenne at its zenith was a camp-meeting compared with Pithole, Petroleum Centre or Babylon. Men and women of every degree of decency and degradation huddled as closely as the pig-tailed Celestials in Chinatown. Millions of dollars were lost in bogus stock-companies. American history records no other such era of riotous extravagance. The millionaire and the beggar of to-day might change places to-morrow. Blind chance and consummate rascality were equally potent. Of these centers of sin and speculation, strange transformations and wild excesses, scarcely a trace remains. Where hosts of fortune-seekers and devotees of pleasure strove and struggled nothing is to be seen save the bare landscape, a growth of underbrush or a grassy field. Sodom was not blotted out more completely than Pithole, the type of many oil-towns that have been utterly exterminated.

The first funeral at Fagundas was a novelty. A soap-peddler, stopping at the Rooling House one night, died of delirium-tremens. He was put into a rough coffin and a small party set off to inter the corpse. Somebody thought it mean to bury a fellow-creature without some signs of respect. The party returned to the hotel with the body, a large crowd assembled in the evening, flowers decorated the casket, services were conducted and at dead of night two-hundred oil men followed the friendless stranger to his grave.

James Bennett, aged nineteen, worked at a well near Petroleum Centre. Tubing was to be drawn one morning and Bennett wished to engage a substitute, because impressed with a premonition of some misfortune, should he leave home. His mother, long an invalid, tried to quiet his fears and he went to work reluctantly. In a few moments the block and tackle fell and killed Bennett, who was on the derrick-floor. The shock of her son's death proved fatal to the sorrowing mother.

Col. W. H. Kinter, of Oil City, a man of kindliest impulses, genial and whole-souled, greeting a neighbor one Sunday evening, remarked : "Goodnight, old boy—no, make it good-bye ; we may never meet again !" He retired in excellent health and spirits. Next morning, feeling drowsy, he asked his wife—a daughter of Hamilton McClintock—to bring him a cup of tea. She returned in a short time to find her husband asleep in death. John Vanausdall, partner of William Phillips in the biggest well on Oil Creek, left his home at Oil City in the morning, took ill at Petrolia and telegraphed for his wife. She hurried on the first train and reached his bedside just as he drew his last breath.

The funeral of Mr. and Mrs. William B. Magee, on May ninth, 1891, was one of the largest ever witnessed in Oil City. Mrs. Magee, aged eighty-three, died on May sixth, from an attack of grip. Mr. Magee, four years older, died on May seventh, eighteen hours after his wife. This singular coincidence, the happy ending of a long pilgrimage, was not anticipated. The venerable pair had "clamb the hill thegither" sixty-one years, thirty of them in Oil City. Mr. Magee was ill only two days. Both retained consciousness to the last, passed away peacefully and were laid side by side in Grove-Hill cemetery.

Judge Keating, who built the Keating furnace in Clarion county and acquired a fortune in iron and oil, died in December of 1880. He was buried at Emlenton, hundreds of old friends attending the funeral. As the service at the church was beginning John Middleton, aged eighty-one, gasped for breath and died in the pew. While descending the hill from the cemetery, after the burial, Wm. McCullough, of Pittsburg, aged seventy-two years, uncle of Mr. Keating, fell and fractured his left arm.

Joseph Wood, proprietor of the St. James Hotel at Paterson, N. J., died on May thirteenth, 1896. He was a wit and story-teller of the best kind, a gallant fighter for the Union and for a year lived at Pithole. A fortune made by operating and speculation he lost by fire in a year. He conducted hotels at Hot Springs, Washington, Chicago and Milwaukee and was one of the famous Bonifaces of the United States. On his business-cards he printed these " religious beliefs :"

" Do not keep the alabaster-boxes of your love and tenderness sealed up until your friends are dead. Fill their lives with sweetness. Speak approving, cheering words while their ears can hear them and while their hearts can be thrilled and made happier by them. The kind things you mean to say when they are gone say before they go. The flowers you mean to send to their coffins send to brighten and sweeten their homes before they leave them. If my friends have alabaster-boxes laid away, full of fragrant perfumes of sympathy and affection, which they intend to break over my dead body, I would rather they would bring them out in my weary and troubled hours and open them, that I may be refreshed and cheered by them while I need them. I would rather have a plain coffin without a flower, a funeral without a eulogy than a life without the sweetness of love and sympathy. Let us learn to anoint our friends beforehand for their burial. Post-mortem kindness does not cheer the burdened spirit. Flowers on the coffin cast no fragrance backward over the weary way."

Let down the bars and enter the field that was once the seething, boiling caldron called Pithole. A poplar-tree thirty feet high grows in the cellar of the National Hotel. Stones and underbrush cover the site of the Metropolitan Theater and Murphy's Varieties. This bit of sunken ground, clogged with weeds and brambles, marks the Chase House. Here was Main street, where millions of dollars changed hands daily. For years the Presbyterian church stood forsaken, the bell in the tower silent, the pews untouched and the pulpit-

Bible lying on the preacher's desk. John McPherson's store and Dr. Christie's house were about the last buildings in the place. Not a human-being now lives on the spot. All the old-timers moved away. All? No, a score or two quietly sleep among the bushes and briars that run riot over the little graveyard in which they were laid when the dead city was in the throes of a tremendous excitement.

> The rate at which towns rose was surely most terrific,
> Nothing to rival it from Maine to the Pacific ;
> The rate at which they fell has never had an equal—
> Woods, city, ruin'd waste—the story and the sequel.

The first "hotel" at Pithole—a balloon-frame rushed up in a day—bore the pretentious title of Astor House. Before its erection pilgrims to the coming city took their chance of meals at the Holmden farm-house. As a guest wittily remarked: "It was table d'hote for men and also table d'oat for horses." The viands were all heaped upon large dishes and everybody helped himself. The Morey-Farm Hotel, just above Pithole, charged twenty-one dollars a week for board, had gas-light, steam-heat, telegraph-office, barber-shop, colored waiters and "spring-mattresses." Its cooking rivalled the best in the large cities. At

THE DINNER HOUR AT WIGGINS'S HOTEL.

Wiggins's Hotel, a three-story boarding-house in the Tidioute field, two-hundred men would often wait their turn to get dinner. This was a common experience in the frontier towns, to which throngs hastened before houses could be erected for their accommodation. E. H. Crittenden's hotel at Titusville was the finest Oildom boasted in the sixties. Book & Frisbee's was notable at the height of the Parker development. A dollar for a meal or a bed, four dollars a day or twenty-eight dollars a week, be the stay long or short, was the invariable rate. Peter Christie's Central Hotel, at Petrolia, was immensely popular and a regular gold-mine for the owner. Oil City's Petroleum House was a model hostelry, under "Charley" Staats and "Jim" White. The Jones House cleared Jones forty-thousand dollars in nine months. Its first guest was a Mr. Seymour, who spent one year collecting data for a statistical work on petroleum. His manuscripts perished in the flood of 1865. The last glimpse my eyes beheld of Jones was at Tarport, where he was driving a dray. Bradford's Riddell House and St. James Hotel both sized up to the most exacting requirements. Good hotels and good restaurants were seldom far behind the triumphant march of the pioneers whose successes established oil-towns.

W. J. Bostford, who died at Jamestown in November of 1895, operated at Pithole in i s palmy days. Business was done on a cash basis and oil-property was paid for in money up to hundreds-of-thousands of dollars. Bostford made a big sale and started from Pithole to deposit his money. A cross-country trip was necessary to reach Titusville. Shortly after leaving Pithole he was attacked by robbers, who took all the money and left him for dead upon the highway. He was picked up alive, with a broken head and many other injuries, which he survived thirty years.

"French Kate," the woman who aided Ben Hogan at Pithole and followed him to Babylon and Parker, was a Confederate spy and supposed to be very friendly with J. Wilkes Booth. Besides his oil-interests at Franklin, the slayer of Abraham Lincoln owned a share in the Homestead well at Pithole. A favorite legend tells how, by a singular coincidence, which produced a sensation, the well was burned on the evening of the President's assassination. It caught fire about the same instant the fatal bullet was fired in Ford's Theater and tanks of burning oil enveloped Pithole in a dense smoke when the news of the tragedy flashed over the trembling wires. The Homestead well was not down until Lincoln had been dead seven weeks, Pithole had no existence and there were no blazing tanks; otherwise the legend is correct. Two weeks before his appalling crime Booth was one of a number of passengers on the scow doing duty as a ferry-boat across the Allegheny, after the Franklin bridge had burned. The day was damp and the water very cold. Some inhuman whelp threw a fine setter into the river. The poor beast swam to the rear of the scow and Booth pulled him on board. He caressed the dog and bitterly denounced the fellow who could treat a dumb animal so cruelly. At another time he knocked down a cowardly ruffian for beating a horse that was unable to pull a heavy load out of a mud-hole. He has been known to shelter stray kittens, to buy them milk and induce his landlady to care for them until they could be provided with a home. Truly his was a contradictory nature. He sympathized with horses, dogs and cats, yet robbed the nation of its illustrious chief and plunged mankind into mourning. To newsboys Booth was always liberal, not infrequently handing a dollar for a paper and saying : "No change ; buy something useful with the money." The first time he went to the Methodist Sunday-school, with "Joe" Simonds, he asked and answered questions and put a ten-dollar bill in the collection-box.

Ben Hogan was one of the motley crew that swarmed to Pithole "broke." He taught sparring and gave exhibitions of strength at Diefenbach's variety-hall. He fought Jack Holliday for a purse of six-hundred dollars and defeated him in seven rounds. Four-hundred tough men and tougher women were present, many of them armed. Hogan was assured before the fight he would be killed if he whipped his opponent. He was shot at by Marsh Elliott during the mill, but escaped unhurt. Ben met Elliott soon thereafter and knocked him out in four brief rounds, breaking his nose and using him up generally. Next he opened a palatial sporting-house, the receipts of which often reached a thousand dollars a day. An adventure of importance was with "Stonehouse Jack." This desperado and his gang had a grudge against Hogan and concocted a scheme to kill him. Jack was to arrange a fight with Ben, during which Hogan was to be killed by the crowd. Ben saw his enemy coming out of a dance-house and blazed away at him, but without effect. The fusillade scared "Stonehouse" away from Pithole and on January twenty-second, 1866, a vigilance committee at Titusville drove the villain out of the oil-region, threatening to hang him or any of his gang who dared return. This committee was organized to clear out a nest of incendiaries and thugs. The vigilants erected a gallows near the smoking embers of E. B. Chase & Co.'s general store, fired the preceding night, and decreed the banishment of hordes of toughs. "Stonehouse Jack" and one-hundred other men, with a number of vile women came under this sentence. The whole party was formed in line in front of the gallows, the "Rogue's March" was played and the procession, followed by a great crowd of people, proceeded to the Oil-Creek Railroad station. The prisoners were ordered on board a special train, with a warning that if they ever again set foot upon the soil of Titusville they would be summarily executed. This salutary action ended organized crime in the oil-region.

Blacklegs, thieves and murderers ran little risk of punishment in the early days of oil-developments, unless they became unusually obstreperous and were brought to a period with a shot-gun. Scoundrels lay in wait for victims at every turn and stories of their misdeeds could be told by the hundred. The McFate farm was one of the first on Cherry Run to be sold at a fancy price. S. J. McFate, one of the brothers owning the property, two weeks after the sale in 1862, walked down to Oil City to draw several-thousand dollars from the bank. He displayed the money freely and left for home late at night. The road was dark and lonely and next morning, in a clump of bushes a mile above Oil City, his lifeless body was discovered. A ghastly wound in the head and the absence of the money explained the tragedy and the motive. No clue to the murderer was ever found, although squads of detectives "worked on the case" and queer fictions regarding the mysterious assassin were printed in many newspapers.

John Henderson, a tall, handsome man, came from the east during the oil-excitement in Warren county and located at Garfield. In a fight at a gambling-house one night George Harkness was thrown out of an upstairs-window and his neck broken. Foul play was suspected, although the evidence implicated no one, and the coroner's jury returned a verdict of accidental death. Harkness had left a young bride in Philadelphia and was out to seek his fortune. Henderson, feeling in a degree responsible for his death, began sending anonymous letters to the bereaved wife, each containing fifty to a hundred dollars. The letters were first mailed every month from Garfield, then from Bradford, then from Chicago and for three years from Montana. In 1893 she received from the writer of these letters a request for an interview. This was granted, the acquaintance ripened into love and the pair were married! Henderson is a wealthy stockman in Montana. In 1867 an English vessel went to pieces in a terrible storm on the coast of Maine. The captain and many passengers were drowned. Among the saved were two children, the captain's daughters. One was adopted by a merchant of Dover, N. H. He gave her a good education, she grew up a beautiful woman and it was she who married George Harkness and John Henderson.

In February of 1880 Conductor W. W. Gaither, of the Clarion Narrow-Gauge Railroad, ejected a peddler named John Clancy from his train, near King's Mills, for refusing to pay his fare. Clancy shot Gaither, who died in a few days from the wound. W. L. Fox, of Foxburg, president of the road, was a warm personal friend of the murdered conductor. He took charge of the pistol and became active in bringing Clancy to punishment. Clancy was placed on trial at Clarion. President Fox was to produce the pistol in court. Leaving home on the early train for Clarion, he had proceeded some distance from Foxburg when he discovered he had forgotten the pistol. He stopped the train and ran back to get the weapon. When he returned he was almost exhausted. W. J. McConnell, beside whom he was sitting, attempted to revive him, but he sank into unconsciousness and expired in the car at the exact spot where his friend Gaither was shot. Clancy was convicted of murder in the second degree and sentenced to eight years in the penitentiary. His wife and twelve-year-old son were left destitute. The boy went to work for a farmer near St. Petersburg. A week later he was crossing a field in which a vicious bull was feeding. The bull attacked him, ripped his side open and tossed him from the field into the road. The boy died in a short time. Besides these fatalities resulting from Clancy's crime, the business of Foxburg was seriously crippled. The village depended on the oil-business of the Fox estate. W. L. Fox, although only twenty-nine years old, was the manager of this estate, comprising three-thousand acres of oil-land, which, only partially developed, yielded twenty-five-thousand barrels a month. He owned the only extensive individual pipe-line in the oil-region and at the time of his death was erecting an immense refinery. He had a capital of three-millions and was completing plans for the construction of other lines of railways, with Foxburg as their center. The refinery was abandoned, the pipe-line was sold and no further development of the Fox property was made. The death of W. L. Fox took the distribution of a million dollars a year from the region around Foxburg. The family erected a splendid church to his memory, but it is seldom used. The bank closed its doors years ago, most of the business has sought other sections and the pretty village is merely a shadow of the past.

> " The massive gates of circumstance
> Are turned upon the smallest hinge,
> And thus some seeming pettiest chance
> Oft gives our life its after tinge."

The narrow-gauge railroad from Foxburg to Clarion was an engineering novelty. It zig-zagged to overcome the big hill at the start, twisted around ravines and crossed gorges on dizzy trestles. Near Clarion was the highest and longest bridge, a wooden structure on stilts, curved and single-tracked.

RAILROAD BRIDGE NEAR CLARION.

One dark night a drummer employed by a Pittsburg house was drawn over it safely in a buggy. The horse left the wagon-road, got on the railroad-track, walked across the bridge— the ties supporting the rails were a foot apart—and fetched up at his stable about midnight. The drummer, who had imbibed too freely and was fast asleep in the vehicle, knew nothing of the drive, which the marks of the wheels on the approaches and the ties revealed next morning. The horse kept closely to the center of the track, while the wheels on the right were outside the rails. Had the faithful animal veered a foot to the right, the buggy would have tumbled over the trestle and there would have been a vacant chair in commercial ranks and a new voice in the celestial choir. That the horse did not step between the ties and stick fast was a wonder. The trip was as perilous as the Mohammedan passage to Paradise over a slack-wire or Blondin's tight-rope trip across Niagara.

The first railroad to enter Oil City was the Atlantic & Great Western, now of the Erie system, in 1866. Its first train crossed the mouth of Oil Creek on a track laid upon the ice. "Billy" Stevens and John Babcock were early conductors. Stevens went to Maine and Babcock died several years ago at Meadville, soon after completing a term as mayor of the city. The Farmers' Railroad was finished in 1867, the Allegheny Valley in 1868 and the Lake-Shore in 1870. A short railroad up Sage Run conveyed coal from the Cranberry mines. On August fourth, 1882, the engineer—Frank Wright—lost control of a train on the down grade, one of the steepest in the state. He reversed the engine to the last notch and jumped, sustaining injuries that caused his death in four days. For two miles the track was torn up and coal-cars were smashed to splinters by running into a train of freight-cars at McAlevy's Mills. Six men were killed outright and five died from their injuries next day.

The popular auditor of the New York Central, W. F. McCullough, was an Oil-City boy. His brother, James McCullough, is traveling-auditor of the New York, New Haven & Hartford ; another brother, E. M. McCullough, is traveling bill-agent for the U. S. Steamship-Railway Company. They are sons of the late Dr. T. C. McCullough, who died at Oil City in 1896.

Hon. Thomas Struthers, of Warren, who died in 1892 at the age of eighty-nine, donated the town a public-library building that cost ninety-thousand dollars. He aided in constructing the Pennsylvania Railroad, built sections of the Philadelphia & Erie and Oil-Creek Railroads and the first railroad in California. He was the first manager of the Oil-Creek road. Frank Thomson, now first vice-president of the Pennsylvania Railroad, was also superintendent of the Oil-Creek. C. J. Hepburn, now residing at Sunbury and permanently disabled as the result of an accident, held the same position for years. He was a thorough railroader, esteemed alike by the employés and the public for his efficient performance of duty. The old-time Oil-Creek conductors were lock-switch, steel-track and rock-ballast clear through. Gleason, postmaster at Corry a term or two, runs the Mansion House at Titusville. "Bill" Miller is on the Pacific coast. Mack Dobbins died at St. Louis and "By" Taylor has made his last trip. Barber lives at Buffalo. "Mike" Silk, who yanked oil-trains from Cherry Run, is a wealthy citizen of Warren. Selden Stone and "Pap" Richards are still on deck, the last of a coterie of as white railroad-men as ever punched pasteboard.

"Never quarrel with a preacher or an editor," said Henry Clay, "for the one can slap you from the pulpit and the other hit you in his paper without your getting a chance to strike back." Col. William Phillips, president of the Allegheny-Valley Railroad, violated the Kentucky statesman's wise maxim by making war on the Oil-City *Derrick*. He was building the Low-Grade division, from Red Bank to Emporium, and the main-line suffered. The track was neglected, decayed ties and broken rails were common and accidents occurred too frequently for comfort. The winter and spring of 1873 were fruitful of disaster. At Rockland an oil-train ran over the steep bank into the river, upsetting the passenger-coach at the rear. The oil caught fire, several passengers were burned to death and others were terribly injured. The railroad officials, acting under orders from headquarters, refused to give information to the crowd of frantic people who besieged the office at Oil City to learn the fate of friends on the train. To the last moment they denied that anything serious had happened, although passengers able to walk to Rockland Station telegraphed brief particulars. At last a train bearing some of the injured reached Oil City. Next

morning the *Derrick* gave full details and criticised the management of the
road severely for the bad condition of the track and the stupid attempt to with-
hold information. The heading of the article—"Hell Afloat"—enraged Col.
Phillips. He and Superintendent J. J. Lawrence prepared a circular to the conductors, instructing them "to take up pass of C. E. Bishop or J. J. McLaurin
whenever presented, collect full fare, prohibit newsboys from selling the Oil-
City *Derrick* on the trains, not allow the paper to be carried except in the mails
or as express-matter, and to report to the General Superintendent." Conductor
Wench, a pleasant, genial fellow, on my next trip from Parker looked perplexed as he greeted me. He hesitated, walked past, returned in a few moments and asked to see my pass. The document was produced, he drew a
letter from his pocket and showed it to me. It was the order signed by Phillips and Lawrence. "That's clear enough, here's your fare," was my rejoinder.
It was agreed at the office to say nothing for a day or two. Doubtless Phillips and Lawrence thought the paper had been scared and would send a flag of
truce. A big wreck afforded the opportunity to open hostilities. For months
the war raged. The paper had a regular heading—"Another Accident on the
Valley of the Shadow Road"—which was printed every morning. Accidents
multiplied and travel sought other lines. Phillips threatened to remove the
shops from South Oil City, his partners wished Bishop to let up, he refused and
they bought his interest. Peace was proclaimed, the road was put into decent
order and the Pennsylvania Railroad eventually secured it. The fight had no
end of comical features. It worried Col. Phillips exceedingly and spread the
reputation of the *Derrick* over the continent. The cruel war is over and Col.
Phillips and Col. Lawrence journeyed to the tomb long years ago.

"Jim" Collins—he ought to be manager—is about the only one of the
early conductors on the Allegheny-Valley Railroad still in the traces. His
record of twenty-seven years shows capable, faithful attention to duty and care
for the comfort and safety of passengers that has gained him the highest popularity. Superintendent "Tom" King, now vice-president of the Baltimore &
Ohio, is among the foremost railroad-officials of the United States. His brother
was crushed to death by the cars. Wench, the Taylors, Reynolds and Bonar
have been off the road many years. Long trains of crude are also missing,
some towns along the route have disappeared and the crowds of operators who
formerly thronged the line between Parker and Oil City have vanished from
the scene.

The United-States Pipe-Line has overcome legal obstructions, laid its tubes
under railroads that objected to its passage to the sea and will soon pump oil
direct to refineries on the Jersey coast. Senator Emery, the sponsor of the
line, is not the man to be bluffed by any railroad-popinjay who wants him to
get off the earth. The National-Transit Line has ample facilities to transport
all the oil in Pennsylvania to the seaboard, but Emery is a true descendant of
the proud Highlander who wouldn't sail in Noah's ark because "ilka McLean
has a boat o' his ain."

"Hell in harness!" Davy Crockett is credited with exclaiming the first
time he saw a railroad-train tearing along one dark night. Could he have seen
an oil-train on the Oil-Creek Railroad, blazing from end to end and tearing
down from Brocton at sixty miles an hour, the conception would have been yet
more realistic. Engineer Brown held the throttle, which he pulled wide open
upon discovering a car of crude on fire. Mile after mile he sped on, thick
smoke and sheets of flame each moment growing denser and fiercer. At last

he reached a long siding, slackened the speed for the fireman to open the switch and ran the doomed train off the main track. He detached the engine and two cars, while the rest of the train fell a prey to the fiery demon. A similar accident at Bradford, caused by a tank at the Anchor Oil-Company's wells overflowing upon the tracks of the Bradford & Bordell narrow-gauge, burned two or three persons fatally. The oil caught fire as the locomotive passed the spot and enveloped the passenger-coach in flames so quickly that escape was cut off.

Do any of the pioneers on Kanawha remember "Dick" Timms's Half-way House? The weather-beaten sign bore the legend, in faded letters: "Rest for the Weary. R. Timms." The exterior was rough and unpainted, but inside was cheery and homelike in its snugness. When travelers rode up to the door "Uncle Dick," in full uniform of shirt and pantaloons, barefooted and hatless, rough and uncouth in speech and appearance, but with a heart so big that it made his fat body bulge and his whole face light up with a cheerful smile, stood ready with his welcome salutation of "Howdy, howdy? 'Light; come in."

Fifteen-thousand wells that produced oil and forty-five-hundred dry-holes have been drilled in Ohio. The average daily yield of the wells now producing is four barrels. The oil is inferior and sells at a trifle above half the price of Pennsylvania crude. The Indiana oil-region is a level country, about forty miles long east and west and three to four wide. The oil, dark green in color and thirty-six gravity, is found in the Trenton limestone, at a depth of a thousand feet. Thirty to a hundred feet of driving-pipe and three-hundred feet of casing are needed in each well. The main belt runs in regular pools and may be considered ten-barrel territory. The aggregate production of the field is ten to twelve-thousand barrels a day. The largest well started at two-thousand barrels and some have records of five-hundred to eight-hundred. The great gas-field, south of the oil-belt, has boomed manufactures and contributed vastly to the wealth of the Hoosiers. Of sixty-one-thousand wells put down in the United States since 1880 dawned fully nine-thousand were dry-holes. One Oil-City broker's customers lost four-million dollars in a year by staying on the wrong side of the market. Speculation and over-production played strong hands in the attempt to bankrupt the oil-business permanently, but the former is dead and the latter can be taken care of should rich fields happen to come in suddenly. The pathway of the oil-producer has not always been strewn with flowers, yet he has kept up his courage and striven hard to land on Easy Street. No part of the country to-day is so prosperous as the oil-region. The boys can wear diamonds when they want to do a turn in the social swim. Yet they don't turn up their noses and claim that the Lord made them out of whole cloth and the rest of mankind out of the leavings.

J. W. Stewart, of Clarion, is in Africa drilling for oil. An English syndicate is behind the enterprise and test-wells are to be drilled in the gold-fields on the southern coast. Stewart writes that it is amusing to see the monkeys climb up a derrick and watch the drillers at work. Just how amused they will be, if Stewart strikes a spouter that drenches the monkeys and the derrick, each must diagram for himself until the result of carrying the petroleum-war into Africa is decided.

Alas for sentiment! Nero proves to have been a humanitarian, a good man who was merely a bad fiddler. Henry the Eighth turns out to be a model husband, rather unfortunate in the loss of wives, but sweetly indulgent and only a trifle given to fall in love with pretty girls. William Tell had no son and shot no arrow at an apple on young Tell's head. Now Charlotte Temple is a myth, the creation of an English novelist, with her name cut on a flat tombstone in Trinity Churchyard over a grave which originally bore a metal-plate supposed to commemorate a man! At this rate some historic sharp in the future may demonstrate that the oil-men were a race of green-tinted people governed by King Petroleum. Colonel Drake may be pronounced a figure of the imagination, the Standard a fiction, the South-Improvement Company a nightmare and the Producers' Association a dream. Then some inquisitive antiquarian may come across a copy of "Sketches in Crude-Oil," stored in a forgotten corner of the Congressional Library, and set them all right and keep the world running in the correct groove with regard to the grand industry of the nineteenth century.

The Salt-Creek oil-field, the first worked in Wyoming, is in the northern part of Natrona and the southern part of Johnson county, fifty miles north of Casper, the terminus of the Fremont, Elkhorn & Missouri Valley Railroad. As known to-day the field is eighteen by thirty miles. It lies along Salt Creek and its tributaries, which drain northward and empty into Powder River, and is a rough country, cut by deep gulches, beneath which there are table-lands of small extent. Vegetation is scanty and timber is found only on the highest bluffs. In 1889 the Pennsylvania Oil-Company, composed of Pennsylvanians and under the management of George B. McCalmont, located on Salt Creek and drilled a well which, early in the spring of 1890, struck oil. Obstacles of no small magnitude were met with. The oil had to be freighted fifty miles by wagon; railroad-freights were controlled by eastern oil-producers and rates that would allow shipment seemed almost impossible, and the oil had to be proved before it could be placed upon the market in competition with well-known brands. In the face of these difficulties the company continued work and in the spring of 1894 succeeded in making arrangements to ship crude-oil. Storage-tanks were erected at the wells and at the railroad and a refinery is now in operation at Casper. The wells vary in depth from nine-hundred to fifteen-hundred feet and three companies are operating. The oil is a valuable lubricant. The transportation of the oil to the railroad is effected in freight-wagons of the ordinary sort. Behind them is a fourth wagon, or the freighter's home, which has wide boards projecting from the sides of the wagon-box over the wheels, making a box of unusual width covered with heavy canvas over the ordinary wagon-bows and provided with a window in the back, a door in front, a bed, cook-stove, table, cupboard and the necessary equipment for keeping house. In this house on wheels the freighter passes the night and in breaking camp he is not bothered with his camp-outfit. This novelty has been recently introduced by Mr. Johnson, the leading freighter for the Pennsylvania Company. With sixteen mules he draws his four wagons with nine tons of oil, over a very sandy road.

John G. Saxe once lectured at Pithole and was so pleased with the people and place that he donated twenty-five dollars to the charity-fund and wrote columns of descriptive matter to a Boston newspaper. "If I were not Alexander I would be Diogenes," said the Macedonian conqueror. Similarly Henry Ward Beecher remarked, when he visited Oil City to lecture, "If I were not pastor of Plymouth church I would be pastor of an Oil-City church." The train conveying Dom Pedro, Emperor of Brazil, through the oil-region stopped at Foxburg to afford the imperial guest an opportunity to see an oil-well torpedoed. He watched the filling of the shell with manifest interest, dropped the weight after the torpedo had been lowered and clapped his hands when a column of oil rose in the air. An irreverent spectator whispered: "This beats playing pedro."

Daniel Fisher, ex-mayor of Oil City and chief of the fire-department, donned a new suit one day when oil-tanks abounded in the Third Ward. Hearing a cry of distress, he mounted a tank and saw a man lying on the bottom, in a foot of thick oil. He dropped through the hatchway, pulled up the victim of gas and with great difficulty dragged him up the small ladder into the fresh air. Of course, the new clothes were spoiled beyond hope of redemption. The man revived, said his name was Green, that he earned a living by cleaning out tank-bottoms and was thus employed when overcome by gas. Next day Fisher met Green, who thanked him again for saving his life, borrowed ten dollars and never repaid the loan or offered to set up a new suit of clothes.

"Brudders an' sistern," ejaculated a colored preacher, "ef we knowed how much de good Law'd knows about us it wud skeer us mos' to deff." A Franklin preacher once seemed to forget that the Lord was posted concerning earthly affairs, as he prayed thirty-six minutes at the exercises on Memorial Day. The sun beat down upon the bare heads of the assembled multitude, but the divine prayed right along from Plymouth Rock to the close of the war. Col. J. S. Myers, the veteran lawyer, presided. Great drops of perspiration rolled down his face, but he was like the henpecked husband who couldn't get away and had to grin and bear it. He summed up the situation in a sentence: "I think ministers ought to take it for granted that the Almighty knows enough American history to get along nicely without having it prayed at Him by the hour!"

Philo M. Clark built an inclined railroad up a hill at Oil City, known as Clark's Summit, designing to lay out a residence-suburb on the stretch of flat land that overlooked the country for miles. A race-track was constructed and a crowd gathered one afternoon to witness an exhibition of speed. Just as the horses started in a running-race Wesley Chambers rode up on a mustang. The mustang had been a racer himself and he dashed upon the track. The others had a big start, but the mustang closed the gap and passed under the wire first, in spite of his rider's efforts to rein him in. The yells of the crowd were heard clear to Franklin.

Judge Trunkey, who presided over the Venango court a dozen years and was then elected to the Supreme Bench, was nearing a case of desertion. An Oil-City lawyer, proud of his glossy black beard, represented the forsaken wife, a comely young woman from Petroleum Centre, who dandled a bright baby of twenty months on her knee. Mother and baby formed a pretty picture and the lawyer took full advantage of it in his closing appeal to the jury. At a brilliant climax he turned to his client and said: "Let me have the child!" He was raising it to his arms, to hold before the men in the box and describe the heinous meanness of the wretch who could leave such beauty and innocence to starve. The baby spoiled the fun by springing up, clutching the attorney's beard and screaming: "Oh, papa!" The audience fairly shrieked. Judge Trunkey laughed until the tears flowed and it was five minutes before order could be restored. That ended the oratory and the jury salted the defendant handsomely. Hon. James S. Connelly, an Associate Judge, who now resides in Philadelphia and enjoys his well-earned fortune, was also on the bench at the moment. Judge Trunkey, one of the purest, noblest men and greatest jurists that ever shed lustre upon Pennsylvania, passed to his reward six years ago.

Colorado counts confidently on a production sufficient to give the Centennial State a solid reputation for oil. Wyoming is blossoming into prominence as a prospective source of supply. The Pennsylvania Oil and Gas Company has tested a wide area and is drilling a number of wells. The product is transported in heavy wagons, drawn by eight, twelve or sixteen horses, but the construction of pipe-lines will soon supplant this ponderous system. P. M. Shannon is president of the company and Judge McCalmont—known all over Venango, Butler and McKean counties as "Barney"—is second in command. Kansas has promising fields, opened in 1894 by Guffey & Galey, who sold to the Standard Oil-Company. One of its branch-companies is drilling twenty wells a month and a refinery has been erected. California is not confined to gold-mines, mammoth trees and luscious fruits. For years developments have

OIL-WELLS AT LOS ANGELES, CALIFORNIA.

been carried on, centering finally at Los Angeles. City-lots are punctured with holes and three-hundred wells have been drilled on two-hundred acres. Samuel M. Jones, formerly of the Pennsylvania oil-region and now president of the Acme Sucker-Rod Company of Toledo, leveled his kodak at the Los Angeles wells in 1895, securing the view printed in the cut. Hon. W. L. Hardison, who operated in the Clarion and Bradford fields and served a couple of terms in the Legislature, has been largely interested in the California field for ten years. Los Angeles wells are seven to nine-hundred feet deep, yield six barrels to seventy-five at the start and employ six-hundred men. The oil is used for fuel and lubrication, produces superior asphaltum and a distillate for stove-burners and gasoline-engines. It cannot be refined profitably for illuminating purposes. Thus the Occident and the Orient clasp hands in the petroleum-column.

Los Angeles is a genuine California town, with oil-wells as an extra feature. Derricks cluster on Belmont Hill, State street, Lakeshore avenue, Second street and leading thoroughfares. A six-inch line conveys crude to the railroads and car-tanks are shipped over the Southern Pacific and Santa Fé routes. At least one of the preachers seems to he drilling "on the belt," if a tourist's tale of a prayer he offered be true. Here it is :

"O. Lord! we pray that the excursion train going east this morning may not run off the track and kill any church-members that may be on board. Thou knowest it is bad enough to run oil-wells on Sunday, but worse to run Sunday excursions. Church-members on Sunday excursions are not in condition to die. In addition to this, it is embarrassing to a minister to officiate at a funeral of a member of the church who has been killed on a Sunday excursion. Keep the train on the track and preserve it from any calamity, that all church-members among the excursionists may have opportunity for repentance, that their sins may be forgiven. We ask it for Christ's sake. Amen."

A thief broke into a Bradford store and pilfered the cash-drawer. Some months later the merchant received an unsigned letter, containing a ten-dollar bill and this explanatory note : "I stole seventy-eight dollars from your money-drawer. Remorse gnaws at my conscience. When remorse gnaws again I will send you some more."

It is not surprising that evil travels faster than good, since it takes only two seconds to fight a duel and two months to drill an oil-well.

D. A. Dennison, the lively editor of the Bradford *Era*, is rarely vanquished in any sort of encounter. A "sweet-girl graduate" wrote a story and wanted him to print it. Thinking to let her down gently, he remarked : "Your romance suits me splendidly, but it has trivial faults. For instance, you describe the heroine's canary as drinking water by 'lapping it up eagerly with her tongue.' Isn't that a peculiar way for a canary to drink water?" "Your criticism surprises me," said the blushing girl in a pained voice. "Still, if you think your readers would prefer it, perhaps it would be better to let the canary drink water with a teaspoon." Dennison wilted like an ice-cream in July, promised to publish the story and the girl walked away mistress of the situation.

"The Producers' Consolidated Land and Petroleum Company," the formidable title over the Bradford office of the big corporation, is apt to suggest to observant readers the days of old long sign.

Seventy-five years ago an honest German erected the first house on what was to be part of a village in Clarion county. His name was a combination which in English meant dog and town. In the course of years other settlers arrived and a little hamlet arose, which naturally came to be called Dogtown in honor of the original pioneer. Among the early villagers was R. Monroe, for whom the place was for a time termed Monroeville. But euphonious Dogtown suited the expressive vernacular of the oilmen, after whose advent the old title held undisputed sway, leaving Monroeville to rust as a memory of the past and Dogtown to go to the dogs eventually.

A dry-joke tickles and a dry-hole scrunches. It's a poor mule won't work both ways, a poor spouter that can't keep its owner from going up the spout, a poor boil in the pot that isn't better than a boil on the neck, a poor chestnut on the tree that doesn't beat a chestnut at a minstrel-show and a poor seed that produces no root or herb or grain or fruit or flower. "Who made you?" the Sunday-school teacher asked a ragged urchin. "Made me? Well, God made me a foot long and I growed the rest !" And so the early operators on Oil Creek made the oil-development "a foot long" and it "growed the rest." The tiny seed is a vigorous plant, the puling babe a lusty giant. Amid lights and shadows, clouds and sunshine, successes and failures, struggles and triumphs, starless nights and radiant days, petroleum has moved ahead steadily. Growth, "creation by law," is ever going on in the healthy plant, the tree, the animal, the mind, the universe. We must go forward if the acorn is to become an oak, the infant a mature man, the feeble industry a sturdy development. Progress implies more of *in*volution than of *e*volution, just as the oak contains much that was not in the acorn, and the oil-business in 1896 possesses elements unknown in 1859. Not to advance is to go backward in religion, in nature and in trade. "An absentee God, sitting idle ever since the first Sabbath, on the outside of the universe, and *seeing* it go," is not a correct idea of the All-Wise Being, working actively in every point of space and moment of time. Stagnation means decay in the natural world and death in oil-affairs. The man who sits in the pasture waiting for the cow to come and be milked will never skim off the cream. The man who wants to figure as an oil-operator must bounce the drill and tap the sand and give the stuff a chance to get into the tanks. Still a youngster in years, the petroleum-colt has distanced the old nags. The sucker-rod is the pole that knocks the persimmons. The oil-well is the fountain of universal illumination. The walking-beam is the real balance of trade and of power. The derrick is the badge of enlightenment. Petroleum is the bright star that shines for all mankind and doesn't propose to be snuffed out or shoved off the grass. Its past is known, its present may be estimated, but what Canute dare fence in its future and say: "Thus far shalt thou come and no farther?"

> If there be friendly readers, as they reckon up the score,
> Who find these random " Sketches " not a burden and a bore
> Too heavy for digestion and too light for solemn lore—
> Who find a grain of pleasure has been added to their store
> By some glad reminiscence of the palmy days of yore,
> Or tender recollection of the old friends gone before—
> Who find some things to cherish and but little to deplore—
> Good-bye, our voyage ended, we must anchor on the shore.
> The last line has been written, all the labor now is o'er,
> The task has had sweet relish from the surface to the core;
> The sand-rock is exhausted, for the oil has drain'd each pore,
> The derrick stands neglected and we cease to tread its floor;
> My feet are on the threshold and my hands are on the door—
> The pen falls from my fingers, to be taken up no more.